STABILITY OF PARTICLE MOTION IN STORAGE RINGS

AIP CONFERENCE PROCEEDINGS NO. 292

PARTICLES AND FIELDS SERIES 54

STABILITY OF PARTICLE MOTION IN STORAGE RINGS

UPTON, NY 1992

EDITORS:
MELVIN MONTH
ALESSANDRO G. RUGGIERO
WU-TSUNG WENG
ASSISTANTS TO THE EDITORS:
MARGARET DIENES
PAMELA MANNING

BROOKHAVEN NATIONAL LABORATORY

American Institute of Physics   New York

Authorization to photocopy items for internal or personal use, beyond the free copying permitted under the 1978 U.S. Copyright Law (see statement below), is granted by the American Institute of Physics for users registered with the Copyright Clearance Center (CCC) Transactional Reporting Service, provided that the base fee of $2.00 per copy is paid directly to CCC, 27 Congress St., Salem, MA 01970. For those organizations that have been granted a photocopy license by CCC, a separate system of payment has been arranged. The fee code for users of the Transactional Reporting Service is: 0094-243X/87 $2.00.

© 1994 American Institute of Physics.

Individual readers of this volume and nonprofit libraries, acting for them, are permitted to make fair use of the material in it, such as copying an article for use in teaching or research. Permission is granted to quote from this volume in scientific work with the customary acknowledgment of the source. To reprint a figure, table, or other excerpt requires the consent of one of the original authors and notification to AIP. Republication or systematic or multiple reproduction of any material in this volume is permitted only under license from AIP. Address inquiries to Series Editor, AIP Conference Proceedings, AIP, 500 Sunnyside Boulevard, Woodbury, NY 11797-2999.

L.C. Catalog Card No. 93-073534
ISBN 1-56396-225-X
DOE CONF-921077

Printed in the United States of America.

Contents

Preface ix
    A. G. Ruggiero

I. INVITED TALKS

Activity in 1992 Related to the Design of the CERN
Large Hadron Collider 3
    W. Scandale
Accelerator Simulation Activities at the SSCL 13
    G. Bourianoff
Beam Dynamics Work at HERA 33
    F. Willeke
Activities at Fermilab Related to Collider
Present and Future 36
    G. P. Goderre and J. Holt
A Summary of Studies of Particle Stability at RHIC 48
    G. Parzen
The Others 58
    É. Forest

II. WORKING GROUP SUMMARIES

Summary for Working Group A on Short-Term Stability 73
    C. Iselin
Summary for Working Group B on Long-Term Stability 85
    S. Peggs
Comments on Working Group C: Methods 90
    A. G. Ruggiero

III. CONTRIBUTIONS: WORKING GROUP A
Short-Term Stability

Remarks on the Differential Algebraic Approach
to Particle Beam Optics by M. Berz 93
    V. Garczynski
Determination of Coupled-Lattice Properties Using
Turn-by-Turn Data 102
    G. Bourianoff, S. Hunt, D. Mathieson, F. Pilat,
    R. Talman, and G. Morpurgo
Chromaticity Correction for the Collider at the SSC 135
    Y. Nosochkov, F. Pilat, and T. Sen

Longitudinal Phase-Space Measurements at IUCF 163
    D. D. Caussyn, M. Ball, B. Brabson, J. Budnick, A. W. Chao,
    J. Collins, V. Derenchuk, S. Dutt, G. East, M. Ellison, T. Ellison,
    D. Friesel, W. Gabella, B. Hamilton, H. Huang, W. P. Jones, S. Y. Lee,
    D. Li, M. G. Minty, S. Nagaitsev, K. Y. Ng, X. Pei, G. Rondeau,
    T. Sloan, M. Syphers, S. Tepikian, Y. Wang, Y. Yan, and P. L. Zhang

Betatron Coupling Correction at the IUCF Cooler,
Leading to Improved Determination of Fourth-Order
Resonance Hamiltonian 170
    M. Ellison, M. Ball, B. Brabson, J. Budnick, D. D. Caussyn, J. Collins,
    S. Curtis, V. Derenchuk, G. East, T. Ellison, D. Friesel, B. Hamilton,
    H. Huang, W. P. Jones, W. Lamble, S. Y. Lee, D. Li, S. Nagaitsev,
    X. Pei, G. Rondeau, T. Sloan, A. W. Chao, S. Dutt, M. Syphers, Y. Yan,
    M. G. Minty, S. Tepikian, K. Y. Ng, and W. Gabella

Success in One-Turn Maps for Dynamic Aperture
Studies—A Brief Review 177
    Y. T. Yan

Tracking Experience with PATRIS at BNL 182
    J. Milutinovic and A. G. Ruggiero

Study of Stability of Beam in the Fermilab
Main Injector 208
    C. S. Mishra and F. Harfoush

An Optimal Procedure for Magnet Sorting 217
    R. K. Koul

Taylor Series Maps and Their Domain of Convergence 230
    D. T. Abell and A. J. Dragt

Is the Momentum Space Optimally Used with
the FODO Lattices? 260
    D. Trbojevic, K. Y. Ng, and S. Y. Lee

IV. CONTRIBUTIONS: WORKING GROUP B
Long-Term Stability

Quasi-Closed Orbit in a Harmonically Perturbed
Magnetic Field 267
    G. V. Stupakov

Dynamic Aperture and Emittance Growth Rates
in HERA Proton Ring 273
    F. Zimmermann

Slow Emittance Growth Due to Modulation Effects
in the Presence of Nonlinear Fields 289
    O. S. Brüning

Symplectic Integrators for Spin Motion 307
    S. R. Mane
The Method of Minimal Normal Forms 310
    S. R. Mane and W. T. Weng
Perturbation Expansion for Particle Distribution
in Hadron Storage Rings.. 325
    J. Shi and S. Ohnuma
A Study of Effects of Tune Modulation in
Nonlinear Maps.. 337
    A. Bazzani, M. Pusterla, and M. Venturini
Exact Physical Model for Magnets in Storage Rings 345
    D. Maletić and A. G. Ruggiero
A Simulation of Modulational Diffusion for the
Fermilab Tevatron 361
    T. Satogata and S. Peggs
The Applicability of Diffusion Phenomenology to Particle
Losses in Hadron Colliders 375
    A. Gerasimov
Invariant Manifolds and Stability: Some Results for 1-D Maps 385
    M. Giovannozzi
Aperture Determination of RHIC92 From Randomly
Generated Initial Coordinates 397
    G. F. Dell
The Effect of Synchrobetatron Coupling on the Dynamic Aperture 409
    G. Parzen

V. CONTRIBUTIONS: WORKING GROUP C
Methods

The Modern Approach to Single-Particle Dynamics
for Circular Rings... 417
    É. Forest, L. Michelotti, A. J. Dragt, and J. S. Berg
    with **Foreword** by J. Bengtsson
Towards C++ Object Libraries for Accelerator Physics 488
    L. Michelotti
Photographs... 492
List of Participants... 497
Author Index.. 501

Preface

The Workshop on the Stability of Particle Motion in Storage Rings was held at Brookhaven National Laboratory, October 19–24, 1992. It was sponsored by BNL, AUI, and DOE, and some funding was provided by the HEP Department of DOE. We would also like to acknowledge the assistance given by the U.S. Particle Accelerator School for the organization of the Workshop and the publication of the proceedings.

The program and the format of the Workshop were set up by the following Program Committee:

> Alex Chao, Superconducting Super Collider Laboratory
> Alex Dragt, University of Maryland
> Étienne Forest, Lawrence Berkeley Laboratory
> Melvin Month, Brookhaven National Laboratory
> Stephen G. Peggs, Brookhaven National Laboratory
> Alessandro G. Ruggiero, Brookhaven National Laboratory
> Walter Scandale, CERN
> David Sutter, U.S. Department of Energy
> Richard Talman, Superconducting Super Collider Laboratory
> Wu-Tsung Weng, Brookhaven National Laboratory
> Ferdi Willeke, DESY

The Local Organizing Committee, responsible for arranging the meetings and the social events, had the following members:

> Pamela M. Manning
> Melvin Month
> Alessandro G. Ruggiero
> Elaine Taylor
> Wu-Tsung Weng

The Workshop was attended by about 80 people, listed under "Participants," from U.S.A. and European laboratories and institutions. These proceedings represent their contributions.

The first day of the meeting was dedicated to the presentation of work at several major laboratories with projects on large hadron colliders. Late in the day, the appointed group leaders presented their working agenda for the rest of the meeting.

The Workshop was divided into three sessions: A and B were run in parallel and C was attended by a general audience. Group A dealt with the topic of short-term behavior of particles in storage rings, typically 1000 revolutions. Group B dealt with long-term behavior and addressed the issue of diffusion and diffusion-like processes. The general session (C), about 2 hours each morning, was comprised of talks and lectures on the most advanced methods of research proper to the investigation of motion stability in storage rings. In conclusion, on the morning of the last day, each group presented its summary.

It has become customary for a group of experts from laboratories and institutions around the world to meet periodically, every two years or so, to discuss the stability of particle motion in storage rings. The fundamental problem is that an electrically charged particle such as a proton, an antiproton, or a heavy ion, has to circulate safely in a storage ring, for high-energy and nuclear physics experiments, over extremely long periods of time, from a few hours to possibly one day. Typically, this corresponds to one billion revolutions.

The following article, which appeared in the CERN Courier, Vol. 33 (Jan./Feb. 1993), best summarizes the goal and the results of the Workshop:

"... Particle beam stability is crucial to any accelerator or collider, particularly big ones, such as Brookhaven's RHIC heavy ion collider and the larger SSC and LHC proton collider schemes. A workshop on the Stability of Particle Motion in Storage Rings held at Brookhaven in October dealt with the important issue of determining the short- and long-term stability of single-particle motion in hadron storage rings and colliders, and explored new methods for ensuring it.

In the quest for realistic environments, the imperfections of superconducting magnets and the effects of field modulation and noise were taken into account.

The workshop was divided into three study groups: Short-Term Stability in storage rings, including chromatic and geometric effects and correction strategies; Long-Term Stability, including modulation and random noise effects and slow-varying effects; and Methods for determining the stability of particle motion. The first two were run in parallel, but the third was attended by everyone.

Each group considered analytical, computational, and experimental methods, reviewing work done so far, comparing results and approaches and underlining outstanding issues. By resolving conflicts, it was possible to identify problems of common interest.

The workshop reaffirmed the validity of methods proposed several years ago. Major breakthroughs have been in the rapid improvement of computer capacity and speed, in the development of more sophisticated mathematical packages, and in the introduction of more powerful analytic approaches.

In a typical storage ring, a particle may be required to circulate for about one billion revolutions. While ten years ago it was only possible to predict accurately stability over about 1000 revolutions, it is now possible to predict over as many as one million turns. If this trend continues, in ten years it could become feasible to predict particle stability over the entire storage period. About 80 participants from the U.S.A. and Europe attended the meeting. ..."

Alessandro G. Ruggiero
Chairman, Workshop

I. INVITED TALKS

Activity in 1992 Related to the Design of the CERN Large Hadron Collider

W. Scandale

CERN, CH-1211 Geneva 23, Switzerland

1 INTRODUCTION

In 1992, the definition of the parameters and the evaluation of the performances of the CERN LHC progressed along three main lines:

The magnetic structure of the lattice was modified in order to increase the filling factor of the main dipoles in the regular cells.

The tolerances on the field quality of the guiding magnet were investigated with long-term computer tracking simulations as well as with experiments performed at the SPS.

Additional studies were performed to speed up the simulations using truncated maps, to propose phenomenlogical descriptions of the diffusive behavior in the transverse phase spaces, and to obtain a deeper insight into the nonlinear motion in the case of dynamical systems as simple as the Hénon map.

Other activities dealing with the design of hardware components, uncorrelated to the issues of the present workshop, will not be reported.

2 MAGNETIC STRUCTURE OF THE LHC LATTICE

The magnetic structure of the LHC lattice has been modified in such a way that there are 24 regular cells per arc and three main dipoles per half-cell. The inner diameter of the dipole superconducting coils is 56 mm. The cell is 102,042 m long and the main dipoles are 13.145 m long. Near the 3.05-m-long main quadrupoles there is a short straight section with beam position monitors, closed-orbit dipoles, tuning quadrupoles, chromaticity sextupoles, and Landau-damping octupoles. Each main dipole contains small sextupole and decapole correctors, located in the shadow of the electrical connections, the role of which is to compensate almost locally the systematic part of the b_3 and the b_5 field-shape imperfection components (in European notation b_3 is a sextupole, etc.).

The insertions have been re-matched to the modified arcs of regular cells. Each dispersion suppressor contains eight cell dipoles, one shorter dipole 8.8 m long, and four quadrupoles. Each long straight section beside the crossing points contains two triplets of quadrupoles and a set of horizontal separation-recombination dipoles. The utility straight sections for the dump and the halo cleaning have a special design.

The orbit functions of the regular cell and of the experimental insertions are shown in Figs. 1 and 2 respectively.

4 Design of the CERN Large Hadron Collider

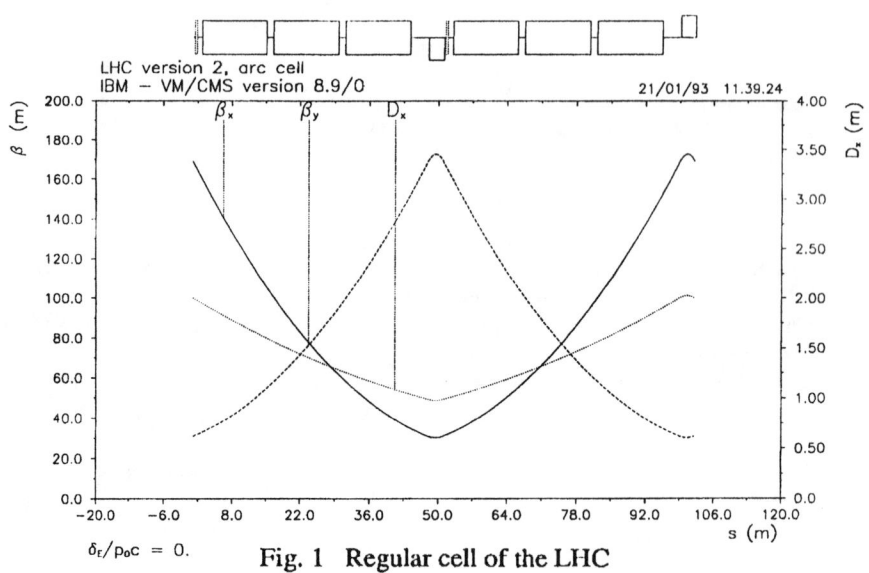

Fig. 1 Regular cell of the LHC

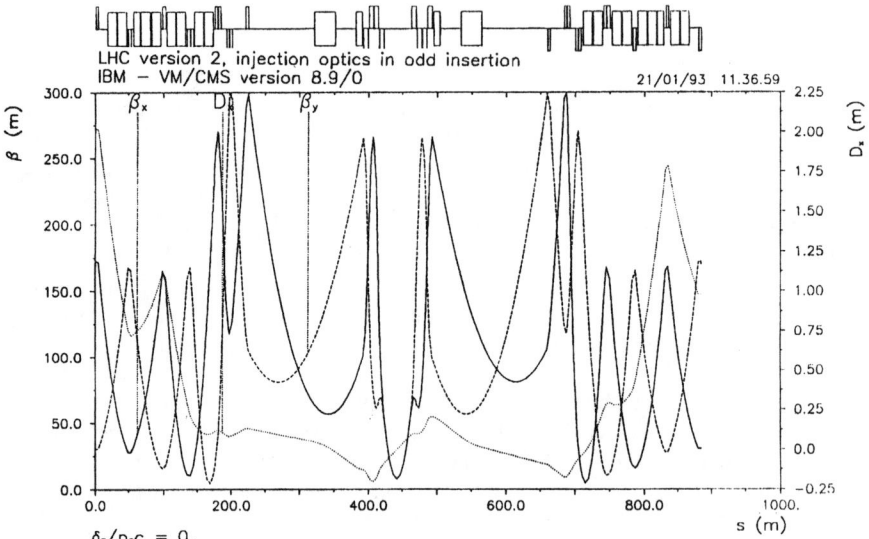

Fig. 2 Experimental insertion with injection optics

3 NONLINEAR CORRECTORS

The systematic components b_3, b_4 and b_5 of the field-shape imperfections are particularly pronounced at injection due to the effect of the persistent currents in the superconducting coils of the main dipoles. The typical values expected for them in the LHC magnets are respectively of the order of -2.5, -0.15, and 0.5 in 10^{-4} units at a reference radius of 1 cm.

Fortunately there is no need to compensate the systematic component b_4, since its value changes sign at each LHC octant and therefore its effect is globally self-compensated. Such a change of sign is inherently related to the side-by-side design of the two-in-one magnets and to the fact that the beam trajectory goes from the inner to the outer ring or vice-versa at each of the eight crossing points of the LHC.

The systematic components b_3 and b_5 have to be compensated using the sextupoles and the decapoles located at the ends of each main dipole. By setting the integrated strength of each of these multipoles equal to the strength of the systematic error integrated along one dipole, an almost local correction can be obtained which is in general sufficient to bring the values of tuneshift close to those of an ideal machine without errors and correctors, as shown in Fig. 3.

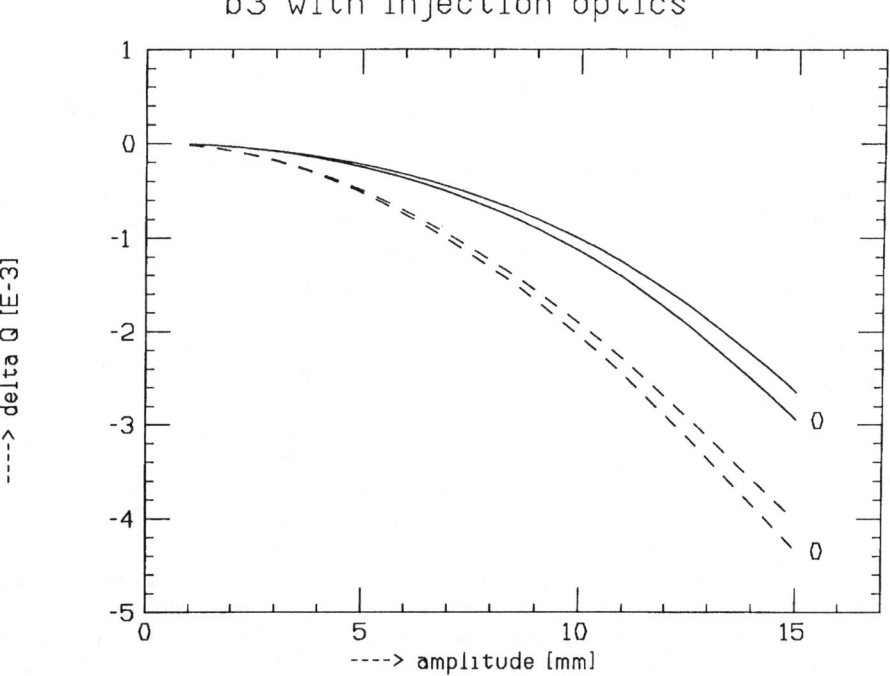

Fig. 3 Detuning with the transverse amplitude.
Solid curve: horizontal detuning.
Dashed curve: vertical detuning.
Curves with the symbol O: lattice without field-shape errors.
Other curves: lattice with field-shape errors and local corrections.

A more sophisticated approach to set the nonlinear correctors is based on use of the normal forms [1]. The tuneshift functions and their dependence on the strength of the errors and of the correcting gradients are computed order by order. A merit function, defined as the average of the tuneshift integrated over the sum of the normal-form invariants, is evaluated and factorized in order to isolate the terms dependent on the sum of the invariants. The residual terms, which describe the dependence of the tuneshift functions on the lattice structure, on the field-shape errors, and on the correcting gradients, are minimized by varying the correcting gradients. For a final confirmation, the tuneshift with amplitude and momentum is computed in the range of interest by tracking the particles with a symplectic integrator. The various coefficients of the tuneshift functions are estimated by using Martin Berz's DA package. Typical results of the nonlinear compensation are shown in Figs. 4 and 5.

4 TOLERANCES OF THE FIELD-SHAPE ERRORS

The tolerances of the field-shape errors in the guiding magnet have been investigated with computer tracking simulations [2]. The heuristic approach proposed for the LHC is based on long-term (up to a few 10^6 turns) simulations with thin-lens approximation and symplectic integrators. The field-shape imperfections, which can be expressed as the sum of two parts, one systematic and the other random, are represented by multipolar expansions stopped at the 11th order. The possible correlations between random multipoles of different order are neglected. The tolerances are more critical at injection, since the beam size is larger. On the other hand, the systematic errors are larger at injection due to the persistent currents.

The large values of the low-order (3rd and 5th) systematic multipoles provoke a sizeable detuning with amplitude and momentum, which can be corrected either locally or by using a clever cancellation of the detuning terms with a robust minimization procedure based on the normal forms, as discussed in the previous section.

Large high-order (7th and 9th) systematic multipoles destabilize the off-momentum particles and have to be minimized by design.

Random imperfections which vary from magnet to magnet due to manufacturing tolerances are the main source of the resonances and the distortion functions. Statistical moments of the distributions can be easily predicted, but are insufficient for complete knowledge of the nonlinear optics, since the resonance strengths depend on the specific sequence of the random errors around the ring rather than on statistical properties. Therefore, criteria for magnet design are to be studied on several nonlinear lattices, with different sequences of random multipoles. Additional parameters to be considered are the residual closed orbit, the linear coupling, and the synchrotron motion.

The beam stability is influenced by the linear parameters such as the tune, the residual linear coupling and the peak β-values in the insertion quadrupoles.

In most of the cases, the dynamic aperture is contained in the vacuum pipe of the LHC. However, sophisticated collimation systems are to be located close to the stable orbit to protect superconducting magnets from losses. For safe operation, careful matching of the physical aperture in the presence of collimators and the stability border is needed. The budget of the physical aperture includes the space required for the beam and that required for the injection oscillations, the magnet misalignment, the residual sagitta of the dipoles, which are curved to follow the beam path, the estimate of the residual closed-orbit and of the β-function modulations, and finally the space required to

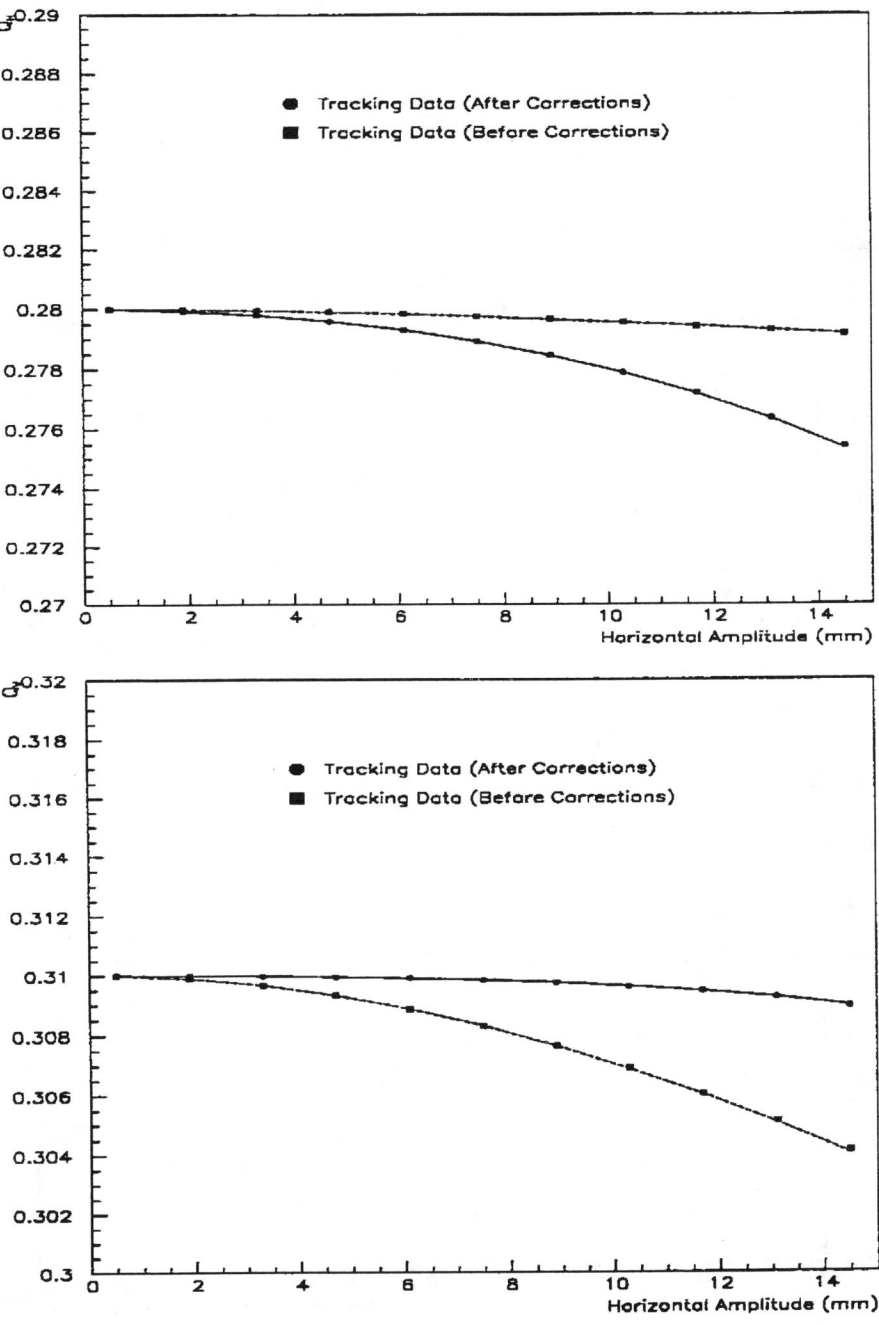

Fig. 4 Detuning with the amplitude

8 Design of the CERN Large Hadron Collider

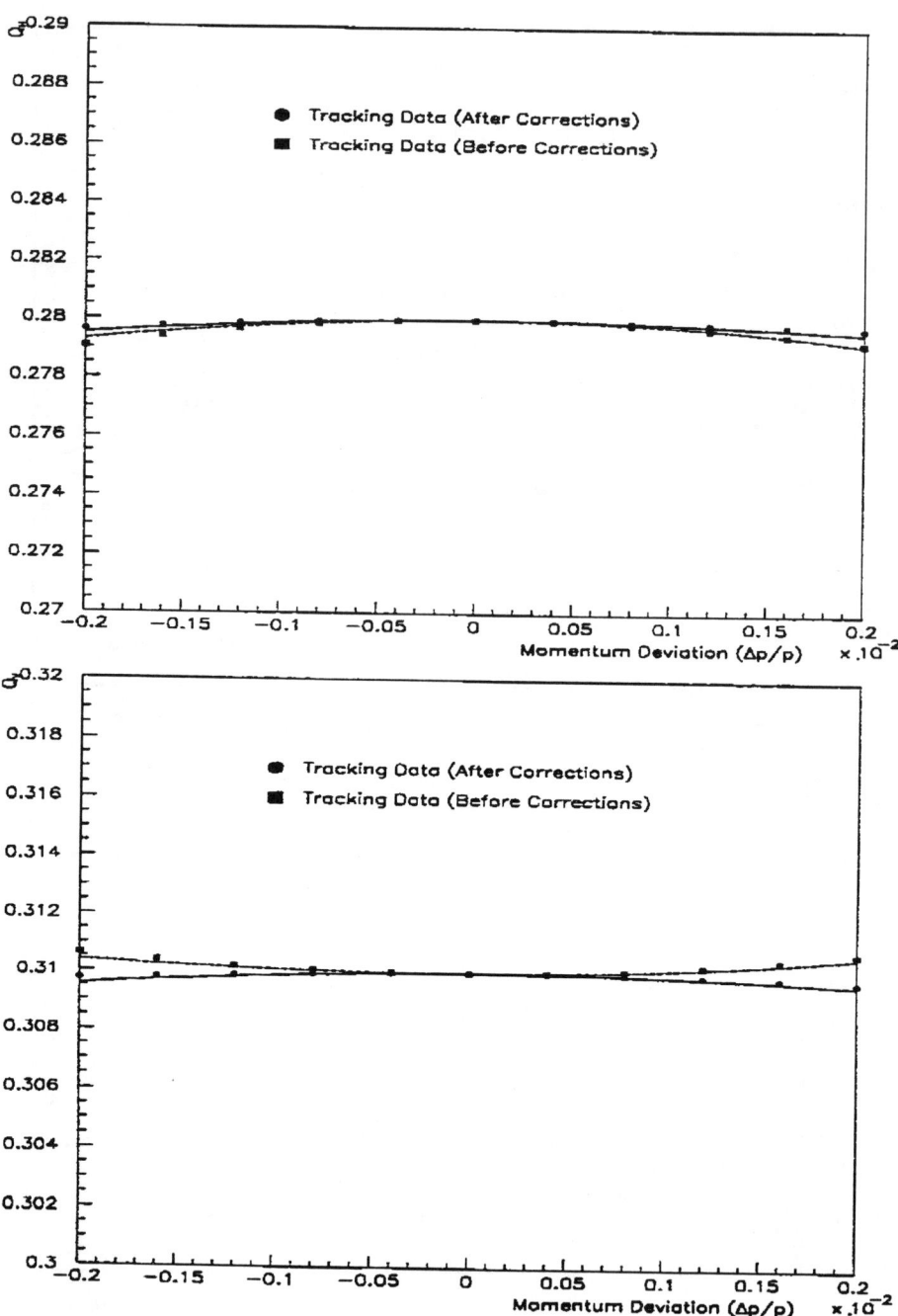

Fig. 5 Detuning with the momentum

devise a safe collimation system. At injection, the inner collimator jaw is expected to be located at 8 mm from the central orbit. In Fig. 6 the computed value of the dynamic aperture, defined as the initial amplitude of a particle surviving at least $5 \cdot 10^5$ turns, is plotted as a function of the inner collimator position. The dashed curve is the chaotic border above which the value of the Lyapunov exponent becomes positive. As a rule of thumb, the field-shape imperfections are considered tolerable when the chaotic limit almost coincides with the nominal position of the collimator, whilst the dynamic aperture is sufficient to accommodate three times the r.m.s. transverse beam dimension augmented by injection oscillations of a reasonable amplitude.

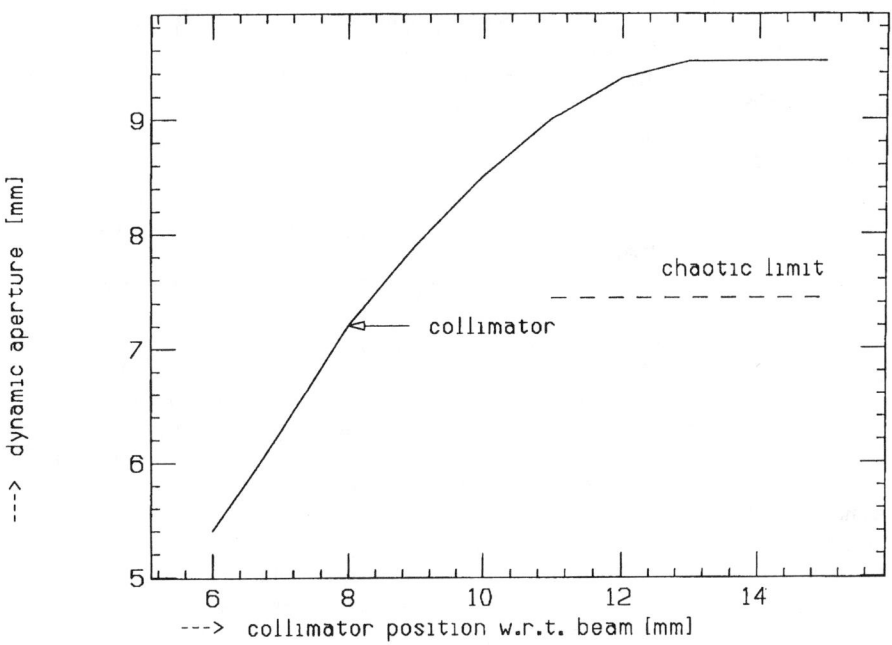

Fig. 6 Dynamic aperture vs physical aperture

5 EXPERIMENTS WITH THE CERN-SPS

Experiments are performed in the CERN-SPS to study the dynamic aperture and the diffusion of circulating particles in a regime as similar as possible to that expected in the LHC [3]. The protons stored in the SPS at a momentum of 120 GeV/c are excited with already existing sextupoles in order to introduce in a controlled fashion nonlinearities in an otherwise linear lattice. To probe large amplitudes, a pencil beam with small emittance and momentum spread is used, to which a large enough coherent deflection is applied. In a few hundred turns, a "hollow" distribution of charges is created around the central orbit due to nonlinear filamentation. Its behavior is observed with several instruments: current transformers record lifetime, Schottky noise detectors give tune and tune-spread, flying wires provide transverse profiles, and orthogonal pairs of position monitors are able to produce a phase-space portrait.

The sextupolar excitation used minimizes the strength of the third integer resonance. Detuning compensation was experimentally tested by using existing octupoles: a 30% increase of dynamic aperture resulted from a factor of ten reduction of tune-spread. This provides experimental guidance in devising correction schemes for large hadron accelerators. However, most of the emphasis was put on the study of slow diffusion induced by tune modulation produced by natural power supply ripple as well as controlled modulation of a special quadrupole. The diffusion coefficient was measured as a function of the amplitude, the modulation frequency and depth, and the tune. It was obtained by scraping the beam tail with horizontal and vertical collimators, retracting them suddenly by a few mm, and observing the beam lifetime to estimate the time taken by the particles to fill the gap created by the retraction. Diffusion immediately sets in when tune modulation is turned on, and there is evidence that a ripple which leads to tune modulation of 10^{-3} cannot be tolerated in a machine with strong nonlinearities. A simultaneous tune modulation at two frequencies is far more destructive than a modulation at a single frequency for the same overall depth. The agreement with numerical simulations is of the order of 20%; however, the dependence of diffusion on modulation depth and frequency is not yet fully explained.

A critical comparison of the experimental results at the CERN-SPS with those of similar experiments at the FNAL-Tevatron introduced a crucial discussion [4]. The results of the latter experiment can be interpreted in terms of fast-growing diffusion obeying to a major degree the following general principle: the initial transverse distribution remains unaffected for betatron amplitudes smaller than a certain time-dependent value, whilst the density is fully depleted for larger betatron amplitudes. The transient region is narrow and independent of the shape of the initial distribution. The characteristic results of the CERN experiment obtained with the scraper retraction show features that are incompatible with the fast-growing diffusion model, since the transition to losses, after collimator retraction, when protons reach the new collimator position, is rather sharp, whereas the following pattern of intensity losses is quasi-linear. On the other hand, the features of a typical survival plot obtained with computer tracking simulations are also incompatible with the diffusion mechanisms à la Fokker-Plank, since the spread of the survival time with the initial conditions is too large. A coherent description of these phenomena is not yet available.

6 TRUNCATED TAYLOR MAPS

The conventional element-by-element tracking performed with a symplectic integrator is in general limited to about 10^6 turns even with the computing facilities available nowadays. To overcome this limitation, the use of high-order truncated Taylor maps in the long-term stability studies for the LHC has been investigated [5]. However, with this sort of transport algorithm the volume in the phase space is no longer an invariant of the motion. The symplecticity of the one-turn matrix has been recovered by using either the well-known kick factorization à la Irving, or a novel procedure called dynamic re-scaling. The latter is quite pedestrian. It consists not of adding higher-order terms to eliminate the high-order violation of symplecticity but, rather, adding a linear transformation. The transformation is staged in three different scale transformations, two in the transverse direction and one in the longitudinal direction, characterized by three different scaling factors chosen in a manner to ensure that Liouville's theorem is obeyed, at least on average. In fact, there is an infinite set of possibilities for the choice of the three scaling factors. The additional criterion to determine their final choice is based on the following arguments. It is assumed that Taylor maps up to 11th order are sufficient for an accurate description of the beam dynamics in the LHC, at least in the region of the phase space where the motion is regular or only weakly chaotic. The beam trajectories are computed both with element-by-element tracking and with Taylor maps with initial values of the scaling factors. The differences between amplitudes in the phase space of the two results are estimated in a few iterations by slightly varying the scaling factors. From that one can optimize the scaling factors in such a way that the amplitude differences between direct tracking and iteration of the Taylor map are constant as a function of the number of turns. The comparison between direct tracking and iteration of truncated maps with and without dynamic re-scaling has been made for a large number of turns: the agreement is two orders of magnitude better with re-scaling that without.

7 HÉNON MAP

The Hénon map has been studied by numerical tracking in order to clarify the role of the trapping phenomenon in the presence of a weak sinusoidal modulation of the tune and to check the possibility of defining a diffusion coefficient [6]. For small values of the frequency modulation, the well-known pulsation of the whole characteristic structure of islands and the periodic beat of the separatrices has been made evident. A critical frequency has been found numerically which has the following property: for frequency modulations of the order of the critical frequency the locking of the particles on the resonance islands can work only in one direction during oscillation of the tune. This partial trapping acts as a very fast transport mechanism towards outer regions of the phase space. The critical frequency has been found to be almost inversely proportional to the depth of the tune oscillation. Such a dependence has also been confirmed analytically with a Hamiltonian approach. Moreover it has been found that there is a strong dependence of the critical frequency on the initial coordinates of the particle in the resonance island.

The diffusion coefficient is heuristically defined as the increase in particle emittance per turn. In a standard Hamiltonian treatment this parameter is expected to be inversely proportional to the tune modulation depth. Numerical simulations with the Hénon map have shown that, under given conditions, the growth of the emittance is linear with time only for a limited number of turns. In this time scale the diffusion coefficient is in fact roughly decreasing with the modulation depth. However, the dependence is almost parabolic and shows a fine structure for small values of the modulation depth.

The domain of stability of the unperturbed Hénon map has been investigated [7] with a novel technique based on numerical estimation of the invariant manifolds of the unstable fixed point. There are two sub-spaces emanating from the hyperbolic fixed point, one characterized by expanding and the other by contracting behavior. These two manifolds are characterized by the following features: they are both invariants of the motion, and they have at least one intersection — the unstable fixed point itself. This in fact implies that the two manifolds have an infinite number of intersections and that they can be constructed by repeatedly iterating an ensemble of the initial conditions of a small part of these manifolds around the hyperbolic fixed point. It turns out that these initial conditions can be chosen on the eigenvalues of the linearized map as long as one stays within a small distance from the fixed point. The invariant manifolds of the unstable fixed point have been compared to the stability regions as computed by direct tracking, and have been found to be practically identical for all values of the tune.

ACKNOWLEDGMENTS

I would like to acknowledge the contributions of J. Gareyte, M. Giovannozzi, T. Risselada, and F. Schmidt to the activity related to LHC design, reported in this paper.

REFERENCES

[1] W. Scandale, F. Schmidt, E. Todesco, "Compensation of the Tune Shift on the LHC using the Normal Form Techniques", Particle Accelerators **35**,53 (1991). M. Giovannozzi, F. Schmidt, Private communication (1992).

[2] Z. Guo, T. Risselada, W. Scandale, "Dynamic Behavior of the CERN Large Hadron Collider (LHC)", Third EPAC Conference, Berlin (1992).

[3] J. Gareyte, W. Scandale, F. Schmidt, "Review of the Dynamic Aperture Experiment at CERN-SPS", Workshop on Nonlinear Problems in Accelerator Physics, Gosen, Germany (1992).

[4] A. Gerasimov, "The Applicability of Diffusion Phenomenology to Particle Losses in Hadron Colliders", Divisional Report CERN SL/92-30 (AP) (1992).

[5] R. Kleiss, F. Schmidt, F. Zimmerman, "The Use of Truncated Taylor Maps in Dynamic Aperture Studies", to be published in Particle Accelerators (1993).

[6] A. Bazzani, M. Pusterla, M. Venturini, "A Study of Effects of Tune Modulation in Nonlinear Maps", These proceedings.

[7] M. Giovannozzi "Invariant Manifolds and Stability: Some Results for 1-D Maps", These proceedings.

Accelerator Simulation Activities at the SSCL

George Bourianoff, SSCL

1 Introduction

This paper will attempt to summarize the activities related to accelerator simulation at the SSC laboratory during the recent past. The work presented here will largely be the work of others and specific contributors are identified in the acknowledgments. The majority of the work was done by the Machine Simulation and Correction Group with contributions from individual machine groups and the Accelerator Theory Group.

The paper will be organized into three basic areas as follows. The first section deals with operational simulations. As the name implies, the results presented here deal with simulating various routine accelerator operations such as injection, extraction and correction. The topic of correction includes many specific operations such as chromatic correction, decoupling, and orbit smoothing as well as corrector failure simulations. Specific examples that will be discussed here are local chromatic correction of the low-beta insertion of the collider, experimental determination of local coupling strengths in LEP, and results of steering-corrector failure simulations in the collider.

The second section deals with performance prediction of a specified lattice design. This involves running the tracking kernel of the code to evaluate the performance of a specific configuration. This is usually done as part of a family of runs to investigate parametric dependencies by incrementally changing one parameter and observing the results. The effects are usually quantified by either the linear aperture or dynamic aperture. The specific results to be discussed here include the effect of higher-order multipoles on linear aperture and the effect of power supply ripple on emittance growth in the collider.

The third area discusses development and application of advanced techniques to particle tracking. The goal is to extend the scope of problems that may be attacked with tracking codes beyond the current range. One approach is to apply the computational power of the current generation of parallel processors to accelerator physics simulation, while the other technique is to utilize high-order maps to extend the range of simulations. Both approaches are being exploited at the SSC. The recent developments in parallel processing will be described here, and mapping techniques are discussed separately in a paper by Yiton Yan in these proceedings.

2 Operational Simulation

2.1 LOCAL CHROMATICITY CORRECTIONS

The first example of operational simulation to be described here is described more fully in Ref. [1], which discusses the local chromatic corrections made to correct for the effect of the strong-focusing quadrupoles in the low-beta insertion and the simulations done to support that effort.

The natural chromaticity of the collider is approximately -170 units of chromaticity from the arcs. For β^* equal to 0.5 m, each IR contributes an additional -50 units of chromaticity and correspondingly -100 units for β^* of 0.25 m. The current design calls for 4 families of sextupole correctors located adjacent to the IR regions to be used to correct part or all of the IR-induced chromaticity. The simulations described here were used to evaluate quantitatively the relative performance of the different operational scenarios that were possible with the available set of sextupoles.

The simulation effort required evaluating a three-dimensional array of parameters and therefore consisted of a great number of individual cases. The possible optics configurations of the low-beta IRs are shown in Table 1. These 6 cases constitute one of the three dimensions mentioned above.

Table 1: Optics Configurations Studied

Case	β_n^*	β_s^*
I	0.25 m	0.25 m
II	0.25 m	0.50 m
IV	0.25 m	8.00 m
V	0.50 m	0.50 m
VII	0.50 m	8.00 m
X	8.00 m	8.00 m

The second dimension of this three-dimensional parameter space specifies the amount of chromaticity corrected by the local correction scheme with the remaining chromaticity corrected by the global scheme. For example, the curves labeled local_50 + 50 indicate that 50 units of chromaticity are compensated by each of the local corrector families. The curves marked local_0+0 do not use the local correctors to correct any first-order chromaticity but use them to balance the chromatic contributions of the two IRs when they are operated asymmetrically. A full discussion and analysis of this situation is contained in Ref. [1].

The last dimension of the parameter space defines the distribution of errors in the lattice. The local chromatic correction scheme requires precise phase relationships between the various members of a correction family and hence is sensitive to the magnitude and distribution of errors in the lattice. The dynamic aperture calculations were done with various combinations of errors in the arcs and IR regions to study this dependence.

The results of this study are summarized in Figures 1 to 4. These show the 1000-turn dynamic aperture as a function of betatron amplitude and momentum offset. This set of curves displays only one of the six optics configurations (0.25m+8.00m). The other cases were calculated but are not shown here. The curves show that the local schemes have a larger momentum aperture relative to the globally corrected case.

The brief summary of this work given here is meant only to give an overview of the scope and content of the effort. A complete analysis and discussion are contained in Ref. [1].

2.2 COUPLING MEASUREMENTS AT LEP

A second example of operational simulations could equally well be discussed under the second section of this report dealing with performance prediction since it deals with both topics. The example to be discussed is the measurement of local coupling coefficients at LEP by analyzing turn by turn BPM data and comparing the measured results to simulation results. The motivation for this effort is that the random skew quadrupole component of the SSC dipoles will cause the betatron motion to be strongly coupled and interfere with routine accelerator operations. Standard techniques of decoupling using the closest tune approach will not be able to control the deviations of the eigenplanes in the arc and it is therefore necessary to detect and correct the coupling locally.

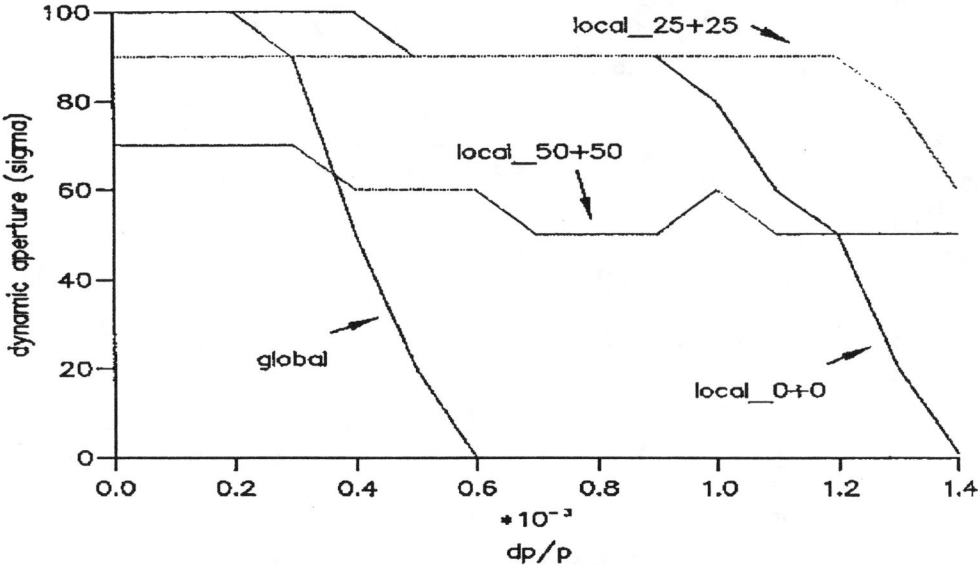

Figure 1. Dynamic Aperture as a Function of Amplitude and Momentum

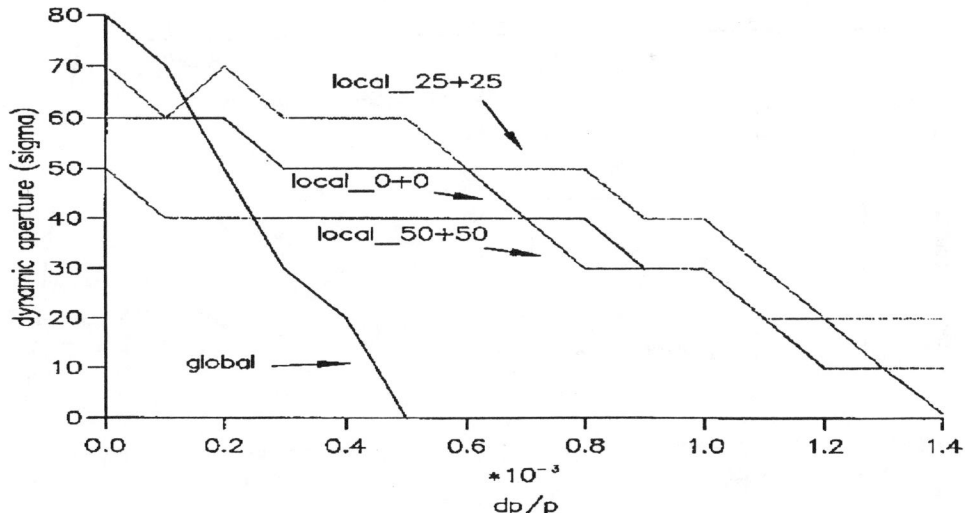

Figure 2. Dynamic Aperture as a Function of Amplitude and Momentum

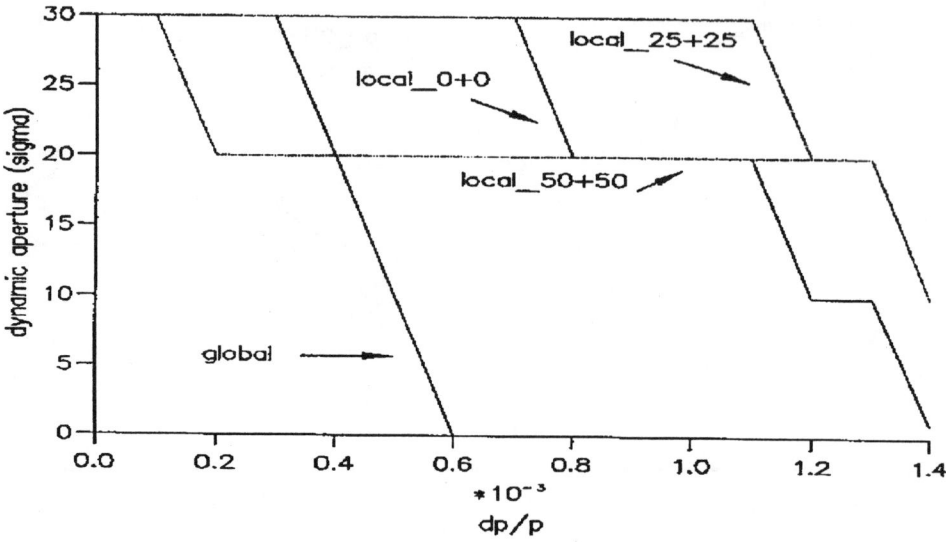

Figure 3. Dynamic Aperture as a Function of Amplitude and Momentum

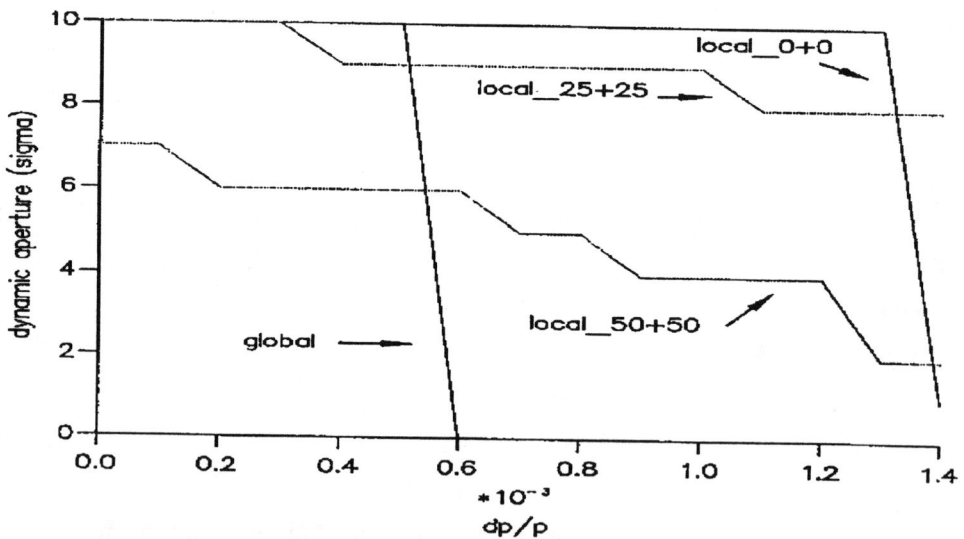

Figure 4. Dynamic Aperture as a Function of Amplitude and Momentum

The theoretical foundation for this procedure is contained in Ref. [2]. However, there existed operational questions about one's ability to measure the needed parameters to the required accuracy in an operational environment. The description of the experimental procedures, signal processing, supporting simulations and measured results are summarized here and more fully described in Ref. [3].

The basic formalism for local decoupling contained in Ref. [2] is summarized here for the sake of completeness. The basic goal is to block diagonalize the once-around transfer matrix M;

$$M = \begin{bmatrix} A & B \\ C & D \end{bmatrix} \qquad (1)$$

where A, B, C, and D are all 2 x 2 submatrices. It has been shown in Ref. [3] that M may be diagonalized by the transformation

$$\begin{bmatrix} \underline{A} & 0 \\ 0 & \underline{D} \end{bmatrix} = g^2 \begin{bmatrix} I & -R_D \\ -R_A & I \end{bmatrix} \begin{bmatrix} A & B \\ C & D \end{bmatrix} \begin{bmatrix} I & -R_D \\ -R_A & I \end{bmatrix} \qquad (2)$$

where symplectic conjugation is indicated by an overbar and

$$R_A = \frac{C + \overline{B}}{\Lambda_A - trD} \, , \qquad R_D = \frac{B + \overline{C}}{\Lambda_D - trA} \, ,$$

$$\Lambda_{A,D} = \left(\frac{trA + trD}{2}\right) \pm \sqrt{\frac{(trA - trD)^2}{4} + det|C + \overline{B}|} \, , \qquad (3)$$

$$g^2 = \frac{|\Lambda_D - trA|}{|\Lambda_D - \Lambda_A|} \, , \qquad \Lambda_{A,D} = 2\cos(\mu_{A,D}) \, . \qquad (4)$$

Λ_A and Λ_D, the eigenvalues of the matrix $M + \overline{M}$, are related to the generalized betatron tunes μ_A and μ_D as shown above. In what follows, the nominally horizontal (vertical) motion will be labeled A (D).

At a fixed point in the lattice of a coupled machine, the A betatron oscillations will be visible in the y motion and the D oscillations will be visible in x. The A betatron motion may be readily distinguished from the D betatron motion on the basis of their characteristic frequencies. The x and y motion at a fixed point and at the A betatron frequency may be written

$$x = g\cos(-\Psi_A) \, , \quad y = ge_A\cos(-\Psi_A + \Phi_A)$$

where

$$e_A^2 = \left[R_{A11} + \left(\frac{\alpha_A}{\beta_A}R_{A12}\right)\right]^2 + \left(\frac{R_{A12}}{\beta_A}\right)^2 \, .$$

$$\Phi_A = -\mathrm{atan}\left(\frac{R_{A12}/\beta_A}{R_{A11} - (\alpha_A/\beta_A)R_{A12}}\right) \tag{5}$$

where Ψ_A is the phase advance per turn of the A betatron motion. It can be seen from Eq. (5) that two constants e_A and Φ_A define the state of local coupling from x to y at a given location in the lattice. Two additional constants could be defined in order to parameterize the coupling from y to x but it will be assumed here that this coupling is essentially reciprocal since the eigenplanes are approximately (but not exactly) perpendicular. Further discussion of this is contained in Ref. [3].

The experimental procedure called for a betatron signal to be generated in the horizontal plane with the injection kickers. The constants e_A and Φ_A were determined by Fourier analyzing the X and Y motion at each BPM. The Fourier transforms of the applied signals may be written as

$$\{X\}_k = \frac{1}{\sqrt{N}}\left(\sum_{j=0}^{N-1} X_j \exp\left(-i\frac{2\pi jk}{N}\right)\right) \tag{6}$$

where $\{X\}$ = Discrete Fourier transform of X, N = number of discrete time samples (turns), and X_j = output of horizontal BPM at turn j.

Let the index k_{xmax} be defined to be the k value at which the maximum magnitude of $\{X\}$ occurs. The noise is defined to be the average value of all Fourier components with the exclusion of $k = k_{xmax}$. The signal to noise ratio of the actual data obtained at LEP is quite good, with 99 percent lying above 10.0.

The complex function $\{X\}_k$ is converted to polar form

$$\{X\}_k = \rho_{x_k} e^{i\phi_{x_k}}, \quad e_A = \frac{\rho_{x_{k_{xmax}}}}{\rho_{y_{k_{xmax}}}}, \quad \Phi_A = \phi_{x_{kmax}} - \phi_{y_{kmax}}, \tag{7}$$

and the amplitude ratio ε_A and phase difference Φ_A are extracted at every BPM. These are the two pieces of information necessary to perform the decoupling calculation based on experimental data and referenced in Eq. (5). The actual matrix elements needed for computing the corrector strengths (R_{A11} and R_{A12}) are obtained by inverting Eq. (5);

$$R_{A12} = e_A^2 \beta_A \sin(\Phi_A), \quad R_{A11} = e_A^2 \cos(\Phi_A) + \frac{\alpha_A}{\beta_A} R_{A12}. \tag{8}$$

The experimental effort consisted of measuring the local coupling parameters and comparing them to simulated values determined from a computational model of LEP. The experimentally measured results are shown in Figures 5 and 6.

The notation of experiment 3 in these figures distinguishes them from other data in which the betatron oscillations were generated by methods different from firing the injection kickers. The functions in the graphs are the amplitude ratio ε_A and phase difference Φ_A defined in Eq. (7).

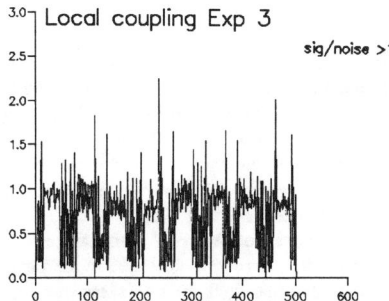
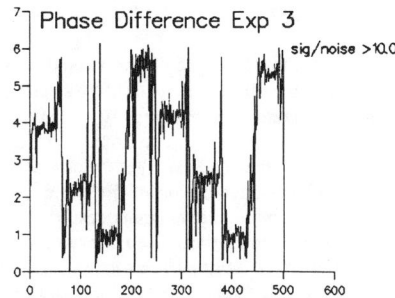

Figure 5. Measured coupling coefficient   Figure 6. Measured phase factors

This situation was simulated using the TEAPOT simulation code modified to produce an output file identical in structure to the output of the Beam Orbit Measurement system of the real experiment. These data were processed in the same fashion and by the same code and used to produce the plots in Figures 5 and 6. The actual experimental field and alignment errors that contribute to the experimentally measured coupling are of course unknown. The simulation code uses a Monte Carlo algorithm to generate an error distribution with what are believed to be the correct statistical averages [4]. In this case, the coupling is produced by a small systematic a_1 component of 0.01 units in the arc dipoles and a random rotation of 2 mrad in the main arc quads; the latter is dominant.

The simulated local coupling coefficients are shown in Figures 7 and 8. A direct comparison of these two figures with Figures 5 and 6 (experimental data) indicates that the coupling characteristics of the simulation agree closely with the coupling characteristics of the experiment. The simulation data are a little "cleaner" than the real data and the peak coupling values in the IR's are approximately 50 percent larger in the experimental case. However, the coupling in the arcs is within 20 percent of the experimental case. Overall, the agreement between experiment and simulation is quite good and we may use it as a base case on which to apply the local decoupling algorithm to estimate its effectiveness. This direct comparison of simulated results to measured results also serves to verify the simulation code in its treatment of basic optics and the treatment of error sources which create the need for operational correction.

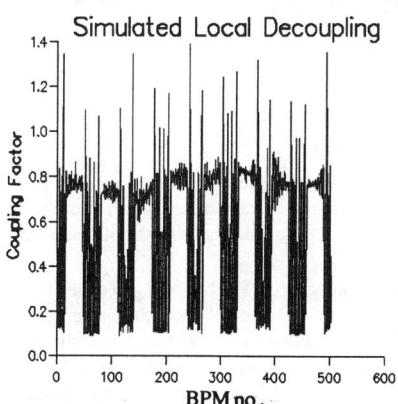
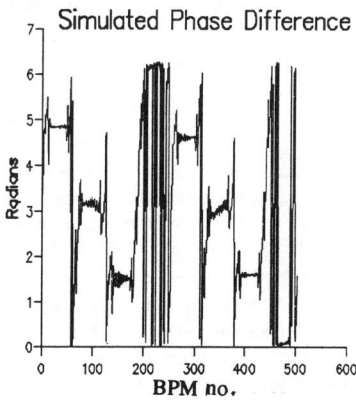

Figure 7. Simulation of Experiment 3   Figure 8. Simulation of Experiment 3

2.3 CORRECTOR FAILURE SIMULATIONS

A third example of operational issues that are being addressed through simulation is the impact of the failure of one or more steering correctors on collider operation. Each ring of the collider will contain approximately 1000 independently powered steering correctors. The specified mean time to failure of one of the thousand corrector power supplies is 29 hours. The question addressed in this study is what the probability is that failure of one corrector will cause beam loss or significant degradation of the beam. Beam loss will probably result in quenching of one or more superconducting magnets, resulting in a down time of perhaps half a day.

The simulations used the working collider lattice as of Aug 11. This lattice had all assigned errors (alignment errors, random and systematic field errors) as specified in the Level 3B Specifications. It also used the full set of correctors as specified, including 44 skew quadrupoles to correct the coupling. The study was done at collision energy (20 TeV) with tunes of 123.285 for v_x and 122.265 for v_y. The distributions of horizontal and vertical steering corrector strengths for this case are shown in Figures 9 and 10. The corrector strengths have a roughly normal distribution with peak values of 1.83 and 2.49 Tesla-meters for the horizontal and vertical cases respectively.

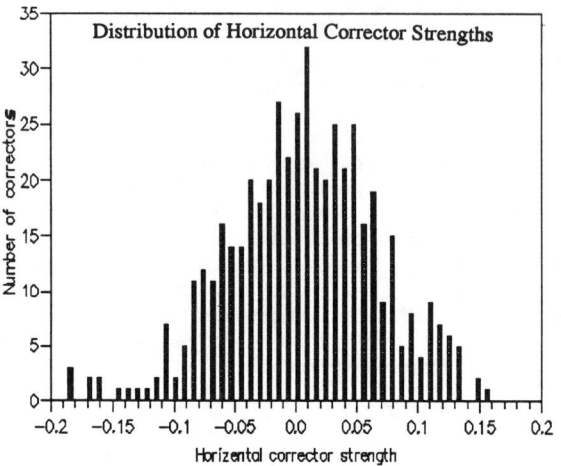

Figure 9. Strength x 0.1 in Tesla-meters

The simulation was done by successively "turning off" correctors beginning at the strongest and proceeding to less strongly powered correctors. Turning off the strongest correctors has the strongest effect. The correctors are grouped in blocks of 10 for this study so the simulations are done by turning off the 11th strongest, 21st strongest, etc.

The results of this process are given in Table 2 in tabular form. The top row gives the standard deviation of the closed orbit, the maximum excursion of the closed orbit and the linear aperture for the reference case (the case where all correctors are functioning). The study was done for both on-momentum and off-momentum particles as indicated by the last two columns in Table 2. These columns contain entries of "all," "yes" or "no" for particles lost. The entry "yes" means that some but not all the particles were lost and is further quantified in Figure 11.

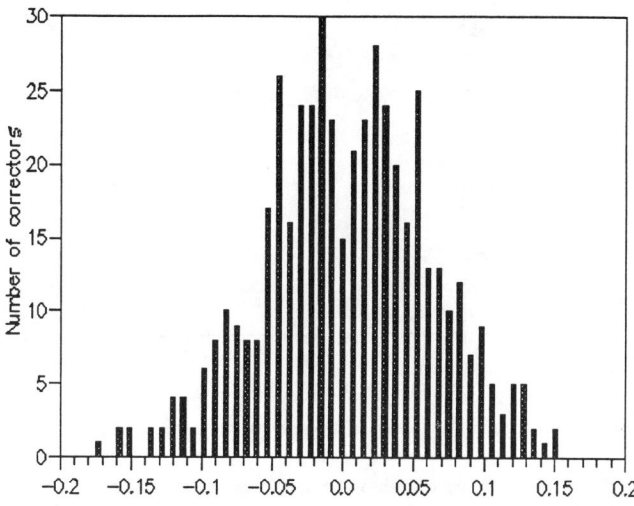

Figure 10. Strength x 0.1 in Tesla-meters

Table 2: Effect of Corrector Failure

case	$\sigma_{co}(y)$	Max (y_{co})	Linear Aperture	Particle lost (dp=0.0)	Particle lost (dp=0.0005)
Reference case	0.1137mm	0.5187mm	0.917mm	No	No
1st				All	All
11th				All	All
21th				All	All
31th	2.201mm	15.49mm		Yes	Yes
41th	1.947mm	13.62mm	0.148mm	Yes	Yes
51th	2.544mm	12.21mm		Yes	Yes
61th	1.73mm	11.86mm	0.17mm	Yes	Yes
71th	1.67mm	9.80mm	0.43mm	Yes	Yes
81th	1.50mm	9.85mm	0.61mm	Yes	Yes
91th	1.42mm	9.48mm	0.81mm	Yes	Yes
100th	1.097mm	7.55mm	0.73mm	No	No

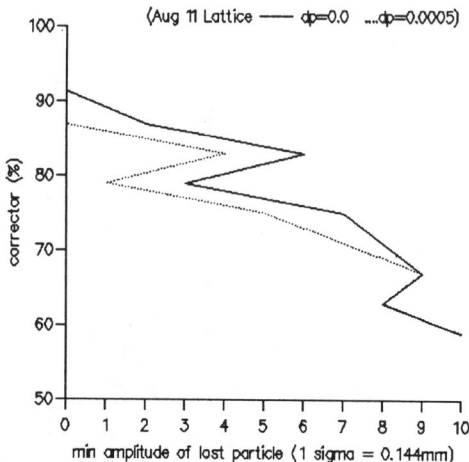

Figure 11. Corrector strength vs. lost particle amplitude

The figure shows the relationship of corrector strength to the minimum betatron amplitude at which particles are lost. The corrector strength is specified by rank in the distribution of strengths and is on the vertical axis. The betatron amplitude is expressed in terms of beam spot size (1 σ = 0.144 mm). From this graph, it is possible to determine what fraction of the correctors will affect the core of the beam (arbitrarily defined here to be made up of those particles with betatron amplitude less than 3σ). It can be seen from Fig. 11 that if any of the strongest 20 percent of correctors fail, particles will be lost in the core of the beam. Since there are 470 correctors, the 20th percentile ranking corresponds to the 47th corrector. From Table 2, it can be seen that failure of this corrector will produce a rms closed-orbit distortion of 2.5 mm and a peak closed-orbit deviation of 12.0 mm.

The functional dependence shown in Figure 11 is characterized by a pronounced dip which makes the function non-single-valued for certain values of the corrector strength. This may be explained by taking into account the beta function at the corrector location. It turns out that the strength of the correctors is correlated with the local beta function by the optical properties of the lattice. If one defines a parameter called effective strength equal to the corrector strength times the square root of beta, the functional relationship can be displayed as in Fig 12. It can be noticed that a peak in the curve occurs at the 80th percentile which is correlated to the observed dip in Fig. 11.

The lost particle amplitude may now be plotted against effective strength as shown in Figure 13. The effective strength corresponding to the top 20 percent of the correctors is indicated and clearly corresponds to the 3σ point. Using effective strength as an independent coordinate eliminates the double valued nature of the function shown in Figure 11.

The figures and discussion here pertain to the vertical motion of the Aug 11 lattice. The analysis has been repeated for the horizontal motion in this lattice and the entire process repeated for two other lattice configurations. The basic result is that there is a 20 percent chance that the failure of 1 corrector will result in loss of all or a significant fraction of the beam. This means that (given the 29-hour MTF) approximately once a week the beam will be lost due to corrector failure. This is unacceptable and indicates that corrector reliability must be improved.

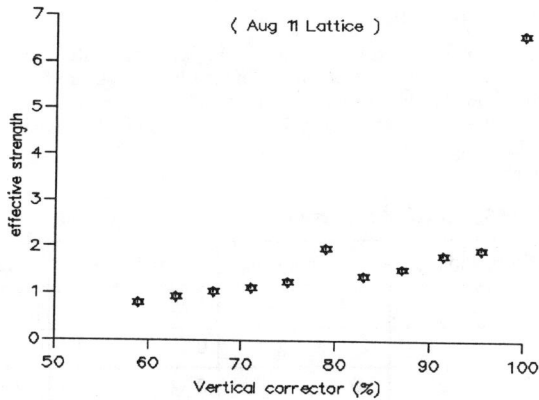

Figure 12. Effective strength vs. corrector ranking

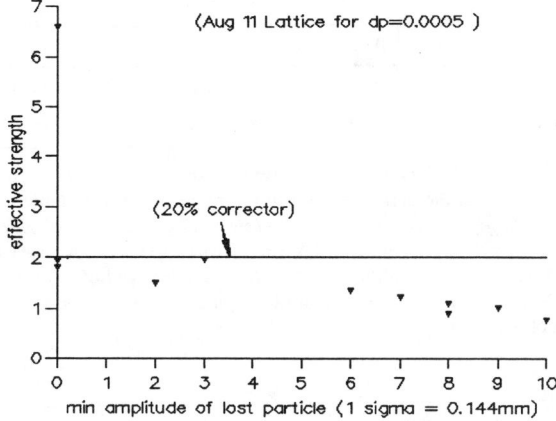

Figure 13. Effective strengths vs. lost particle amplitude

3 Performance Prediction

The second major area of activity is performance prediction. As the name implies, this activity will simulate the performance of a given lattice configuration, normally comparing the performance of a family of lattices differing only in one parameter. The simulation does not in general require any of the correction capabilities built into the TEAPOT code described in Ref. [5] and can be done using only the tracking kernel of TEAPOT on a parallel processor or other mainframe computer. These simulations tend to be much more computationally intensive than the simulations described in the first section because a set of lattices must be tracked for a sufficient number of random seeds for a sufficient number of turns. As an example, the calculation of the effect of higher-order multipoles on collider performance required ten seeds for five cases to generate the required statistics. Each run required approximately 20 hours on an HP70 RISC station leading to a total 1000 hours of computation on a high performance workstation array.

3.1 HIGHER-ORDER MULTIPOLES

The first example to be discussed is the impact of higher-order multipoles on collider performance. The problem is framed in terms of calculating the linear aperture for 5 separate cases which correspond to the HOM levels exceeding the level 3B specifications by factors of 2 or 4. The specified levels of random and systematic field errors for the dipole magnets in the collider are contained in the level 3b specifications and shown in Table 3. The strengths in the table are Tesla times 10^{-4} measured at 1 cm.

Table 3: Specified Higher-Order Multipole Strengths

Order	Systematic a_n	Systematic b_n	Random a_n	Random b_n
1	0.04	0.04	1.25	0.50
2	0.032	-2.0	0.35	1.15
3	0.026	0.026	0.32	0.16
4	0.02	0.08	0.05	0.22
5	0.016	0.016	0.05	0.02
6	0.013	0.02	0.01	0.02
7	0.01	0.01	0.01	0.01
8	0.008	0.02	0.0075	0.0075

The simulation consisted of increasing the values of systematic b_3 and b_4 by factors of 2 and 10 relative to the entries in the table. The lattice used was the Mar 31 lattice with tunes of 123.765 and 122.2791 for v_x and v_y respectively. Two sets of 17 particles were loaded with initial betatron amplitudes ranging from 0.4 mm to 8.4 mm (1σ to 21σ). The two sets had different momenta with one set being on momentum and the other having a $\delta p/p$ of 0.0005. The particles were tracked for 1024 turns, and tune versus amplitude plots such as shown in Figure 14 were produced for each set of particles.

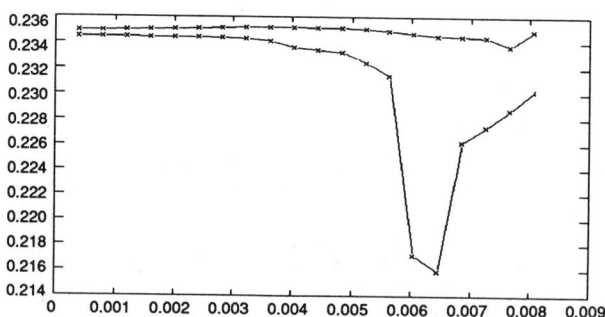

Figure 14. Horizontal tune versus amplitude for Oct X 2 case

The linear apertures for the 5 cases based on 10 random seeds each were computed. The results for one case are shown in Table 4.

Table 4: Octupole X 2 Case

seed	$\delta v_x < .005$	$\delta v_y < .005$	smear <5% @$\delta p = 0.0$	smear < 5% @$\delta p = .0005$	Linear Aperture
1015	7.7 mm	5.8 mm	6.86 mm	6.86 mm	5.8 mm
2701					
3455	7.7 mm	5.8 mm	6.05 mm	5.65 mm	5.65 mm
4011	7.6 mm	6.3 mm	1.6 mm	2.8 mm	1.6 mm
5176	6.1 mm	5.1 mm	5.65 mm	5.24 mm	5.1 mm
6869	6.5 mm	5.6 mm	5.24 mm	5.24 mm	5.24 mm
7531	7.3 mm	6.1 mm	6.45 mm	6.05 mm	6.05 mm
8999	7.7 mm	6.4 mm	6.45 mm	7.26 mm	6.4 mm
9204	>8.0 mm	6.0 mm	5.65 mm	2.82 mm	2.82 mm
10975	> 8.0 mm	6.1 mm	6.86 mm	6.05 mm	6.05 mm
MEAN	7.4 mm	5.9 mm	5.64 mm	5.33 mm	4.96 mm
STD DEV	0.66 mm	0.39 mm	1.61 mm	1.57 mm	1.64 mm

The smear is loosely defined to be the turn to turn variation of a quantity proportional to the linear invariant expressed as a percentage (an exact quantitative definition can be found in Ref. [6]). It is therefore a measure of the nonlinearity of the motion and is hence an increasing function of amplitude. The entry in the table defines the radius at which the smear exceeds 5 percent, which is somewhat arbitrarily defined as the point at which nonlinearity limits machine performance. The linear aperture is defined to be the smallest of the 4 radii shown in Table 4. A mean and standard deviation for the 10 seeds is computed for each of the 5 cases studied.

The results, shown in Figure 15, indicate that the machine performance, as quantified by the linear aperture, would not be affected if the dipole magnets exceeded the specifications by a factor of 10 in the decapole component and a factor of 2 in the octupole component.

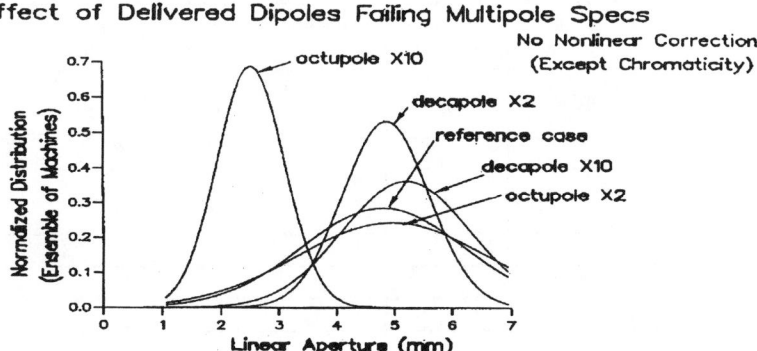

Figure 15. Linear Aperture Distributions

A similar calculation was performed to assess the impact of b_5 and b_9 components in all quadrupoles and separately in the triplet quads located in the low-beta insertions. The emphasis on b_5 and b_9 is due to two factors, the first being that they are by far the largest of the multipole components and the second being that they are "allowed" multipoles and can in principle be improved by rearranging conductors.

This set of simulations was carried out on a lattice that includes closed-orbit steering dipoles in the IR region that produce a crossing angle of 135 μrad. This is relevant to the present simulation since the nonzero crossing angle will cause the beam to traverse the triplet quads off axis and therefore see stronger nonlinear fields, adversely affecting the dynamic aperture.

Table 5 shows the dependence of short-term dynamic aperture on the presence or absence of b_5 in different sets of quadrupoles.

Table 5: Effect of b_5

systematic b_5	$\delta p/p = 0$	$\delta p/p=10^{-4}$	$\delta p/p=4\ 10^{-4}$
$b_5 =1.4(0.573)$ in all 4(5)cm quads	9	10	9
$b_5=0$ in triplets	12	12	12
$b_5=0$ in triplets and 5cm quads	12	12	11
$b_5=0$ in all quads	12	12	11

The representation in Table 5 is based on classifying all quadrupoles into 1 of 3 groups. There are 4-cm arc quads, 5-cm tuning quads in the IR regions and 5-cm triplet quads in the IR region. The first row has b_5 in all quads at the specified level. By comparing the first and second rows of Table 5 to the other rows, it may be deduced that the b_5 component in the triplet quads has the 3σ impact on the dynamic aperture. Comparing the other rows shows that the b_5 component is insignificant in the other quadrupoles.

The impact of b_5 in the triplet may be further quantized by varying this parameter incrementally between 0.0 and the specified value of 0.574 and looking for existence of a threshold. The results of this investigation are shown in Table 6. The value of b_9 was set to zero for the pur-

Table 6: Effect of varying b_5 in the triplet quadrupoles

sys b_5 (triplets)	dynamic aperture dp/p=0	linear aperture dp/p=0	dynamic aperture dp/p=10^{-4}	linear aperture dp/p=10^{-4}	dynamic aperture $\delta p/p=4 \times 10^{-4}$	linear aperture $\delta p/p=4 \times 10^{-4}$
0.	14	4	12	5	13	0
0.1	14	4	14	4	12	0
0.2	13	4	12	4	11	0
0.3	13	4	11	4	11	0
0.4	11	3	11	4	11	0
0.574	11	3	11	4	10	0

poses of this study. The table reveals a threshold at a b_5 value of about 0.3 units. The existence of such nonlinear thresholds is not surprising but their exact location in parameter space is in general not known and their mapping constitutes one of the most important results of computer simulation of nonlinear systems.

The fact that the linear aperture goes to zero in some cases above is indicative of resonance behavior and could probably be remedied by shifting the tune slightly. This remains to be demonstrated

3.2 RIPPLE

A second important example of performance prediction deals with the calculation of the effect of power supply ripple on beam emittance. This is a very difficult problem to simulate directly because very small effects accumulate over very long times in a very nonlinear system.

There are 10 power feed points in each main collider ring at which power is transmitted from the surface to the ring. The superconducting dipoles and quadrupoles are connected as a series string on superconducting buses. These components all have inductance, capacitance and resistance and form a transmission line impedance to the propagation of the AC current components (the ripple). The detailed analysis of this transmission line will not be described here but the results of that analysis have been included in the performance prediction as follows.

The strongest AC ripple components of the dipole *magnetic* field expressed in terms of $\delta B/B$ are shown in Table 7 as a function of frequency.

The ripple amplitudes are given at injection energy field amplitudes and collision energy field amplitudes. The field amplitudes are exponentially damped as a function of distance from the feedpoint. The exponential damping length expressed in units of cell lengths is given in the fourth column of Table 7. The column marked relative influence is simply the product of decay length and the amplitude (at injection energy) normalized to the most influential frequency. The frequency sensitivity of the particle motion in the accelerator has not been included in this numerical factor.

Table 7: Ripple Amplitudes (max) and half-widths for Injection and Collision Energies

Freq.Hz	Injection $\delta B/B$	Collision $\delta B/B$	Decay Length-cells	Rel. Influence
60	1.00e-7	2.76e-8	99	3.85e-1
120	4.35e-7	2.52e-8	59	1.
180	5.43e-7	1.41e-8	46	9.73e-1
240	9.48e-8	1.26e-8	38	1.40e-1
300	1.80e-8	2.28e-9	34	2.38e-2
360	1.24e-7	8.66e-9	27	1.30e-2
420	4.89e-9	9.33e-10	27	5.13e-3
480	1.55e-8	1.50e-9	26	1.57e-2
540	1.81e-9	5.34e-10	25	1.76e-3
600	4.97e-9	6.22e-10	23	4.45e-3
660	4.57e-10	1.35e-10	22	3.91e-4
720	1.70e-7	1.08e-8	22	1.46e-1
1440	2.00e-8	1.26e-9	15	1.17e-2
2880	2.75e-9	1.72e-10	11	1.18e-3

The ripple simulation was done using a superposition of components at 4 frequencies at 120, 720, 1440 and 2880 Hz. These frequencies were selected by considerations of beam dynamics. The number of frequencies was limited to 4 by the memory available on the Intel ISPC/860 parallel computer where the calculations were performed

The study included 3 cases consisting of one without ripple and without synchrotron oscillations, one without ripple but with synchrotron oscillations, and one with ripple and with synchrotron oscillations. The ripple amplitude was increased by a factor of 10 over the values shown in Table 7 in order to produce an observable effect.

Figure 16 shows the emittances of a very fat beam as smaller than those of the simulated beam by a factor of approximately 6. This was also done to exaggerate the effect and make it observable. It can be seen from Figure 16 that there is a barely observable emittance growth after 50,000 turns for the exaggerated simulation case. The present simulation used 256 particles in the Mar_31_lattice.

The difficulties associated with simulating the effect of power supply ripple demonstrate one of the most difficult aspects of simulating the SSC collider operation, namely the accumulation of very small errors (of the order of 1 part in 10^9) over many turns (order of 10^8). The magnitude of the physical effect is just 4 orders of magnitude larger than round-off error and hence 10^8 turns would be the absolute maximum number of turns possible to simulate assuming that errors accumulate like the square root of the number of turns. In addition to round-off errors, one must consider the effect of numerical errors which could have a value greater than the ripple current as well as other physical effects of comparable magnitude being left out of the simulation.

Based on these considerations, it must be concluded that a direct simulation of the emittance blowup due to power supply ripple is beyond the capabilities of present day general simulation codes and computers. The most fruitful numerical approach appears to be to write a special purpose code to analyze this effect in the absence of all others.

4 Advanced Techniques

The discussion above illustrates one of the limits to tracking codes. A more common limit is imposed by the CPU time and elapsed time required to calculate a given result. Using a widely accepted element by element tracking code such as TEAPOT on a large lattice such as the SSC collider will require 15 hours of CPU time on a large mainframe supercomputer to track a reasonable number of particles for 100,000 turns. One such run will typically require a week to complete in a time-share environment. It is very desirable to be able to perform many such runs with a much quicker turn-around time. Hence new hardware and software methods must be pursued to accomplish this.

4.1 SPACE-CHARGE CODE

A decision was made in 1990 to apply the power of Massively Parallel Processing to the applications discussed here. To this end, the SSC acquired a 64-node ISPC/860 distributed memory parallel computer manufactured by Intel. The basic experience and performance of this machine for particle tracking without space-charge effects is contained in Ref. [7]. The principle result is that the tracking calculation can be done at approximately 10 double precision MFOPS per node at very high parallel efficiency for a rate slightly faster than with a Cray YMP if all 64 nodes are used in one calculation.

The basic tracking code has been combined with an electrostatic particle in a cell module to include the space-charge effects in a self-consistent manner. This represents a substantial extension to the physical domain of the problem. The physical model and some tracking results for the LEB are described in Ref. [8].

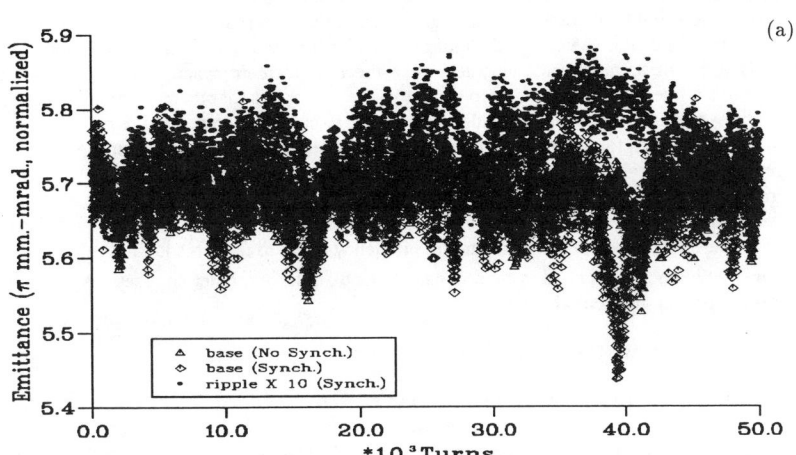

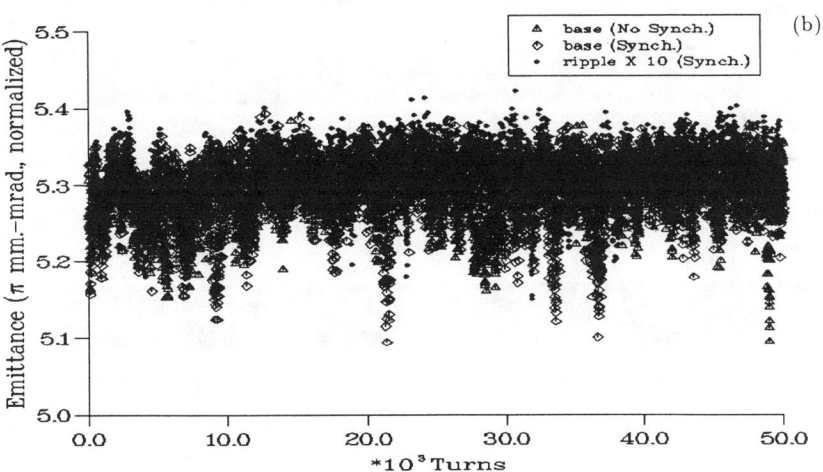

Fig 16. Emittance growth due to power supply ripple

Implementation of the space-charge algorithm in a distributed memory processor is considerably more difficult than implementation of the basic tracking routines since the treatment of collective effects necessitates a great deal more inter-node communication than in the case of non-interacting particles. The space-charge tracking code has been successfully written and tested on the parallel processor and is described in Ref. [9].

Execution of the space-charge code is very time-consuming for several reasons. One reason is that many particles must be tracked in order to produce acceptable fluctuation levels in the electrostatic field calculation. A second reason is that the space-charge calculation must be done at intervals that are determined by numerical stability requirements of the PIC solver. These intervals turn out to be considerably less than the inter-element spacing for the LEB and for planned beam intensities. A third reason for slower execution is that there are simply many more calculations to do. The forgoing considerations apply to any computer, serial or parallel. An additional consideration for distributed memory parallel machines is that the space-charge code requires a great deal more inter-node communication than the non-interacting particle code. Although exact comparison is difficult, the existing implementation of the space-charge code is approximately a factor of 10 slower than the non-interacting particle version. This implementation of the space-charge code achieves a parallel efficiency of 50 to 60 percent on 32 nodes, which is similar to efficiencies obtained in fluid dynamic calculations and other applications involving the solution of partial differential equations.

4.2 PARTICLE VISUALIZATION SYSTEM

The primary new development related to parallel processing is a high performance, interactive graphical interface known as the particle visualization system (PVS). This system can be operated synchronously with the space-charge simulation so that it is possible to step the simulation time step by time step or element by element displaying the results at each iteration. The PVS is written for a Silicon Graphics Crimson workstation in C^{++}. The control panel of the PVS is shown in Figure 17.

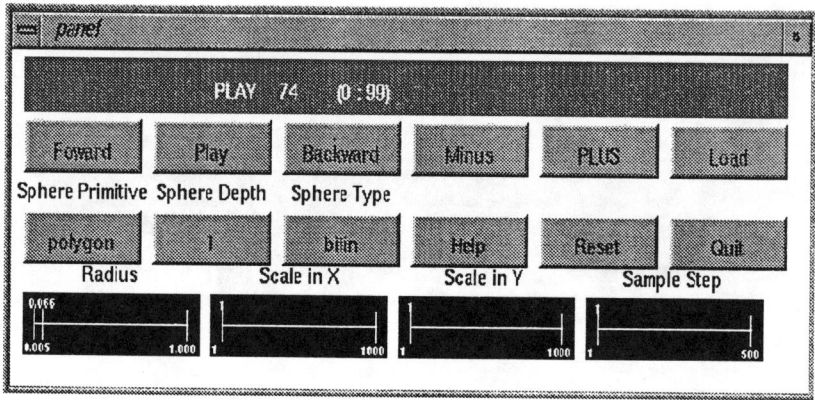

Figure 17. Control panel for Particle Visualization System

The button marked PLUS will advance the simulation and display one frame whereas the button marked MINUS will cause only the display to back up one frame. The button marked FORWARD causes the simulation to proceed at its fastest rate until the end of file is encountered. Various parameters of the display can be adjusted from the slider bars at the bottom of the control panel.

The PVS can display several windows simultaneously giving different viewpoints. Figure 18 shows the particle bunch viewed from inside and outside the ring. Within each window, the viewpoint orientation and zoom range can be adjusted with the mouse. In addition, it is possible to select a viewing frame that is stationary with respect to the ring, translating with the bunch or rotating at a specified rate with respect to either of the above reference frames. There are also several options for how much of the beamline is displayed along with the particle bunch. Figure 19 shows an interior view of the bunch internal to the LEB with and without beamline elements. In either case, one can display the inset window giving detailed specifications of the magnetic element through which the particle is tracking. The amount of information displayed in a given frame determines the speed at which displays can be generated, and it can be adjusted to the needs of the particular simulation.

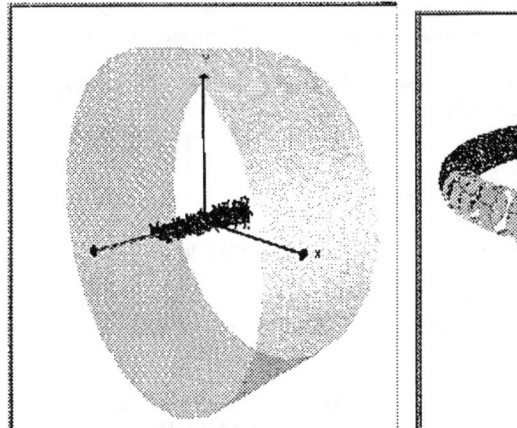

Figure 18. Exterior views of a bunch in the LEB

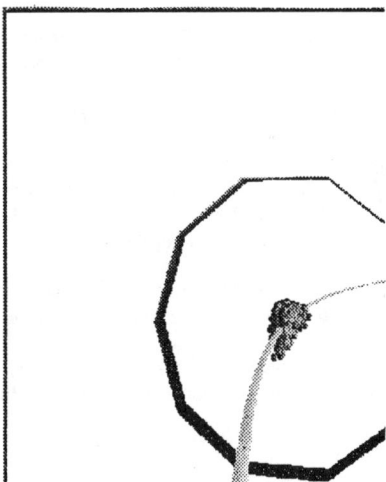

Figure 19. Interior views of a particle bunch in the LEB

Acknowledgments

The author would specifically like to acknowledge the contributions of the following people for the pieces of work indicated. Fulvia Pilat was responsible for the simulations of local chromaticity correction in the IR's and for analyzing the effects of b_5 and b_9 errors in the triplet quadrupoles. Xiaofeng Fang was primarily responsible for the work on corrector failures. Ben Cole provided the work on emittance growth caused by power supply ripple. Long Chang did the work on the particle visualization system. Richard Talman provided many helpful insights and suggestions.

The author would like to acknowledge the support of the Defense Advanced Research Projects Agency. Some of the support was was a joint DARPA/DOE program in the area of high performance computing.

The SSCL is operated by Universities Research Association, Inc., for the U.S. Department of Energy under Contract No. DE-AC35-89ER40486.

References

1. Y. Noschokov, F. Pilat, T. Sen, Chromaticity Correction for the Collider at the SSC, *This Proceeding*.

2. R. Talman, Single Particle Motion, *Notes for Joint US-CERN School on Beam Observation, Diagnosis and Correction*, Capri, Italy, 1988.

3. G. Bourianoff, S. Hunt, D. Mathieson, F. Pilat, R. Talman, G. Morpurgo, Determination of Coupled-Lattice Properties Using Turn-by-Turn Data. *SSCL Preprint #181*, Dec. 1992.

4. A. M. Fauchet, T. Fieguth, J. P. Koutchouk, T. Risselada, Coupling in LEP(II): Azimuthal Distribution in Octants 2,3 & 4, *LEP Commissioning Note 18*, Dec. 1989.

5. L. Schachinger, R. Talman, Teapot: Thin-Element Accelerator Program for Optics and Tracking, *Particle Accelerators* 22, 35-56 (1987).

6. D. Bintinger et al., *The Compensation of SSC Lattice Optics in the Presence of Dipole Field Errors*, SSC-SR-1038, Feb. 1989, p. 11.

7. G. Bourianoff, B. Cole, Parallel Processing at the SSC: The Fact and the Fiction, *Advanced Beam Dynamics Workshop on Effects of Errors in Accelerators, Their Diagnoses and Corrections, AIP Conference Proceedings No. 255*, Corpus Christi, Texas, 1991.

8. S. Machida, G. Bourianoff, Y. Huang, N. K. Mahale, Tracking Study of Hadron Boosters, *XV International Conference on High Energy Accelerators*, 1992, Hamburg, Germany.

9. L.C. Chang, G. Bourianoff, B. Cole, S. Machida, A parallel Implementation of Particle Tracking with Space Charge Effects on an Intel IPSC/860, SSCL (1992).

Beam Dynamics Work at HERA

F. Willeke
Deutsches Elektronen Synchrotron
Notkestr.85, 2000Hamburg 52, Germany

The following is a short summary of a talk given at the workshop on single-particle beam dynamics held at BNL in October 1992. For more detailed descriptions of the topics mentioned, see the references.

HERA, the first electron-proton collider [1,2,3,4] was commissioned in 1991 [6,7] and started to provide colliding beams and luminosity for the two experiments ZEUS and H1 in 1992 [8,9,10,11] Beam dynamics issues have played an important role in the design of HERA. Strong nonlinear field imperfections caused by persistent currents affect the proton beam stability at injection energy, as predicted by extensive simulation studies [12]. The commissioning confirmed that the extensive correction system of HERA, consisting of long sextupole coils inside the dipoles that provide almost local compensation of the persistent current sextupole, and the short multipole (12-pole and 10-pole) coils, is necessary and adequate. The dynamic aperture at injection of HERA is not very large. It is only approximately one third of the physical aperture, or five sigma of the injected beam. However, this turned out to be sufficient for stable beam operation at low energy. It agrees quite well with simulations (see also [14]). Before HERA came into operation, no experience was available on collision of different species of particle beams. The lifetime of the proton beam has always been a subject of concern considering that coherent motion is always present in the electron beam. The first *e-p* collisions (Oct. 20, 1991), however, already showed that the most harmful impact on the lifetime of the proton beam is caused by the electron beam size at the collision point being smaller than the proton beam size. This also confirms similar observations made earlier at the SPS *p-p* collider [13]. Since different beam sizes have been avoided by appropriate choice of the β-functions of the proton and electron beam respectively, the proton beam lifetime exceeded 50 h in collision even with beam-beam tuneshift values around $\Delta\nu = 0.002$, which is close to the design values.

To achieve a good proton beam lifetime it is also important to centre the two beams carefully with respect to each other at the interaction point. Precision of the order of 10 μm, which corresponds to roughly 10% of the transverse beam size (sigma), is required. Furthermore it is important to choose the working point close to the coupling resonance $Q_x - Q_y = -1$ near a tune of 0.3. Once the two beams are brought into collision, only occasional small corrections in tunes and orbit are required to maintain the specific luminosity close to the design value of $L_{\text{spec}} \simeq 4 \cdot 10^{29}$ cm^{-2} s^{-1} (mA)$^{-2}$ and a good proton beam lifetime. For more details on the initial experience with electron-proton collisions, see Refs. [15,16].

Beam dynamics studies on HERA are underway to improve our understanding of the machine behaviour. Very useful information is obtained by scraper experiments, which can measure the amplitude growth for large-amplitude particles. This provides insight into the mechanism of particle loss in the presence of nonlinear fields. So far, the motion of the particles near the scraper jaws is characterized well by diffusion-like behaviour [17]. Such experiments are particularly important for a possible parasitic fixed-target experiment in the HERA proton ring, which is fed by the particles in the beam halo [18].

Considerable effort [19] has been made to understand the dynamic aperture limiting effects by analytic studies. A pump diffusion model has been used to investigate the interference between several tune modulations with different frequencies. Experiments at the SPS [20] can be at least qualitatively explained by this model.

Local diffusion coefficients have been calculated for HERA [14] using the theory of emittance growth by resonance crossing [21]. It is remarkable that the underlying structure of the phase space has been calculated for HERA taking into account all measured multipole components of the magnets. This is an example of the use of new efficient methods such as automatic differentiation [22] combined with Lie-algebra and normal-form techniques [23]. The critical amplitude for the onset of diffusion agrees quite well with the dynamic aperture.

Some work is also being done on stochastic mapping of nonlinear accelerator systems which provides particle distributions in phase space and may be used to calculate beam lifetimes in the presence of nonlinearities [24]. Also, I should at least mention the fruitful effort on longitudinal electron spin polarization in the HERA electron ring [25].

References

[1] HERA, A Proposal for a Large Electron Proton Colliding Beam Facility at DESY, DESY HERA 81-10 (1981).

[2] G.-A. Voss, in Proc. 1st European Part. Accel. Conf., Rome, June 1988.

[3] B.H. Wiik, in Proc. 1989 Part. Accel. Conf., Chicago, March 1989.

[4] H. Kumpfert and M. Leenen, in Proc. 14th Int. Conf. on High Energy Accelerators, Tsukuba, 1989.

[5] B. H. Wiik, in Proc. 2nd European Part. Accel. Conf., Nice, June 1990.

[6] B.H. Wiik, in Proc. 1991 Part. Accel. Conf., San Francisco, May 1991.

[7] D. Degele, in Proc. 3rd European Part. Accel. Conf., Berlin, March 1992.

[8] F. Willeke, in Proc. 15th Int. Conf. on High Energy Accelerators, Hamburg, July 1992.

[9] B. Wiik, in Proc. Rochester Conf., Dallas, August 1992.

[10] R. Brinkmann, in Proc. 3rd Int. Conf. on Calorimetry in High Energy Physics, Corpus Christi, Sept. 1992.

[11] B.H. Wiik, in Proc. 1993 Part. Accel. Conf., Washington, May 1993.

[12] R. Brinkmann and F. Willeke, in Proc. 2nd Advanced ICFA Beam Dynamics Workshop, Lugano, April 1988, CERN 88-04, p. 203.

[13] L. Evans and J. Gareyte, CERN 82-8 (DI-MST) (1982).

[14] F. Zimmermann, in These Proceedings.

[15] F. Zimmermann et al, in Proc. Workshop on B-Factories: The State of the Art in Accelerators, Detectors, and Physics, Stanford, April 1992.

[16] S. Herb and F. Zimmermann, in Proc. 15th Int. Conf. on High Energy Accelerators, Hamburg, July 1992.

[17] M. Seidel, ibid.

[18] H. Albrecht et al, Letter of Intent, DESY-PRC 92-04.

[19] O. Brüning, in These Proceedings.

[20] W. Scandale, in These Proceedings.

[21] A. Schoch, CERN 57-21 (1957); G. Guignard CERN 70-24 (1970).

[22] M. E. Forest, M. Berz, and J. Irwin, Part. Accel. 24, 91 (1989).

[23] M. Berz, Part. Accel. 24, 109 (1989).

[24] A. Pauluhn et al, in Proc. 15th Int. Conf. on High Energy Accelerators, Hamburg, July 1992.

[25] M. Boege et al, in Proc. 1993 Part. Accel. Conf., Washington, May 1993.

Activities at Fermilab Related to Collider Present and Future

G. P. Goderre and J. Holt

Fermi National Accelerator Laboratory[*]

1 INTRODUCTION

The long-range Fermilab program requires fully capitalizing on the world's highest energy accelerator, the Tevatron, throughout the decade of the 90's. The program calls for increasing the collider luminosity with each successive run until peak luminosities of $> 5 \times 10^{31}$ cm^{-2} s^{-1} and integrated luminosities > 100 pb^{-1} per run are achieved, effectively doubling the mass range accessible for discovery. If, as appears likely, the top quark lies at the upper range of the mass reach of the Tevatron, then increasing the energy of the collider operation could prove to be a crucial factor in the future program as well. In order to achieve these goals, we present a highly challenging upgrade of the present accelerator complex, called Fermilab III. During the 1989 collider running period the maximum luminosity attained was slightly in excess of the design goal of the Tevatron I project which was set at 10^{30} cm^{-2} s^{-1}. In order to increase this performance level by a factor of 50, many changes are needed. Such a plan, of necessity, has modifications in almost all areas of the accelerator as the present system is reasonably optimized.

Fermilab III places emphasis on collider operation since productively searching new physics domains requires a continual increase in integrated luminosity. Fixed-target physics intensity improvements are also part of the overall considerations. The increased proton intensities needed for both collisions and antiproton production will also substantially benefit the fixed-target program.

2 PRESENT

During phase I of the this upgrade, there have been major modifications to the Tevatron. These modifications were commissioned at the start of this collider run and include the installation of electrostatic separators to separate the orbits and new low-beta insertions at both experiment interaction regions. These modifications have already enabled the Tevatron to achieve a record peak luminosity of 7.45 x 10^{30} cm^{-2} sec^{-1}

---

[*] Operated by the University Research Association under contract with the U. S. Department of Energy

and a record weekly integrated luminosity of 1.477 pb^{-1}. Figure 1 shows a comparison of the first four hundred days between this collider run and the last collider run of the initial luminosity as a ten times running average. Figure 2 is a comparison of the weekly luminosity and total luminosity.

The goal for the present run is 5×10^{30} cm^{-2} sec^{-1}. In the 1989 collider run the record peak luminosity was 2.07×10^{30} cm^{-2} sec^{-1}. The Tevatron was operated with six bunches colliding head-on in all locations (twelve collision points). One of the luminosity-limiting factors was a maximum sustainable tune shift of 0.025 due to the beam-beam interaction. A similar value has also been achieved at CERN with a different working point in the tune diagram and with different bunch parameters. In addition to a tune shift, the beam-beam interaction causes a tune spread across the beam and enhances the strength of various destructive nonlinear resonances. In the Tevatron, resonances up to twelfth order must be avoided. An orbit separation scheme was developed to eliminate the unnecessary collision points (there are only two experiments) as well as total separation during injection, acceleration and low-beta squeeze. The beams are brought into collision at only two points when low beta is reached.

In contrast with LEP and Cornell, the Tevatron uses separators in both the horizontal and vertical planes to produce a helical orbit. A helical orbit is accomplished by creating a betatron oscillation in the horizontal and vertical plane such that the phase between the two oscillations is n times π over 2 where n in an odd integer. The location of the separators in the Tevatron is shown in Figure 3. Helical orbits were chosen to keep the beams separated everywhere in the ring so that when the position of the bunches was cogged from the injection location to the collision location the beams remain separated.

The design goals were a minimum separation of 5σ and a maximum separation of 15 mm (beam-beam center to center). This goal has been met. During operation we have run with separators as small as 3σ with no problems. The separators have been very reliable. There has been only one separator spark during operations to date (1000 hours). The observed spark did not appreciably affect the store in progress.

During injection, acceleration and the low-beta squeeze, separation is achieved using the horizontal separators at B17 and vertical separators at C17. During the injection process the horizontal separator at B11 is used to adjust the phase of the helix through the injection Lambertson. Finally, the beams are brought into collision at the B0 and D0 insertions and kept apart everywhere else with local electrostatic three bumps in each plane. One pair of bumps creates a helical orbit from B11 to C49. The other pair of bumps keeps the beam apart from D11 to A49. This results in 6.5

Activities Related to Collider Present and Future

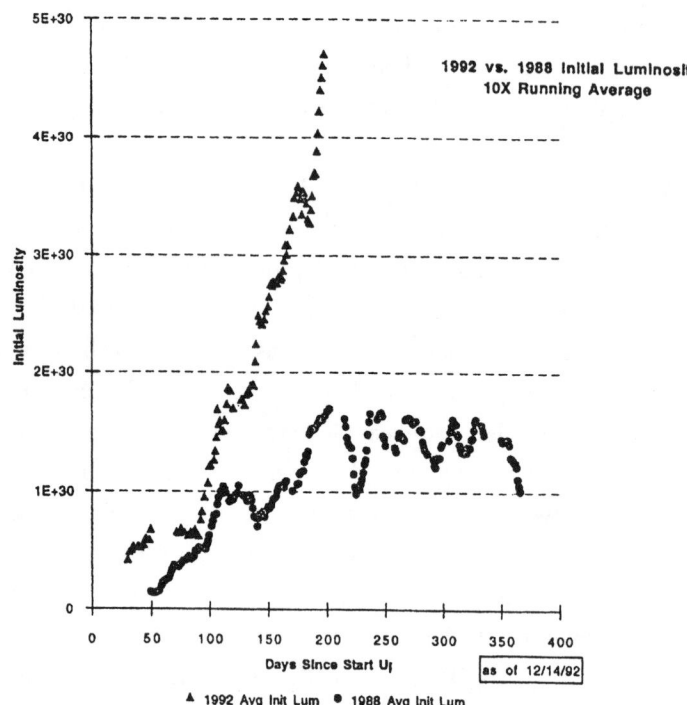

Figure 1. Comparison of the weekly luminosity and total luminosity for this collider run and last collider run.

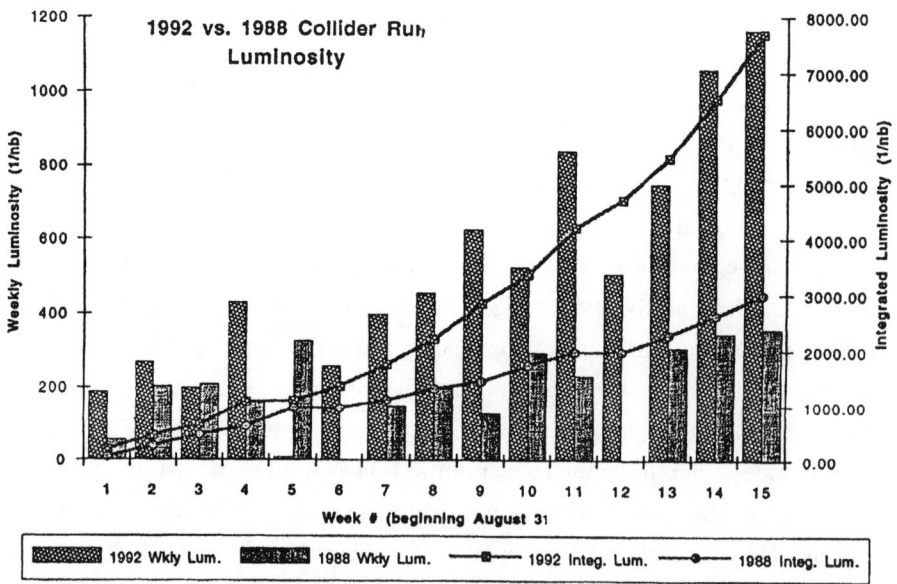

Figure 2. Comparison of the weekly luminosity and total luminosity for this collider run and last collider run.

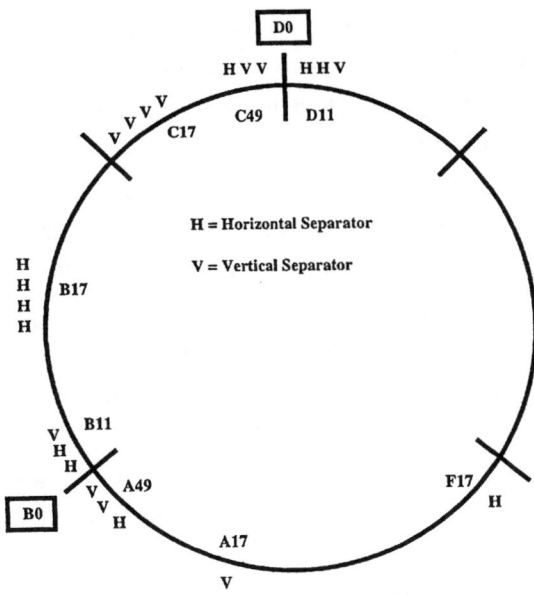

Figure 3.   Location of Separators in the Tevatron.

betatron oscillations in the third of the ring between B0 and D0 and 13 betatron oscillations in the remaining two thirds of the ring. Since the number of betatron oscillations in these bumps is only approximately an integer or half-integer, the local bumps require 3 elements. The location of the middle element of the three bumps in the vertical plane is C17, A17, and B17, F17 in the horizontal plane.

Since the protons and antiprotons are traveling on different orbits they experience different nonlinear fields and therefore have different tunes, coupling, and chromaticities. These differential effects were measured and it was found that differential tune and coupling that were produced could be explained by persistent currents in the main bending dipoles (b2) and the chromaticity sextupoles. However, no differential chromatic effects were observed. A correction scheme was designed to correct these differential effects using existing sextupoles in the secondary correction spool packages in the Tevatron. The idea is to create three circuits, two for adjusting the tunes and one to correct coupling. The correction scheme consists of 46 sextupoles distributed around the ring, 16 normal sextupoles and 30 skew sextupoles. The tune-adjusting sextupoles are connected in pairs such that their chromatic effects cancel. The three correction circuits are configured by controls software. It is necessary to reconfigure the 23 hardware circuits into three software circuits whenever the lattice changes, i.e. at each step in the squeeze. The system (46 sextupoles) has more than enough strength to correct any differential effect for intensities through the Main Injector era. The system was designed to compensate differential tune shifts up to 0.020. However, it is not known if such strong sextupole fields will cause problems related to dynamic aperture effects. Part of the system has been tested, and it was shown that a field strength can be achieved which allows correction of differential tune shifts of 0.010. The partial system is currently being used to control the antiproton tunes at energies up to 500 GeV.

The construction of a colliding beam facility at the D0 long straight section of the Tevatron, coupled with the presence of the CDF detector at the B0 straight section, has produced the need for a low-beta insertion that, unlike the old system, permits the simultaneous and essentially independent operation of more than one interaction region. The new low-beta insertion enables simultaneous operation of a multiple of such systems by matching each insertion to the arcs of the machine in betatron and momentum space. Matched insertions in collider mode are independent except for the need to maintain a constant tune with distributed tune correction quadrupoles in the rest of the accelerator. The addition of each low-beta insertion to the accelerator lattice raises the tune of the accelerator approximately a half unit unless compensated. The operating point of the collider with two collision regions has vertical and horizontal tunes of 20.576 and 20.585 respectively. By comparison, the lattice used for fixed target has no low-beta inserts and has a tune of 19.4.

The insertions at both experimental regions are optically identical. The original low-beta region at B0 was deliberately unmatched and produced a large beta and dispersion wave in the rest of the accelerator, and it was replaced. Each new low-beta insertion is composed of 18 high-gradient quadrupoles that are physically located symmetrically around the straight section region and in the arcs. The magnetic gradients are antisymmetric relative to the center. A field-free region, 15.24 m long, is available for each detector between the final quadrupoles. The lattice design is a relatively conventional one; the low-gradient quadrupoles are used to provide the matching into the arcs, the high-gradient ones provide the strong focussing close to the interaction point to give the small beam size. Twelve independent circuits are used to vary the insertion optics. The inclusion of an extra circuit beyond the minimum of eight results in a more efficient insert by allowing some of the quads to run at less than their maximum values. The β^* at injection is 170 cm. Currently the β^* at the end of the squeeze is 50 cm. The magnets are capable of going to a β^* of 25 cm.

Both the separators and the low-beta insertions were commissioned at the beginning of the current collider run, which started in May 1992. Both systems have performed reliably. Further upgrades to the Fermilab accelerator complex include upgrading the Linac to 400 MeV from 200 MeV, improvements to the antiproton source, and construction of the Main Injector which will replace the Main Ring. These improvements are expected to yield another factor of ten improvement in the luminosity that can be delivered by the Tevatron collider.

The Antiproton Source has been performing exceptionally well in the collider run. Figure 4 shows a comparison between this collider run and the last run of the weekly antiproton production, and the integrated antiproton total for the first 15 weeks. The record antiproton stacking rate now stands at 3.83 mA/hour, just short of the goal of 4 mA/hour for collider run Ia. The stacking record previous to the 1992 collider run was 2.5 mA/hour. This improvement is due primarily to increased delivered Main Ring intensity from improvements in the Booster on antiproton production cycles, and to major upgrades to all the stochastic cooling systems in the Antiproton Source. This includes the core cooling system (upgraded from 2 to 4 GHz bandwidth to 4 to 8 GHz bandwidth since the last collider run), the stacktail cooling system (improved pickup arrays, plus many other modifications), the stacktail betatron system (newly

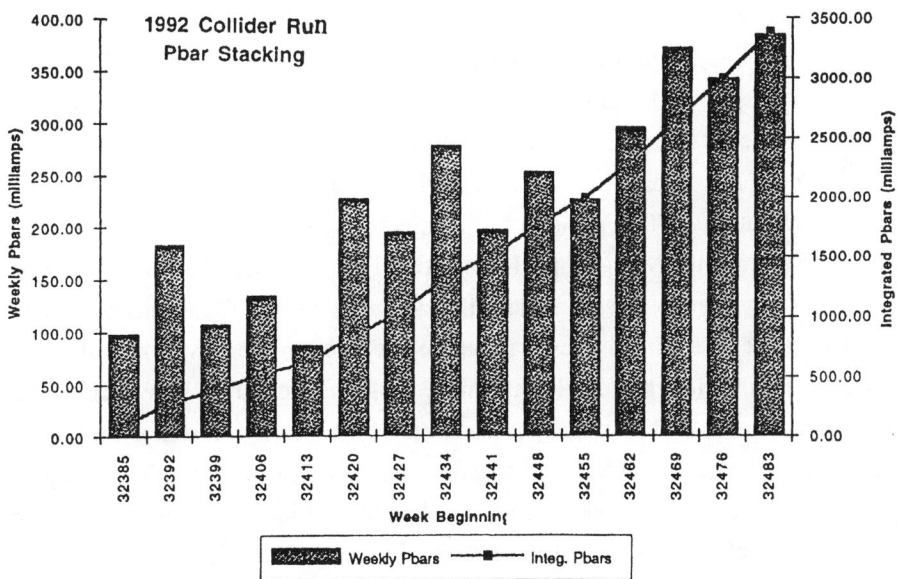

Figure 4. Comparison of the weekly antiproton production and integrated total for the first fifteen weeks of this collider run and last collider run.

commissioned), and the Debuncher cooling systems (increased power capability). With several more planned modifications and some fine tuning, we anticipate that we will exceed this run's goal of 4 mA/hour stacking rate.

The efficiency of transferring antiprotons from the Accumulator Ring to the Tevatron has increased dramatically, from 25% to over 60%, during the course of the present collider run. This is a complicated process involving beam manipulations in the Accumulator, Main Ring, and Tevatron. In November the ion clearing system in the Accumulator Ring was upgraded from 100 V to 600 V, eliminating the trapped ions which are a source of beam instability to the antiproton stack. This has had three beneficial effects. First, transverse beam emittances can be made smaller, greatly improving the antiproton transfer efficiency through the limited aperture of the Main Ring to the Tevatron. Second, transfers can be made from larger stacks (we now routinely transfer from 80+ mA stacks). Third, the 4 to 8-GHz core momentum stochastic cooling system can now be used during transfers, reducing the beam momentum width and thereby further increasing the fraction of antiprotons transferred to the Tevatron. These numbers are expected to continue to increase as the collider run progresses.

3 FUTURE

The Fermilab Main Injector (FMI) project is the centerpiece of Fermilab's program for the 1990s, Fermilab III. It is designed to support a luminosity of at least 5×10^{31} cm^{-2} sec^{-1} in the Tevatron collider while providing the potential for development of new capabilities at Fermilab in the realm of rare neutral K decays and neutrino oscillations. The Fermilab Main Injector is an 8 to 150-GeV synchrotron designed to replace the Main Ring. The existing Fermilab Main Ring represents, with the changes required to accommodate the demands of the Tevatron era, a serious limitation in the beam intensities that can be provided to the Tevatron and onto the antiproton production target. The proposed Main Injector will overcome these limitations. The FMI project has passed several significant milestones over the past several months and is now proceeding rapidly towards the initiation of physical construction. The project received a $11.65M appropriation in FY1992 and $15M for the current fiscal year. Through the Energy Systems Acquisition Advisory Board process the Department of Energy has authorized the application of funds to construction of the underground enclosure and service building at the point of tangency between the Main Injector and the Tevatron (MI-60), and to the preparation of bid packages for all remaining project construction.

Earlier this summer, on July 24, wetland mitigation work was initiated. The construction of the FMI requires the filling of approximately six acres of wetlands. Permission to do this has been granted by the U.S. Army Corps of Engineers contingent upon the creation of nine acres of new wetlands in close proximity to those destroyed. The wetland mitigation project includes the immediate filling of approximately four acres of wetland and creation of the full nine-acre mitigation area. Earth moving is now completed, with only spring plantings and five years of monitoring remaining to complete the mitigation.

The accelerator will be constructed of 344 new conventional dipole magnets, but using quadrupoles, accelerating rf cavities and instrumentation from the Main Ring. The lattice features two types of cells: normal (34.6 m) cells in the arcs and straight sections, and dispersion-suppressor (25.9 m) cells adjacent to the straight sections to reduce the dispersion to zero in the straight. The tighter focussing and smaller dispersion result in physically smaller beams than in the Main Ring, and an acceptance over three times as large. The standard cell consists of a FODO lattice containing two dipoles in each half-cell; the dipole length is 6 m. The straight section cells are the same length as the normal cells. The dispersion suppressor cells require special length quadrupoles and dipoles; again, the lattice is a simple FODO array with two 4-m dipoles between quadrupoles.

The dipole magnet has been designed and two prototypes have been constructed and measured. Both magnets exhibit field quality well described by computer models and within the performance specification. The magnets have four turns per pole of conductor with dimensions 2.54 cm x 10.16 cm, with a peak current of 9375 A, and a peak power of 75 kW. The poletip gap is 5 cm, and the good-field region ($\Delta B/B < 1 \times 10^{-4}$) exceeds ±4.4 cm at injection. At the peak field (1.72 T) there is significant saturation producing a sextupole field which determines the required strength of the chromaticity-controlling sextupole magnets. Twelve production prototype dipoles are to be built in 1993 with magnet production starting late in the year.

Four new beamlines are required to connect the FMI to the Fermilab accelerator complex: a 760-m 8-GeV line to connect to the Booster, two 260-m beamlines to connect to the Tevatron (one for protons and one for antiprotons), and a beamline to allow transport of 120-GeV protons to the antiproton production target or to the experimental areas. This last beamline utilizes the Main Ring remnant that will remain in F-sector of the Tevatron enclosure. Designs exist for all of these beamlines. The 8-GeV line has been designed with a lattice strongly resembling the Main Injector lattice, except for the matching section at the Booster end of the line. The beamline utilizes existing Main Ring dipoles for most of the bending elements. The two 150-GeV lines are almost mirror images of one another. They transfer beams to the Tevatron at a point 13 m downstream (in the proton direction) of the center of the F0 straight section with common Lambertson magnets to place the beam onto the Tevatron vertical closed orbit. These beamlines utilize Main Ring magnets for all of the dipoles and quadrupoles. Beam transfers to the Main Ring remnant utilize the same beamline as for proton transfers to the Tevatron. The Lambertson magnets at the Tevatron are turned off, allowing the beam to continue upwards to the Main Ring elevation.

Extensive tracking studies are underway. To date, the most complete studies have been done at the injection energy of 8 GeV. These studies include the measured multipoles from the prototype Main Injector dipoles, and from the Main Ring quadrupoles. Random errors have been included, with distributions based on the more extensive data from Main Ring dipoles. Misalignment errors have been included, with rms position errors of 0.25 mm assumed in both x and y for dipoles and quadrupoles, and rms roll angles of 0.5 mrad for the dipoles. The tracking studies include chromaticity-correcting sextupole magnets which correct the chromaticites to −5 in both planes (+5 for tracking at energies above transition). The rf voltage is turned on at the nominal injection value, and synchrotron oscillations corresponding to the maximum expected momentum offset are included. The tracking studies reveal an uncorrected closed orbit error of 4 to 8 mm, varying with the seed of the random distribution, and survival for the full 35,000 turns (the nominal injection dwell time for injecting six Booster batches) for particles with initial amplitudes of < 20 mm. This corresponds to an admittance of almost 60 π mm-mrad, a full factor of two larger than the largest (95%) beam emittances anticipated. It appears that the admittance at 8 GeV is limited by the amplitude-dependent tune shift from the octupole component on the Main Ring quads. The inclusion of octupole correction elements (which can be recovered from the Main Ring) increases the admittance even further. Limited tracking studies have been done at transition, 120 GeV and 150 GeV. While the work done so far has yielded no surprises, much more work remains, particularly with regard to the resonant extraction of the beam at 120 GeV.

R&D work in support of the project is also well advanced. In addition to the dipole effort discussed earlier, R&D on the 1000-V/10,000-A power supply required to power the dipoles has also been initiated. Twelve such supplies are required in the accelerator. The first prototype will be assembled next spring in the area that was most recently home to the SSC string test. Finally, significant progress has been made on the

200-kW rf power amplifier required for the project. Eighteen units will be required ultimately. Work will continue through 1993 on this critical component.

Construction of the Main Injector is being coordinated by the Fermilab Accelerator Division with significant contributions expected from all divisions of the laboratory. Much work has been accomplished in technical design, in project management and in permit applications. Fermilab was awarded a grant from the State of Illinois, with which the architect/engineering firm Fluor-Daniel was contracted to provide advance conceptual design work for the civil construction and site mitigation required for the project. The state money was also used to prepare an Environmental Assessment which recommended a Finding of No Significant Impact (FONSI) for the project. Following the determination by the DOE that this document was acceptable, the FONSI was published in the Federal Register on July 6, allowing the mitigation work to proceed. All other construction permits have been secured and it is expected that construction of the MI-60 underground enclosure will start early in January 1993.

An important parameter throughout the accelerator chain is ε, the transverse invariant beam emittance, a measure of beam brightness. The initial value of ε is established in the Linac and its pre-injector. At each step in the acceleration process, ε can only become larger through dilution processes. The first place where this occurs is in transferring the protons into the Booster. The most straightforward way in which to preserve the small emittance of the Linac is to raise the energy at which the Linac beam is injected into the Booster.

The Linac currently provides 0.7×10^{10} H$^-$/bunch/turn into the Booster with $\varepsilon = 7$ to 8π mm-mrad (95% normalized). Theoretical calculations and experimental measurements have established that the tune-spread caused by space-charge, Δv, degrades transverse emittance of the proton beam at injection into the Booster at intensities corresponding to two turns or greater injected from the Linac. It has been shown that the degradation of the transverse beam emittance is determined by a space-charge tune-shift limit of $\Delta v = 0.37$ in the Booster at the start of acceleration after the beam has been bunched. As the space-charge tune shift is inversely proportional to a relativistic factor $\beta\gamma^2$, and proportional to N/ε, it should be possible to reduce the emittance in the space-charge-dominated intensity regime by increasing the injection energy from 200 MeV to 400 MeV, a change in $\beta\gamma^2$ of 1.75. Normalized emittances of 10π for bunch intensities of 3.5×10^{10} particles out of the Booster could be expected instead of the currently measured 18π numbers. Conversely, transmitted intensities in the Booster and Main Ring, which are limited by the transverse emittance growth, could be expected to increase by approximately the same 1.75 factor for the emittance that is now being used.

Most of the construction for the 400-MeV Linac Upgrade has been completed. Beam commissioning plans have been started in preparation for the laboratory shutdown in the summer of 1993. All of the new Linac is in the temporary position adjacent to the operating drift tube Linac inside the present Linac enclosure. The seven new side-coupled accelerator modules are under power without beam to verify overall system reliability. All rf power systems in the Linac gallery now operate 24 hours a day with little intervention. All major components for the 400-MeV transfer line to the Booster are complete. Only the four orbit-bump magnets for the Booster Injection Girder are still in fabrication, and these should be complete by March 1993.

The operating energy of the Tevatron is defined by the ability of the superconducting cable to carry the necessary current density in the presence of high magnetic fields; the so-called cable short sample limit. The 1000 magnets in the ring form an ensemble with a distribution of quench currents with a full width of ± 60 GeV; the machine operating energy is determined by the lowest quench current of any magnet in the ring. There are two possible methods of increasing the machine energy: identifying and replacing the weak magnets with higher quench current elements, or lowering the temperature of the magnets, which increases the critical current. Since the quench current of each magnet was measured prior to installation in the ring we can estimate with good accuracy that ~40% of the magnets would need replacement to reach 1000 GeV at the present cryogenic temperature. Since weak magnets can only be identified and replaced serially this option is not viable, independent of financial considerations. We have therefore decided to implement lower temperature operation. For the Tevatron magnets the enhanced current-carrying capability amounts to about 15% per degree Kelvin.

The present operating threshold of the Tevatron is 935 GeV with a peak coil temperature of 4.9 K. A coil temperature reduction of 1.0 K is expected with the new system. Assuming that no mechanical limits are reached in the magnets, this will result in an energy of 1075 GeV. In reality, mechanical limits and weak splices will be encountered between 935 GeV and 1075 GeV which will require a small number of magnet changes. The 1075-GeV energy level represents the operational limit of several major subsystems of the Tevatron: the cryogenic system capacity, power supply and power lead current, and magnet collar strength.

Lower temperature will be achieved by pumping on the magnet two-phase helium circuit with cold vapor compressors. A subcooling dewar will be located between the refrigerator and the magnet strings to buffer oscillations. The dewar is sized to help minimize the transients caused by the AC losses during ramp turn on/off in Fixed-Target Physics.

Control of the system will be achieved by replacing the 10-year-old multibus I, Z80 based refrigerator control system. This system is currently at the limit of input channels, controlled devices, and control software. A new Multibus II and Intel 80386 based system will handle the additional hardware and software necessary and allow for expansion to more elaborate system control and tuning.

The Fermilab satellite refrigeration system and the Central Helium Liquefier (CHL) provide cooling for approximately 1000 superconducting magnets and assorted cryogenic components that make up the Tevatron. The current system is capable of maintaining magnet temperatures at about 4.9 K, allowing the Tevatron to operate reliably at a beam energy of 900 GeV. The cryogenic system upgrade requires that the cold compressors, with an accompanying phase separator return dewar, be fitted into existing refrigerator building piping. The phase separator dewar protects the cold compressor from possible damage due to liquid surges and provides a source of stored refrigeration available from the liquid in the dewar.

This fall, Fermilab took delivery of 27 centrifugal cold compressors. A centrifugal type cold compressor was small enough to fit into the existing system, requiring no further floor space in an already crowded satellite refrigerator building. Seven of the 25 replacement satellite valve boxes have been delivered. Continued shipment is expected from the first of 1993. Nearly all subassemblies have been completed.

The satellite refrigerator controls are being upgraded to accommodate the added requirements of the low temperature upgrade as well as to allow for future expansion. A prototype system had been operational in an auxiliary satellite refrigerator since June. Installation is currently underway in one of the eight satellite compressor buildings during collider operations. It is planned to install the new controls system in as many compressor buildings as possible prior to the next shutdown.

A SUMMARY OF STUDIES OF PARTICLE STABILITY AT RHIC*

G. Parzen

Brookhaven National Laboratory, Upton, NY 11973, USA

1 INTRODUCTION

This paper summarizes some studies of particle stability done at RHIC. The topics reported on include

1. Long term tracking and the dynamic aperture
2. Linear coupling effects and their correction
3. Tune spreads in the beam due to field multipoles in the magnets and their correction.

2 LONG-TERM TRACKING AND THE DYNAMIC APERTURE

Those working on this problem include J. Claus, G.F. Dell, H. Hahn, M. Harrison, D. Maletić, J. Milutinovic, G. Parzen, S. Peggs, A.G. Ruggiero, S. Tepikian, D. Trbojevic and J. Wei. Tracking programs used included ORBIT, PATRICIA, PATRIS, and TEAPOT.

2.1 Tracking Results without Synchrotron Oscillations

Particles were tracked[1,2] for about 10^6 turns. Error field multipoles were present in each magnet; random multipoles and systematic multipoles up to the tenth order were included. The momentum of the particle was fixed; no synchrotron oscillations were present. Ten different distributions of field multipoles were studied.

Figure 1 shows a survival plot for a RHIC lattice with six $\beta^* = 6$ insertions. The survival time in turns is plotted against the initial betatron amplitude for $\Delta p/p = 0$ particles and for 10 different distributions of field errors. The particles are started with $x' = y' = 0$ and $\epsilon_y = \epsilon_x$. The figure also shows a rough linear extrapolation to 3×10^9 turns which is about a 10-hour survival time in RHIC. In this case the long-term effects are small. The dynamic aperture, A_{SL}, does not vary much with the survival time.

Figure 2 shows a similar plot for a RHIC lattice with six $\beta^* = 2$ insertions. In this case the dynamic aperture, A_{SL}, depends appreciably on the survival time. It decreases by about 20% as the required survival time increases from 400 turns to 10^6 turns. The linear extrapolation to 3×10^9 turns is not to be trusted very much. One may note that this decrease of A_{SL} with the survival time is present here in the absence of synchrotron oscillations.

The $\beta^* = 2$ results are shown also in Fig. 3, where the survival time is plotted against initial betatron amplitude for four distributions of field errors. In one case, seed No. 8, the particle starting with $x_0 = 6$ mm was tracked for about 8×10^6 turns where it went unstable. In this case, there is some

---

* Work performed under the auspices of the U.S. Department of Energy.

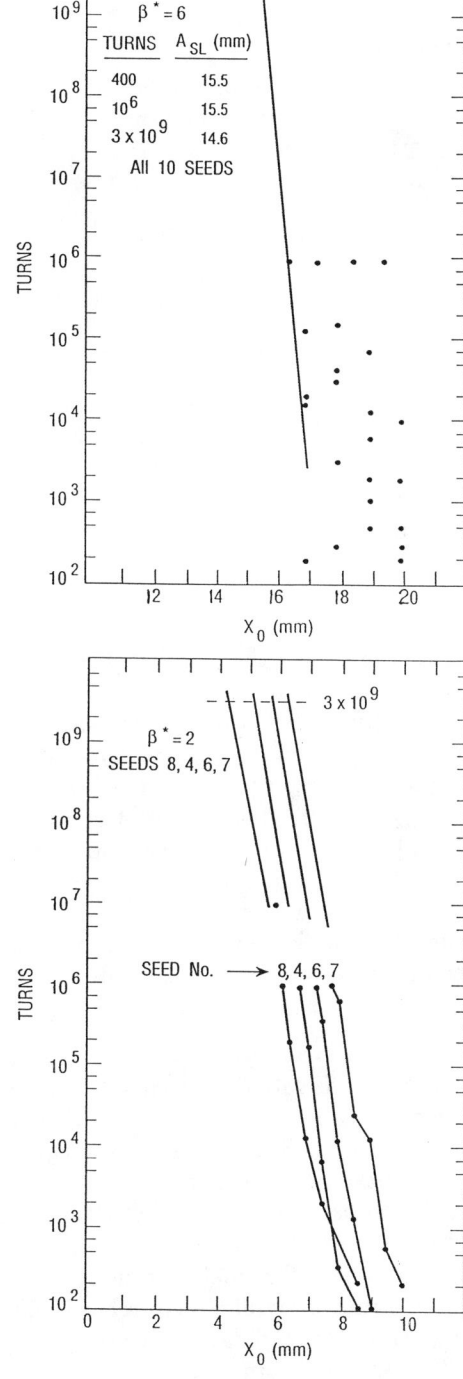

Fig. 1:
Survival time in turns versus the initial betatron amplitude for 6 $\beta^* = 6$ insertions.

Fig. 2:
Survival time versus the initial betatron amplitude for 6 $\beta^* = 2$ insertions.

Fig. 3:
Survival time versus the initial betatron amplitude for 6 $\beta^* = 2$ insertions for four distributions of field errors.

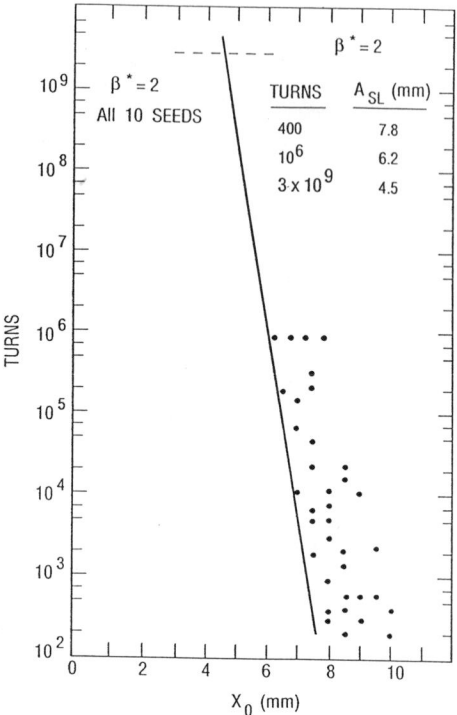

indication that the dependence of A_{SL} on the survival time is starting to flatten out.

In the above tracking studies, the particles were started with initial conditions $x' = y' = 0$, $x \neq 0$, $y \neq 0$, $\epsilon_x = \epsilon_y$. A multiparticle approach was explored[3] where particles are started with different x, x', y, y' with $\epsilon_x = \epsilon_y$ and $x' \neq 0$, $y' \neq 0$. Figure 4 compares the results for the dynamic aperture found using the one-particle approach with those found using the multiparticle approach. Figure 4 shows histograms which indicate the largest initial x that is stable for each of 10 different distribution of field errors in the magnets. For the multiparticle case, an effective initial x is assigned to each set of starting conditions according to the rule

$$2 x^2/\beta_x = (\epsilon_x + \epsilon_y)_{\text{initial}}. \tag{1}$$

Figure 4 appears to show that the multiparticle approach, tracking 100 particles for 1000 turns for each set of field errors, gives about the same dynamic aperture as the one-particle approach, tracking one particle with $x' = y' = 0$ for 10^6 turns. Studies on the multiparticle approach are continuing.

Long-term tracking studies are being done including the effect of synchrotron oscillations. The ORBIT program has been modified[4] to use point magnets in order to obtain symplectic tracking when synchrotron oscillations are present. The methods used are similar to those of the TEAPOT program with some differences regarding the choice of the reference orbit. Tracking studies including synchrotron oscillations are also being done[5] with the TEAPOT program.

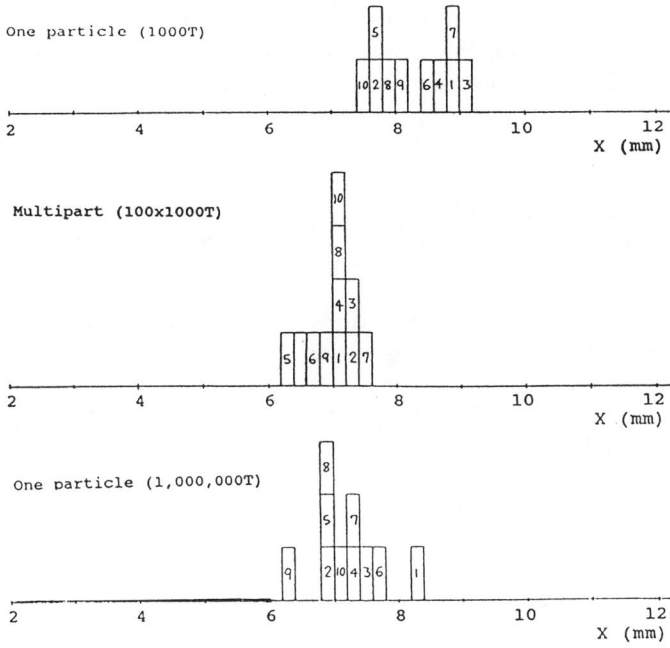

Fig. 4: Histograms showing the largest stable initial x for 10 distributions of field errors and 6 $\beta^* = 2$ insertions.

2.2 Some On-Going Long-Term Tracking Studies

The following describes studies on several aspects of long-term tracking. The effects of round-off error on the accuracy were studied.

Using[6] the PATRIS program and the RHIC AGR lattice, with no sextupoles and no field errors, studies indicate that 14-figure accuracy is reduced to 7-figure accuracy after 10^6 turns.

Using[7] a one-cell lattice, and tracking runs for up to 10^9 cells, the loss of accuracy was studied using a procedure that allows the round-off error to be varied continuously.

Studies are being done[8] on understanding the dynamic aperture by doing detailed studies of the motion near the stability boundary and attempting to correlate the results with resonances.

Studies are being done[9] to estimate the effects of the magnetic field not being Maxwellian by adding the field terms required to make the fields Maxwellian and noting the change in tracking results.

Studies are being done[10] to see the effects of doing tracking using a quantum mechanical description of the particle motion.

2.3 Gradient Ripple Effects

A study was done[11] of the effect of a gradient ripple in the quadrupoles of RHIC. The gradients in the quadrupoles were changed with time according to

$$G = G_0 \left(1 + \frac{\Delta G}{G} \sin(2\pi t/T)\right). \qquad (2)$$

The corresponding tune ripple is given by

$$\Delta\nu = 80 \Delta G/G \qquad (3)$$

for a RHIC lattice with 6 $\beta^* = 2$ insertions. In Fig. 5, the dynamic aperture for 10^6 turns is plotted against the magnitude of the ripple, $\Delta G/G$, for a 60-cycle ripple. Limiting the dynamic aperture loss to 5%, one finds a tolerance for $\Delta G/G$ of $\Delta G/G \leq 0.5 \times 10^{-5}$ corresponding to tune ripple of $\Delta\nu \leq 0.4 \times 10^{-3}$. There are no synchrotron oscillations in this study, and $\Delta p/p = 0$.

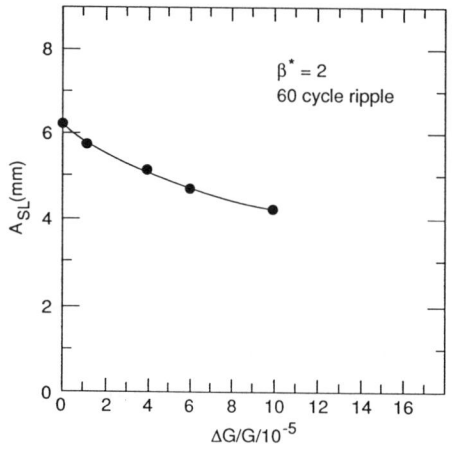

Fig. 5: Dynamic aperture versus the gradient ripple for 6 $\beta^* = 2$ insertions.

3 LINEAR COUPLING EFFECTS

Linear coupling effects have become more important for proton colliders. The reasons for this include the following:

1. Superconducting magnets tend to give rise to larger skew quadrupole error fields.
2. The tune is close to the coupling resonance in order to avoid higher-order nonlinear resonances.
3. Low-beta insertions have higher beta magnets which can cause significant coupling effects.

The most important linear coupling effect is probably the tune splitting driven by the nearby difference resonance. This effect is roughly linear in a_1, the field multipole that describes the skew quadrupole field. For this effect the important harmonic of a_1 is the $\nu_x - \nu_y$ harmonic. This effect can be largely corrected with a 2-family skew quadrupole correction system.

There are also a number of effects driven mostly by the harmonics of a_1 close to $\nu_x + \nu_y$ which include the following:

$\nu_x + \nu_y$ effects

1. Residual higher-order tune shift that remains after correction with a 2-family correction system. This tune shift is roughly quadratic in a_1.
2. Possible large shifts in the beta-functions.
3. Emittance increase at injection.
4. Normal-mode rotation angle $\neq 0$, after correction with a 2-family correction system.

Those working on these problems include, on theory, V. Garczynski, G. Parzen, S. Peggs, A.G. Ruggiero, and S. Tepikian; and on simulation and correction, G.F. Dell, H. Hahn, M.A. Harrison, S.Y. Lee, G. Parzen, S. Peggs, and S. Tepikian. Programs used include ORBIT and PATRICIA.

Two approaches have been used in the theoretical treatments. These may be described as follows:

1. Find the change in the 4×4 one-turn transfer matrix due to the presence of the a_1 field, using point a_1 errors. Use the result for the transfer matrix to find orbit parameters.[12,13]
2. Find the 4 eigenfunctions of the transfer matrix in the presence of the a_1 field. Use the results for the eigenfunctions to find the orbit parameters.[14]

The theory and its results are fairly complicated. A few results are useful for the insight they provide in designing a correction system for the linear coupling effects. The eigenfunction approach leads to the following driving terms[14] for the linear coupling effects.

The driving terms for the tune splitting due to the nearby difference resonance, for the case when this difference resonance is $\nu_x - \nu_y = 0$, is given by

$$\Delta \nu = \frac{1}{4\pi \rho} \int ds \, (\beta_x \beta_y)^{\frac{1}{2}} a_1 \exp\left[i\bar{\nu}(\theta_x - \theta_y)\right], \tag{4}$$

$$\theta_x = \psi_x/\nu_x, \quad \theta_y = \psi_y/\nu_y, \quad \bar{\nu} = (\nu_x + \nu_y)/2.$$

For the four $\nu_x + \nu_y$ effects, the higher-order tune shift, the shift in the beta-functions, etc., the driving terms are

$$b_n = \frac{1}{4\pi \rho} \int ds \, (\beta_x \beta_y)^{\frac{1}{2}} a_1 \exp\left[i((n - \nu_y)\theta_x + \nu_y \theta_y)\right],$$

$$c_n = \frac{1}{4\pi \rho} \int ds \, (\beta_x \beta_y)^{\frac{1}{2}} a_1 \exp\left[i(\nu_x \theta_x + (n - \nu_x)\theta_y)\right]. \tag{5}$$

The b_n, c_n driving terms usually occur multiplied by the resonance factor $1/(n - \nu_x - \nu_y)$. The b_n, c_n for n near $\nu_x + \nu_y$ are the important driving terms.

The four $\nu_x + \nu_y$ effects listed above have the same driving terms. Correcting one of them, like the residual higher-order tune shift, may be expected to correct the other effects to a considerable extent. This was observed in a simulation study.

A simulation study has been done[15] which computes the effects of linear coupling and the performance of a proposed correction system to correct these effects. Some of the results of this simulation study are given in Table 1 and Table 2. The simulation study indicates that 4 or 5 families of a_1 correctors may be able to correct all the linear coupling effects.

Table 1: Results for the correction of the tune splitting for a $\beta^* = 2$ RHIC lattice using a 2-family tune-splitting correction system set to make $\Delta \nu = 0$; $a_1 = 0$, $\nu_1 = 0.826$, $\nu_2 = 0.821$.

Error Field Dist.	Uncorrected			Corrected		
	ν_1	ν_2	$|\nu_1 - \nu_1|/10^{-3}$	ν_1	ν_2	$|\nu_1 - \nu_2|/10^{-3}$
1	0.796	0.854	59	0.828	0.823	6
2	0.707	0.935	228	0.838	0.819	19
3	0.869	0.783	86	0.825	0.829	4
4	0.772	0.883	111	0.831	0.823	7
5	0.779	0.872	93	0.836	0.820	16
6	0.848	0.805	43	0.832	0.821	11
7	0.840	0.847	7	0.852	0.834	18
8	0.742	0.895	153	0.838	0.818	20
9	0.785	0.866	81	0.828	0.823	6
10	0.749	0.891	142	0.823	0.827	5

Table 2: Results for the beta-functions for a $\beta^* = 2$ RHIC lattice before and after correction with a 2-family tune-splitting correction system set to make $\Delta \nu = 0$. With $a_1 = 0$, $\beta_{max} = 50$.

Error Field	Uncorrected		Corrected	
	$\beta_{1,max}$ at QF(m)	$\beta_{2,max}$ at QD(m)	$\beta_{1,max}$ at QF(m)	$\beta_{2,max}$ at QD(m)
1	68	75	60	58
2	138	95	53	58
3	89	78	55	66
4	83	74	65	62
5	69	65	82	64
6	65	67	58	61
7	76	82	100	80
8	78	79	68	61
9	80	78	57	61
10	77	87	67	74

4 BEAM TUNE SPREAD DUE TO FIELD MULTIPOLES

Field multipoles may be present due to errors, iron saturation, coil design, etc., which produce a tune shift that depends on $\Delta p/p$, ϵ_x and ϵ_y. The resulting beam tune spread may be comparable to the width of the resonance-free box in tune space in which RHIC will operate.

Those working on this problem include J. Claus, G.F. Dell, H. Hahn, M.A. Harrison, G. Parzen, S. Peggs, M. Rhoades-Brown, A.G. Ruggiero, D. Trbojevic, and J. Wei.

The relative importance of the various sources of this tune spread depends on the size of the field multipoles, which changes with time. At this time it appears that a large beam tune spread for Au ions may occur at $\gamma = 30$ that is due primarily to the random b_2, b_3, b_4 in the dipoles. This assumes $\beta^* \geq 2$. Tune spreads for $\beta^* < 2$ require special consideration. In the following, it is usually assumed that $\beta^* \geq 2$.

By random multipoles, one means those multipoles that are not intended to be present in the design. Although they are called "random," they may not have a truly random distribution.

4.1 b_3, b_4, a_3, a_4 Multipoles

It has been found[16-19] that the tune spread due to the random b_3 and b_4 is generated by the average b_3, b_4 in the dipoles. Thus the tune spread depends on how well one can control the average b_3, b_4 in the dipoles. One choice for the achievable average b_3, b_4 would give a worst-case tune spread of about $\Delta \nu = 15 \times 10^{-3}$ in RHIC.

The tune spread due to the random b_3, b_4 can be adequately corrected with b_3 and b_4 correctors at the center and ends of the dipoles. Correctors at the dipole center are essential if the tune spread is large, but this is not possible in RHIC, which has just one dipole in each half-cell. Studies have been done[19,20] putting the center corrector in the insertion region where β_x, β_y, X_p are about the same as in the center of the dipole. These lumped b_4 correctors raise some concern as they may generate a strong 1/5 resonance. Tracking studies done[20] so far show no loss in aperture due to the lumped b_3, b_4 correctors in the insertion region.

4.2 b_2, a_2 Multipoles

The tune spread due to the random b_2, a_2 is quadratic in the strength of the b_2 multipole. Nevertheless, one observes[17] field error distributions that produce an appreciable tune spread. One worse case was seen[17] that produced a tune spread of $\Delta \nu = 6 \times 10^{-3}$. This tune spread may be partly correctable using the chromaticity sextupoles.

4.3 Some Comments

If $\beta^* \leq 1$ then b_5 near the high-beta quadrupoles may become important.[21] The tune spread due to b_k goes like some high power of β_x, β_y for large k. Thus for $\beta^* \leq 1$, the b_k, a_k in the high-beta quadrupole triplet become important.

It may be helpful that the tune spread due to b_k, a_k and the tune spread due to beam-beam interaction are not directly additive. The first is larger at large betatron amplitudes and the second is larger at small betatron amplitudes. Using analytical results[22] for the tune shifts, the tune spreads due to all the sources – field multipoles, beam-beam interaction, space charge – have been added together to produce results like those shown in Fig. 6.

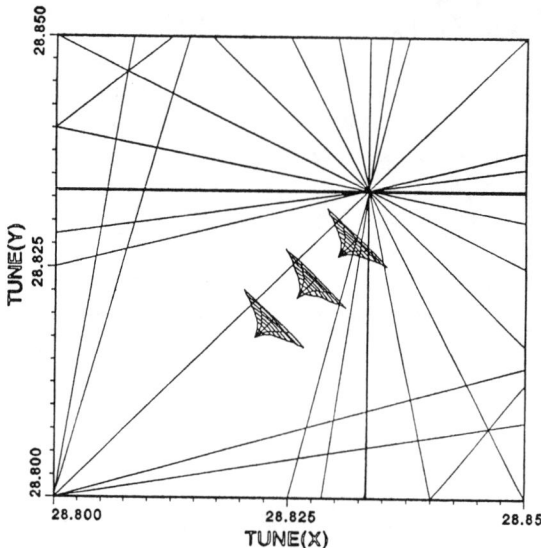

Fig. 6:
The beam tune spread in the beam due to various sources, for Au ions at $\gamma = 100$ with emittances of 40π and for $\Delta p/p = 0, \pm 2 \times 10^{-3}$.

5 REFERENCES

1. G. Parzen, Long Term Stability in RHIC, BNL Report AD/RHIC-85 (1991).
2. G.F. Dell and G. Parzen, Aperture Determination by Long Term and Multiparticle Tracking, BNL Report AD/RHIC-95 (1991).
3. G.F. Dell, private communication (1992).
4. G. Parzen, private communication (1992).
5. S. Tepikian, private communication (1992).
6. J. Milutinovic and A.G. Ruggiero, Proc. of Workshop on the Stability of Particle Motion in Storage Rings, BNL, October 1992.
7. D. Maletić, private communication (1992).
8. J. Milutinovic and A.G. Ruggiero, Long Term Tracking with PATRIS, Part 3, BNL Report AD/AP-44 (1992).
9. D. Maletić and A.G. Ruggiero, Exact Physical Model for Magnets in Storage Rings, BNL Report AD/AP-50 (1992).
10. D. Maletić, private communication (1992).
11. G. Parzen, private communication (1990).

12. V. Garczynski, Effects Due to Linear Coupling,to Second Order in the Skew Quadrupole Strengths, BNL Report AD/AP-39 (1992).
    Beta Function Distortion Due to Linear Coupling, BNL Report AD/AP-24 (1991).
13. A.G. Ruggiero, Exact Analysis to Any Order of the Linear Coupling Problem in the Thin Lens Model, in Proc. Advanced Beam Dynamics Workshop, AIP Conf. Proc. No. 255 (1991), p. 139.
14. G. Parzen, Theory of the Tune Shift Due to Linear Coupling, BNL Report AD/RHIC-100 (1991).
    Theory of the Beta Function Due to Linear Coupling, BNL Report AD/RHIC-102 (1991).
    Beta Functions, the Normal Mode Rotation Angle, and Eigenfunctions in the Presence of Linear Coupling, BNL Report AD/AP-49 (1992).
15. G. Parzen, Tune Shifts and Beta Functions Due to Linear Coupling, in AIP Conf. Proc. No. 255 (1991), p. 131.
    Tune Splitting in the Presence of Linear Coupling, in Proc. 1991 IEEE PAC, p. 1615 (1991).
    Beta Functions in the Presence of Linear Coupling, Ibid., p. 1872.
16. A.G. Ruggiero, Comments on the Effects of Random Errors, BNL Report AD/RHIC/AP-86 (1989).
17. G. Parzen, ν-Spread Due to Random Field Errors, BNL Report AD/RHIC-61 (1990).
    Tune Spread and Tune Splitting Due to Random Field Errors, BNL Report AD/RHIC/AP-102 (1987).
18. G. Parzen, G.F. Dell, H. Hahn and A.G. Ruggiero, Beam ν- Spread Due to Field Errors in RHIC, in Proc. EPAC90, p. 1512 (1990).
19. J. Claus, G.F. Dell, H. Hahn, G. Parzen, M.J. Rhoades-Brown and A.G. Ruggiero, The Decapole Correction Scheme for RHIC, Ibid., p. 1467.
20. G. Parzen, The Beam ν-Spread Due to the Random b_3 and the Random b_4 and its Correction, BNL Report AD/RHIC/RD-30 (1991).
21. J. Wei, private communication (1992).
22. J. Wei and M. Harrison, Tune Spread Due to Magnetic Multipoles in RHIC, in Proc. HEACC92 (1992), to be published.

THE OTHERS

Etienne Forest
Center for Beam Physics
Lawrence Berkeley Laboratory, Berkeley, CA 94720

Abstract

In this paper we present in an informal way the work done in other fields which has an impact on circular ring dynamics. In particular, we put a strong emphasis on celestial mechanics.

1 Introduction

When I was asked by Sandro to give a talk on "Others...," it was loosely defined by Sandro as "Talk about stuff done outside the Laboratories." Since then I have redefined the objective of this talk and I hope that Sandro will excuse the scientific licence I had to exercise.

My talk has three objectives:

1. To classify the others not in terms of their <u>physical location</u>, but in terms of their <u>scientific objectives</u>. I will contrast their fields with ours.

2. To discuss how the work of "Others..." may impact on the topic of this workshop.

3. To leave you with a conclusion, or to be more precise, with a plea, which is addressed primarily to those amongst you who are involved in theoretical investigations and in teaching.

I divide the "Others..." into three categories:

1. Those who deal with beam lines and track particles through them. What they have in common with us is mainly superficial:

    <u>$F = ma$ in magnets and similar hardware problems.</u>

    They do not face beam dynamics stability issues. Nevertheless, there are certain tools and techniques we ought to learn from them.

2. The celestial dynamicists: they do not share the hardware with us. Planets and magnets rhyme but that is about it! However, they are interested in stability issues in complex periodic systems. I will talk mainly about them but also about plasma physicists, chemists, and mathematical physicists.

3. Mathematicians: we can learn a lot from mathematicians— but first, their work has to be digested by theoretical physicists with a strong bend towards applications. Otherwise, it is just like the tower of Babel.

2 Beam Line Physicists

Who are the "Others..."?

Let us start with an obvious possibility: the beam line physicists. I list a few fields:

1. Spectrometer design
2. Electron microscopy
3. Lithography
4. Linacs
5. Experimental beam lines
6. etc. ...

All the individuals in these fields deal with the problem of tracking a charged particle through an electromagnetic field, just like us.

Unfortunately, they differ from us in a very fundamental way which is central to the topic of this workshop. They do not circulate a beam over and over again through the same device: hence they are not interested in stability issues. This has a profound and qualitative impact on the type of approximations they use.

Despite this fact, there are some ideas which we can borrow from the single-pass community. I will list a few:

* Expressing the tracking codes in terms of local quantities (dubbed the LEGO approach by Hirata and yours truly)[1]. Why? Single-pass people (lithography for example) have to deal with the horrible details of each magnet. They must worry about fringe fields and very high-order effects. Given a horribly complicated magnet, the only way to push a particle through correctly, is to concentrate on a description which suits this magnet. Hence it must be local.

  We must learn to think this way. To think "modules" as Karl Brown would say. For example, our relative inability to write a "small machine code" is a direct consequence of our desire for global coordinates. (Closed orbit, Frenet coordinates, curvature, torsion etc. ...all concepts which are not necessary.)

* They analyse the motions by using modules, i.e. finite s-maps.

Of course, if you don't have a nice global Hamiltonian, you are forced to rephrase everything in terms of maps. This is very easy for them to do. The plain Taylor series expansion is usually what they want. It is very easy to see why: take an electron microscope. You come in with parallel rays and you try to focus them on a small spot. All of this is perfectly well defined in terms of the usual (x, x', y, y') coordinates and the transfer matrix of the microscope.

Now, one can be fancy and use Lie operators or generating functions to exploit some symmetries, but, in the final analysis, the position and the angle of the rays is what matters[2].

Not surprisingly, single-pass scientists invented the so-called "matrix codes," which are Taylor series expansions around a design orbit. And they mold their theories around such a representation.

Our theories are geared towards the computation of global quantities which are the result of our attempt to understand the iteration of the map. They are the usual dynamical concepts: apertures, resonances, tune shifts, equilibrium emittances, beta functions, etc. ...

Theoretically they are extracted from some beautiful Hamiltonian in action-angle variables. What do we do if we don't have it anymore???? The answer is a "pseudo-Hamiltonian" for the one-turn map, i.e. Lie operators. Once you circulate through a periodic system, you are naturally forced to consider pseudo--Hamiltonians.

* What else about the single-pass gang? We know they like finite s-maps or black box maps, so they pioneered automatic differentiation to extract the Taylor series approximation of a map. Berz, who came out of the spectrometer group of Wolnik, introduced his "DA-Package" in our community. It has no intrinsic physics: it is an automatic differentiation utility.

The moral of the single-pass story is:
> STAY LOCAL WHEN TRACKING RAYS, MAPS OR MOMENTS
> ANALYSE MAPS WHEN DOING PERTURBATION THEORY
> DON'T BE AFRAID OF TAYLOR SERIES WHEN ANALYZING THE TRACKING DATA

The morass of the single-pass story is:

In the final analysis, our tools and techniques are strongly influenced by the iteration of the map. Hence, despite the hardware similarities (i.e. magnets, beam pipes, etc.), the mathematical tools and techniques are very different. Those guys don't give a damn about iterating a map (*incidently, by map I mean everything from experimental data to approximate analysis of a ring through perturbation theory*). They have it real easy on that count. Unfortunately, this is the real theoretical problem addressed in this workshop.

So, we are back to our search. Who are the people interested in the stability of Hamiltonian systems and what can we learn from them? I found, to my surprise, that the folks in celestial mechanics have a lot to teach us. Not only do they use tools which are similar to ours but, in my view, they have a great theoretical understanding of where they stand and where they are going.

3 Celestial Mechanics: a Field to Emulate

In contrast, our field is not unified. It has been said in previous workshops (such as the Lugano meeting) that they are the "tracking" people on one hand and the "map" people on the other hand. I can tell you as an organiser of this workshop that this is still the belief our community. There is a strong tendency for us to view the world into trackers and mappers because it reflects the "political" divisions within our field. I find this division to be totally false and to be a reflection of our lack of theoretical unity. Anyone familiar with my work will notice that it covers the topics of symplectic integration (i.e. kick codes) [5,4,3], perturbation theory on maps [6,7] and even tracking with approximate maps. If the world is really divided between "trackers" and "mappers" then I suffer from split personality. When I find myself agreeing with Dick Talman on the issue of the so-called "exact tracking" by kick codes and at the same time, agreeing with Alex Dragt or Bob Warnock, this is not split personality on my part. Instead, it is the result of my effort to spend time understanding the unifying concepts in our field. Some would say this is a waste of time, but I disagree. In section C, Chairman Bengtsson will guide an attempt to dot the "i's" and cross the "t's" on the pseudo-issue of trackers and mappers, so I will stop here.

But in the meantime, allow me to use celestial mechanics as a field where greater theoretical unity is present; a field where a faithful "tracker" (13th-order Störmer, really "exact"[8]) and an insane "mapper" (tracking with a crazy integer map for example [9]) is in fact the same individual, none other than our distinguished guest, Professor Scott Tremaine of the Canadian Institute of Theoretical Astrophysics. How can this be?

Before going into the achievements of celestial mechanicists, I want to point out that the mathematical distance between their pure tracking studies and their approximate maps is even greater than ours. Why? First of all, in addition to symplecticity, long-term problems in celestial mechanics are very tricky because the underlying Hamiltonian has several constants of the motion which are non-trivial to preserve during the integration process. Hence, while it is easy for us to replace the exact Hamiltonian with an approximate one (i.e. kick codes) and declare the new model "exact" as Dick Talman would say, it is not so immediate in celestial mechanics. For example, the Hamiltonian is conserved: what does it imply for symplectic integration (kick codes)? The answer is troubling for symplectic integrators. Take the pendulum:

$$H = \frac{p^2}{2} + k[1 - \cos(x)].$$

Now, integrate H exactly for a time t. The solution is free from any kind of chaos because we have one constant of the motion, namely the energy H. Now, submit this H to a special kind of torture called first-order symplectic integration:

Really exact $\qquad\qquad$ Exact in the Talman Sense
$\qquad\downarrow\qquad\qquad\qquad\qquad\qquad\downarrow$
$\exp\left(:-tH:\right) \quad\cong\quad \exp\left(:-t\frac{p^2}{2}:\right)\exp\left(:-t\,k[1-\cos(x)]:\right).$

What is the problem? The exact solution is the regular pendulum while the "kick code" produces the standard map. For all values of t, no matter how small, the standard map produces some chaos!!

I am sure that you can understand now why Professor Tremaine may need a very high-order integrator to be "really exact" and why this is important to the members of his field. When the celestial mechanicists discover a source of chaos in the motion of planets through a less than absolutely exact method of integration, they must always worry about the potential artificial nature of the chaos as my pendulum example suggests. Let me quote Sussman and Wisdom[10] on this topic:

> Altogether, the evidence for the chaotic behavior in these long-term planetary integrations is very convincing, but there remains the logical possibility that exponential divergence is a subtle numerical artifact. To positively conclude that the chaos observed in these long-term planetary integrations is not the result of numerical artifacts requires an unambiguous identification of a physical mechanism and an analytical evaluation to determine that the mechanism actually accounts for the observed chaos.

Let me emphasize that this very cautious and lucid remark comes from the scientist who introduced mapping methods in celestial mechanics: Prof. Jack Wisdom of MIT.

Fortunately, we do not have any nontrivial constant of the motion. Our accelerators are sufficiently crappy that it may be appropriate for us to be "exact" in a Talman or a modified Talman sense. So Dick is right when he says that I do agree with him after all. The important point I am trying to make, is that it is far more difficult to adopt the "Talman" point of view in celestial mechanics. Let us continue with the planetary guys. We want to study the evolution of the whole solar system. What do we have?

1. For 3 Myr or less: direct integration of the exact kind (no impact on us because we can be sloppy, as I said).

2. For 100 Myr:
Symplectic n-body Mapping Integration of Wisdom and Holman[11]

$$H = H_{\text{Kepler}} + H_{\text{interaction}}.$$

Here, this can be viewed as a non-trivial application of the Ruth-Forest symplectic integration method since is not a simple "drift." I suggested this half jokingly to Prof. Martin Duncan not knowing that Wisdom had done it already (1990). Where does it come in accelerator physics?

It comes in the different types of split we can use in symplectic integration. We are not restricted to the drift/kick setup:

$$H_{\text{small machine}} = H_{\text{ideal bend}} + H_{\text{multipole}}.$$

This integrator was first used by Oide at KEK. Note that artificial chaos can be present in Wisdom's Hamiltonian, but it is unlikely in ours. All our chaos is probably real.

3. For a few Gyr:
Here, Duncan, Quinn and Tremaine found it necessary to use a mapping approach first concocted in 1982 by Wisdom to investigate the asteroid gaps. Essentially, it follows this logic:

$$\begin{array}{c} \text{Real Hamiltonian} \\ \downarrow \\ H_{simpler} = \sum_{i=1}^{N} H_i \\ \downarrow \\ \text{Map} = \prod_{i=1}^{N} \exp\left(: -tH_i :\right) \end{array} \quad \text{physical intuition}$$

Note that the last line is first-order symplectic integration on $H_{simpler}$. The first line, which I dubbed "physical intuition," is a complex sequence of approximations which would give nightmares to the average accelerator physicist.

Again, let me remind you that Tremaine and Duncan are "trackers" even more than we are [i.e. symmetric multistep methods for the numerical integration of planetary orbits, Quinlan and Tremaine, or, 3-million-year integration of Quinn, Tremaine and Duncan (1991), etc. ...].

So, what do they have to say about the results of the mapping methods? Let me quote from "The Long-Term Evolution of Orbits in the Solar System: A Mapping Approach"[12]:

> Our results suggest several speculations about the history of the Solar System: (i) there may still be stable bands in the outer Solar System (beyond Saturn) containing residual planetesimals from the proto-planetary disk; it is likely that these bands are sources of the short-period comets; (ii) Pluto may have begun in a near-circular orbit and have reached its present high-inclination, high-eccentricity, resonant orbit as a result of chaotic motion caused by Uranus and Neptune; (iii) the depletion of asteroids in near circular orbits beyond $a = 3.3AU$ may be due to long-term instabilities of orbits in this region.
>
> *(This is real physics and you don't get it by staying "exact," even in the Talman sense! But, here is what I like the best and greatly admire...)*
>
> ....The mappings presented here involve many approximations, including the neglect of mutual planetary perturbations, the neglect of planetary inclinations and eccentricities, the approximation that the test particle eccentricity is small compared to its relative separation from the planet, and the approximation that all perturbations are localized at conjunction (*i.e. delta function kicks I believe*). Clearly the mapping is not a satisfactory substitute to numerical integration. Nevertheless, the experiments presented here are the first to follow a dynamical system resembling the Solar System for a time equal to the age of the Solar System...

I believe that we have a great deal to learn from these gentlemen. In our field, Dragt, Irwin and Warnock have advocated the kind of approximations performed by Tremaine and his colleagues. Despite many more pitfalls and subtleties, the celestial mechanicists perform the necessary studies. We ought to draw a lesson from this.

4. What about tracking beyond the ability of the computer itself? Indeed, even if we are symplectic, we get killed by truncation errors. I am very suspicious of tracking results in the chaotic layer just below

the million turns dynamic aperture. What can we do about it? Again the answer is found in astronomical circles. The idea is to replace a floating point map by an integer map:

$$\begin{array}{c} \text{Real map} \\ R^{2N} \quad \rightarrow \quad R^{2N} \\ \Downarrow \text{ Replaced by integer map} \\ Z^{2N} \quad \rightarrow \quad Z^{2N} \end{array}$$

This work was pioneered by Rannou. Recently, Tremaine studied the standard map and a billiard map of Tennyson using lattice maps. The results are extremely interesting. In short, the maps are exactly symplectic with no truncation errors and appear qualitatively correct at $t = \infty$.

More interesting for us: these lattice maps can be created within the standard large machine "kick codes" using the explicit integrator of Ruth (this was noticed first by Clint Scovel of Los Alamos). And, finally, the "jolt factorized maps" of Irwin, which Dragt described during the workshop, can also be put on a lattice. It remains to be seen whether or not the fitted maps of Warnock and Berg can be thrown unto Z^{2N}. The work of Tremaine, which is not easily applicable to celestial mechanics, is immediately applicable to circular hadron machines. We ought not to ignore this.

Now I would like to turn to symplectic integration and the recent advances in this field.

4 Symplectic Integration

Enough about long-term stability. Let me address the related topic of symplectic integration. This is a topic of great pedagogical value because it provides a remarkable example of the unity found in theoretical science.

After a quiet beginning in 1956 due to De Vogelaere at Notre Dame, symplectic integration stayed dormant until 1983. As you know, in 1983 Ron Ruth of SLAC derived an explicit integrator for the Hamiltonian

$$H = A(\vec{p}) + V(\vec{q}).$$

To a great extent it also died in 1983 as far as the vast majority of accelerator physicists are concerned. The "Others..." took it over and generalized Ruth's method beyond his wildest dreams. In fact, it could not have been done without Lie methods. How did this happen? Ruth's derivation relied on the use of generating functions and on the particular form of H, namely the drift/kick split. His derivation was very complex and did not reveal the incredible generality of his find. For example, none of the small machine application were apparent in the original paper of Ruth.

Later, the Lie methods reveal the immense potential of Ruth's algorithm. In fact, one can show that the Ruth integrators are intimately connected to Lie groups. As I pointed out, the first application in our field belongs to the area of small machine modelling. In celestial mechanics it can be applied to the n-body maps of Wisdom. In plasma physics, the Lie generalization was used by Cary[13], providing a more efficient way to compute the motion of an electron through a plasma. In our field, Cary's algorithm can be used in coasting-beam simulations. As far as derivations are concerned, the Lie group connection was exploited by myself in deriving a 6th-order integrator as well as three special integrators for celestial mechanics. These integrators were tested by Gladman and Duncan.

The greatest triumph of Lie methods in explaining and expanding the Ruth integrators is due to Haruo Yoshida[14], another celestial mechanicist. Ruth's original derivations had 5 equations containing hundreds of terms. It was a miracle that Ruth succeeded in getting an exact solution. Yoshida's proof has two lines! Furthermore it shows how Ruth's method is the beginning of a continuous series of higher and higher order integrators. The generality of the method is astonishing: Prof. Masuo Suzuki, of Tokyo University, had derived Yoshida's results in the context of path integral calculations. He also proved a conjecture of Ruth: one needs negative s steps for symplectic integrators beyond the second-order leap-frog.

Almost simultaneously, André Bandrauk and Hai Shen, of the Chemistry Department of Sherbrooke University, independently derived Yoshida's results under the title "An Improved Exponential Split Operator Method for Solving the Time-Dependent Schrödinger equation."

Finally, Suzuki completely classifies the minimal symplectic integrators using some very elegant methods based on time ordering and Kubo cumulants [15].

The story does not end... Forest, Bengtsson and Reusch [3] noted that the entire Yoshida scheme applies also to implicit integration. Scovel, using our result, pointed out its applicability to maps of the form

$$Map(t) = \varphi(\frac{t}{2}) \circ \varphi(-\frac{t}{2})^{-1}.$$

In addition, the Ruth algorithm has been applied to Lie-Poisson integrators. In classical dynamics, it is possible to reduce the dimensionality of the phase space whenever integrals of the motion are present. In this reduced space, the Poisson bracket looks very different: it is called a Lie-Poisson bracket. If one can preserve the Lie-Poisson bracket, the original Poisson bracket will also be preserved. This does not have obvious application in our field since, as I said before, we have no subtle invariants. It might have applications in tracking the evolution of moments. Since moments are functionals of the maps, the Poisson bracket induces a Lie-Poisson bracket of the moments.

There is a vast and complex literature on this subject. But, again, I want to emphasize the unity of the theoretical tools.

5 The Mathematicians

Critical Function

Finally, we are left with the mathematicians. Recently, while visiting the National Observatory of Japan where the Yoshida of symplectic integration works, I was lucky to attend a seminar by Professor Ian Percival of Queen Mary College, London. He described his work done in collaboration with N. Buric and F. Vivaldi. By the way, Vivaldi attended the Capri workshop. How many amongst you remember what he talked about? Probably nobody. Mathematical work has a pico-second retention rate in the brains of accelerator physicists, yours truly included. Incidentally, Vivaldi's contribution was about the use of integers to study chaotic dynamics, as Tremaine did in his recent Physica D paper. Let me continue the story anyway.

Percival discussed the concept of the critical function[16]. Consider a completely integrable system, for example, a map for a ring with detuning only. Add to this a perturbation, for example a beam-beam kick. We can quantify the beam-beam kick in terms of the "tune shift" parameter ξ. The critical function $\xi(\nu)$ gives the maximum value of ξ for which the torus of tune n still exists. This is a very bizarre function: it is zero on all rational and it is continuous at every rational. It is discontinuous at all irrational. Percival and his collaborators succeeded in approximating this fractal function by smooth functions. In addition they calculated it using finite perturbation theory for some simple problems like the standard map and a problem containing two islands[17]. This two-island problem can be viewed as a quasi-rigorous analysis of the two-resonances problem. The method can be extended to any Hamiltonian containing a finite number of first-order resonances.

Constant Tune Perturbation Theory

It is to be noted that the kind of perturbation theory needed here is a constant tune perturbation expansion. This is rarely used in accelerators.

I had an undergraduate student, Ken Lamon, who did use it to jump over islands. Also it is done by default, in the calculation of invariant tori which Bob Warnock uses to put a lower bound on number of stable turns in a circular ring [19,20].

Group Theory

There are other topics where mathematicians can help. For example, what representations of $Sp(4)$ or $Sp(6)$ are to be found in a monomial expansion of a given function? This is intimately connected to the problem of the Irwin factorization. Also, along the same line, we need a better understanding of the Lie-Poisson structure of moments: is this useful for beam-beam simulations as Chanell suggested or is it a glorious waste of time? There is a lot of work in other fields on this issue, for example:

→ Ye, Morrison and Crawford, Poisson Bracket for the Vlasov Equation on a Symplectic Leaf.

→ Holm and Wolf, Lie-Poisson Description of Hamiltonian Ray Optics.

As I said before the Lie-Poisson structure is very useful in the presence of invariants. Another application can be found in the field of moment algebra: For single pass systems, Dragt, Gluckstern and Rangarajan have derived sequences of conserved 2nd-order and higher-order moments under $Sp(6)$. These are generalizations of the emittance used in single pass systems [18],

$$\text{Invariant} = \sqrt{\langle x^2 \rangle \langle p^2 \rangle - \langle xp \rangle^2}.$$

This is also connected to the Lie-Poisson structure of the moments.

Enough mathematics! Mathematics has to be translated and digested before it has a chance to impact our field. We ought to be aware of progress and encourage the more mathematical practitioners of our trade to apply its power and beauty to our problems.

6 Conclusion

The purpose of this talk was not to turn you into celestial mechanicists, mathematical physicists or electron microscopists. After all, I have been pontificating on topics which are not my specialty.

As a matter of fact, I am not even a real accelerator physicist, I can't build the damned thing. If you say the word magnet, the first thing coming to my mind is the plastic banana on my wife's refrigerator. So, I don't want you to think that I am here trying to teach you anything about the real world of accelerators and magnets design.

My presence here is a metaphor of my career. People ask me and even pay me to help them with narrow theoretical problems:

How do you compute tune shifts?
How do you integrate through a small ring?
How about radiation integrals?
How about nonlinear momentum compaction?
What is "DA"?
etc. ...

Why is it easy for me to answer these apparently disconnected questions? The answer lies in the fabric which holds this talk together:

Theoretical Unity

When faced with a problem, it is part of my devious nature to react like an academic person. Instead of jumping into the problem, I take one step backward and I look at the entire field, just like a child looking at his toys. Then I try to solve it because I do not lose track of the reason why people pay my salary. Most of you do not have the luxury to play the "professor" and fool around with crazy concepts. So, by design, our field lacks the cultural biases which force its practitioners to seek a cultural, pedagogical and theoretical unity. As a community, we pride ourselves on being able to build some of the most complex technological instruments ever conceived. There is no debate on this issue. We have to build our own solar system, but with all due respect, Scott Tremaine and his colleagues have it already made and delivered!

In the end, I am absolutely convinced that I can do certain calculations because I took the time to take an academic look at the field of single-particle dynamics in rings. My plea to you today is to allow yourself to look into the progress made in the last 10 years in rephrasing and ordering many of the topics relevant to single-particle dynamics. This will involve:

1. a boring exercise in semantics: we must simply define the words we use,

2. an intellectual effort to free oneself from old-fashioned approaches.

We owe it to ourselves and to the students of our field to take a fresh look at the theoretical tools available just as Courant and Snyder did so beautifully 40 years ago.

References

[1] E. Forest and K. Hirata, KEK Report 92-12, August 1992

[2] A.J. Dragt and E. Forest, Advances in Electronics and Electron Physics **67**, 65 (1986)

[3] E. Forest, J. Bengtsson and M. Reusch, Phys. Lett. A **158**, 99 (1991)

[4] E. Forest, J. Comp. Phys. **99** (2), 209 (1992)

[5] E. Forest and R.D. Ruth, Physica D **43**, 105 (1990)

[6] E. Forest, J. Math. Phys. **31**, 1133 (1990)

[7] E. Forest and J. Irwin, Workshop Proceedings, Capri, 1990, W. Scandale and G. Turchetti editors, World Scientific, 46 (1990)

[8] G.D. Quinlan and Scott Tremaine, Astron. J. **100**, 1694 (1990)

[9] D.J.D. Earn and S. Tremaine, Physica D **56**, 1 (1992)

[10] G.J. Sussman and J. Wisdom, Science **257**, 56 (July 1992)

[11] J. Wisdom and M. Holman, Astron. J. **102**, 1528 (1991)

[12] M. Duncan, T. Quinn and S. Tremaine, Icarus **82**, 402 (1989)

[13] J. R. Cary and I. Doxis, to appear in J. Comp. Phys.

[14] H. Yoshida, Phys. Lett. A **150**, 190 (1990)

[15] M. Suzuki, Phys. Lett. A **165**, 387 (1992)

[16] N. Buric, I.C. Percival and F. Vivaldi, Nonlinearity **3**, 21 (1990)

[17] O. Piro, N. Buric and I.C. Percival, Phys. Lett. A **165**, 320 (1992)

[18] A.J. Dragt, F. Neri and G. Rangarajan, Phys. Rev. A **45**, 2572 (1992)

[19] K. Lamon, J. Phys. A-Mathematical and General **23** (17), 3875 (1990)

[20] R. L. Warnock, Phys. Rev. Lett. **66** (14), 1803 (1991)

II. Working Group Summaries

SUMMARY FOR WORKING GROUP A ON SHORT-TERM STABILITY

Christoph F. Iselin
CERN, CH-1211 Geneva 23, Switzerland

ABSTRACT

Seventeen papers were presented during the ten sessions of the working group A on short-term tracking. This paper summarizes the author's conclusions on the working group's discussions and comments on the state of the art in short-term tracking methods.

1 INTRODUCTION

Seventeen papers were presented during the ten sessions of the working group A on short-term tracking.[1-17] The reader is referred to their printed versions in these proceedings. A few remarks will also be made in the following sections.

The next section is the author's view on how opinions on particle tracking in circular accelerators and storage rings evolved over the last ten years. It is followed by the working group's recommendations on correction algorithms. Next come some clarifications on coupled motion, followed by an attempt to demonstrate some advantages of the object-oriented approach. The final section contains the author's conclusions and recommendations.

2 HISTORICAL REVIEW

2.1 Symplecticity

Mainly due to the lack of computing power, tracking programs available in the early 1980's would not track more than a few hundred turns. It is striking that one tracking program even contained DIMENSION statements which limited tracking to 400 turns. Most storage rings being built at that time used classical separate function magnets and exhibited a rather linear behaviour. In these machines short-term tracking could not show any adverse effects of non-symplectic tracking. It was thus not generally accepted that only symplectic tracking could produce meaningful results for long-term stability studies. Many accelerator physicists believed that symplectic tracking would cause an unnecessary slow-down in tracking.

The 1982 Brookhaven workshop finally stressed the importance of symplectic tracking. The available computing power had drastically increased, allowing many more turns to be tracked, and results were shown which clearly indicated beam blow-up for non-symplectic tracking programs. Nevertheless it took several years until this was generally accepted, but fortunately today there is no doubt left about the necessity of symplectic tracking.

© 1994 American Institute of Physics 73

2.2 Lie Algebra

Lie algebraic tools were first presented[18-20] as a tool for symplectic tracking. The authors hoped that composition of transfer maps for individual elements could produce a one-turn map, and that by repetitive squaring of the one-turn map they could obtain a many-turn map, which would yield a very fast simulation method for long-term tracking studies.

Soon it was recognized that unavoidable truncations of high-order effects during composition could hide important non-linear effects, and that composition could only be used for short sequences of elements, typically for a single cell or even a half cell.

However, Lie algebraic tools have proven to be an excellent tool for analyzing transfer maps. Normal form and resonance analysis of the one-turn map provides insight on the behaviour of the machine, and quantities like chromaticity, tune shift with amplitude, distortion functions of moderate order, resonance terms, etc. can easily be obtained. These techniques have also been used with success to analyze maps of high orders generated by composition of truncated Taylor maps.

2.3 Thin-Lens Tracking

Another method for symplectic tracking, using thin lenses, was offered in specialized programs like PATRICIA,[21,22] RACETRACK,[23,24] PATRAC,[25,26] PATRIS,[27] TEAPOT,[28] SIXTRACK,[29] and SSCTRK.[30] Thin-lens maps are exactly symplectic, but they do not represent the accelerator lattice accurately. The thin-lens method has been called the exact solution to a wrong problem.

Thin-lens approximations provide a very high tracking speed, and they have been applied with success to many large storage rings.

2.4 Differential Algebra

Differential algebra[11,31] packages have been proposed to permit map evaluation and composition to orders typically as high as 10 to 15. These packages provide truncated Taylor expansions ("jets") of the desired transfer maps. The truncated maps are not symplectic, but the error can be made small by using sufficiently high orders. However, the convergence problem has not been completely solved.

Truncated Taylor maps can be analyzed directly or using Lie algebraic techniques to find quantities like chromaticities and tune shifts with energy or with amplitude.

Differential algebra can easily represent the dependence of a map on various machine parameters. The effect of parameter variations can then be determined, studied, and corrected.

Differential algebra methods come in two flavours. The hard way evaluates the transfer map for each element up to a given order. The resulting maps are then concatenated to form a one-turn map. Evaluation of element maps to high orders is an error-prone process, and composition of maps requires a lot of computing work. For these two reasons this approach is not recommended.

It is easier to start with input phase-space coordinates for the first element defined as a vector of differential algebra variables. After tracking this vector once around the ring, the phase-space coordinates at the exit of the last element represent a differential algebra map for one turn. This method requires no evaluation of maps beyond what is available in a tracking program; all one has to do is replace every arithmetic operation in tracking by the corresponding differential algebraic operation.

When used for long-term tracking, a differential algebra map must be replaced by an equivalent symplectic map. To this aim one has to convert the truncated Taylor series to a representation which agrees with the former in terms up to its order.

Differential algebra also has its merit for analyzing transfer maps. It has been used with success for extracting all kinds of global quantities.

2.5 Generating Function Methods

In many cases transfer maps are available as Taylor series. For practical reasons the series must be truncated, associated with a loss of symplecticity.

A canonical transformation generated by a generating function always produces a symplectic map. This fact can been used for the evaluation of truncated Taylor maps. Convenient generating functions are the mixed-variables ones:

$$F = F(q_1, p_2) \quad \Rightarrow \quad q_2 = \frac{\partial F(q_1, p_2)}{\partial p_2}, \quad p_1 = \frac{\partial F(q_1, p_2)}{\partial q_1}. \quad (1)$$

Or

$$F = F(q_2, p_1) \quad \Rightarrow \quad q_1 = \frac{\partial G(q_2, p_1)}{\partial p_1}, \quad p_2 = \frac{\partial G(q_2, p_1)}{\partial q_2}. \quad (2)$$

Programs like MARYLIE[20,32] and MAD[33,34] convert the Lie transformation to a generating function which expanded to a Taylor series matches the Lie series up to the truncation order. A similar approach is also suited for evaluating differential algebra maps.

The generating function is not unique, as the conversion adds terms above the truncation order, which may become rather large. This can result in large computational errors. Since the equations are implicit for either the positions or the canonical momenta, an iterative method is required to solve for these quantities.

2.6 Cremona Map Methods

A Cremona map is a symplectic map the Taylor expansion of which terminates at a finite order. Simple examples are those maps which can be described by generating functions of the form

$$F(q_1, p_1) = q_1 p_2 + G(p_2) \quad \Rightarrow \quad q_2 = q_1 + \frac{\partial G(p_2)}{\partial p_2}, \quad p_2 = p_1. \quad (3)$$

Or
$$F(q_2, p_1) = q_2 p_1 + H(q_2) \quad \Rightarrow \quad q_2 = q_1, \quad p_2 = p_1 + \frac{\partial G(q_2)}{\partial q_2}, \quad (4)$$
where G and H are polynomials. The first type resembles a drift map, e. g.
$$x_2 = x_1 + p_{x1}(1 - p_{t1}/\beta)L \quad (5)$$
and the second type looks like a thin-lens kick map. Compositions of such maps also produce Cremona maps, if no terms are truncated.

This method ("kick factorization") has been proposed[35,36] as a tool to replace truncated Taylor maps by a symplectic map composed of simple Cremona maps. The procedure builds a Cremona map which agrees up to the truncation order with the given map. Once the Cremona map is known, it can be evaluated by subsequently applying its "factors." This is usually much faster than the iterative procedure required to apply a generating function, and introduction of spurious higher-order terms can be controlled more easily.

3 CORRECTION ALGORITHMS

The working group recommends the following correction procedure:
- Define quantities to be corrected, e.g.:
    - Tune shift with momentum,
    - Variation of lattice functions with momentum,
    - Orbit displacement,
    - Tune shift with amplitude,
    - etc.
- Translate to a mathematical criterion.
- Identify the parameters to change, e.g.:
    - Ordering of magnets,
    - Position of corrector magnets,
    - Excitation of corrector magnets,
    - etc.
- Set up an algorithm and execute it.

4 COUPLING

It is understood that the most complete information about linear coupling can be obtained from the normal form analysis. However, for certain purposes the formulations below may be useful.

4.1 Transverse Linear Coupling

Transverse coupling has been analyzed by several authors.[6,15,37] Consider the linear transfer map **M** in two degrees of freedom partitioned into four 2×2 blocks:
$$\mathbf{M} = \begin{pmatrix} m_{11} & m_{12} & m_{13} & m_{14} \\ m_{21} & m_{22} & m_{23} & m_{24} \\ m_{31} & m_{32} & m_{33} & m_{34} \\ m_{41} & m_{42} & m_{43} & m_{44} \end{pmatrix} = \begin{pmatrix} A & B \\ C & D \end{pmatrix}. \quad (6)$$

The 4-dimensional phase-space vector shall also be partitioned according to the horizontal and vertical planes. Edwards and Teng introduced a "symplectic rotation"

$$\mathbf{R} = \begin{pmatrix} I\cos\phi & \overline{R}\sin\phi \\ -R\sin\phi & I\cos\phi \end{pmatrix}, \qquad (7)$$

where R is a 2×2 matrix with unit determinant, and $\overline{R}$ denotes its symplectic conjugate:

$$R = \begin{pmatrix} a & b \\ c & d \end{pmatrix}, \quad |R| = \begin{vmatrix} a & b \\ c & d \end{vmatrix} = 1, \quad \overline{R} = \begin{pmatrix} d & -b \\ -c & a \end{pmatrix}. \qquad (8)$$

This leaves three free parameters for the elements of R, and a fourth free parameter ϕ. Edwards and Teng determine $\mathbf{R}$ such that $\mathbf{M}$ conjugated with $\mathbf{R}$ becomes block diagonal:

$$\mathbf{RMR}^{-1} = \begin{pmatrix} E & 0 \\ 0 & F \end{pmatrix} \Rightarrow R = \frac{1}{\sqrt{|B+\overline{C}|}}(B+\overline{C}). \qquad (9)$$

If $|B+\overline{C}| < 0$ both ϕ and all elements of R become imaginary. This can be avoided by redefining

$$\mathbf{R} = \frac{1}{\sqrt{1+|S|}}\begin{pmatrix} I & \overline{S} \\ -S & I \end{pmatrix} \qquad (10)$$

where

$$S = R\tan\phi, \quad \text{and} \quad \cos\phi = \frac{1}{\sqrt{1+|S|}}. \qquad (11)$$

All four elements of S are free parameters. The solutions is

$$S = -\left(\frac{1}{2}(\operatorname{Tr} A - \operatorname{Tr} D) + \operatorname{sign}(|B+\overline{C}|)\sqrt{|B+\overline{C}| + \frac{1}{4}(\operatorname{Tr} A - \operatorname{Tr} D)^2}\right)^{-1}$$
$$\times (B+\overline{C}),$$
$$E = A - BS, \qquad (12)$$
$$F = D + \overline{S}C.$$

The block diagonalized matrix can be parametrized as usual. From the eigenvectors of the conjugated system

$$V_1 = \begin{pmatrix} \sqrt{\beta_1} & 0 \\ \frac{\alpha_1}{\sqrt{\beta_1}} & \frac{1}{\sqrt{\beta_1}} \end{pmatrix}, \quad V_2 = \begin{pmatrix} \sqrt{\beta_2} & 0 \\ \frac{\alpha_2}{\sqrt{\beta_2}} & \frac{2}{\sqrt{\beta_2}} \end{pmatrix} \qquad (13)$$

one finds the eigenvectors of the coupled system:

$$W_1 = \frac{1}{\sqrt{1+|S|}}\begin{pmatrix} V_1 \\ \overline{S}V_1 \end{pmatrix}, \quad W_2 = \frac{1}{\sqrt{1+|S|}}\begin{pmatrix} -SV_2 \\ V_2 \end{pmatrix}. \qquad (14)$$

4.2 Tracking the Edwards-Teng Functions

For tracking the coupled lattice functions we assume that the transfer matrix for one element is partitioned as above:

$$\mathbf{R}_e = \begin{pmatrix} A_e & B_e \\ C_e & D_e \end{pmatrix}. \tag{15}$$

The symplectic rotation at element entrance changes the diagonal blocks to

$$E_e = (A_e - B_e S_1)/\sqrt{|A_e - B_e S_1|}, \qquad F_e = (D_e + \overline{S}_1 C_e)/\sqrt{|A_e - B_e S_1|}, \tag{16}$$

and the new coupling matrix at exit becomes

$$S_2 = -(C_e - D_e S_1)\overline{(A_e - B_e S_1)}/|A_e - B_e S_1|. \tag{17}$$

We may track the decoupled lattice functions using the matrices E_e for mode 1 and F_e for mode 2. The above formalism has been extended to three dimensions by Teng.[38]

4.3 Mais-Ripken Lattice Functions

An alternative way to handle coupling has been given by Mais and Ripken.[39,40] Below an alternate derivation is given. A straightforward description of the three eigenmodes is contained in the three eigenvectors of the one-turn transfer matrix $R(0, C)$:

$$V_k(0) = \begin{pmatrix} v_{1,k} \\ v_{2,k} \\ v_{3,k} \\ v_{4,k} \\ v_{5,k} \\ v_{6,k} \end{pmatrix}, \qquad k = 1, 2, 3. \tag{18}$$

They are ordered such that the first one is mainly horizontal, the second one mainly vertical, and the third one mainly longitudinal, and normalized such that

$$\sum_{m=1}^{3} \left(\Re v_{2m-1,k} \Im v_{2m,k} - \Im v_{2m-1,k} \Re v_{2m,k}\right) = 1, \qquad k = 1, 2, 3. \tag{19}$$

The real and imaginary parts of the eigenvectors, placed as columns in a matrix, then form a symplectic matrix. The general initial condition for the eigenmode k is

$$Z_k(0) = \Re\left[V_k(0) \exp(-2\pi i \mu_k(0))\right]$$
$$= \Re V_k(0) \cos[2\pi \mu_k(0)] + \Im V_k(0) \sin[2\pi \mu_k(0)], \qquad k = 1, 2, 3, \tag{20}$$

where $\mu_k(0)$ is the initial principal phase for this mode. The eigenvectors behave like ordinary trajectories, thus they can be tracked by

$$V_k(s) = R(0, s) V_k(0). \tag{21}$$

The phase for the projection of mode i on plane k is

$$\mu_{i,k}(s) = \arctan \frac{\Im v_{2k-1,i}}{\Re v_{2k-1,i}}. \tag{22}$$

All three projections of an eigenmode have the same phase advance Q_k for one turn around the machine. However, the projections of the eigenmode on the three planes need not all start at the same phase. Mais and Ripken use the projections of the Courant-Snyder lattice functions on the three planes:

$$\begin{aligned} \beta_{i,k}(s) &= v_{2k-1,i} v_{2k-1,i}* \\ \gamma_{i,k}(s) &= v_{2k,i} v_{2k,i}*, \qquad k, i = 1, 2, 3. \\ \alpha_{i,k}(s) &= v_{2k-1,i} v_{2k,i}* \end{aligned} \tag{23}$$

Here the index i refers to the eigenmode, and the index k to the plane. Note that the equation

$$\beta_{i,k}\gamma_{i,k} - \alpha_{i,k}^2 = \left(\beta_{i,k}\frac{d\mu_{i,k}}{ds}\right)^2 \quad \text{but} \quad \beta_{i,k}\frac{d\mu_{i,k}}{ds} \neq 1 \tag{24}$$

holds, where $\mu_{i,k}$ is the phase advance for the projection on plane k of the eigenmode i.

4.4 Relations between Different Formalisms

Expressing the beam widths in Mais-Ripken form and in Edwards-Teng form one obtains the following relations:

$$\begin{aligned} \beta_{1,x} &= \beta_1 \cos^2 \phi, & \beta_{1,y} &= \beta_1 \sin^2 \phi, \\ \beta_{2,x} &= \beta_2 \sin^2 \phi, & \beta_{2,y} &= \beta_2 \cos^2 \phi, \end{aligned} \tag{25}$$

where $\beta_{i,k}$ is the projection of β_i on the plane k. The apparent emittances for each mode are

$$\begin{aligned} E_{1,x} &= J_1 \beta_1 \cos^2 \phi, & E_{1,y} &= J_1 \beta_1 \sin^2 \phi, \\ E_{2,x} &= J_2 \beta_2 \sin^2 \phi, & E_{2,y} &= J_2 \beta_2 \cos^2 \phi. \end{aligned} \tag{26}$$

As expected the total transverse energy remains constant for each mode.

5 OBJECT-ORIENTED PROGRAMMING

Object-oriented methods have been around for at least ten years.[33,41] The concept was emerging from the evolution of new computer languages. Then the object-oriented languages were not yet generally available, so the concept could only be implemented with inadequate means. Dynamic memory management had to be implemented in FORTRAN, and PASCAL or C structure had to be simulated using PARAMETER and EQUIVALENCE statements.

In the meantime compilers for object-oriented languages like C++,[42] Smalltalk, Eiffel, and others are generally accessible. This has greatly simplified the situation, and sets of classes for beam optics calculations could be implemented.[11]

The object-oriented approach allows to build "plug compatible" software modules which may be interchanged without affecting any existing code. Assume we have defined three C++ classes to be used with a tracking program:

```
// virtual base class for beam elements
class element {
public:   ...
// track phase space vector:
virtual void propagate(vector);
...
};
// class for drift spaces:
class drift:  public element {
public:   ...
// track phase space vector:
virtual void propagate(vector);
...
private:
double length;
};
// class for sequences of elements:
class line:  public element {
public:   ...
// track phase space vector:
virtual void propagate(vector);
...
private:
// description of sequence:
...
};
```

We have also defined a function to implement propagation through a sequence:

```
void line::propagate(vector v)
{
// for all members of the sequence do
...
member->propagate(v);
}
```

Having written, compiled and tested this code we may easily add a new element class:

```
// class for Siberian snakes:
class snake:  public line {
public:   ...
```

```
virtual void propagate(vector);
...
private:
// additional data:
...
};
void snake::propagate(vector v)
{ ... }
```
No recompilation of existing code is required. Nevertheless, whenever a `snake` occurs in a `line`, the function `propagate` for the `line` knows that it should call `snakes` propagate.

6 CONCLUSIONS AND RECOMMENDATIONS

6.1 Map Methods versus Integrals

In the past global quantities like tune shifts or resonance widths were invariably evaluated using single or multiple integrations around the ring. Important terms could easily get lost in this process. An instructive example has been given by Jäger and Möhl,[43] when the chromaticity of a small ring (LEAR) was calculated with several programs and compared. Most programs which use integral methods ignore some terms which are important for this type of ring, while all programs using map methods give correct results.

Integral methods are considered to be dangerous and, if available, map methods should be preferred.

6.2 Survival Plots

Sometimes the dynamic aperture determined by a survival plot shows a sharp limit, and sometimes it depends logarithmically on the number of turns tracked. The general consensus of the working group is that with non-symplectic tracking it almost certainly exhibits a slope, while with symplectic tracking both cases are possible. In the latter case the Lyapunov exponent may help decide on stability of particles.

6.3 Comparison of Programs

One often hears statements like "We have run this problem with the three programs ..., and found that ...". Since several programs may make the same erroneous assumption, this approach is dangerous, and that it is by far better to put some effort to understand one program thoroughly. This allows to asses its validity for the problem to be solved, and have greater confidence in the results.

If programs are compared one should also take into account the different definitions used. Frequent sources of misunderstandings include:
- Different numbering of multipole components in Europe and in the USA.
- Different sign conventions for magnetic fields.
- Including or excluding numeric denominators like $k!$ in multipole coefficients.

- Use of different variables: (x, x', y, y') versus (x, p_x, y, p_y).
- Dropping factors like $\beta = v/c$ or ignoring terms containing $\gamma^{-1} = m/E$ in programs for electron machines.

6.4 Convergence Problems

Truncated Taylor maps are often applied without giving consideration to the convergence problem.[44] At best it is observed that a certain optimum order gives the best accuracy, but a precise convergence criterion is not stated.

Knowing the poles of the developed function, the convergence radius of a Taylor series in one variable is defined by the magnitude of the pole closest to the origin. A Taylor map with one degree of freedom uses two variables, and the convergence radius in the first variable depends on the value of the second variable. Finding its convergence domain requires study of the locus of the poles in a plane with two complex coordinates. This is only feasible when the exact map is known. The paper by Abell[1] is an example of such a study, it gives good estimates for an example.

A Taylor map in n degrees of freedom yields a series in $2n$ variables. One would expect one has to study the locus of the poles in a $2n$-dimensional space with complex coordinates, which may be a formidable task. A lot of research remains to be done in this domain.

6.5 Measuring Transfer Maps

It has been shown[3,4,10,15] that it is possible to determine most terms in the coupled linear one-turn map from measurement on the machine. It is important to use the symplecticity of the map together with the results of Fourier analysis. This allows to check the consistency of measurements and to compute some coefficients which are otherwise not available. Even with large measurement errors a reasonable accuracy can be obtained and used for correcting coupling effects.

6.6 Object-Oriented Programming

For large machines it is vital that both the computer programs and their data are properly structured. Only under these conditions can the computations be organized in a flexible way, permitting to analyze the whole machine and parts of the machine with equal ease. It has been shown by various authors[11,33,34,41] that an object-oriented approach promises the most flexible organization.

7 REFERENCES

1. D. Abell, Convergence of Taylor Series Maps. Workshop on Stability of Particle Motion in Storage Rings, Brookhaven National Laboratory, October 1992.
2. Y. Cai, Correction Scheme for the SSC. Ibid.
3. D. Caussyn, Experimental Longitudinal Tracking with RF Phase Modulation. Ibid.

4. M.J. Ellison, Experimental Measurement of 1-D and 2-D Poincaré Maps. Ibid.
5. V. Garczynski, Derivations on the Differential Algebra Underlying COSY-Infinity. Ibid.
6. V. Garczynski, Effects Due to Linear Coupling, to the Second Order in the Skew Quadrupole Strength. Ibid.
7. M. Giovanozzi, LHC Correction Scheme. Ibid.
8. R. Koul, Lie Algebraic Methods Revisited. Ibid.
9. R. Koul, Magnet Sorting. Ibid.
10. S.Y. Lee, Future Nonlinear Beam Dynamics Experiments at IUCF Cooler. Ibid.
11. L. Michelotti, Object-Oriented Tools. Ibid.
12. J. Milutinovic, Tracking Experiments with PATRIS at BNL. Ibid.
13. C.S. Mishra, Study of Stability in the Fermilab Main Injector. Ibid.
14. F. Pilat, The Dynamic and Momentum Aperture of the SSC Collider Ring in Presence of the Interaction Region's Local Chromaticity Correction. Ibid.
15. R. Talman, Symplecticity in Real Life. Ibid.
16. D. Trbojevic, Are the Present Lattice Designs Using Momentum Space Optimally? Ibid.
17. Y. Yan, Tracking Short Term (10^6 turns) Symplectic One-Turn Maps of Moderate (≥ 4) Order. Ibid.
18. A.J. Dragt, Lectures on Nonlinear Orbit Dynamics. 1981 Summer School, High Energy Particle Accelerators, FNAL, July 1981.
19. D.R. Douglas, Lie Algebraic Methods for Particle Accelerator Theory. Dissertation, University of Maryland, 1982, unpublished.
20. A.J. Dragt and D.R. Douglas, Lie Algebraic Method for Charged Particle Beam Transport. Workshop on Accelerator Orbit and Particle Tracking Programs, BNL, May 1982.
21. H. Wiedemann, User's Guide for PATRICIA. SLAC, PTM-230, Feb. 1981.
22. H. Wiedemann, User's Guide for PATRICIA Version 85.5. Stanford Synchrotron Radiation Laboratory internal report ACDE-NOTE 29, May 1985.
23. A. Wrülich, Tracking Studies in HERA. Workshop on Accelerator Orbit and Particle Tracking Programs, BNL, May 1982.
24. A. Wrülich, RACETRACK, A Computer Code for the Simulation of Nonlinear Particle Motion in Accelerators. DESY 82/026, 1984.
25. P. Faugeras et al., PATRAC: Particle Tracking Program. Workshop on Accelerator Orbit and Particle Tracking Programs, BNL, May 1982.
26. A. Hilaire and A. Warman, A Program for Single Particle Tracking. CERN internal report SOS/AMS 88-8, 1988.

27. J. Milutinovic and A.G. Ruggiero, Comparison of Accelerator Codes for a RHIC Lattice. AD/AP Technical Note No. 9, 1988.
28. L. Schachinger and R. Talman, TEAPOT — A Thin Element Accelerator Program for Optics and Tracking. SSC Central Design Group internal report SSC-52, Dec. 1985.
29. F. Schmidt, Sixtrack: A Single Particle Tracking Code. Workshop on Nonlinear Problems in Future Particle Accelerators, Capri, Italy, April 1990.
30. D. Ritson, SSCTRK: A Particle Tracking Code for the SSC. Ibid.
31. M. Berz, Symplectic Tracking through Circular Accelerators with High Order Maps. Ibid.
32. A.J. Dragt et al., MARYLIE 3.0, A Program for Charged Particle Beam Transport Based on Lie Algebraic Methods. University of Maryland internal document, user's guide, 1989.
33. F.C. Iselin, The "MAD" Program (Methodical Accelerator Design). Workshop on Accelerator Orbit and Particle Tracking Programs, BNL, May 1982.
34. H. Grote and F.C. Iselin, MAD Version 8.10, User's Reference Manual. CERN/SL/90-13(AP), Revision 3 to be published.
35. A.J. Dragt, Method for Symplectic Tracking. Workshop on Nonlinear Problems in Future Particle Accelerators, Capri, Italy, April 1990.
36. J. Irvin, A Multi-Kick Factorization Algorithm for Nonlinear Maps. SSC-228, 1989.
37. D.A. Edwards and L.C. Teng, Parameterization of linear coupled motion in periodic systems. IEEE Trans. Nucl. Sci. 20:885, 1973.
38. L.C. Teng, Concerning n-Dimensional Coupled Motion. FN 229, FNAl, 1971.
39. Gerhard Ripken, Untersuchungen zur Strahlführung und Stabilität der Teilchenbewegung in Beschleunigern und Storage- Ringen unter strenger Berücksichtigung einer Kopplung der Betatronschwingungen. DESY internal Report R1-70/4, 1970.
40. H. Mais and G. Ripken, Theory of Coupled Synchro-Betatron Oscillations. DESY internal Report, DESY M-82-05, 1982.
41. J. Niederer and B. Morris, LILA: The Long Island Lattice Analogue. Workshop on Accelerator Orbit and Particle Tracking Programs, BNL, May 1982.
42. B. Stroustrup, The C++ Programming Language, Second Edition, Addison-Wesley, New York, 1991.
43. J. Jäger and D. Möhl, Comparison of Methods to Evaluate the Chromaticity in LEAR. CERN PS/DL/LEAR/Note 81-7.
44. E. Forest and K. Hirata, A Contemporary Guide to Beam Dynamics, KEK Report 92-12, Aug. 1992.

Summary for Working Group B on Long-Term Stability

Stephen G. Peggs
BNL, Upton, NY 11973, USA

1 Introduction

A total of 36 workshop participants attended at least one session of the Long-Term Stability working group. We avoided turning these sessions into a specialized seminar series by meeting in two subgroups, loosely labeled Analysis and Diffusion & Tracking, so that working discussions among a reasonably small number of people were possible. Nonetheless, no attempt is made to categorize the 13 group B papers according to original subgroup.

A similar workshop, the Workshop on Accelerator Orbit and Particle Tracking Problems, was held almost exactly 10 years ago at Brookhaven. It is interesting to see how many of the participants in the photograph of that workshop appear again in the photograph at the front of these proceedings. Fortunately, it is not correct to infer that little progress has been made in the last decade, or that the average age of the participants has increased significantly. Rather, the recent photograph has many more, younger, faces than its predecessor. This attests to the ongoing interest and vigorous activity in an area of central importance to accelerator physics.

2 Ten Years After

While a conclusive remark made during the workshop correctly states that "major breakthroughs have been made in the rapid improvement of computer capacity and speed, in the development of more sophisticated mathematical packages, and in the introduction of more powerful analytic approaches," it is easy to overlook the profound philosophical shift that has adiabatically accompanied these advances, and that underpins them. In the last decade, our attitude towards the legitimacy and role of numerical methods (computers) has changed from a rather grudging acceptance of the inevitable, to enthusiastic investigation of the possible. The average level of computer literacy has risen far. The following two examples illustrate this point.

The motion of particles in an accelerator is inherently discrete — first one magnet is traversed, and then an entirely different one. Although one may argue about reducing nonlinear magnets of interest to thin elements of vanishing length, it is clear that the underlying mathematical model is closer to one of **difference** equations, than of **differential** equations. Unfortunately, it is usually impossible to solve nonlinear difference equations analytically. For this reason, classical dynamicists, in this century and the last, developed differential Hamiltonian methods that are most appropriate for continuous systems. (Note, however, that Poincaré knew of the existence of chaos.) By contrast, it is natural to model difference systems such as accelerators using iterative computer codes. A decade or so ago, when computers began to be used intensively to investigate map dynamics, Hamiltonian descriptions of accelerators were attempted by incorporating time-dependent delta-function perturbations. Today, such Hamiltonian approaches are generally accepted as having limited applicability -- especially to the problem of long term-stability — although they have had some important successes. Numerically derived results are more rigorous than before and enjoy a higher degree of confidence. Ten years ago, we were still trying to understand round-off errors, and still learning to enforce symplecticity.

The second example is the change of paradigm that lies behind the advent of differential algebra. A decade ago, the unequivocal object of interest was the 4- or 6-dimensional phase-space vector. One wrote subroutines that input a phase-space vector in order to output it one turn later. In the interim, the propagation of high-order Taylor expansions of maps has gone from being a formal mathematical curiosity to being a routine numerical operation. Object-oriented languages now allow the manipulation of such large analytically unwieldy structures as pyramids of polynomial coefficients, using natural mathematical operators such as + and * in code, through the miracle of function overloading. The expansion variables need not be limited to 6-phase-space coordinates, but may also include lattice quantities such as sextupole corrector strengths, so that the **parametric** behavior of the map can now be investigated directly.

The study of modulation intrinsic to the beam, or caused by ripple in magnet power supplies, has been a growth industry. Having been mentioned in only 2 of the 20 papers in the 1982 proceedings, it is important in 8 of the 13 papers from this working group. One reason for this is that modulation effects are important, but another is that the time scales under study have vastly increased with the computing power available. Loosely speaking, we used to track for 10^3 turns, but now 10^6-turn studies are routine. Since the periods for most significant modulation sources are in the range of 10^2 to 10^3 turns, it could be said that we still track for 10^3 time units, but that now the natural unit of time is a modulation period, and not an accelerator turn.

Last but not least, it is crucial to note how much more operational experience we now have. Then, the SPS was the only hadron storage ring in routine operation. Now, not only have the Tevatron, HERA, and IUCF joined the SPS, but well thought out dynamics experiments have resulted in an important body of high quality data being made available. In ten years' time we hope and expect that RHIC, the SSC, and LHC will also join the ranks. And we expect it to be hard to remember how little we knew, and what our perspectives were, in 1992.

3 Contemporary Themes

Without pretending to be comprehensive, this section outlines some of the themes that appear in the working group B papers.

Although almost all tracking codes are kick codes that represent nonlinear magnets by vanishingly thin elements, the opinion of Maletic and Ruggiero is that "... in order to predict the stability of the motion over very long periods of time it is mandatory that all kinematic terms be properly included in the model. Neglecting some of them may invalidate the results of very time-consuming exercises on the computer." Nonetheless, the other authors who report tracking results all use kick codes. They use a spectrum of accelerator tracking models, from a minimalist 1-D Henon map without tune modulation [Giovannozzi], to a maximalist model including all measured magnet multipoles in HERA [Zimmerman].

Stupakov introduces the concept of a quasi-closed orbit that closes on itself if the modulation period is an integer number of turns, in order to analytically investigate allowable levels of ripple in SSC dipole magnets. He concludes that "its envelope allows one to determine the range of particle deviations at any position of the ring for a given amplitude of the [modulation]."

The Henon map consists of a rotation in phase-space coordinates followed by a single sextupole kick. It is parameterized by a single quantity, the tune, since the sextupole strength can be normalized by rescaling phase space. The difficulty of determining the border of stability of even this simple system, using traditional analytic tools, is noteworthy. One reason for this is the need to go to high-order perturbation theory, making simple Hamiltonian approaches unsuitable. Giovannozzi describes one analytic method, and attempts to extend it to more general 1-D maps. The papers by Brüning and by Bazzani and Pusterla go on to add tune modulation to the Henon map. Brüning makes a connection with experiments at the SPS, in which extrinsic tune modulation was introduced at one or two frequencies, with variable parameters. Bazzani and Pusterla present analytical and numerical results, culminating in the dependence of a 1-D diffusion coefficient on the tune modulation amplitude.

Tune modulation due to finite chromaticity and synchrotron motion is an essential ingredient in the independent tracking models used by Dell, and by Parzen, to simulate RHIC. Parzen confirms the conventional wisdom that it makes a considerable difference to the dynamic aperture whether or not a particle launched with a momentum offset of $\Delta p/p = 0.005$ undergoes synchrotron oscillations. He goes on to make the phenomenological observation that "synchrobetatron coupling becomes important when the particle transverse displacement due to $\Delta p/p$ is about equal to the betatron oscillation amplitude." Dell compares the dynamic aperture results obtained by tracking 10^2 particles for 10^3 turns with those found by tracking 1 particle for 10^6 turns. He launches 100 particles on a surface of constant total initial action by randomly selecting the remaining free initial coordinates. Ten different magnetic multipole seeds are used in each method. He concludes that, while "apertures determined by the worst case values from 10^6-turn runs and multiparticle runs are essentially equal ... [the] use of 10^2 particles tracked for 10^3 turns requires approximately one tenth the computer time needed for 10^6-turn runs and thus enables more varied studies for a given computer budget."

In 1-D, the quantitative agreement between the analytic theory of tune modulation effects, simulation results, and accelerator experiments such as E778, is well established — when a single isolated resonance model is valid. The situation is conveniently encapsulated in a plot in the space of tune modulation depth and modulation tune, showing 4 distinct dynamical phases. Unfortunately the real world has 2 transverse degrees of freedom. Satogata and Peggs attempt to extend the agreement between analysis and simulation to 2-D, by investigating thick-layer diffusion. Here, the chaos caused by overlapping synchrobetatron sidebands of a horizontal primary resonance feeds into the vertical motion via a nearby secondary resonance. The conventional Hamiltonian model tortuously derives a diffusion coefficient for the vertical motion that exhibits steps when plotted versus the distance from the working point to the secondary resonance. In their paper, Satogata and Peggs observe that the vertical diffusion is best described by an exponential chaotic random divergence, and not by root time diffusion. The exponent that they observe also exhibits steps as a function of the distance to the secondary resonance, but as yet there is no quantitative agreement between simulation and analytical prediction.

Zimmerman presents an impressively detailed theoretical and numerical model of HERA, derived from his recently published thesis. Among other topics, he considers the use of a Lyapunov exponent method in forecasting long-time scale behavior, and derives a complementary version of the tune modulation dynamical phase diagram, emphasizing the width of the stochastic layer between sidebands rather than the behavior of particles near the center of the sideband islands. After predicting diffusion rates for HERA and comparing them with operational observations, he concludes "the observed dynamic aperture ... in the HERA proton ring at 40 GeV is in good agreement with the results of computer simulations and analytical predictions which include a modest tune modulation and the effect of the measured nonlinear field errors of all superconducting magnets."

Experiments both at the Tevatron and at the SPS have investigated diffusion effects. Quite different approaches were used, but Gerasimov nonetheless attempts to compare results from the superconducting storage ring with results from the normal conducting ring. "In view of the incompatibilities of the diffusion models with the CERN diffusion experiment data and tracking survival data one could naturally ask why the Fermilab diffusion experiment data were basically quite successfully fitted with diffusion models." Shi and Ohnuma develop a technique to study the evolution of particle distribution as a function of oscillation amplitude. "Using perturbation expansion with multiple scales, we have solved the Vlasov equation and the Fokker-Planck equation in the time domain." Unfortunately, "when the system is close to major resonances, the perturbation expansion ... may not converge ... large storage rings, however, are generally operated far from all major resonances."

In a short paper, Mane presents two simple examples of second-order symplectic integrators for spin motion. He claims that "the use of Lie group theory makes the generalization [from orbit symplectic generators] to include spin motion relatively straightforward." In a lengthy paper, Mane and Weng review normal form methods for solving nonlinear differential equations, and compare the relative merits of three ways to evaluate normal forms. They conclude that "...the superiority of the minimal normal form method ... for ordinary nonlinear autonomous differential equations ... has been demonstrated. The minimal normal form method has also been extended, to treat discrete maps. The application to the evaluation of one-turn maps for accelerators yields mixed results, hence the superiority of the minimal normal form is not as clearly visible."

4 Working Group Opinions

The last session of the working group was held with both subgroups, to see if we could form a group consensus on some issues. The 32 attendees were informally polled on two questions:

<u>Is there a dynamic aperture?</u> Survival plots, showing the logarithm of the number of turns before loss plotted versus the launch amplitude, have become commonplace. Since data on these plots typically extend only to about 10^6 turns, it is natural to speculate on how to extrapolate to the 10^9-turn time scale of interest. One school of thought claims that there is a minimum initial amplitude — the dynamic aperture — below which all particles are unconditionally stable (neglecting Arnol'd diffusion). The survival plot will be vertical at the dynamic aperture. On the other hand, some claim that all particles eventually make their way out to the vacuum chamber wall. When the participants were asked if they expect the survival plot to become vertical, demonstrating an unambiguous dynamic aperture, 24 said YES, 1 said NO, and 7 abstained.

<u>What time scales do we understand?</u> Although we routinely track for 10^6 turns, some argue that we don't use a complete enough model to trust the results. Others argue that all the important physics is in the model, and that we can already trust one-turn map techniques for very long times. Of the 18 people who responded

to the question, 14 believed that we understand the 10^6- to 10^7-turn time scale, but 2 pessimists trust only in 10^3- to 10^4-turn results. The most optimistic opinion was that we can trust our present models even beyond 10^{10} turns.

<u>Is there anything better, or faster, or more reliable than brute force single-particle tracking?</u> We discussed three alternative approaches to tracking, without voting:

1) Lyapunov exponent methods. There was a loose consensus that the Lyapunov method, based on about 10^4 turns' worth of tracking data, is "a conservative stability test that gives a lower bound limit on the aperture for 10^5 to 10^6 turns."

2) One-turn truncated concatenated maps. After much discussion, we agreed to disagree. Notable comments from different ends — and dimensions — of the spectrum included:

"The reliability of one-turn map methods is case dependent, and so one must always test results with brute force codes anyway, removing the advantage of speed."

"There is no need to compare one-turn map and brute force tracking results. One should instead compare one-turn map tracking results with different truncation orders."

"There is no advantage in speed below about 20,000 turns."

"A larger source of uncertainty is the validity of the mathematical model, and the accuracy of the magnet data. How well do we model or measure colliders anyway?"

3) More particles for fewer turns. Discussion focused on the issue of whether accelerator motion is ergodic, so that a particle launched with equal horizontal and vertical displacements (for example) will adequately sample all parts of phase space. The consensus was that the motion is non-ergodic, so that many particles with different initial amplitude ratios should be launched to get truly reliable aperture results. However, there was no consensus on the optimum trade-off between number of particles and number of turns, for a fixed number of particle-turns.

Acknowledgement

I would like to thank Dusan Maletic for the unstinting and invaluable help he provided as secretary of working group B.

Comments on Working Group C: Methods

Alessandro G. Ruggiero
Chairman of the Workshop

Because Working Group C dealt with the subject of Methods, which is of general interest, it was arranged that all workshop attendees could participate in it.

Introductory talks were given on the first day by Etienne Forest and by the chairman of the Working Group, Johan Bengtsson. Presentations were then given during the two-hour long plenary sessions in the morning of each of the following days. On the last day of the workshop, Johan Bengtsson gave a summary of the deliberations of this Group, which appears as the foreword to the contribution by Forest, Michelotti, Dragt and Berg in these proceedings.

The purpose of this Working Group was to provide a coherent presentation of the state of the art of the methods used in determining the stability of particle motion in storage rings. It has produced a consistent and comprehensive documentation of the field which is now available to all who want to familiarize themselves at this subject.

III. Contributions Working Group A

SHORT-TERM STABILITY

REMARKS ON THE DIFFERENTIAL ALGEBRAIC APPROACH TO PARTICLE BEAM OPTICS BY M. BERZ[*]

V. Garczynski

Brookhaven National Laboratory, Upton, NY 11973, USA

ABSTRACT

The underlying mathematical structure of the differential algebraic approach of M. Berz to particle beam optics is isomorphic to the familiar truncated polynomial algebra. Concrete examples of derivations in this algebra, consistent with the truncation operation, are given.

1 INTRODUCTION

The differential equations of beam optics

$$z' = F(z, \delta),$$
$$z(s_i) = z_i \quad (\delta - \text{parameters}), \tag{1.1}$$

can be solved iteratively to any order in z_i [2-9]. The solution can be stated as a mapping $\mathcal{M}$ between the initial variables z_i and the final ones $z_f = z(s_f)$,

$$z_f = \mathcal{M}(z_i, \delta). \tag{1.2}$$

The derivatives

$$\frac{\partial^k \mathcal{M}}{\partial z_i^k}, \quad k = 1, 2, \ldots, n, \tag{1.3}$$

and

$$\frac{\partial \mathcal{M}}{\partial \delta} \tag{1.4}$$

are of interest as they yield the transfer matrix if $k = 1$, the aberrations if $k > 1$, and the sensitivities of a system to the external parameters δ. They can be quickly computed, without recourse to the definition of a derivative as a (computer inconvenient) limit,

$$f'(x) = \lim_{\Delta x \to 0} \frac{f(x + \Delta x) - f(x)}{\Delta x}. \tag{1.5}$$

To find all the derivatives, through the nth order, one evaluates the function $f(x)$ in an algebra $D(n, 1)$ (described in Section 2) which yields [1,5]

$$f(xe + d) = \left[f(x), f'(x), \ldots, f^{(n)}(x) \right], \tag{1.6}$$

---

[*] Work performed under the auspices of the U.S. Department of Energy.

where

$$e = (1, 0, \ldots, 0) - \text{unit element}, \qquad (1.7)$$
$$\underset{\text{0th place}}{\uparrow}$$

and

$$d = (0, 1, 0, \ldots, 0) - \text{differential in } D(n, 1). \qquad (1.8)$$
$$\underset{\text{1st place}}{\uparrow}$$

In case of a function of several independent variables $x_1, \ldots, x_\nu$ the formula (1.6) generalizes as follows

$$f(x_1 e + d_1, \ldots, x_\nu e + d_\nu) = (f, \nabla f, \ldots, \nabla^n f)(x_1, \ldots, x_\nu), \qquad (1.9)$$

where

$$\nabla = \left(\frac{\partial}{\partial x_1}, \ldots, \frac{\partial}{\partial x_\nu} \right), \qquad (1.10)$$

and where the unit element e and the differentials d_k are

$$e = (1, 0, \ldots, 0), \qquad (1.11)$$
$$\underset{\text{0th place}}{\uparrow}$$

$$d_1 = (0, 1, 0, \ldots, 0), \qquad (1.12)$$
$$\underset{\text{1st place}}{\uparrow}$$

$$\ldots\ldots\ldots\ldots$$

$$d_\nu = (0, 0, \ldots, 0, 1, 0, \ldots, 0). \qquad (1.13)$$
$$\underset{\nu\text{th place}}{\uparrow}$$

The dimensionality of the above elements is [1]

$$\dim(e) = \dim(d_k) = N(n, \nu) = \frac{(n+\nu)!}{n!\nu!}, \quad k = 1, \ldots, \nu. \qquad (1.14)$$

The array (1.6), (1.9) can be easily implemented on a computer and evaluated in parallel. The task of computing derivatives is reduced to (computer convenient) arithmetic operations on the N-tupoles of real numbers which are described in Section 2. It turns out that the algebra $D(n, 1)$ is a commutative algebra [9]. It is also a differential algebra [10] because it does admit derivations. This means that it is possible to find some linear operations ∂ acting on the N-tupoles, and obeying the Leibnitz rule

$$\partial (A \cdot B) = (\partial A) \cdot B + A \cdot \partial B. \qquad (1.15)$$

Essentially this kind of approach to particle optics was proposed by Berz [1]. His formulation culminated in the computer code package – COSY INFINITY – which can handle aberrations of any order seemingly in any optical system [5]. The simplest, in our opinion, formulation of the basic assumptions and rules of this approach is based on the familiar truncated polynomial algebra [7,9].

The case of a real function $f(x)$ of a single real variable, and how to find its derivatives effectively through the nth order by evaluating it in the algebra

$D(n,1)$ will be shown. Examples of the derivations in this algebra will be constructed. More detailed treatment, also containing concrete examples of functions, can be found in [12].

After the Berz works it became clear that the differential algebraic techniques are useful and overcome the well known restrictions on computations of higher-order aberrations inherent in the TRANSPORT [13] or MAD codes [14]. In that respect the COSY INFINITY supersedes earlier codes like TRIO [15], GIOS-BEAMTRACE [16], COSY 5.0 [17] and HAMILTON [18]. It competes successfully with the more established MARYLIE 3.0 code [19], which becomes exceedingly complicated in higher orders.

2 THE TRUNCATED POLYNOMIAL ALGEBRA $D(n,1)$

Let us consider the set $D(R)$ of infinitely differentiable functions of a single, real variable $x \in R$. Taylor expansion of a function $a(x)$ from this set, through the order n,

$$a(x) = a_0 + a_1 \frac{x}{1!} + \cdots + a_n \frac{x^n}{n!} + o(x^n), \qquad (2.1)$$

is fully characterized by the array A of real $n+1$ numbers

$$A = \{a_0, a_1, \ldots, a_n\} \in D(n,1). \qquad (2.2)$$

Conversely, given A, one recovers an equivalence class $A = [a]$, of functions which can differ in the terms beyond the nth order. These terms form an ideal $I_n(R)$ in $D(R)$ in the sense that:

1. a linear combination of terms $o(x^n)$ is again of order $o(x^n)$, viz.,

$$o(x^n) + o(x^n) = o(x^n), \qquad (2.3)$$

2. the product of any function $a(x) \in D(R)$, with an element from $I_n(R)$, belongs to the ideal, viz.,

$$a(x) o(x^n) = o(x^n). \qquad (2.4)$$

These properties are essential for mathematical correctness of the algebraic operations on the equivalence classes. Results of an addition and a multiplication do not depend on the particular choices of representatives of these equivalence classes when they are satisfied.

Any operation which is defined on the functions $a \in D(R)$ induces the corresponding operation on the coefficients $a_0, a_1, \ldots, a_n$ if the division onto the polynomial and the terms of higher orders is respected. One may tolerate some "spill" from the polynomial part into the ideal $I_n(R)$, but the reverse should be prevented. For instance, when multiplying $a(x)$ with x one loses the terms $a_n (x^{n+1}/n!)$ which goes now to the ideal. Differentiation, however, is dangerous in the sense that it brings unspecified term $a_{n+1}(x^n/n!)$ from the ideal into the "clean" polynomial area. Note that the operation $x\,(d/dx)$ preserves the ideal and nothing spills back to the polynomial part. The same is true for the operations $x^\alpha\,(d/dx)$, $\alpha = 1, 2, \ldots, n$. Higher α's render trivial operations destroying the polynomial part entirely.

One may write the function $a(x)$ as a scalar product, up to the higher-order terms

$$a(x) = (A, X) + o(x^n), \tag{2.5}$$

where the vector X is composed of the basic monomials, normalized by the factorials

$$X = \left(1, \frac{x}{1!}, \ldots, \frac{x^n}{n!}\right). \tag{2.6}$$

A truncation operation T_n is introducted which preserves all the terms through the nth order:

$$T_n\{a(x)\} = (A, X), \tag{2.7}$$

and

$$T_n\{I_n(R)\} = 0. \tag{2.8}$$

Hence, after the truncation, one is dealing with the nth-order polynomials. Note that the algebraic operations in $D(R)$ induce the corresponding operations in $D(n, 1)$:

1. Scalar multiplication

$$\lambda a(x) = \sum_{k=0}^{n} \lambda a_k \frac{x^k}{k!} + o(x^n), \tag{2.9}$$

2. Addition

$$a(x) + b(x) = \sum_{k=0}^{n} (a_k + b_k) \frac{x^k}{k!} + o(x^n), \tag{2.10}$$

3. Multiplication

$$\begin{aligned}a(x) \cdot b(x) &= [(A, X) + o(x^n)][(B, X) + o(x^n)] \\ &= (A, X)(B, X) + o(x^n) \equiv (A \cdot B, X) + o(x^n) \\ &= c(x) = (C, X) + o(x^n)\end{aligned} \tag{2.11}$$

where

$$(C, X) = T_n\{(A, X)(B, X)\}. \tag{2.12}$$

The following arithmetic rules for $D(n, 1)$ are obtained:

1. Scalar multiplication

$$\lambda(a_0, a_1, \ldots, a_n) = (\lambda a_0, \lambda a_1, \ldots, \lambda a_n), \tag{2.13}$$

2. Addition
$$(a_0, a_1, \ldots, a_n) + (b_0, b_1, \ldots, b_n) = (a_0 + b_0, a_1 + b_1, \ldots, a_n + b_n), \quad (2.14)$$
3. Multiplication
$$(a_0, a_1, \ldots, a_n)(b_0, b_1, \ldots, b_n) = (c_0, c_1, \ldots, c_n), \quad (2.15)$$
$$c_k = k! \sum_{\substack{0 \leq i,j \leq n \\ (i+j=k)}} \frac{a_i b_j}{i! j!}, k = 0, 1, \ldots, n. \quad (2.16)$$

The multiplication is commutative, and is distributive across the addition:
$$A \cdot B = B \cdot A, \quad (2.17)$$
$$A \cdot (B + C) = A \cdot B + A \cdot C. \quad (2.18)$$

Hence, the set $D(n, 1)$ is a commutative algebra [10] which is the quotient algebra
$$D(n, 1) \sim D(R)/I_n(R). \quad (2.19)$$

It is also a differential algebra since it does admit derivations [11]. For example, the derivations on the algebra $D(R)$ of differentiable functions given by
$$\mathcal{D}_\alpha = x^\alpha \frac{d}{dx}, \quad \alpha = 1, 2, \ldots, n, \quad (2.20)$$

generate the corresponding derivations ∂_α on the algebra $D(n, 1)$ as follows:
$$(\partial_\alpha A, X) = T_n \{(A, \mathcal{D}_\alpha X)\}, \quad (2.21)$$

where $\mathcal{D}_\alpha$ acts on every component of the vector X.

To understand this, note first that the derivations $\mathcal{D}_\alpha$ satisfy the condition
$$\mathcal{D}_\alpha I_n(R) \subset I_n(R), \text{ or } T_n \{\mathcal{D}_\alpha I_n(R)\} = 0, \quad \alpha = 1, 2, \ldots, n. \quad (2.22)$$

Since the derivations $\mathcal{D}_\alpha$ obey the Leibnitz rule, one has the equality
$$\mathcal{D}_\alpha (ab) = (\mathcal{D}_\alpha a) b + a \mathcal{D}_\alpha b \quad (2.23)$$

where, due to the condition (2.22), one has
$$\mathcal{D}_\alpha a(x) = (A, \mathcal{D}_\alpha X) + o(x^n) \\
\equiv (\partial_\alpha A, X) + o(x^n). \quad (2.24)$$

Applying the truncation operation T_n to both sides of the equality (2.23) and using the formulae (2.11) and (2.23), one gets, for any X, the equality
$$[\partial_\alpha (A \cdot B), X] = [(\partial_\alpha A) \cdot B, X] + (A \cdot \partial_\alpha B, X). \quad (2.25)$$

According to the main theorem of algebra, it is possible only when coefficients on both sides of this equation agree [10]. This implies that the Liebnitz rule is satisfied by ∂_α

$$\partial_\alpha (A \cdot B) = (\partial_\alpha A) \cdot B + A \cdot \partial_\alpha B, \quad \alpha = 1, 2, \ldots, n. \tag{2.26}$$

Hence, ∂_α are derivations in the algebra $D(n,1)$.

Using the definition (2.21) and the explicit form (2.20) of the derivations $\mathcal{D}_\alpha$, one finds components of $\partial_\alpha A$

$$(\partial_\alpha A)_k = \begin{cases} k!/(k-\alpha)! \, a_{k-\alpha+1} & k \geq \alpha \\ 0 & k < \alpha \end{cases} \quad k = 0, 1, \ldots, n. \tag{2.27}$$

It is clear that for $\alpha \geq n+1$ the derivations ∂_α trivialize to zero. More derivations on the algebra $D(n,1)$ are obtained by taking linear combinations of ∂_α

$$\partial = \sum_{\alpha=1}^{n} \lambda_\alpha \partial_\alpha, \tag{2.28}$$

where λ_α are arbitrary real numbers.

Finally let us note that the usual derivative d/dx violates the condition (2.22), and thus does not generate a derivation on the algebra $D(n,1)$; cf. [2], formulae (5), (6).

Some elements of the algebra $D(n,1)$ are special. Apart from the zero and unit elements

$$\begin{aligned} 0 &= (0, 0, \ldots, 0), \\ e &= (1, 0, \ldots, 0), \end{aligned} \tag{2.29}$$

the differential elements are considered:

$$\begin{aligned} d_1 &= (0, 1, 0, \ldots, 0) \equiv d, \quad \text{1st differential}, \\ d_n &= (0, 0, \ldots, 0, 1), \quad n\text{th differential}. \end{aligned} \tag{2.30}$$

It follows from the multiplication rule, (2.15), that

$$d^k = k! d_k, \quad k = 1, 2, \ldots, n, \tag{2.31}$$

$$d^{n+1} = 0, \quad \text{nilpotency property.} \tag{2.32}$$

Any element $A \in D(n,1)$ can then be decomposed as a linear combination of the unit element and the differentials:

$$\begin{aligned} A &= (a_0, a_1, \ldots, a_n) = (a_0, 0, \ldots, 0) + (0, a_1, 0, \ldots, 0) + \\ &\quad \cdots + (0, 0, \ldots, 0, a_n) = a_0 e + a_1 d_1 + \cdots + a_n d_n \\ &= a_0 e + a_1 \frac{d}{1!} + \cdots + a_n \frac{d^n}{n!} = a(d). \end{aligned} \tag{2.33}$$

The differentials can be consistently ordered as follows [1,2]:

$$\lambda_0 e > \lambda_1 d_1 > \cdots > \lambda_n d_n > 0, \qquad (2.34)$$

for any positive numbers $\lambda_0, \lambda_1, \ldots, \lambda_n$. The first components, when not equal, decide which of the two elements is larger. Only when they coincide do we compare the second components, etc. The element λe can be viewed as an extension of the real number λ to the algebra $D(n,1)$.

One may also extend a function $f(x) \in D(R)$ to the algebra, as well. One replaces any constant c, appearing in f, by its extension ce, and the argument x is replaced by the element $xe + d \in D(n,1)$. Its formal Taylor expansion terminates because of the nilpotency of the differential and it yields [cf. (1.6)]

$$\begin{aligned} f(xe+d) &= f(xe) + f'(xe)\frac{d}{1!} + \cdots + f^{(n)}(xe)\frac{d^n}{n!} \\ &= f(x)e + f'(x)d_1 + \cdots + f^{(n)}(x)d_n \qquad (2.35) \\ &= \left(f, f', \ldots, f^{(n)}\right)(x). \end{aligned}$$

The above extension is obviously very useful as it yields all the derivatives through the (computer convenient) arithmetic operations on $D(n,1)$. More general extensions are also possible by using higher order differentials.

Examples:

1. $\exp(A) = \exp(a_0, a_1, \ldots, a_n) = \exp\left[(a_0, 0, \ldots, 0) + \underbrace{(0, a_1, \ldots, a_n)}_{b}\right]$

$$= \exp(a_0 e)\exp(b) = \exp(a_0)\sum_{k=0}^{n}\frac{b^k}{k!}. \qquad (2.36)$$

2. $A^{-1} = (a_0, a_1, \ldots, a_n)^{-1} = \dfrac{1}{a_0 e + b}$

$$= \frac{1}{a_0}e\left[1 - \frac{b}{a_0} + \left(\frac{b}{a_0}\right)^2 - \cdots + (-1)^{n+1}\left(\frac{b}{a_0}\right)^n\right], \quad a_0 \neq 0. \qquad (2.37)$$

3. $\sqrt{A} = \sqrt{a_0 e + b} = \sqrt{a_0}e\sqrt{1 + \dfrac{b}{a_0}}$

$$= \sqrt{a_0}e\left[1 + \sum_{k=1}^{n}(-1)^{k-1}\frac{1\cdot 3\cdots(2k-1)}{2^k k!}\left(\frac{b}{a_0}\right)^k\right], \quad a_0 > 0. \qquad (2.38)$$

In all these cases finite amounts of the algebraic operations are needed in order to compute the extended functions.

The whole scheme generalizes to any number of independent variables, at some expense of notational inconvenience, and yields the formula (1.9) [2,6,12].

3 ACKNOWLEDGMENTS

I would like to thank Dr. M. Berz of Michigan State University, East Lansing, for several clarifying and helpful conversations concerning his approach to particle beam optics, for providing copies of his papers on the subject, and for his code COSY INFINITY. I also thank Dr. Chih-Han-Sah and Dr. V. Schechtman from the Department of Mathematics of SUNY at Stony Brook for useful discussions about mathematical issues related to the differential algebraic approach. I would like to thank Dr. Johannes Claus, Dr. M. Month and Judy Colman for their interest in this work, for critical remarks and for correcting my English. Special thanks go to Dr. Leo Michelotti for his illuminating comments and for providing his code MXYZPTLK.

4 REFERENCES

1. M. Berz, The method of power series tracking for the mathematical description of beam dynamics. Nucl. Instrum. Methods A258, 431-436 (1987).
2. M. Berz, Differential algebraic description of beam dynamics to very high orders. Part. Accel. 24, 109 (1989).
3. M. Berz, Arbitrary order description of arbitrary particle optical systems. Nucl. Instrum. Methods in Physics Research A 298, 426-440 (1990).
4. M. Berz, Automatic Differentiation as Nonarchimedean Analysis, MSUCL-809, March 1992.
5. M. Berz, Computer Program COSY INFINITY Version 5 – User's Guide and Reference Manual, MSU, East Lansing, 1992 (and references given there).
6. L.B. Rall, The arithmetic of differentiation, Math. Magazine 59, 275 (1986).
7. L.B. Rall, Automatic differentiation: techniques and applications, Lecture Notes in Computer Science 120, Springer, 1981.
8. G. Corliss and L.B. Rall, Automatic generation of Taylor series in Pascal-SC: Basic applications to ordinary differential equations, Trans. First Army Conf. on Applied Math. and Computing, ARO Report 84-1, 1984.
9. Y.F. Chang, Automatic solution of differential equations, in Lecture Notes in Mathematics 430, pp. 61-94, Springer, 1974.
10. O. Zarisky and P. Samuel, Commutative Algebra, Vol. 1, 2, Van Nostrand, 1958.
11. J.F. Ritt, Differential Algebra, American Mathematical Society, Washington, DC, 1950.
12. W. Garczynski, On the Differential Algebra Underlying the COSY INFINITY Computer Code Due to M. Berz, BNL AD/AP Technical Note No. 47, 1992.

13. K.L. Brown, C.D. Carey, C. Iselin and F. Rothacker, TRANSPORT: A Computer Program for Designing Charged Particle Beam Transport Systems, SLAC-91, 1973; Rev. NAL-91; CERN 80-04.
14. C. Iselin, The MAD program, in Lecture Notes in Physics 215, Springer, 1984.
15. T. Matsuo and H. Matsuda, Computer program TRIO for third order calculations of ion trajectories. Mass Spectrometry 24, (1976).
16. H. Wollnik, J. Brezina and M. Berz, GIOS-BEAMTRACE, a computer code for the design of ion optical systems including linear or nonlinear space charge, Nucl. Instrum. Methods A258, 408 (1987).
17. M. Berz, H.C. Hofmann and H. Wollnik, COSY 5.0, the fifth order code for corpuscular optical systems, Nucl. Instrum. Methods A258, 402 (1987).
18. M. Berz and H. Wollnik, The program HAMILTON for the analytic solution of the equations of motion in particle optical systems through fifth order, Nucl. Instrum. Methods A258, 364 (1987).
19. A.J. Dragt, L.M. Healy, F. Neri and R. Ryne, MARYLIE 3.0 – a program for nonlinear analysis of accelerators and beamlines, IEEE Trans. Nucl. Sci. NS-3, 2311 (1985).

DETERMINATION OF COUPLED-LATTICE PROPERTIES USING TURN-BY-TURN DATA

G. Bourianoff, S. Hunt, D. Mathieson, F. Pilat, R. Talman
SSC Laboratory, Dallas, TX, USA
G. Morpurgo
CERN, Geneva, Switzerland

ABSTRACT

A formalism for extracting coupled betatron parameters from multiturn, shock excited, beam position monitor data is described. The most important results are nonperturbative in that they do not rely on the underlying ideal lattice model. Except for damping, which is assumed to be exponential and small enough to be removed empirically, the description is symplectic. As well as simplifying the description, this leads to self-consistency checks that are applied to the data. The most important of these is a "magic ratio" of Fourier coefficients that is required to be a lattice invariant, the same at every beam position monitor. All formulas are applied to both real and simulated data. The real data were acquired June 1992 at LEP as part of decoupling studies, using the LEP beam orbit measurement system. Simulated data, obtained by numerical tracking (TEAPOT) in the same (except for unknown errors) lattice, agrees well with real data when subjected to identical analysis. For both datasets, deviations between extracted and design parameters and deviations from self-consistency can be accounted for by noise and signal-processing limitations. This investigation demonstrates that the LEP beam position system yields reliable local coupling measurements. It can be conservatively assumed that systems of similar design at the SSC and LHC will provide the measurements needed for local decoupling.

1 INTRODUCTION

The formulas in this report come from, and expand upon, those in a paper by Talman,[1] which generalizes formulas of Courant and Snyder.[2] Except for rectifying a couple of unfortunate choices of symbols, or where new quantities are introduced, the notation is the same as in those papers. Formulas needed for this report are copied without proof. Also various explanations from those papers are assumed to be understood or are repeated only cursorily.

Only free oscillations following a transverse beam kick are analysed. Description of the equally important case of steady state response to a sinuisoidally shaken beam would be essentially similar, though the likely simultaneous presence of both transient and steady state responses is likely to complicate that situation in practice. The analysis is heavily dependent on tune domain analysis, both for isolating eigenmotions and for noise suppression.

There is no discussion of methods that use adjacent BPM's to determine trajectory slopes, or of other methods that rely heavily on the ideal lattice model in the analysis. This also makes the formulas applicable to the analysis of feedback systems that sense beam properties at a single point. For the same reasons we defer multiplying the horizontal and vertical signals by factors $\sqrt{1/\beta_x}$ and $\sqrt{1/\beta_y}$ respectively, even though, to the extent these quantities are unperturbed, that "normalizes" the geometry appropriately, as will be shown. Some results can be obtained only by feeding in information from the lattice model, but they are inherently perturbative, and we judge it valuable to see how much can be extracted in a model-independent way. The data are analysed in terms of eigenmodes. The motion in one, nominally horizontal, mode labeled A, wobbles around the horizontal x-plane. The nominally vertical mode is labeled D. (The letters A and D come from the Courant-Snyder[2] notation for the block diagonal elements of the lattice transfer matrix.)

It is assumed that the BPM (beam position monitor) data acquisition system is perfect, so that digital values of the horizontal coordinate x and the vertical coordinate y correctly reflect the beam centroid every turn, for at least several hundred turns. Nonlinear amplitude response of the BPM's is assumed absent or compensated for, and it is assumed that all BPM's have accurate calibrations that make the length scales equal at all lattice locations and in both planes.

The analysis assumes betatron amplitudes sufficiently small for the motion to be purely linear though, emphasizing the tune domain as it does, the method may be somewhat tolerant of nonlinearity. The lattice functions β_x, β_y, α_x, and α_y of the (presumably decoupled) lattice are known and are available for use in perturbative calculations, but such use will be deferred as long as possible.

Much of this report is devoted to deriving and using relationships among three "representations" of the motion:

- Coefficients extracted by Fourier analysis from the lists of displacements (x, y) measured at all BPM's. There are six independent coefficients at each BPM. Directly derivable from these are

- transfer functions relating the x and y motions at each BPM, for each of two independent "modes." Each transfer function is parameterized by an amplitude ratio and a phase.

- Coordinates of successive turns at each BPM are related by the sixteen matrix elements of a once-around transfer matrix. Especially because only coordinates x and y are measured, while slopes p and q are not, not all matrix elements are uniquely derivable from the data.

In oversimplified terms, the two mode-tunes ν_A and ν_D are obtained by locating peaks in Fourier spectra, the two mode-emittances ϵ_A and ϵ_D are obtained from magnitudes of Fourier coefficients, and three of four off-diagonal transfer matrix elements are determined by ratios of Fourier coefficients. An important off-diagonal determinant (equivalent to the minimum tune split) constrains the Fourier coefficients but appears not to be derivable from turn-by-turn data at a single BPM in a model-independent way. As well as ν_A, ν_D, ϵ_A, and ϵ_D, the other parameters that are extracted in a model-independent way are β_A, β_D, one of the off-diagonal matrix elements, R_{A12}, and the "magic ratio" $\mathcal{R}_m$ (which is equal to ϵ_D/ϵ_A). This ratio is deserving of the name magic because it is (or should be) a lattice invariant, even in the face of the nasty uninvited effects, coupling in the betatron motion and noise in the actual BPM system.

2 APPLICATION OF THE FORMULAS TO REALISTIC DATA

In this report, formulas useful for measuring lattice parameters are tested on two data sets. Each of these data sets consists of nominally horizontal free oscillations yielding measured (x,y) pairs for 1024 turns at each of 504 BPM detectors of LEP. One data set, labeled 14_18, was acquired during a June 1992 run on the nominally decoupled LEP accelerator. The other was obtained by TEAPOT[3] tracking for the same LEP lattice with the strengths of skew error elements having been increased until the coupling badness (defined below) was approximately equal to that of the actual data. The skew elements introduced to do this were random quad rolls (r.m.s. roll angle = 2 mr) and a (relatively unimportant) systematic skew quadrupole multipole in all dipoles ($a_1 = 0.01$ "units"; "units" refers to the SSC convention of parts per 10^4 at 1 cm.) For both real and simulated data, even though excitation of the "wrong-plane" eigenmotion was unintentional, there is enough signal to extract the most important parameters of both eigenmotions.

In almost all cases in the following the analyses of simulated data and actual data are completely parallel and remarkably similar. For that reason and to save space because there are a large number of figures, all plots are given in pairs, with simulated data on top and measured data on the bottom.

The first such pair, in Fig. 1, shows superimposed Poincaré (x,y projection) plots at two adjacent BPM's. This particular pair of BPM's was chosen because of the near equality of their lattice function values: $\beta_x = 59$ m and $\beta_y = 64$ m. These BPM's are symmetrically placed ±14 m from an intersection region. Being equal-beta points, the transverse distributions can be regarded as "natural" without either axis having been magnified or demagnified by the lattice optics. Though the motion was intended to be purely horizontal and the lattice was "globally decoupled" (as will be reviewed below), it can be seen that vertical amplitudes are comparable to horizontal, and the orientation of the plane of oscillation can change rapidly. These deviations are great enough that a purely perturbative description of the cross-plane coupling is likely to be inadequate. With the much greater coupling strengths expected in the SSC or LHC (because of the much smaller magnet aperture and longer circumference) this will be all the more true for those accelerators. That is one reason the present non-perturbative description was developed.

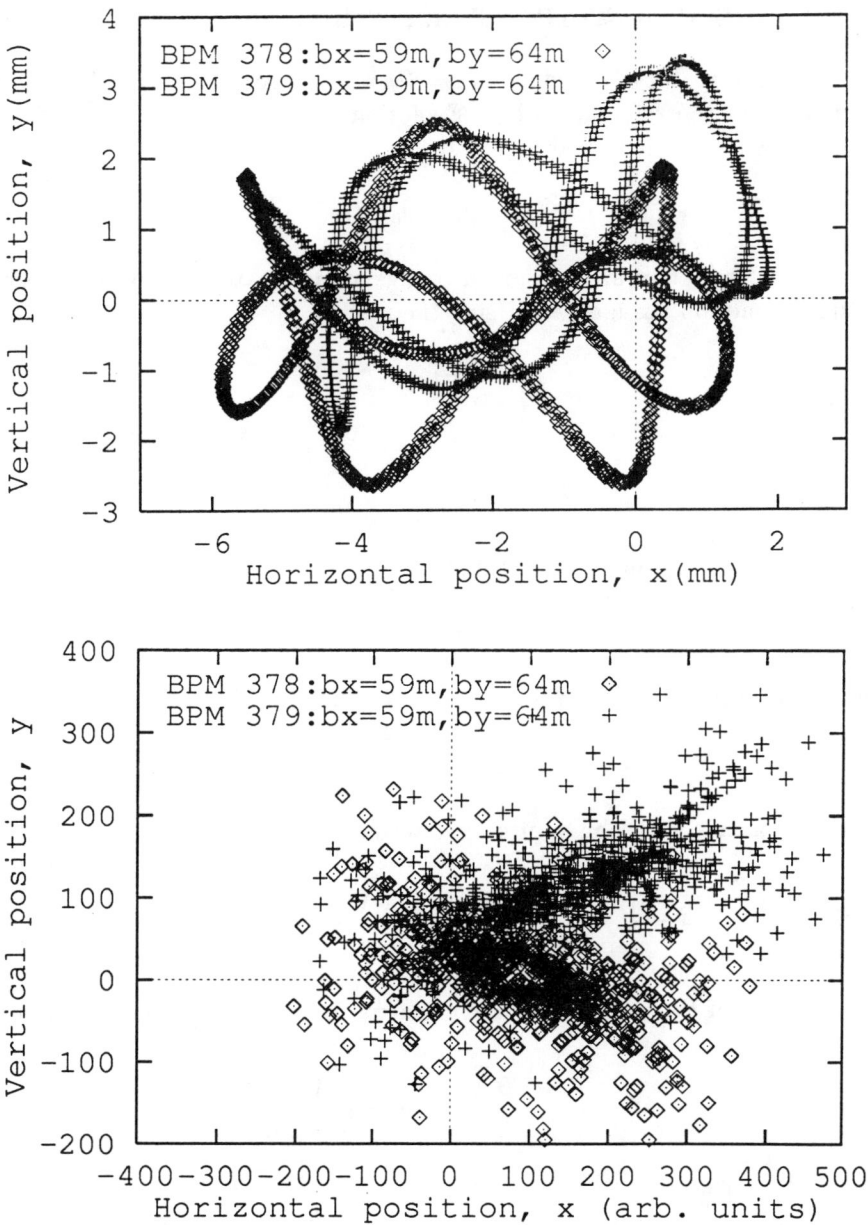

Figure 1. Two superimposed Poincaré plots (x,y projections) at adjacent BPM's for which β_x and β_y are roughly equal. This shows what would be observed on a (nondestructive) phosphor screen inserted at those points. Like all figures in this report the upper figure shows simulated results and the lower figure shows measured results.

3 THE ONCE-AROUND TRANSFER MATRIX

The column vector of coordinates $x = (x, p, y, q)^T$ represents small transverse deviations from the reference orbit. Evolution of a vector x from longitudinal coordinate s_0 to s is described by a transfer matrix M,

$$x(s) = M(s, s_0)x(s_0). \tag{1}$$

A 4×4 transfer matrix M is assumed to give an adequate description; effects of longitudinal motion are ignored. Using the matrix

$$S = \begin{pmatrix} 0 & -1 & 0 & 0 \\ 1 & 0 & 0 & 0 \\ 0 & 0 & 0 & -1 \\ 0 & 0 & 1 & 0 \end{pmatrix}, \tag{2}$$

the symplectic condition M must satisfy is

$$M^T S M = S. \tag{3}$$

"Symplectic conjugation" of any matrix A is defined by

$$\bar{A} = -S A^T S. \tag{4}$$

For a 2×2 matrix with non-vanishing determinant

$$\bar{A} = \overline{\begin{pmatrix} a & b \\ c & d \end{pmatrix}} = \begin{pmatrix} d & -b \\ -c & a \end{pmatrix} = A^{-1} \det |A|; \tag{5}$$

2×2 elements of the partitioned 4×4 matrix M and its symplectic conjugate are defined by

$$M = \begin{pmatrix} A & B \\ C & D \end{pmatrix}, \quad \bar{M} = \begin{pmatrix} \bar{A} & \bar{C} \\ \bar{B} & \bar{D} \end{pmatrix}. \tag{6}$$

Because M is symplectic

$$\bar{M} = M^{-1}. \tag{7}$$

The eigenvalues λ of M satisfy $\det|M - \lambda I| = 0$. The combination

$$M + \bar{M} = M + M^{-1} = \begin{pmatrix} A + \bar{A} & B + \bar{C} \\ C + \bar{B} & D + \bar{D} \end{pmatrix} = \begin{pmatrix} \text{tr}A & E \\ E & \text{tr}D \end{pmatrix} \tag{8}$$

turns out to have simpler properties than M, where the off-diagonal matrix E

and its determinant $\mathcal{E}$ are defined by

$$E = C + \bar{B} = \begin{pmatrix} e & f \\ g & h \end{pmatrix}, \quad \det|E| = eh - fg \equiv \mathcal{E}. \tag{9}$$

This matrix $M + M^{-1}$ exploits the fact that the eigenvalues of M come in reciprocal pairs so that its eigenvalues are two doubly-degenerate real values, of the form $\Lambda = \lambda + \lambda^{-1}$. These sums are real even though the individual eigenvalues λ are complex. The eigenvalue equation is

$$\det \begin{vmatrix} (\text{tr}A - \Lambda)I & \bar{E} \\ E & (\text{tr}D - \Lambda)I \end{vmatrix} = \text{tr}A \, \text{tr}D - (\text{tr}A + \text{tr}D)\Lambda + \Lambda^2 - \mathcal{E} = 0 \tag{10}$$

whose solutions are

$$\Lambda_{A,D} = (\text{tr}A + \text{tr}D)/2 \pm \sqrt{(\text{tr}A - \text{tr}D)^2/4 + \mathcal{E}}, \tag{11}$$

where $A(D)$ goes with the $+(-)$ sign if $\text{tr}A - \text{tr}D$ is positive and vice versa. This choice assures, for weak coupling, that A will correspond to x (this will be called the "nominally horizontal" eigenmotion) and D will correspond to y (the "nominally vertical" eigenmotion.) The eigenvalues satisfy simple equations

$$\Lambda_A + \Lambda_D = \text{tr}A + \text{tr}D, \quad \Lambda_A \Lambda_D = \text{tr}A \, \text{tr}D - \mathcal{E}. \tag{12}$$

In the physically important case, the magnitudes of the eigenvalues of M do not exceed 1, so that there are real angles $\mu_A = 2\pi\nu_A$ and $\mu_D = 2\pi\nu_D$ satisfying

$$\Lambda_A \equiv \lambda_A + 1/\lambda_A = \exp(i\mu_A) + \exp(-i\mu_A) = 2\cos\mu_A,$$
$$\Lambda_D \equiv \lambda_D + 1/\lambda_D = \exp(i\mu_D) + \exp(-i\mu_D) = 2\cos\mu_D. \tag{13}$$

In the special uncoupled case, for which B and C vanish, these angles degenerate into the horizontal and vertical phase advances μ_x and μ_y which satisfy

$$\Lambda_{A,D} = \text{tr}A, D = 2\cos\mu_{x,y} = 2\cos\mu_{A,D}. \tag{14}$$

The determinant $\mathcal{E}$ has special significance since an accelerator is "globally decoupled" if and only if $\mathcal{E} = 0$. This is achieved operationally by adjusting skew quad correction elements so that the two eigentunes can be made to coincide as nearly as possible when $(\text{tr}A - \text{tr}D)$ is scanned through zero using erect quads. That this will work can be seen by manipulating Eqs. (12) and (14) to get

$$(\cos\mu_A - \cos\mu_D)^2 = \frac{1}{4}(\text{tr}A - \text{tr}D)^2 + \mathcal{E}. \tag{15}$$

The left-hand side is small close to either "sum" or "difference" resonances. The former case must be avoided since in that case it turns out that $\mathcal{E} < 0$ and the motion is unstable; in the latter case $\mathcal{E} > 0$ and the tune separation (or rather the squared difference of cosines) cannot be less than $\mathcal{E}$. The minimum tune separation is given by

$$|\nu_D - \nu_A|_{\min} = \frac{\sqrt{\mathcal{E}}}{\pi(\sin\mu_A + \sin\mu_D)}. \qquad (16)$$

The routine accelerator operation called global decoupling consists of adjusting skew quads (typically wired in two families) to minimize this minimum tune separation. This operation was performed on LEP just before the data set labeled 14_18 was acquired.

4 DETERMINATION OF THE EIGENVECTORS

To determine the eigenvectors of $M + M^{-1}$ it is useful to represent a displacement within the x phase space by $\chi^T = (x, p)$ and similarly $\xi^T = (y, q)$. For eigenvalue Λ it is easy to check that the vectors

$$X = \begin{pmatrix} \chi \\ \frac{E}{\Lambda - \mathrm{tr} D}\chi \end{pmatrix}, \qquad Y = \begin{pmatrix} \frac{\bar{E}}{\Lambda - \mathrm{tr} A}\xi \\ \xi \end{pmatrix} \qquad (17)$$

satisfy the equations

$$(M + M^{-1})X = \Lambda X, \qquad (M + M^{-1})Y = \Lambda Y \qquad (18)$$

for either eigenvalue Λ and arbitrary χ or ξ. By picking Λ_A in defining X and Λ_D in defining Y, the nominally horizontal motion labelled A is close to pure x motion, and the nominally vertical motion is close to pure y motion. Toward this end we define 2×2 matrices R_A and R_D by

$$R_A = \frac{E}{\Lambda_A - \mathrm{tr} D}, \qquad R_D = \frac{\bar{E}}{\Lambda_D - \mathrm{tr} A} \qquad (19)$$

in terms of which independent basis vectors can be written as (proportional to)

$$X = \begin{pmatrix} \chi \\ R_A \chi \end{pmatrix}, \qquad Y = \begin{pmatrix} -\bar{R}_A \xi \\ \xi \end{pmatrix} \qquad (20)$$

where a result that follows from Eq. (12),

$$R_D = -\bar{R}_A \qquad (21)$$

has been used.

5 A DIGRESSION CONCERNING DATA PROCESSING STRATEGY

The ideal goal of this analysis would be to extract all 16 elements of M from the turn-by-turn data, but it will not be possible to do this completely. One reason for this is that only coordinates (x, y) are measured; the slope variables (p, q) are either unknown or must be inferred indirectly. Certain functions of the matrix elements, such as tunes, traces, or determinants, may be obtainable more directly than individual elements, but the introduction of such quantities increases, at least temporarily, the number of parameters to be determined. It is sensible to work to eliminate redundant variables, in order to identify a minimum set, and to develop optimal extraction procedures for those. In this spirit Eq. (21) has been used to eliminate R_D. It will be shown below that three, but not all four, of the elements of R_A are directly measurable. The fact that in preparation for taking the data the lattice was globally decoupled, implies that the determinant $\det|R_A|$ is expected to be small at every BPM location. But it would be "cheating" to exploit that in determining the fourth element of R_A; it would violate the spirit of this analysis. Rather we reserve this as a potential consistency check. Obtaining the elements of matrix E can be regarded as an especially important goal since, by Eq. (9), they are directly related to matrix elements of M whose determination is our main purpose. From Eq. (19) it can be seen that matrix R_A is proportional to matrix E. Unfortunately the constant of proportionality will not be obtained directly from the data, and for that reason neither R_A nor E will be eliminated in favor of the other for the time being.

The eigenvalues Λ_A and Λ_D are the most directly and accurately obtainable parameters. By Eq. (13) they are simple functions of the eigentunes, and can be found by locating peaks in the Fourier tune spectrum (as well as by another method to be described below). With no coupling, or even if there is coupling but $\mathcal{E} = 0$, by Eq. (12), $\mathrm{tr}A$ and $\mathrm{tr}D$ would be equal to Λ_A and Λ_D respectively. As above we decline (at least for the time being) to "cheat" in this way.

We introduce various symbols:

$$\mathcal{E} = \zeta(1-\zeta)\Delta_\Lambda^2, \quad \det|R_A| = \frac{\zeta}{1-\zeta},$$

$$\mathcal{G} = \sqrt{\frac{|\Lambda_A - \mathrm{tr}D|}{|\Lambda_A - \Lambda_D|}} = \sqrt{1-\zeta}, \quad E = -\Delta_\Lambda(1-\zeta)R_A, \qquad (22)$$

where

$$e = -\Delta_\Lambda(1-\zeta)R_{A11}, \quad f = -\Delta_\Lambda(1-\zeta)R_{A12},$$
$$g = -\Delta_\Lambda(1-\zeta)R_{A21}, \quad h = -\Delta_\Lambda(1-\zeta)R_{A22}, \qquad (23)$$

and where

$$\Delta_\Lambda = \Lambda_D - \Lambda_A, \quad \mathrm{tr}A = \Lambda_A + \zeta\Delta_\Lambda, \quad \mathrm{tr}D = \Lambda_D - \zeta\Delta_\Lambda. \qquad (24)$$

The quantity ζ is a "small" parameter. When ζ is negligible, all formulas simplify greatly. The quantity $\mathcal{G}$, used below, replaces a different symbol that was multiply defined in reference 1.

6 TRANSFORMATION TO AN EIGENBASIS

In order to define Twiss parameters in a coupled lattice it is necessary to perform a linear transformation from the x, y basis to an eigenvector basis.

In a two-component space basis vectors can be expressed as

$$\hat{\chi}_1 = \begin{pmatrix} 1 \\ 0 \end{pmatrix}, \quad \hat{\chi}_2 = \begin{pmatrix} 0 \\ 1 \end{pmatrix}. \tag{25}$$

These can be used to define an x, y basis in the four-component space:

$$\hat{\imath}^{(1)} = \begin{pmatrix} \hat{\chi}_1 \\ 0 \end{pmatrix}, \quad \hat{\imath}^{(2)} = \begin{pmatrix} \hat{\chi}_2 \\ 0 \end{pmatrix}, \quad \hat{\imath}^{(3)} = \begin{pmatrix} 0 \\ \hat{\chi}_1 \end{pmatrix}, \quad \hat{\imath}^{(4)} = \begin{pmatrix} 0 \\ \hat{\chi}_2 \end{pmatrix}. \tag{26}$$

Similarly, from (20), a basis of eigenvectors is

$$\hat{I}^{(1)} = \mathcal{G}\begin{pmatrix} \hat{\chi}_1 \\ R_A\hat{\chi}_1 \end{pmatrix}, \hat{I}^{(2)} = \mathcal{G}\begin{pmatrix} \hat{\chi}_2 \\ R_A\hat{\chi}_2 \end{pmatrix}, \hat{I}^{(3)} = \mathcal{G}\begin{pmatrix} R_D\hat{\chi}_1 \\ \hat{\chi}_1 \end{pmatrix}, \hat{I}^{(4)} = \mathcal{G}\begin{pmatrix} R_D\hat{\chi}_2 \\ \hat{\chi}_2 \end{pmatrix} \tag{27}$$

where $\mathcal{G}$ is a numerical factor yet to be determined. These bases are related by a linear transformation

$$\hat{I}^{(k)} = G_{ki}\hat{\imath}^{(i)} \tag{28}$$

where summation is implied. A general vector can be expressed in terms of either basis, yielding the equality

$$x_i \hat{\imath}^{(i)} = X_k \hat{I}^{(k)} = X_k G_{ki} \hat{\imath}^{(i)} \tag{29}$$

and from this the coordinates are related, in component and in matrix notation, by

$$x_i = X_k G_{ki}, \quad x = G^T X. \tag{30}$$

By substituting from Eq. (27) into Eq. (28) one obtains

$$G^T = \mathcal{G}\begin{pmatrix} I & R_D \\ R_A & I \end{pmatrix}. \tag{31}$$

The value of $\mathcal{G}$ given above in Eq. (22) is such as to make $\det |G^T| = 1$,

$$(G^T)^{-1} = \mathcal{G}\begin{pmatrix} I & -R_D \\ -R_A & I \end{pmatrix}, \tag{32}$$

and
$$\bar{G}^T = (G^T)^{-1} \tag{33}$$

which shows that G is symplectic.

In the x, y basis the one-turn map relating turns t and $t+1$ is given by

$$x_{t+1} = M x_t. \tag{34}$$

Substituting from Eq. (30) one gets

$$G^T X_{t+1} = M G^T X_t \tag{35}$$

which means that the transfer matrix in the transformed basis is

$$\underline{M} = (G^T)^{-1} M G^T = \mathcal{G}^2 \begin{pmatrix} I & -R_D \\ -R_A & I \end{pmatrix} \begin{pmatrix} A & B \\ C & D \end{pmatrix} \begin{pmatrix} I & R_D \\ R_A & I \end{pmatrix} \equiv \begin{pmatrix} \underline{A} & 0 \\ 0 & \underline{D} \end{pmatrix} \tag{36}$$

from which it follows that

$$\underline{A} = \mathcal{G}^2(A + BR_A - R_D C - R_D D R_A),$$
$$\underline{D} = \mathcal{G}^2(-R_A A R_D - R_A B + C R_D + D). \tag{37}$$

From $\underline{A}$ and $\underline{D}$ the Twiss parameters in the eigenbasis can be extracted. The determinants $\det |\underline{A}|$ and $\det |\underline{D}|$ must both be unity since they are equal to the product of eigenvalues, which is one. As a result $\underline{A}$, for example, can be written in "Twiss form"

$$\begin{pmatrix} \underline{A}_{11} & \underline{A}_{12} \\ \underline{A}_{21} & \underline{A}_{22} \end{pmatrix} = \begin{pmatrix} \cos \mu_A + \alpha_A \sin \mu_A & \beta_A \sin \mu_A \\ -\gamma_A \sin \mu_A & \cos \mu_A - \alpha_A \sin \mu_A \end{pmatrix} \tag{38}$$

where

$$\mu_A = \arccos(\mathrm{tr}\underline{A}/2). \tag{39}$$

It is assumed here that any sign ambiguity has already been resolved in Eq. (11). In Eq. (38) the Twiss parameters are obtained from element-by-element comparison; they are

$$\beta_A = \underline{A}_{12}/\sin \mu_A, \quad \gamma_A = -\underline{A}_{21}/\sin \mu_A, \quad \alpha_A = (\underline{A}_{11} - \underline{A}_{22})/(2 \sin \mu_A), \tag{40}$$

and similarly for $\underline{D}$.

There is a quadratic form which is invariant under the application of the transfer map Eq. (1). Defining $(X, P, Y, Q) \equiv (X_1, X_2, X_3, X_4)$, it and its D counterpart are given by

$$\epsilon_A = \gamma_A X^2 + 2\alpha_A XP + \beta_A P^2,$$
$$\epsilon_D = \gamma_D Y^2 + 2\alpha_D YQ + \beta_D Q^2, \qquad (41)$$

which are the generalizations of the Courant-Snyder invariants. They could be called eigeninvariants or eigenemittances, with the former to be preferred because it tends less to perpetuate the confusion between a single-particle parameter and a beam parameter that comes from the use of the symbol ϵ for both purposes. Because both names are so ugly we will use the term mode invariant instead. A given vector x will, in general, have non-vanishing components in both of the eigenbases. The corresponding mode invariants can be evaluated using the inverse of Eq. (30) to obtain X, followed by substitution into Eq. (41).

Finally we wish to characterize each of the eigenbases by a spatial orientation. The points in a Poincaré plot $(x, y$ projection) are not restricted to a single line; rather they moves along an elliptical trajectory. It is reasonable to characterize the orientation of the $\underline{A}$-eigenbasis by the angular deviation, θ_A, of the major principle axis of the ellipse, away from the x-axis, and similarly for $\underline{D}$.

Observing at a fixed point in the lattice where the lattice functions are β_A and α_A the eigenmotion can be described in the "pseudoharmonic" form

$$X = \sqrt{\beta_A \epsilon_A} \cos \psi_A, \quad P = \sqrt{\beta_A \epsilon_A}(\sin \psi_A - \alpha_A \cos \psi_A)/\beta_A. \qquad (42)$$

Here ψ_A advances by an angle equal to $-\mu_A$ on each turn, and eventually, modulo 2π, takes on all values between 0 and 2π. The factor $\sqrt{\beta_A \epsilon_A}$ (which was not included in the corresponding equation of reference)[1] can initially be regarded as an inessential common factor, but it will be of importance both in combining A and D eigenmotions and in comparing signals at different BPM's. Substitution into Eq. (30) permits the motion to be expressed in the form

$$x = \mathcal{G}\sqrt{\beta_A \epsilon_A} \cos \psi_A, \quad y = \mathcal{G} e_A \sqrt{\beta_A \epsilon_A} \cos(\psi_A + \Phi_A), \qquad (43)$$

where

$$e_A^2 = [R_{A11} - (\alpha_A/\beta_A)R_{A12}]^2 + (R_{A12}/\beta_A)^2,$$
$$\Phi_A = -\arctan \frac{R_{A12}/\beta_A}{R_{A11} - (\alpha_A/\beta_A)R_{A12}}. \qquad (44)$$

It can be shown that the angle of orientation of the ellipse is given by

$$\tan 2\theta_A = -\frac{2[R_{A11} - (\alpha_A/\beta_A)R_{A12}]}{1 - [R_{A11} - (\alpha_A/\beta_A)R_{A12}]^2 - (R_{A12}/\beta_A)^2}. \qquad (45)$$

The orientation of the other mode axis can be found similarly. In general, the two axes are not orthogonal. Normally, since ideal behaviour would have the eigenaxes exactly horizontal and vertical, the deviations of these angles from zero can be regarded as a measure of the seriousness of the coupling.

When two oscillatory quantities are related as x and y are in Eqs. (43), it is natural to relate them by a "transfer function." Since they oscillate at the same frequency, their relationship is completely specified by a ratio of amplitudes, in this case e_A, and a phase difference, in this case Φ_A. The term transfer function is appropriate because the two functions are causally related (otherwise they would not have the same frequency), but one need not "cause" the other. Because the two signals are coherent, the signal-to-noise ratio of their ratio can probably be improved by averaging measurements made at different times. The transfer functions will be obtained by Fourier series analysis in a later section.

7 DIFFERENCE EQUATIONS SATISFIED BY THE MOTION

Using the fact that M^{-1} can be used to propagate backwards in time, relation (8) can be used to obtain four third-order, coupled difference equations that relate the coordinates on three successive turns (labeled $-, 0, +$):

$$\begin{aligned}
x_+ - \mathrm{tr}A\, x_0 + x_- &= hy_0 - fq_0, \\
p_+ - \mathrm{tr}A\, p_0 + p_- &= -gy_0 + eq_0, \\
y_+ - \mathrm{tr}D\, y_0 + y_- &= ex_0 + fp_0, \\
q_+ - \mathrm{tr}D\, q_0 + q_- &= gx_0 + hp_0.
\end{aligned} \quad (46)$$

It is possible to uncouple these equations. Start by squaring Eq. (8):

$$(M + M^{-1})^2 = MM + 2I + M^{-1}M^{-1} = \begin{pmatrix} \mathrm{tr}^2 A + \mathcal{E} & (\Lambda_A + \Lambda_D)\bar{E} \\ (\Lambda_A + \Lambda_D)\bar{E} & \mathrm{tr}^2 D + \mathcal{E} \end{pmatrix}. \quad (47)$$

From this one obtains fifth-order equations linking five successive turns

$$\begin{aligned}
x_{++} + (2 - \mathrm{tr}^2 A - \mathcal{E})x_0 + x_{--} &= (\Lambda_A + \Lambda_D)(hy_0 - fq_0), \\
p_{++} + (2 - \mathrm{tr}^2 A - \mathcal{E})p_0 + p_{--} &= (\Lambda_A + \Lambda_D)(-qy_0 + eq_0), \\
y_{++} + (2 - \mathrm{tr}^2 D - \mathcal{E})y_0 + y_{--} &= (\Lambda_A + \Lambda_D)(ex_0 + fp_0), \\
q_{++} + (2 - \mathrm{tr}^2 D - \mathcal{E})q_0 + q_{--} &= (\Lambda_A + \Lambda_D)(gx_0 + hq_0).
\end{aligned} \quad (48)$$

Substituting from Eq. (46) into Eq. (48) and keeping only the equations for the measured quantities x and y yields

$$\begin{aligned}
x_{++} + x_{--} - (\Lambda_A + \Lambda_D)(x_+ + x_-) + (2 + \Lambda_A \Lambda_D)x_0 &= 0, \\
y_{++} + y_{--} - (\Lambda_A + \Lambda_D)(y_+ + y_-) + (2 + \Lambda_A \Lambda_D)y_0 &= 0.
\end{aligned} \quad (49)$$

What makes these equations truly remarkable in that, in full generality with arbitrary coupling, and with coefficients that are lattice invariants, the same at every BPM, they constrain the output of each plane of every BPM. Since any spurious superimposed signal will not satisfy this constraint, these equations can give a measure of the importance of noise or distortion due to nonlinear electronics in the signal processing or distortion due to nonlinear motion. Eqs. (49) generalize the following difference equations for uncoupled motion:

$$x_+ - \Lambda_x x_0 + x_- = 0, \quad y_+ - \Lambda_y x_0 + y_- = 0. \qquad (50)$$

It can be shown that with no coupling Eqs. (49) and (50) are consistent. The first of Eqs. (50) can be regarded as a "trigonometric" identity relating, in 2D(x, x') phase space, projections of a radius vector that advances through a constant angle μ_x each turn. Eqs. (49) relate similar projections in 4D phase space.

If there were no noise or distortion, and if the x and y measurements were arbitrarily accurate, Eqs. (49) could be used to determine both eigentunes to arbitrarily high accuracy from as few as five turns worth of data.

The eigentunes can be obtained by applying Eq. (49) to actual data. Define the expectation value $< f >$ of N samples f_i by $\sum_i^n f_i/N$. Multiplying the two equations of (49) by x_0 and y_0 respectively, taking expectation values, and rearranging to express as equations for Λ_A and Λ_D yields

$$\begin{pmatrix} <(x_+ + x_-)x_0> & -<x_0^2> \\ <(y_+ + y_-)y_0> & -<y_0^2> \end{pmatrix} \begin{pmatrix} \Lambda_A + \Lambda_D \\ \Lambda_A \Lambda_D \end{pmatrix}$$
$$= \begin{pmatrix} <(x_{++} + x_{--})x_0> + 2<x_0^2> \\ <(y_{++} + y_{--})y_0> + 2<y_0^2> \end{pmatrix}. \qquad (51)$$

When this equation is used to extract eigentunes for the simulated data, the values agree well with peak locations obtained from Fourier analysis: ν_A(diff.eq.) = ν_A(Fourier) $\pm$ 0.00005; ν_D(diff.eq.) = ν_D(Fourier) $\pm$ 0.003. The equation is less accurate when it is applied to measured data : ν_A(diff.eq.) = ν_A(Fourier) $\pm$ 0.001; ν_D(diff.eq.) = ν_D(Fourier) $\pm$ 0.004. It is not suprising that the accuracy for ν_D is not as good as for ν_A, since vertical oscillations should by design not be present at all. The fact that the accuracy is better for simulated data than for actual data is due to noise or signal-processing distortion. While Fourier analysis yields optimal filtering, the difference equation estimates are likely to be biased by noise signals. Our analysis procedure is to use the difference equation to obtain tentative tune values. From this an improved value is obtained by obtaining Fourier amplitudes at points on a fine nearby grid and interpolating to find the maximum. Tunes determined in that way are shown in Fig. 2. The way error bars are determined will be described below. In most cases in these plots they are too small to be visible. These tune values were used in evaluating the Fourier coefficients defined in the next section. Most relationships will be expressed in terms of Fourier coefficients since that tends to incorporate the benefits of filtering in the frequency domain.

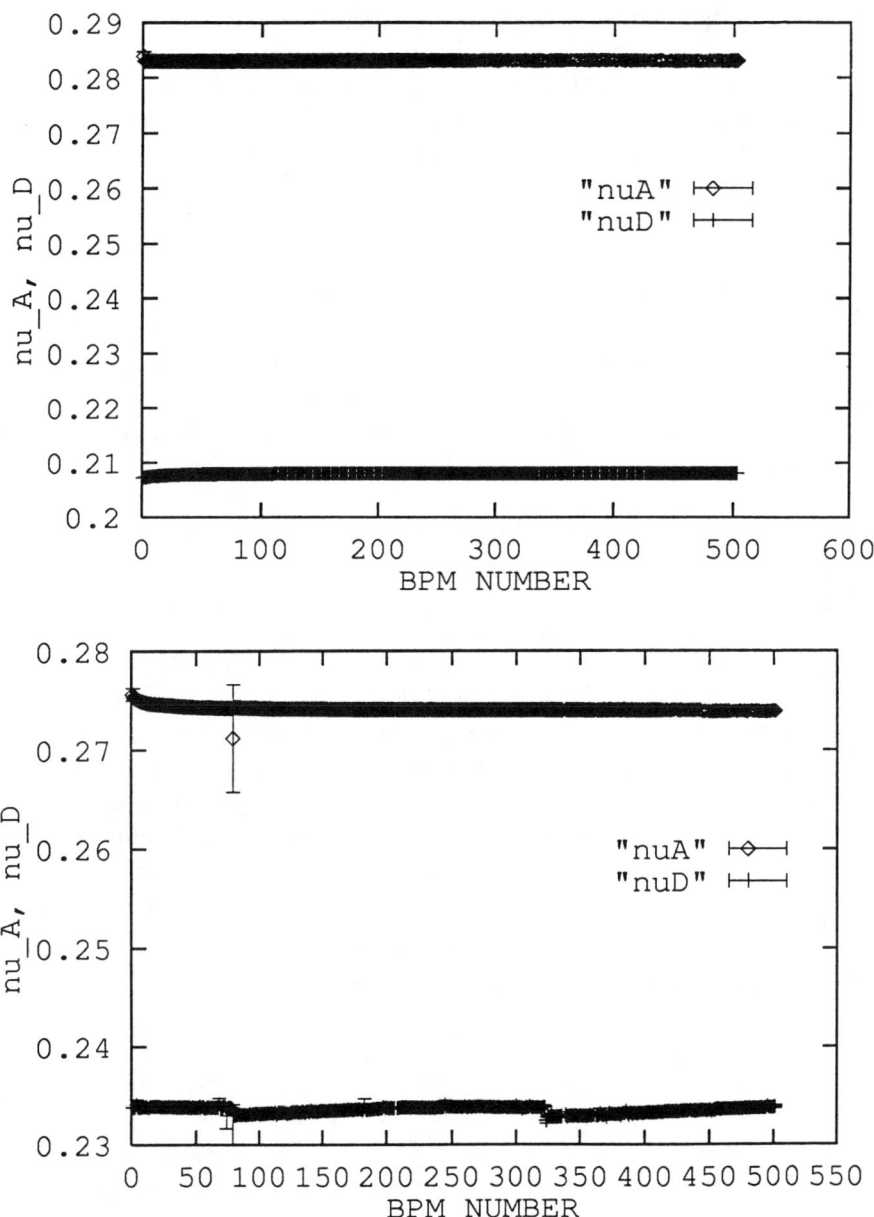

Figure 2. Mode tunes ν_A and ν_D, as determined at all 504 BPM's. Upper figure is from simulation, lower figure from measured data. The discontinuities are artifacts of the Fourier interpolation procedure.

8 EXTRACTION OF FOURIER COEFFICIENTS

In complete generality linear x and y motions can be expressed as superpositions of A-type and D-type motions. The simplest possible description is in "normalized eigencoordinates" $\tilde{X} = (\tilde{X}, \tilde{P}, \tilde{Y}, \tilde{Q})^T$ for which the phase-space motion is circular:

$$\tilde{X} = \epsilon_A^{1/2} \cos \mu_A t, \qquad \tilde{P} = \epsilon_A^{1/2} \sin \mu_A t;$$
$$\tilde{Y} = \epsilon_D^{1/2} \cos \mu_D t, \qquad \tilde{Q} = \epsilon_D^{1/2} \sin \mu_D t. \qquad (52)$$

Here, and wherever it appears in this report, the symbol t can be regarded either as time, measured in units of the revolution period, or as a turn index that increments by 1 each turn. In laboratory x, y coordinates the motion manifests itself in the form

$$x_t = A_x \cos \mu_A t + D_{xc} \cos \mu_D t + D_{xs} \sin \mu_D t,$$
$$y_t = D_y \cos \mu_D t + A_{yc} \cos \mu_A t + A_{ys} \sin \mu_A t. \qquad (53)$$

Here the subscript x or y distinguishes between x or y motion, subscript c or s goes with the cos or sin function, and the capitalized symbols distinguish between A and D eigenmotions. It is assumed that μ_A and μ_D are incommensurate (i.e. not rationally related; i.e. non-resonant) which means there is a time at which the nominally dominant coordinates for the two eigenmotions are simultaneously maximum. Such a time origin has been chosen in (53) [different in general from that in (52)] so that the sin term is missing for x in the A eigenmotion and for y in the D eigenmotion. For that reason the subscript c has been left off A_x and D_y. Also we can assume A_x and D_y are positive without loss of generality.

All six of the coefficients in Eq. (53) are easily and accurately extracted from the turn-by-turn data by Fourier analysis.

For brevity we also define coefficients

$$A_y = \sqrt{A_{yc}^2 + A_{ys}^2}, \qquad D_x = \sqrt{D_{xc}^2 + D_{xs}^2}. \qquad (54)$$

In terms of these coefficients, the transfer function ratios and phases of Eqs. (44) are

$$e_A = \frac{A_y}{A_x}, \qquad \Phi_A = -\arctan \frac{A_{ys}}{A_{yc}};$$
$$e_D = \frac{D_x}{D_y}, \qquad \Phi_D = -\arctan \frac{D_{xs}}{D_{xc}}. \qquad (55)$$

It has been implicitly assumed that there is no damping. In fact, for the actual LEP data there is damping with damping decrement δ (fractional loss per turn) of a few parts per thousand. Furthermore it is found that the damping is different for the two modes. Because of damping, the coefficients A and D, which would otherwise be constant, exhibit time dependence:

$$A(t) = A(0)e^{-\delta_A t}, \quad D(t) = D(0)e^{-\delta_D t}. \tag{56}$$

For the LEP data set 14_18, with $t = 1$ revolution,

$$e^{-\delta_A} = 0.9982 \pm 0.0002, \quad e^{-\delta_D} = 0.9970 \pm 0.0003. \tag{57}$$

The fits of the form (56) are not perfect, and the error assignments in (57) are only rough, but since the damping is so weak it seems legitimate simply to divide the Fourier coefficients by the exponential factors in (56) and then proceed as if there were no damping. That is what has been done and we have chosen to do it implicitly rather than cluttering the equations of this report with explicit $e^{-\delta t}$ factors.

Because the damping in the D mode is appreciably greater than in the A mode (the reason is not known, but the result is consistent with "head-tail damping" that would be expected to be greater in the vertical than in the horizontal plane), and because the D amplitudes are initially smaller than the A amplitudes, the D/A ratio has become really very small after 1000 turns. For this reason the entire analysis reported here is restricted to the first 600 turns. In order to estimate errors for the various parameters these 600 turns were broken into 4 sets of 150 turns each. Identical and completely independent analyses were applied to each 150-turn set; values and errors quoted or plotted are the means and r.m.s. values of the four sets. These errors indicate the variation of these four determinations; they do not indicate the measurement error of the best parameter determination that could be extracted from the data sample.

The transfer function parameters listed in Eqs. (55), evaluated from the coefficients defined in Eqs. (53), with errors calculated as just described, are plotted in Figs. 3 to 6. Though these functions are rather jerky and discontinuous, it will be seen below that appropriate combinations of them are much more regular.

9 RELATIONS BETWEEN FOURIER COEFFICIENTS AND MATRIX ELEMENTS

(Some formulas in this section have already been given at the end of Section 6. They are repeated here because the simultaneous analysis of both A and D modes is essential.)

The normalized eigencoordinates $\tilde{X}$ introduced in the previous section are related to the eigencoordinates $X = (X, P, Y, Q)^T$ of Eq. (30) by a transformation

$$X = \begin{pmatrix} \mathcal{B}_A^{-1} & 0 \\ 0 & \mathcal{B}_D^{-1} \end{pmatrix} \tilde{X} \tag{58}$$

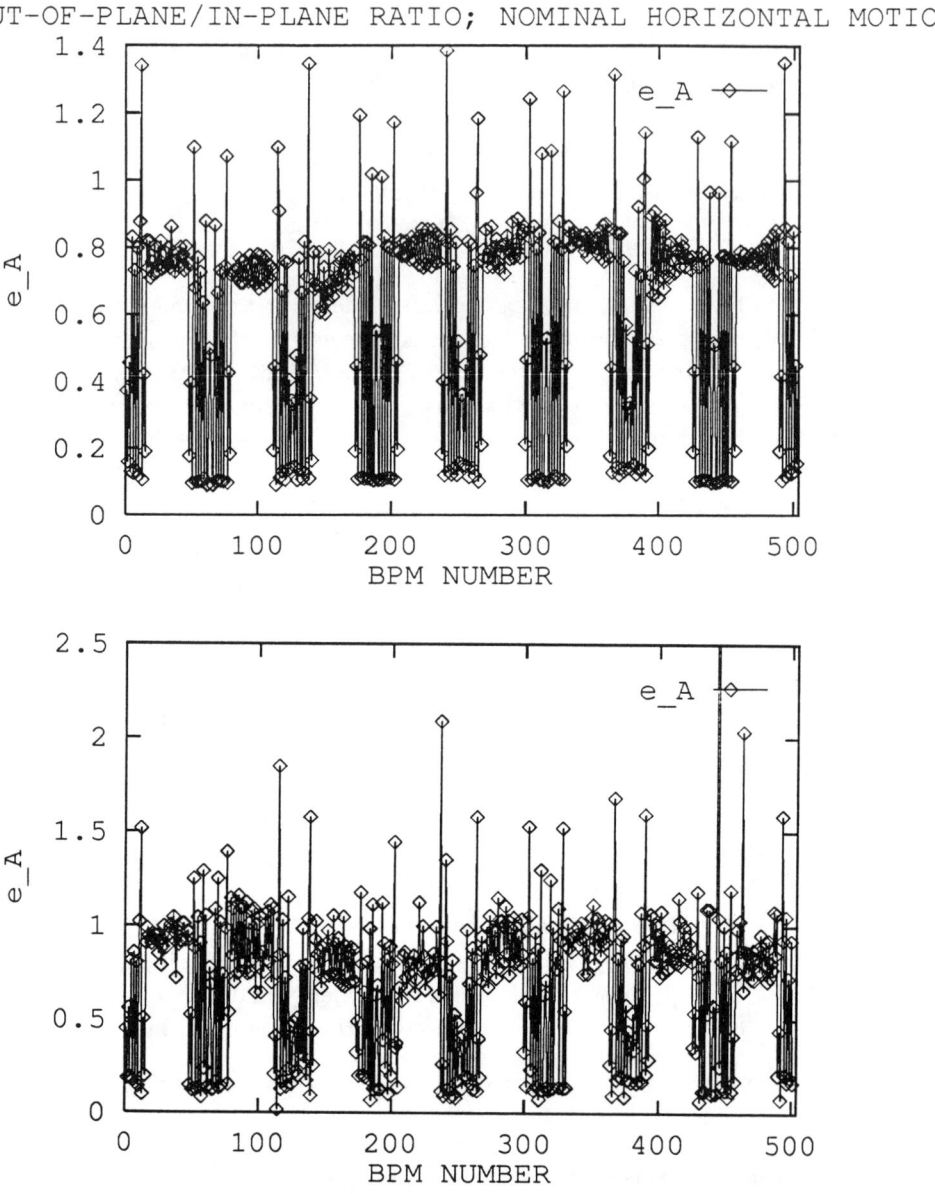

Figure 3. A-mode transfer function amplitude ratio e_A evaluated at every BPM. Upper figure is from simulation, lower figure from measured data.

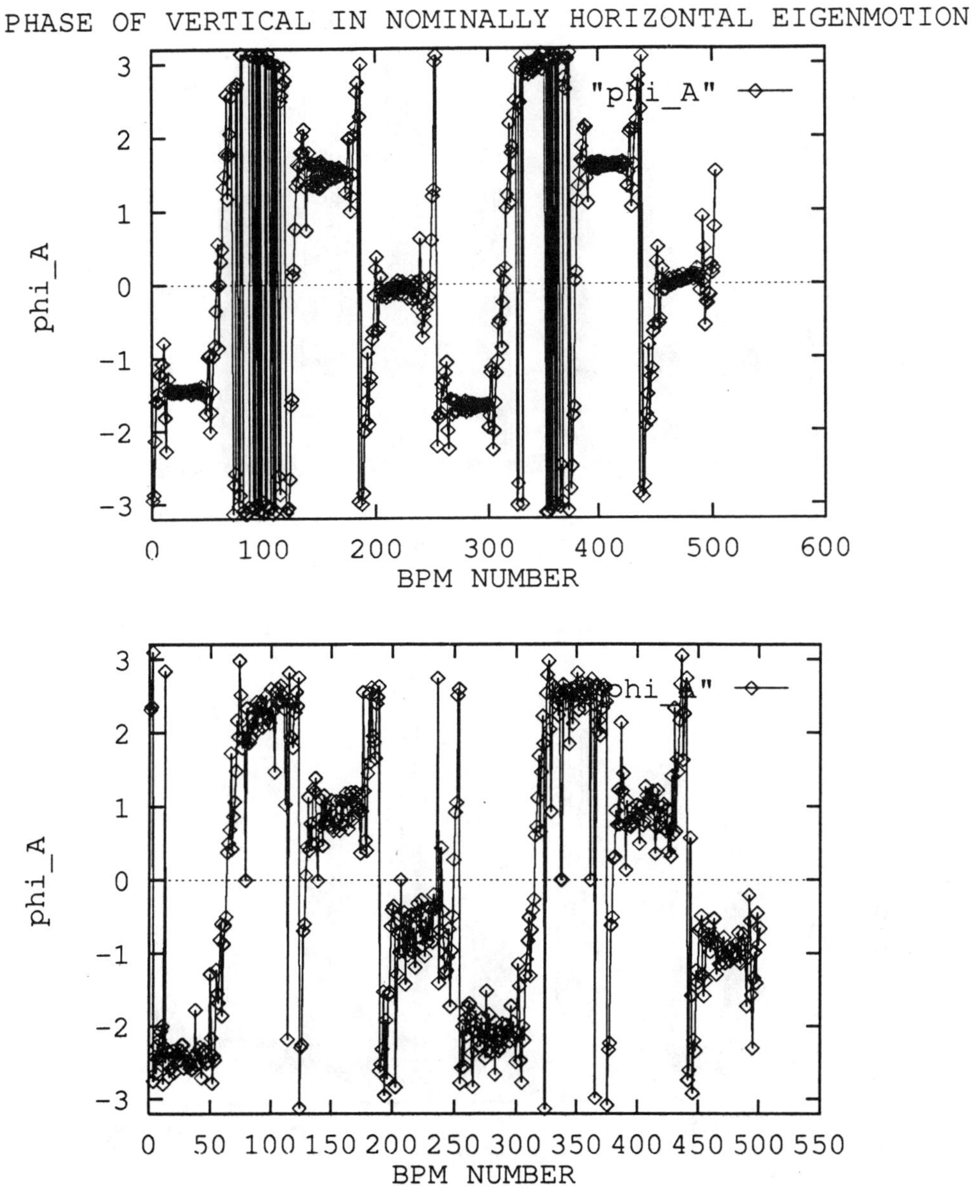

Figure 4. A-mode transfer function phase shift Φ_A evaluated at every BPM. Upper figure is from simulation, lower figure from measured data. Apparent discontinuities, for example near BPM 100 in the upper graph, are caused by dithering near $\pm\pi$ where the phase wraps around.

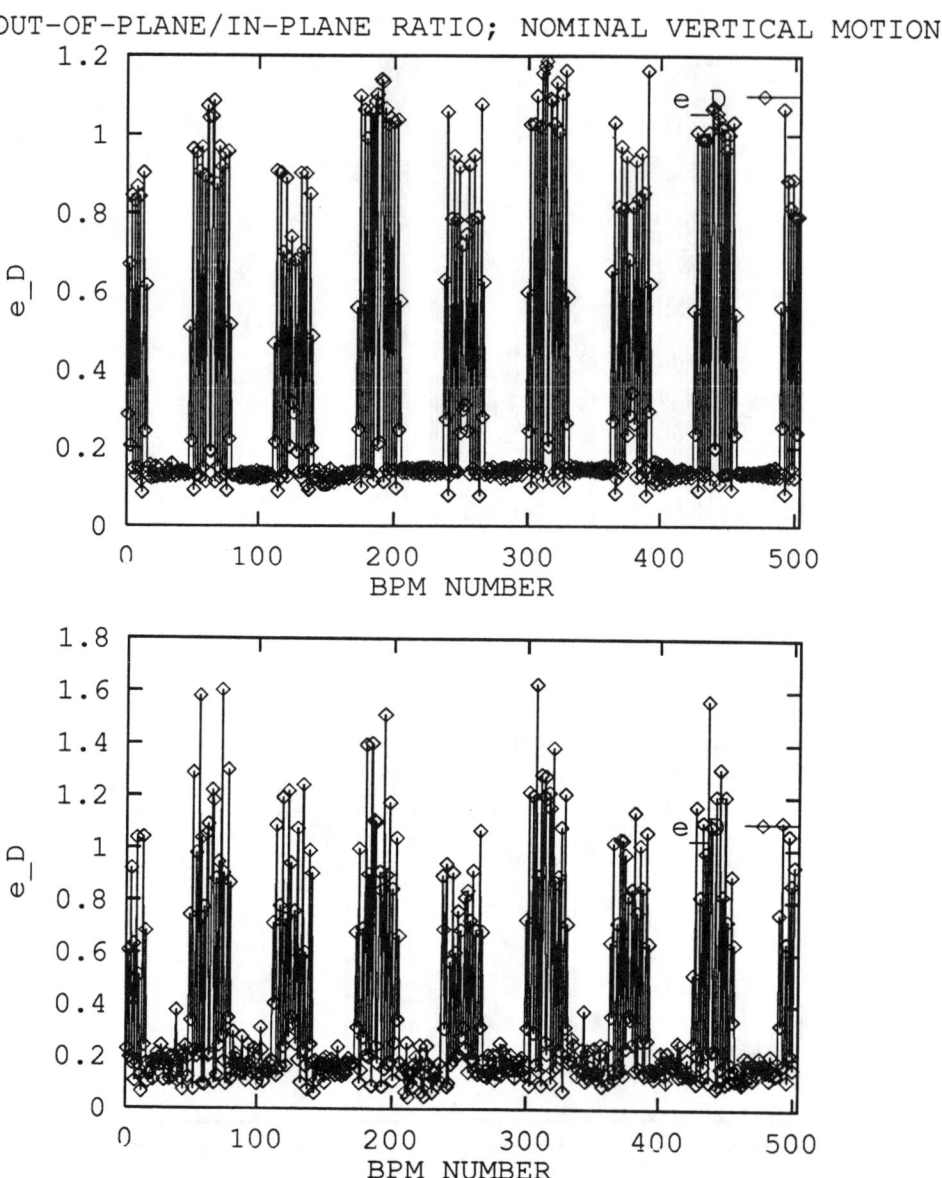

Figure 5. D-mode transfer function amplitude ratio e_D evaluated at every BPM. Upper figure is from simulation, lower figure from measured data.

PHASE OF HORIZONTAL IN NOMINALLY VERTICAL EIGENMOTION

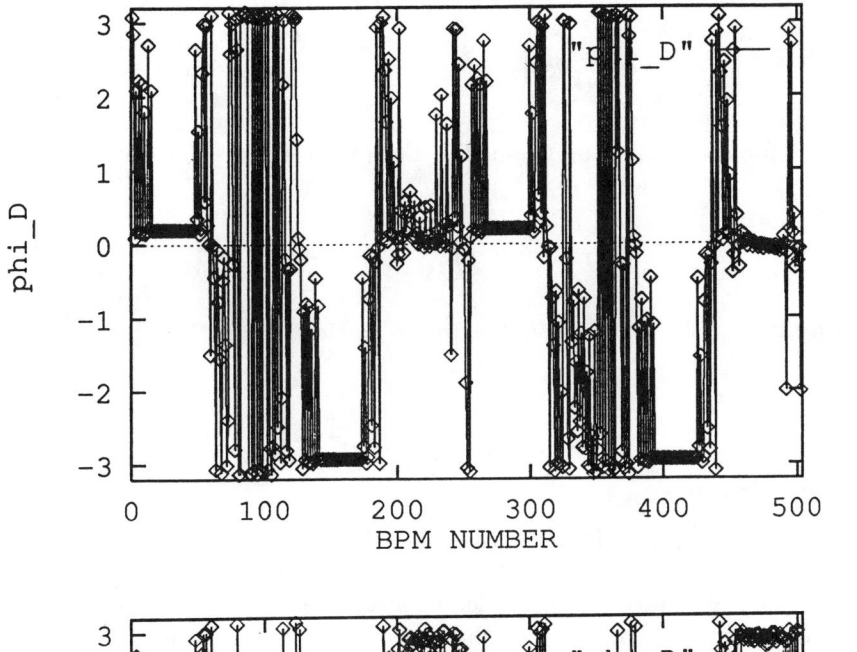

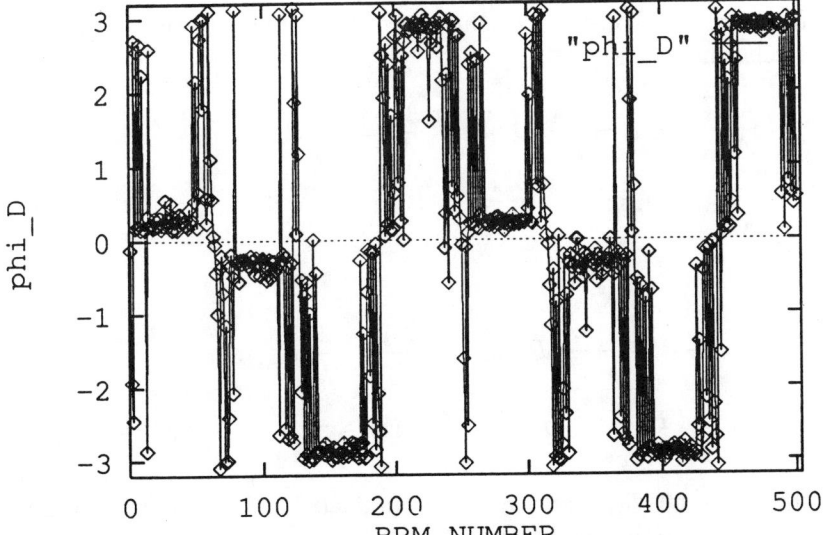

Figure 6. D-mode transfer function phase shift Φ_D evaluated at every BPM. Upper figure is from simulation, lower figure from measured data.

where

$$\mathcal{B}_A = \begin{pmatrix} \beta_A^{-1/2} & 0 \\ \alpha_A \beta_A^{-1/2} & \beta_A^{1/2} \end{pmatrix}, \quad \mathcal{B}_A^{-1} = \begin{pmatrix} \beta_A^{1/2} & 0 \\ -\alpha_A \beta_A^{-1/2} & \beta_A^{-1/2} \end{pmatrix}, \tag{59}$$

and similarly for D. Combining these with Eq. (31) yields

$$x = \mathcal{G} \begin{pmatrix} I & -\bar{R}_A \\ R_A & I \end{pmatrix} \begin{pmatrix} \mathcal{B}_A^{-1} & 0 \\ 0 & \mathcal{B}_D^{-1} \end{pmatrix} \tilde{X}. \tag{60}$$

Substituting from (52), (58), and (59) into (60) yields

$$x_t = \mathcal{G}\sqrt{\beta_A \epsilon_A} \cos \mu_A t$$
$$+ \mathcal{G}(\beta_D^{1/2} R_{A22} - \alpha_D \beta_D^{-1/2} R_{A12})\epsilon_D^{1/2} \cos \mu_D t + \mathcal{G}\beta_D^{-1/2} R_{A12} \epsilon_D^{1/2} \sin \mu_D t,$$
$$y_t = \mathcal{G}\sqrt{\beta_D \epsilon_D} \cos \mu_D t$$
$$+ \mathcal{G}(\beta_A^{1/2} R_{A11} - \alpha_A \beta_A^{-1/2} R_{A12})\epsilon_A^{1/2} \cos \mu_A t + \mathcal{G}\beta_A^{-1/2} R_{A12} \epsilon_A^{1/2} \sin \mu_A t. \tag{61}$$

We are now in a position to evaluate selected matrix elements from the measured Fourier coefficients. By identifying coefficients of Eqs. (61) with those of Eq. (53) we obtain A-relations

$$\frac{A_{yc}}{A_x} = R_{A11} - \frac{\alpha_A}{\beta_A} R_{A12}, \quad \frac{A_{ys}}{A_x} = \frac{1}{\beta_A} R_{A12};$$
$$R_{A12} = \beta_A \frac{A_{ys}}{A_x}, \quad R_{A11} \simeq \frac{A_{yc}}{A_x} + \alpha_x \frac{A_{ys}}{A_x}; \tag{62}$$

and D-relations

$$\frac{D_{xc}}{D_y} = -R_{A22} - \frac{\alpha_D}{\beta_D} R'_{A12}, \quad \frac{D_{xs}}{D_y} = \frac{1}{\beta_D} R'_{A12};$$
$$R'_{A12} = \beta_D \frac{D_{xs}}{D_y}, \quad R_{A22} \simeq -\frac{D_{xc}}{D_y} - \alpha_y \frac{D_{xs}}{D_y}; \tag{63}$$

where the equations for R_{A11} and R_{A22} are indicated only as approximations, reflecting the fact that the replacements $\alpha_A \to \alpha_x$ and $\alpha_D \to \alpha_y$, valid only perturbatively, have had to be made since we will not succeed in extracting α_A and α_D from the data. (One could hope for the terms containing α_A and α_D to be small, but unfortunately in general the two terms of R_{A11} and R_{A22} have comparable numerical magnitudes.) A prime has been added to R'_{A12} in the D-equations to indicate that, although in the formalism it is the same as R_{A12}, its determination from data using Eqs. (63) is independent of its determination using Eqs. (62). Until β_A and β_D have been extracted this redundancy check is not quite ideal since they have to be approximated by the ideal values β_x and

β_y. In a later section β_A and β_D will be obtained directly from the data in an almost model-independent way and we will use those values for the present analysis. It is possible for both R'_{A12} and R_{A12} to vanish or be small, and at such points the ratio is indeterminate or poorly determined. For that reason, in Fig. 7, paired values of R_{A12} as abscissa and R'_{A12} as ordinate are plotted. In this plot the redundancy check reduces to the requirement that all points lie on a straight line. Simulated data are in the upper figure, measured data in the lower. There is near-perfect agreement for the simulated data, nicely satisfying the redundancy check. The ratios also cluster around the best-fit straight line for the actual data; deviations could perhaps be used for estimating the errors in the matrix element determinations.

The equations of this section partially substantiate the statement made previously that three of the four elements of R_A are directly measurable, but that R_{A21} is not. The determination cannot be completed in a model-independent way until the perturbed Twiss functions are determined. Unfortunately, since slope variables p and q are not measured, we will not succeed in extracting α_A and α_D. Also the presence of noise could cause the derived values of β_A and β_D to be even less valid than β_x and β_y but, "planning for success," we hope this is not the case.

10 DETERMINATION OF MODE INVARIANTS AND THE "MAGIC RATIO"

As in Eq. (43), the measured displacements (x, y) can be be expressed in terms of the mode invariants defined in Eqs. (41) and the measured transfer function parameters of Eqs. (55):

$$\begin{aligned} x_t &= \mathcal{G}\sqrt{\beta_A \epsilon_A} \cos \mu_A t + \mathcal{G} e_D \sqrt{\beta_D \epsilon_D} \cos(\mu_D t + \Phi_{Dx}), \\ y_t &= \mathcal{G} e_A \sqrt{\beta_A \epsilon_A} \cos(\mu_A t + \Phi_{Ay}) + \mathcal{G}\sqrt{\beta_D \epsilon_D} \cos \mu_D t, \end{aligned} \quad (64)$$

where the two newly introduced phases will not be important. These equations are the same as Eqs. (53) except for the symbols used for coefficients. We equate the coefficients, explicitly indicating now that the Fourier coefficients and the β-functions depend on longitudinal coordinate s while the mode invariants ϵ_A and ϵ_D do not:

$$\sqrt{\epsilon_A} = \frac{A_x(s)}{\mathcal{G}\sqrt{\beta_A(s)}}, \quad \sqrt{\epsilon_D} = \frac{D_y(s)}{\mathcal{G}\sqrt{\beta_D(s)}}, \quad (65)$$

or, expressing these as formulas for the unknown β-functions,

$$\beta_A(s) = \frac{A_x^2(s)}{\mathcal{G}^2 \epsilon_A}, \quad \beta_D(s) = \frac{D_y^2(s)}{\mathcal{G}^2 \epsilon_D}. \quad (66)$$

124 Coupled-Lattice Properties Using Turn-by-Turn Data

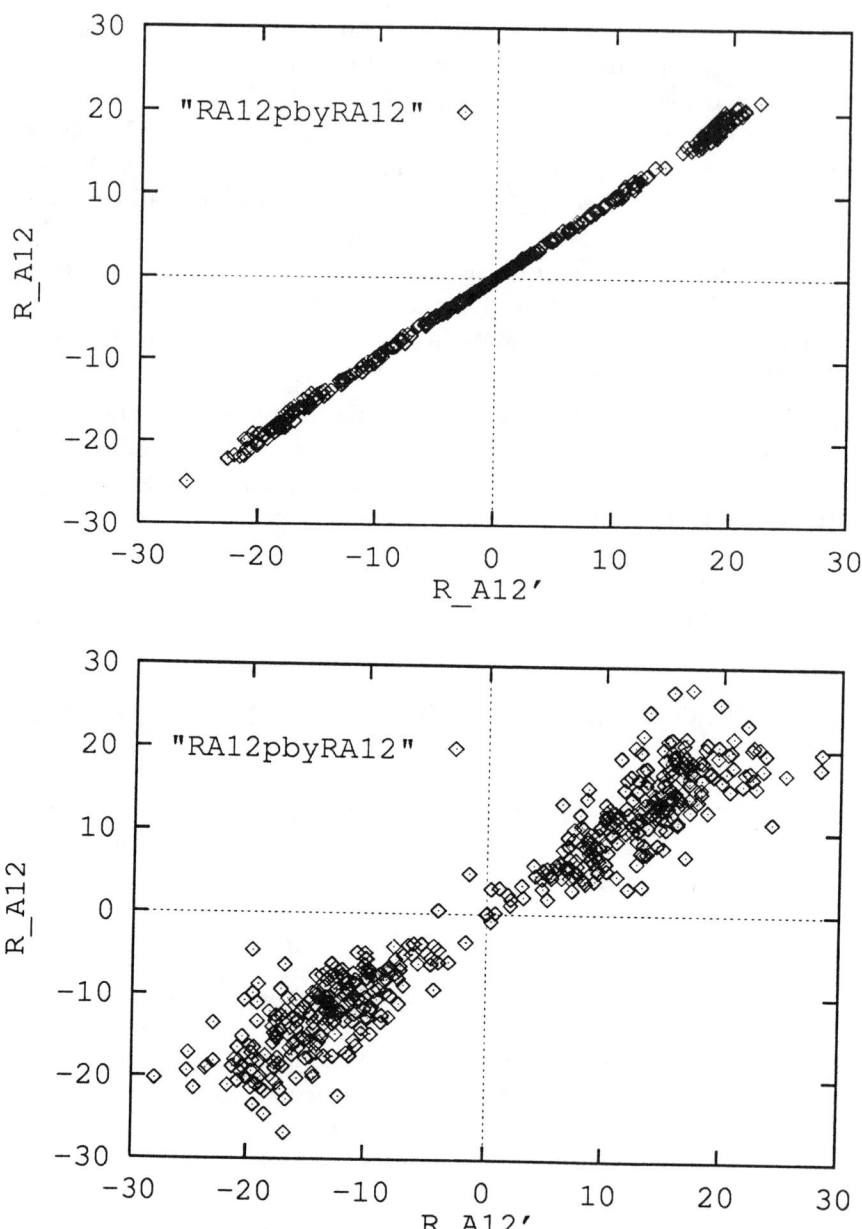

Figure 7. A point is plotted at every BPM with R_{A12} as abscissa and R'_{A12} as ordinate. Since these are two independent determinations of the same matrix element, all points should lie on the main diagonal. Upper figure is from simulation, lower figure from measured data.

We are now in position to derive the advertised "magic ratio." Consistency of Eqs. (62) and (63) requires $R_{A12} = R'_{A12}$ or

$$\beta_A \frac{A_{ys}}{A_x} = \beta_D \frac{D_{xs}}{D_y}, \qquad (67)$$

which can be used to determine the ratio β_A/β_D. Setting this equal to the ratio of Eqs. (66) yields the invariant

$$\mathcal{R}_m \equiv \frac{\epsilon_D}{\epsilon_A} = \frac{D_{xs} D_y}{A_{ys} A_x} \qquad (68)$$

where the ratio $\mathcal{R}_m$, none other than the ratio of mode invariants, is "magic" because it is the same for all BPM's and is obtained directly from readily and accurately measurable Fourier coefficients.

There is a tiniest of potential blemishes on this attractive form. The denominator factor $A_{ys} A_x$ is capable of vanishing, which can make the ratio at best indeterminate and at worst divergent. Analytically the ratio is indeterminate but for actual data the presence of noise or other distortion makes it possible for the ratio to become very large. The ratio has no practical use at positions where this indeterminacy is serious. For this reason, in Fig. 8, we plot numerator factor $D_{xs} D_y$ versus denominator factor $A_{ys} A_x$, and make a linear regression fit to the resultant dot plot. The simulated data are almost perfectly consistent with the constancy of $\mathcal{R}_m$. The actual 14_18 data are also nicely consistent with the linear fit. Except in the indeterminate region near the origin, deviations from constancy of $\mathcal{R}_m$ are mostly less than 20%.

Once $\mathcal{R}_m$ has been made available as the slope of a linear fit in Fig. 8, it is possible to write the formulas of Eq. (66) in an improved form with only a single unknown common factor $(\mathcal{G}^2 \epsilon_A)^{-1}$:

$$\beta_A(s) = \frac{1}{\mathcal{G}^2 \epsilon_A} A_x^2(s), \qquad \beta_D(s) = \frac{1}{\mathcal{R}_m} \frac{1}{\mathcal{G}^2 \epsilon_A} D_y^2. \qquad (69)$$

To this point lattice functions β_A and β_D have been unknown and, in a pure sense, they still are because the factor $(\mathcal{G}^2 \epsilon_A)^{-1}$ is not yet known. But it is so valuable to know the perturbed β-functions that we now reduce our standards slightly and assume (A_x being better determined than D_y)

$$\mathcal{G} \epsilon_A^{1/2} = < \frac{A_x(s)}{\beta_x^{1/2}(s)} > . \qquad (70)$$

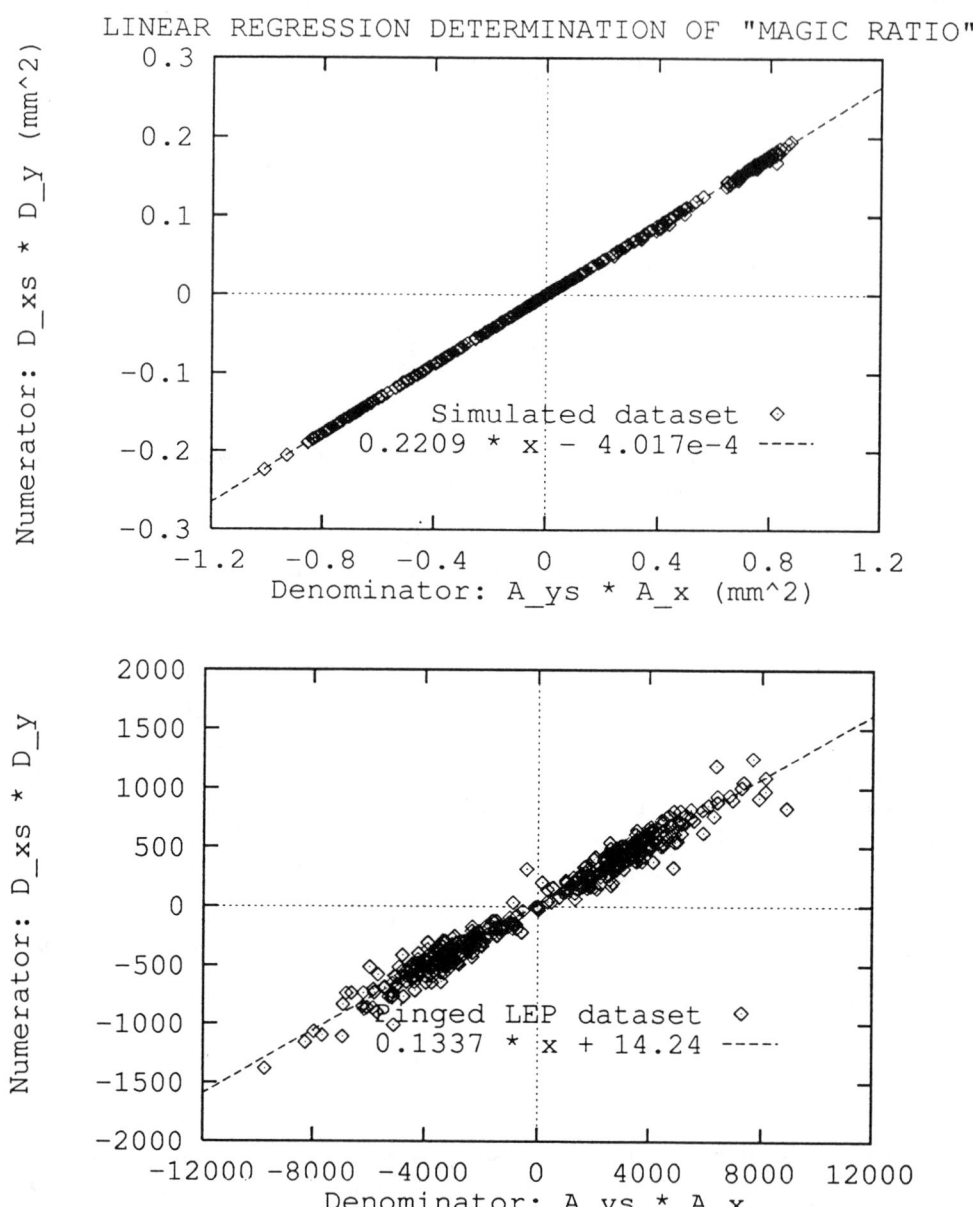

Figure 8. Magic ratio determination. Upper figure is from simulation, lower figure from measured data.

Taken with Eq. (69), this is similar to assuming

$$< \beta_A^{-1} > = < \beta_x^{-1} >; \tag{71}$$

i.e. the average inverse β-function is unperturbed, which is based on preserving the equality of $\int ds/\beta_x(s)$ and the full (integer part plus fractional part) tune of the lattice; that tune is almost surely unperturbed to any relevant accuracy. But relation (71) is compromised by the fact that the BPM distribution around the ring is not uniform. This is especially true for the actual LEP data being analysed, since LEP's BPM's are intentionally placed at points in the lattice where β_y is maximum. Also $< \beta_A^{-1} >$ can be strongly biased by a single bad BPM that yields an anomalously small value of β_A because A_x is small. Anyway Eq. (70) is at least as good as any perturbative relation, and we adopt it tentatively. In any case an error here causes only an overall distance scale error.

Plotted in Fig. 9 are β-functions determined this way. In this case the simulated data (top graph) and real data (bottom graph) are processed somewhat differently. For the simulated data, since the true perturbed functions are known we plot $\sqrt{\beta_{A,D}/\beta_{A,D}(\text{true})}$. For the measured data the perturbed β-function is unknown and we plot $\sqrt{\beta_{A,D}/\beta_{x,y}(s)}$. The deviations from 1 are much greater for the actual data than for the simulated data. It is not known whether that is due to noise or to true deviations of the β-functions.

11 ATTEMPTED FURTHER PARAMETER DETERMINATION

We substitute all presently known parameters into Eqs. (60) and (53) to obtain

$$\begin{pmatrix} x_t \\ p_t \\ y_t \\ q_t \end{pmatrix} = \mathcal{G} \begin{pmatrix} 1 & 0 & \frac{D_{xc}}{D_y} + \alpha_D \frac{D_{xs}}{D_y} & \beta_A \frac{A_{ys}}{A_x} \\ 0 & 1 & R_{A21} & -\frac{A_{yc}}{A_x} - \alpha_A \frac{A_{ys}}{A_x} \\ \frac{A_{yc}}{A_x} + \alpha_A \frac{A_{ys}}{A_x} & \beta_A \frac{A_{ys}}{A_x} & 1 & 0 \\ R_{A21} & -\frac{D_{xc}}{D_y} - \alpha_D \frac{D_{xs}}{D_y} & 0 & 1 \end{pmatrix}$$

$$\times \begin{pmatrix} \beta_A^{1/2} \epsilon_A^{1/2} \cos \mu_A t \\ -\alpha_A \beta_A^{-1/2} \epsilon_A^{1/2} \cos \mu_A t + \beta_A^{-1/2} \epsilon_A^{1/2} \sin \mu_A t \\ \beta_D^{1/2} \epsilon_D^{1/2} \cos \mu_D t \\ -\alpha_D \beta_D^{-1/2} \epsilon_D^{1/2} \cos \mu_D t + \beta_D^{-1/2} \epsilon_D^{1/2} \sin \mu_D t \end{pmatrix}$$

$$= \begin{pmatrix} A_x \cos \mu_A t + D_{xc} \cos \mu_D t + D_{xs} \sin \mu_D t \\ \cdots \\ D_y \cos \mu_D t + A_{yc} \cos \mu_A t + A_{ys} \sin \mu_A t \\ \cdots \end{pmatrix}. \tag{72}$$

128 Coupled-Lattice Properties Using Turn-by-Turn Data

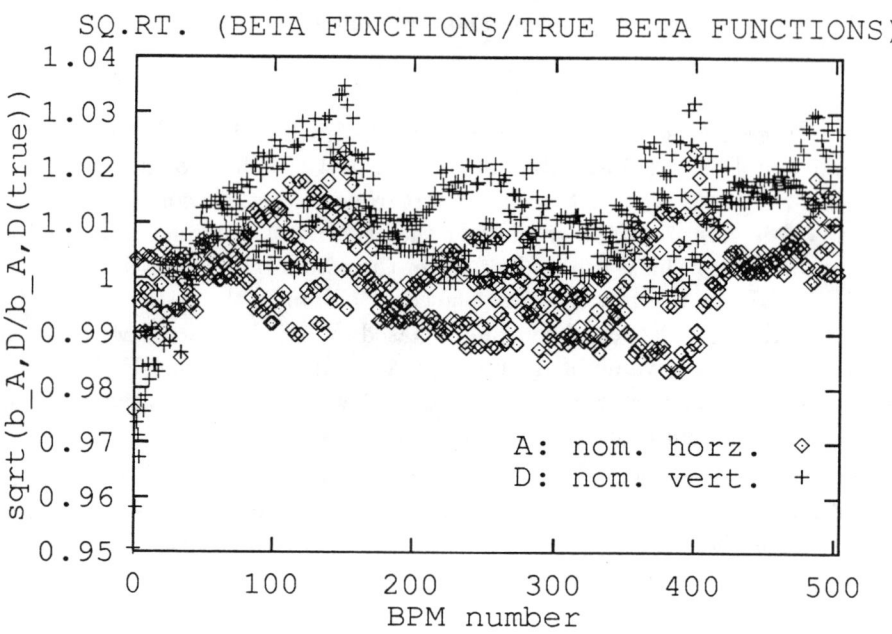

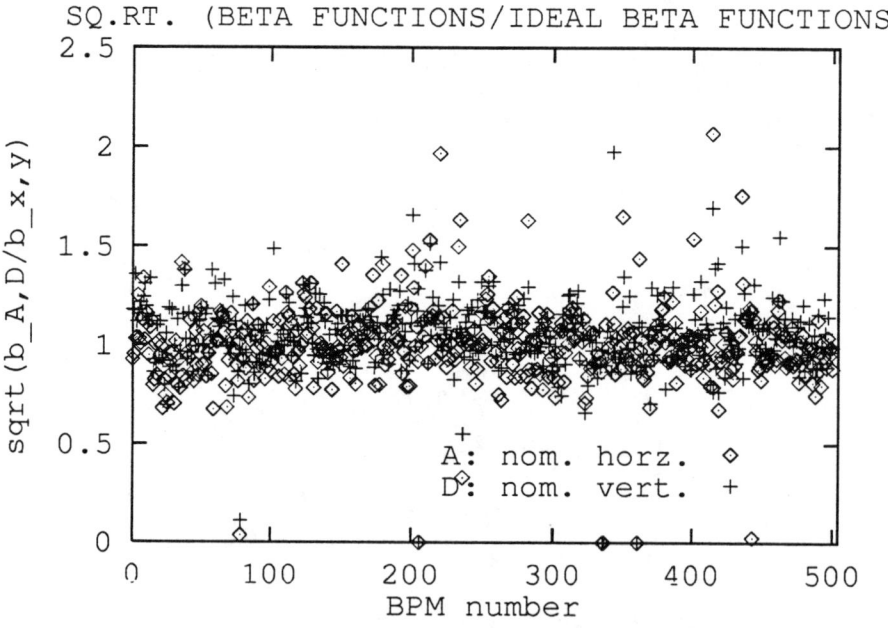

Figure 9. Measured/ideal β-function ratios evaluated at every BPM. Upper figure is $\sqrt{\beta_{A,D}/\beta_{A,D}(\text{true})}$ from simulation, lower figure is $\sqrt{\beta_{A,D}/\beta_{x,y}(s)}$ from measured data.

One can check that these equations for x_t and y_t, as well as not depending on R_{A21}, are identically satisfied independent of α_A and α_D, and hence cannot be used to determine those quantities. This should perhaps not be suprising for the following reason. A coordinate transformation, $x \to x, p \to p' = p - \xi x$, transforms M to

$$M' = \begin{pmatrix} 1 & 0 \\ -\xi & 1 \end{pmatrix} \begin{pmatrix} \cos\mu + \alpha\sin\mu & \beta\sin\mu \\ -\gamma\sin\mu & \cos\mu - \alpha\sin\mu \end{pmatrix} \begin{pmatrix} 1 & 0 \\ \xi & 1 \end{pmatrix} \quad (73)$$

which is equivalent to replacing α by $\alpha + \xi\beta$. Since only the unmeasured "momentum" or "slope" components p and q are influenced by the α parameters, those parameters cannot be inferred by measuring only x and y. (As stated in the introduction we are intentionally refraining from inferring slope coordinates by using adjacent BPM values.) It can be shown that the transformation (73) leaves $\mathcal{E} = \det|C + \bar{B}|$ (but not e, f, g and h) invariant.

Should one tentatively attempt to exploit this invariance to simplify the equations by setting $\alpha_A = \alpha_D = 0$, one obtains, from Eq. (72), expressions for the unmeasured components:

$$p'_t = \mathcal{G}\beta_A^{-1/2}\epsilon_A^{1/2}\sin\mu_A t + \mathcal{G}R_{A21}\beta_D^{1/2}\epsilon_D^{1/2}\cos\mu_D t - \mathcal{G}\frac{A_{yc}}{A_x}\beta_D^{-1/2}\epsilon_D^{1/2}\sin\mu_D t,$$

$$q'_t = \mathcal{G}R_{A21}\beta_A^{1/2}\epsilon_A^{1/2}\cos\mu_A t - \mathcal{G}\frac{D_{xc}}{D_y}\beta_A^{-1/2}\epsilon_A^{1/2}\sin\mu_A t + \mathcal{G}\beta_D^{-1/2}\epsilon_D^{1/2}\sin\mu_D t, \quad (74)$$

where the primes indicate that p'_t and q'_t are not necessarily even approximately equal to the "correct" slope coordinates that might, for example, be obtained from processing adjacent BPM outputs. Having obtained expressions for p'_t and q'_t, one can substitute into the Eqs. (46) in the forlorn hope of obtaining the as-yet-undetermined quantities tr$A$ and tr$D$. This yields sixteen equations: 2(trig functions) times 2(tunes) times 4(equations). Since all these equations turn out to be satisfied identically, they cannot be used to determine any other parameters. For checking these identities one needs to use the result

$$\frac{-\zeta}{1-\zeta}D_y A_x = A_{yc}D_{xc} + (\beta_A R_{A21})A_{ys}D_y. \quad (75)$$

It can be shown using Eqs. (22) that this is simply equivalent to the condition $\det|R_A| = \zeta/(1-\zeta)$. This can be used to determine ζ from R_{A21} or vice versa, but they are not determined independently. The expense of having checked Eqs. (46) can be charged against the QA (quality assurance) account, since self-consistency has been tested. Should we choose to trust that the global decoupling operation has been performed just before the data were obtained, we would substitute $\zeta = 0$

to obtain

$$\beta_A R_{A21} = -\frac{A_{yc}D_{xc}}{A_{ys}D_y} \quad \text{(if and only if } \mathcal{E} = 0\text{).} \tag{76}$$

It seems that the best that can be done to derive the slopes p_t and q_t from the measurements at a single BPM is to expand the second and fourth rows of Eq. (72) using α_x and α_y in place of α_A and α_D and to use Eq. (76) for R_{A21}.

12 QUANTIFYING THE DEGREE OF LOCAL COUPLING

The ultimate goal of our investigation is to "locally decouple" accelerator lattices. A practical algorithm for achieving this is described elsewhere.[1] Here we contemplate two candidate definitions of local coupling, justify the particular choice adopted, and supply qualitative interpretation. As part of the discussion it is necessary to define the degree of local coupling in order to be able to specify a limit that separates acceptable and unacceptable local coupling and to define a global r.m.s. average coupling.

The angle θ_A defined in Eq. (45), when evaluated at local position s has a clear geometric interpretation as the major axes of the ellipse that would be visible on a (non-destructive) phosphor screen placed at s, when only the A-mode is excited. In TEAPOT documentation this is known as the A-eigenangle. Its deviation from zero is a candidate for use in quantifying local coupling. For example, this angle might be required to be small, say less than 0.1. However, this quantity has an undesirable feature. Because of the alternating-gradient optics, the beam profile is strongly distorted as s varies. At vertical focusing quads the vertical axis is stretched and the horizontal compressed. This has the effect of magnifying $\tan \theta_A$, proportional to the stretch ratio $\sqrt{\beta_y/\beta_x}$. The distortion is reversed at horizontally focusing quads. Requiring θ_A to be less than 0.1 at a vertical focusing quad location requires it to be much less than 0.1 at a horizontal focusing quad, which is probably more conservative than intended. To compensate for this effect one can define

$$\text{normalized } A\text{-eigenangle} = \theta_A \sqrt{\frac{\beta_x}{\beta_y}}. \tag{77}$$

Plots of this quantity are shown in Fig. 10. All this discussion can be repeated for the D-mode. The factor $\sqrt{\beta_x/\beta_y}$ has the desired effect of moderating the variation with s, but it is *ad hoc* and the normalized eigenangle cannot be expected to behave gracefully for large eigenangles, because of its definition in terms of a transcendental function.

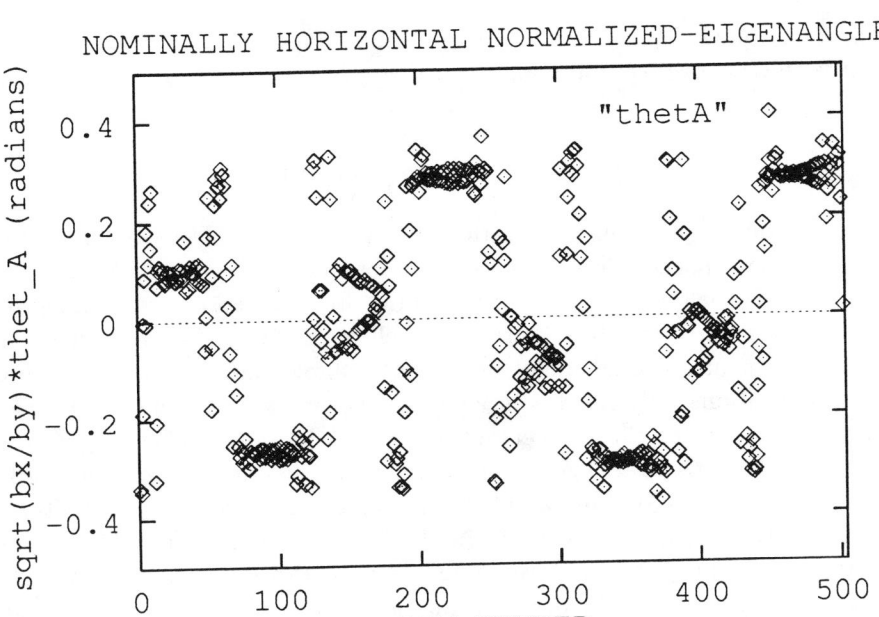

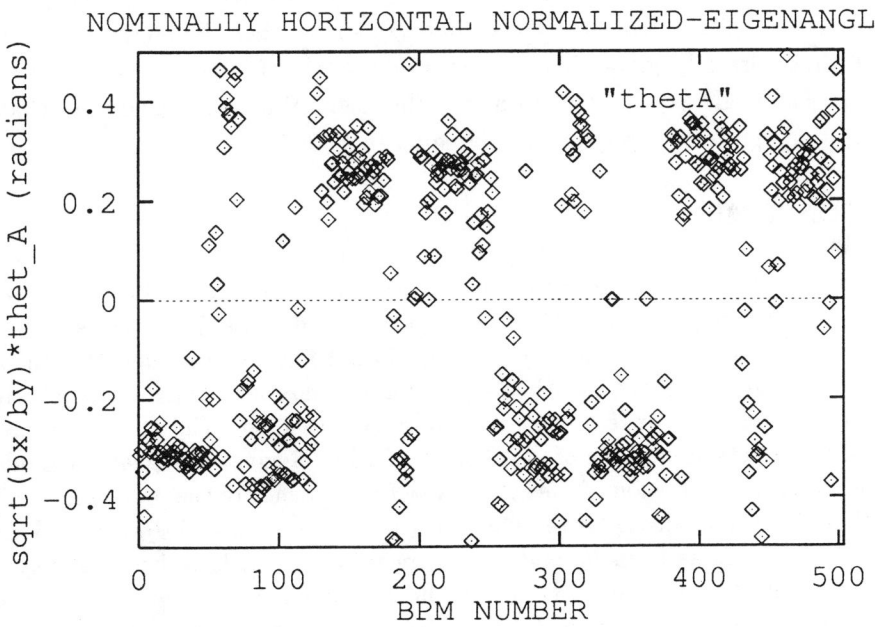

Figure 10. Normalized eigenangle $\theta_A \sqrt{\beta_x/\beta_y}$ evaluated at every BPM. Upper figure is from simulation, lower figure from measured data.

The decoupling algorithm in TEAPOT quantifies coupling by the quantity $\mathcal{B}^C(A)$, called the local coupling badness, where

$$\sqrt{\mathcal{B}^C(A)} = e_A \sqrt{\frac{\beta_x}{\beta_y}} \qquad (78)$$

and e_A is the out-of-plane/in-plane ratio defined in Eq. (43). The same discussion as in the previous paragraph can be used to motivate the inclusion of the square root factor. Because e_A is a simple y/x ratio, Eq. (78) correctly compensates the scale distortions described there. This choice is further supported by the following argument. In what might be a highly coupled lattice, consider a short sector that contains only ideal components with no skew elements. Where it enters this ideal sector the A-eigenmotion can be decomposed into (x, x', y, y') components. Though we are describing nominally horizontal motion, it is possible for the y amplitudes to be comparable to or even greater than, the x amplitudes, because of strong coupling elsewhere in the lattice. Through this ideal sector these amplitudes propagate according to standard two-dimensional formalism. In particular, the Courant-Snyder invariants $\epsilon_x(A) = \gamma_x x^2 + 2\alpha_x x x' + \beta_x x'^2$ and its y counterpart $\epsilon_y(A)$ are conserved quantities. Realizing from Eqs. (43) that $e_A = y(\max)/x(\max)$ it can be seen that $\mathcal{B}^C(A)$ is conserved in sectors that contain no coupling elements. Note that $\sqrt{\beta_x/\beta_y}$ rather than $\sqrt{\beta_A/\beta_D}$ is the factor that correctly gives this conservation. Plots of $\sqrt{\mathcal{B}^C(A)}$ are shown in Fig. 11. The angle $\sqrt{\beta_x/\beta_y}\theta_A$ can have either sign. When it is much less than 1 radian its magnitude is approximately equal to $\sqrt{\mathcal{B}^C(A)}$.

13 CONCLUSIONS

We consider the agreement between theory and observation to be good, simultaneously validating the data, the codes, and our understanding of the situation. The measured magic ratio data of Fig. 8 could not fit so well without the theory being essentially correct and the beam position measurement and data acquisition system yielding more or less what they were supposed to. The good agreement between two determinations of R_{A12} shown in Fig. 7 gives independent model-independent corroboration of the inherent self-consistency of the procedures. The fact that this agreement is better when measured values β_A and β_D are used says that their deviations from β_x and β_y are meaningful and have been reliably (if not accurately) determined. Naturally the simulated data, being free of noise and distortion, give cleaner results, but they are qualitatively similar. They are indicative of the quality of results which can, in principle, be obtained with an ideally functioning BPM system. It does not seem unreasonable to anticipate data of this general quality in the near future at LEP since all the data analysed in this paper were acquired in ten minutes without having expended much time at optimizing conditions.

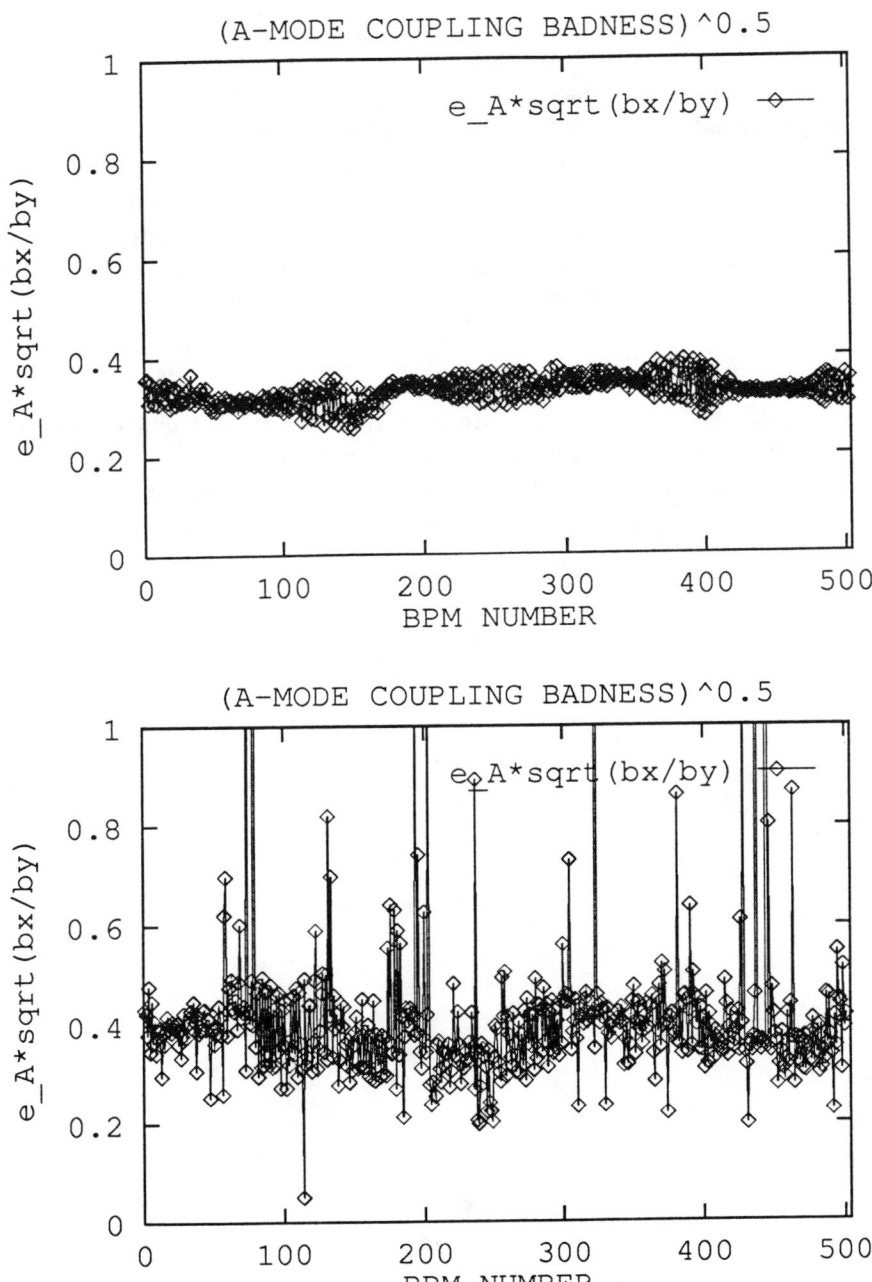

Figure 11. Square root of A-mode local coupling badness $\sqrt{\mathcal{B}^C(A)} = e_A\sqrt{\beta_x/\beta_y}$ evaluated at every BPM. Upper figure is from simulation, lower figure from measured data.

This investigation was undertaken largely to anticipate the performance of beam-position data-acquisition designs of future colliders SSC and LHC. This investigation has shown that a system like that at LEP will reliably yield the coupling information needed for locally decoupling these large colliders. At LEP, coupling measurements are available only at every second half-cell. The SSC design calls for similar instrumentation, with 10^4 turn, x, y digitized pairs available for readout, on demand, at every fifth half-cell. Since cross-coupling effects are expected to be somewhat more serious for SSC than for LEP, the possibility of instrumenting every quadrupole should perhaps be considered.

14 REFERENCES

1. R. Talman, Single Particle Motion, in *Frontiers of Particle Beams; Observation, Diagnosis and Correction*, M. Month and P. Bryant, Editors, US-CERN School on Particle Accelerators, Capri, Springer-Verlag, 1989.
2. E. Courant and H. Snyder, *Ann. Phys.* **3**, 1 (1958).
3. L. Schachinger and R. Talman, *Particle Accelerators* **22**, 35 (1987).

Chromaticity Correction for the Collider at the SSC

Y. Nosochkov, F. Pilat, T. Sen
SSC Laboratory, Dallas, Texas 75237

1 INTRODUCTION

Control of the chromaticity is important in order to avoid collective instabilities like the head-tail effect and to avoid crossing resonances. This is specially so for the collider where the linear chromaticity is spectacularly large, e.g reaching -470 units in the vertical plane at collision, for the design values of the luminosity at the four interaction points. Correcting the linear chromaticity is relatively simple however, and does not require much discussion. In this paper, we present the proposed scheme for reducing the nonlinear chromaticity of the collider in the collision mode. Section 2 discusses briefly the optics of the collider lattice with special emphasis on the Interaction Regions (IRs) which are the main sources of chromaticity. Section 3 is devoted to a theoretical calculation of the second-order chromaticity in a storage ring. Section 4 focuses on the second-order tune shift due to the triplets, the dominant sources of chromaticity in the IRs. Section 5 discusses, mostly in pictures, the theory behind our proposed scheme for correcting higher-order chromaticity. In Section 6 we evaluate the chromatic performance of the proposed scheme and in Section 7 we discuss the effect of the chromaticity-correcting sextupoles on the dynamic aperture. We summarize our conclusions in Section 8.

2 THE COLLIDER LATTICE: SOURCES OF CHROMATICITY

The racetrack shaped collider lattice consists basically of 2 arcs located on the North and South sides and 2 clusters placed on the West and on the East. Each arc contains 196 identical FODO cells with the phase advance across a cell being 90 degrees and the length of each cell is 180 m. The lattice of each cluster includes 2 Interaction Regions, the utility section and the interconnect sections between them. It is intended that the IRs have similar configuration except for the free space length reserved for the detectors.

The arcs occupy about 81% of the lattice and therefore provide a significant contribution to the chromaticity of the machine. They mostly determine the collider chromaticity at injection conditions. However, in the collision mode, the IRs have a larger chromaticity than the arcs. This additional chromaticity comes from the final focussing quadrupoles which at collision are located in the region of extremely high values of the beta functions. The locality of the chromaticity sources in the IRs makes them more dangerous than the chromaticity contributed by the arcs.

The general scheme of an IR is shown in Figure 1. F and D quadrupoles are drawn above and under the beam lines, and the vertical bends are centered on the beam lines. The two rings are separated vertically by a distance of 90 cm everywhere except in the IRs. Within each IR, the beams are brought into collision at the Interaction Point (IP) in two steps by a set of vertical dipoles.

The optics of the IR consists of three main parts. The quadrupoles located in the region where the separation is 90 cm form the tuning section. These quadrupoles are used to match

© 1994 American Institute of Physics

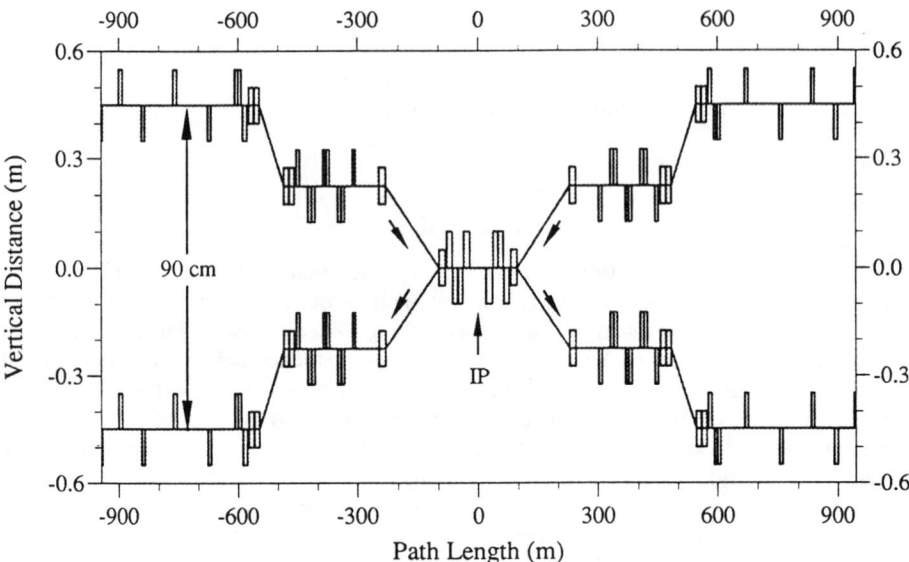

Figure 1: Elevation View of the Low-Beta IR.

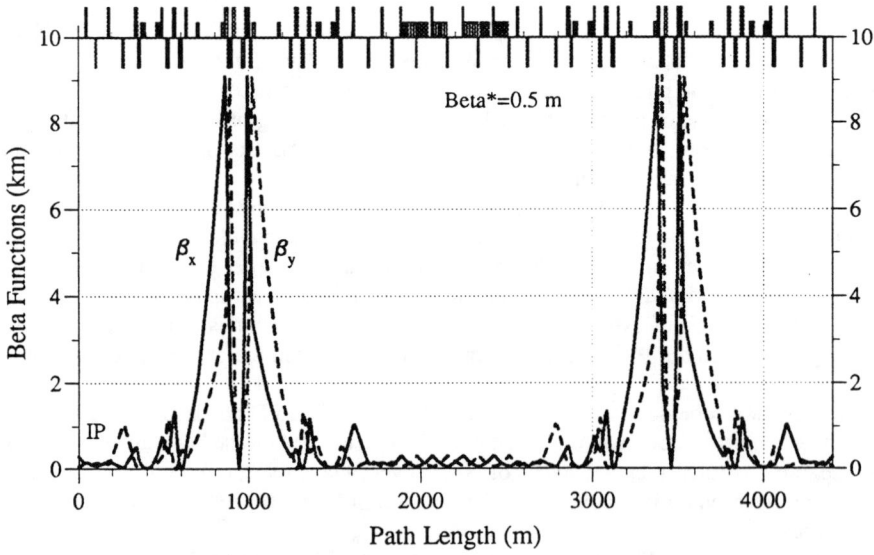

Figure 2: Beta Functions in the Low-Beta IRs.

the optics and vary the β^* at the IP. The second part consists of the vertical dipoles and the quadrupoles located in the 45-cm separation region. The dipoles are used to bring the beams into collision. The phase advance across the quadrupoles is π and collectively the quadrupoles form the $M=-I$ section, whose role is to correct the vertical dispersion locally. The last part of the optics includes the final triplet quadrupoles which focus the beams to the IP. These quads are common to both rings and the beams share the same beam pipe inside them. At high luminosity conditions the beta functions go up rapidly inside the triplets which become a source of a large chromaticity. Figure 2 shows the particular beta functions in 2 IRs for the design value of $\beta^*=0.5$ m. Each IP is surrounded by 2 triplets where the beta function is 30 times larger than in the arcs. The contribution to the 1st-order chromaticity from major sources is listed in Table 1. L* denotes the free space reserved on either side of each IP for the detectors.

SOURCE	horizontal chromaticity	vertical chromaticity
2 arcs	-124	-123
1 low β IR, L*=20.5m, β^*=0.50m 1 low β IR, L*=20.5m, β^*=0.25m	-51 -101	-51 -101
1 medium β IR, L*=90m, β^*=1.95m	-45	-45
Maximum allowed sextupole component in dipoles ($b_2=0.8*10^{-4}$ m^{-2})	160	-136
Complete collider lattice: 2 low β IRs (β^*=0.50m) 2 medium β IRs (β^*=1.95m) $b_2=0.8*10^{-4}$ m^{-2} in dipoles	-171	-469

Table 1: Major sources of chromaticity in the collider lattice

3 TUNE SHIFT AND CHROMATICITY TO 2nd ORDER

Consider a storage ring and label two points on it as 1 and 2. Let μ_o be the global phase advance around the ring and (β_1, α_1, γ_1) the Twiss functions at point 1. The periodic transfer matrix at point 1 is

$$M_1 = \begin{bmatrix} \cos\mu_o + \alpha_1\sin\mu_o & \beta_1\sin\mu_o \\ -\gamma_1\sin\mu_o & \cos\mu_o - \alpha_1\sin\mu_o \end{bmatrix} \equiv M(2 \to 1) \times M(1 \to 2) \quad (3.1)$$

where $M(2 \to 1)$ is the transfer matrix from point 2 to 1 etc. Let μ_1 and μ_2 be the phase advances at points 1 and 2 respectively with respect to an arbitrary reference point and define $\mu_{21}=|\mu_2-\mu_1|$. We now introduce two infinitesimally thin quads of strengths $q_1=k_1\Delta s_1$ and $q_2=k_2\Delta s_2$ at points 1 and 2 respectively. Their perturbations to the transfer matrix are described by the matrices

$$P_1 = \begin{bmatrix} 1 & 0 \\ -k_1 \Delta s_1 & 1 \end{bmatrix}, \qquad P_2 = \begin{bmatrix} 1 & 0 \\ -k_2 \Delta s_2 & 1 \end{bmatrix}. \qquad (3.2)$$

These quad errors change the cyclic transfer matrix at point 1 to $\overline{M}_1$,

$$\overline{M}_1 = M(2 \to 1) \times P_2 \times M(1 \to 2) \times P_1 \equiv M_1 + \Delta M_1. \qquad (3.3)$$

The elements of ΔM_1 (the change in the transfer matrix) are found using this equality. Let $\Delta \mu$ be the change in the global phase advance around the ring. We scale the quad errors by an arbitrary parameter ε, i.e.

$$k_1 \to \varepsilon k_1 \qquad k_2 \to \varepsilon k_2$$

and expand $\Delta \mu$ as a power series in ε,

$$\Delta \mu = \varepsilon \Delta \mu_1 + \varepsilon^2 \Delta \mu_2 + \dots . \qquad (3.4)$$

In keeping with our declared aim, our calculation will be to second order in ε. The new global phase advance $\overline{\mu}_o = \mu_o + \Delta \mu$ is to be found from

$$\cos \overline{\mu}_o = \frac{1}{2} Tr \overline{M}_1 = \cos \mu_o + \frac{1}{2} Tr \Delta M_1. \qquad (3.5)$$

We also have

$$\cos \overline{\mu}_o = \cos \mu_o \cos \Delta \mu - \sin \mu_o \sin \Delta \mu.$$

Substituting Equation (3.4) into the above and equating it to the expression for $\cos \overline{\mu}_o$ given by Equation (3.5), we have

$$-\varepsilon \sin \mu_o \Delta \mu_1 - \varepsilon^2 \left[\sin \mu_o \Delta \mu_2 + \frac{1}{2} \cos \mu_o (\Delta \mu_1)^2 \right] + O(\varepsilon^3) = \frac{1}{2} Tr \Delta M_1. \qquad (3.6)$$

To obtain the corrections to μ_o order by order, we equate the coefficients of like powers of ε on both sides of the above equation:

$$\Delta\mu_1 = \frac{1}{2}(k_1 \Delta s_1 \beta_1 + k_2 \Delta s_2 \beta_2),$$

$$\Delta\mu_2 = \frac{1}{4\sin\mu_o} k_1 \Delta s_1 \beta_1 k_2 \Delta s_2 \beta_2 [\cos\mu_o - \cos(\mu_o - 2\mu_{21})] - \frac{\cot\mu_o}{2}(\Delta\mu_1)^2. \quad (3.7)$$

Generalizing to the case when there are N quad errors distributed around the ring,

$$\Delta\mu_1 = \frac{1}{2}\sum_{i=1}^{N} k_i \beta_i \Delta s_i,$$

$$\Delta\mu_2 = \frac{1}{4\sin\mu_o}\sum_{i=1}^{N-1}\sum_{j=i+1}^{N}(k_i \beta_i \Delta s_i)(k_j \beta_j \Delta s_j)[\cos\mu_o - \cos(\mu_o - 2\mu_{ji})]$$

$$-\frac{\cot\mu_o}{2}(\Delta\mu_1)^2. \quad (3.8)$$

We go now to the limit of infinitesimally thin quads distributed around the ring of circumference C. In this limit,

$$\Delta\mu_1 = \frac{1}{2}\int_0^C k(s)\beta_o(s)\,ds,$$

$$\Delta\mu_2 = \frac{1}{4\sin\mu_o}\int_0^C k(s)\beta_o(s)\,ds \int_s^C k(s')\beta_o(s')[\cos\mu_o - \cos(\mu_o - 2|\mu(s') - \mu(s)|)]\,ds'$$

$$-\frac{\cot\mu_o}{2}(\Delta\mu_1)^2. \quad (3.9)$$

Here we have let $\beta_o(s)$ denote the unperturbed β function at the point s. In the equation for $\Delta\mu_2$ we convert the integral over part of the ring to one over the complete ring and obtain

$$\Delta\mu_2 = -\frac{1}{8\sin\mu_o}\int_0^C k(s)\beta_o(s)\,ds \int_s^{s+C} k(s')\beta_o(s')\cos[\mu_o - 2|\mu(s') - \mu(s)|]\,ds'. \quad (3.10)$$

The gradient error changes not only the tune of the machine but also the β function around the ring. We recall that the change in β to first order in the gradient errors is given by [1]

$$\frac{\Delta\beta_1(s)}{\beta_o(s)} = -\frac{1}{2\sin\mu_o} \int_s^{s+C} k(s')\beta_o(s') \cos[\mu_o - 2|\mu(s') - \mu(s)|] \, ds'. \qquad (3.11)$$

Recognizing that the term on the right-hand side occurs within Equation (3.10) we obtain the rather simple expression

$$\Delta\mu_2 = \frac{1}{4}\int_0^C k(s) \Delta\beta_1(s) \, ds. \qquad (3.12)$$

This important relation tells us that the first-order distortion in the β function propagating around the machine gives rise to the second-order tune shift. The total phase shift to second order in the gradient errors is (after putting the arbitrary parameter ε to unity)

$$\Delta\mu = \frac{1}{2}\int_0^C k(s)\beta_o(s)\,ds + \frac{1}{4}\int_0^C k(s)\Delta\beta_1(s)\,ds + O(k^3). \qquad (3.13)$$

The gradient perturbations of interest here are those seen only by particles off the design momentum. The chromatic error introduced by the quads is corrected by placing sextupoles at places of non-zero dispersion. Assuming that only the horizontal dispersion D_x is non-zero, the effective quadrupole strength in the horizontal plane for a particle with relative momentum deviation $\delta = \Delta p/p_o$ is

$$K^{eff} = K(s,\delta) + S(s,\delta)D(s,\delta)\delta. \qquad (3.14)$$

As functions of δ,

$$K(s,\delta) = \frac{K_o(s)}{1+\delta} = K_o(s)[1 - \delta + \delta^2 + \ldots],$$

$$S(s,\delta) = \frac{S_o(s)}{1+\delta} = S_o(s)[1 - \delta + \delta^2 + \ldots], \qquad (3.15)$$

where K_o and S_o are the nominal quad and sextupole strengths experienced by a particle on momentum. The gradient errors also change the dispersion function around the ring, making it a function of δ as well. We expand D as a power series in δ,

$$D(s,\delta) = D_o(s) + \Delta D_1^C(s)\delta + \Delta D_2^C(s)\delta^2 + \ldots \qquad (3.16)$$

and similarly the β function,

$$\beta(s, \delta) = \beta_o(s) + \Delta\beta_1^C(s)\delta + \Delta\beta_2^C(s)\delta^2 + \ldots \qquad (3.17)$$

where the superscript C denotes a chromatic expansion. Hence the gradient error in the horizontal plane for the off-momentum particles is

$$k(s) \equiv K^{\textit{eff}} - K_o = [-K_o + S_o D_o]\delta + [K_o + S_o(-D_o + \Delta D_1^C)]\delta^2 + O(\delta^3) \;.$$
$$(3.18)$$

Substituting into Equation (3.13) and writing the tune shift in terms of the first- and second-order chromaticity ξ_1 and ξ_2 respectively,

$$\Delta\nu \equiv \frac{1}{2\pi}\Delta\mu = \xi_1\delta + \xi_2\delta^2 + O(\delta^3), \qquad (3.19)$$

we obtain

$$\xi_1 = \frac{1}{4\pi}\int_0^C \beta_o(s)\,[-K_o + S_o(s)D_o(s)]\,ds,$$

$$\xi_2 = \frac{1}{8\pi}\int_0^C [-K_o + S_o(s)D_o(s)]\,\Delta\beta_1^C(s)\,ds + \frac{1}{4\pi}\int_0^C \beta_o(s)\,S_o(s)\,\Delta D_1^C(s)\,ds - \xi_1\;.$$
$$(3.20)$$

We recall that the first-order changes in β and D are given by

$$\frac{\Delta\beta_1^C(s)}{\beta_o(s)} = \frac{-1}{2\sin\mu_o}\int_s^{s+C} \beta_o(s')F_o(s')\cos[\mu_o - 2|\mu(s') - \mu(s)|]\,ds',$$

$$\Delta D_1^C(s) = \frac{-\sqrt{\beta_o(s)}}{2\sin\frac{\mu_o}{2}}\int_s^{s+C} \sqrt{\beta_o(s')}D_o(s')F_o(s')\cos\left[\frac{\mu_o}{2} - |\mu(s') - \mu(s)|\right]ds'$$

with $F_o(s')$ defined as

$$F_o(s') = -K_o(s') + S_o(s')D_o(s')\;.$$

Ignoring the phase factors for the moment, we see that $\Delta\beta_1^C$ which contains factors of $\beta_o(s)$ rather than $\beta_o(s)^{1/2}$, as occurs in $\Delta D_1^C(s)$, will dominate the contribution to the second-order chromaticity. This situation can change if we choose the phase advances between the major chromatic error sources appropriately. For example, two sources of equal strength $\pi/2$

apart in phase will produce β waves exactly out of phase so there will be no resultant β wave. The dispersion waves produced by the same two sources will add in quadrature. Alternatively, if we want to cancel the net dispersion wave, the two sources should be π apart in phase. In this case the β waves will add exactly in phase.

Returning to the issue of chromaticity correction, the sextupole strengths $S_o(s)$ are usually chosen to make the linear chromaticity ξ_1 vanish, i.e. $S_o(s)$ is obtained by solving

$$\int_0^C S_o(s) \beta_o(s) D_o(s) \, ds = \int_0^C \beta_o(s) K_0(s) \, ds \qquad (3.22)$$

and ξ_2 is then given by

$$\xi_2 = \frac{1}{8\pi} \int_0^C [-K_o + S_o(s) D_o(s)] \Delta\beta_1^C(s) \, ds + \frac{1}{4\pi} \int_0^C \beta_o(s) S_o(s) \Delta D_1^C(s) \, ds. \qquad (3.23)$$

Hence to reduce the second-order chromaticity, the first-order changes in β and also in the dispersion D should be minimized. Conversely, the regions where $\Delta\beta_1^C$ is large (e.g. the triplets in the IRs) will contribute the most to the second-order chromaticity. The above expression also exhibits the variation of ξ_2 with the global tune. Since the first-order β wave diverges at integer and half-integer tunes, ξ_2 will be amplified as ν_o approaches 0 or 0.5 and will be a minimum at $\nu_o=0.25$.

4 BETA WAVE AND CHROMATICITY DUE TO IR TRIPLETS

The perturbation to the periodic β function from chromatic errors in the IR quads can be calculated as a power series in δ,

$$\frac{\Delta\beta(s)}{\beta_o(s)} = \frac{\Delta\beta_1^C(s)}{\beta_o(s)} \delta + \frac{\Delta\beta_2^C(s)}{\beta_o(s)} \delta^2 + O(\delta^3) \qquad (4.1)$$

where $\beta_o(s)$ is the unperturbed β function. Let

$$Q_i \beta_i \equiv \int_{i\text{th triplet}} K\beta \, ds.$$

The first-order change due solely to the IR triplets is

$$\frac{\Delta\beta_1^C(s)}{\beta_o(s)} = \frac{1}{2 \sin\mu_o} \sum_{i=1}^4 Q_i \beta_i \cos[2\mu_{is} - \mu_o] \qquad (4.2)$$

where μ_{is} is the phase advance from the point s to the ith triplet. $i=1$ labels the first triplet after the point of observation s when going along the ring in a specified direction.

From earlier considerations we have seen that the first-order β wave (and hence the second-order chromaticity) from the IR quads is minimized when the phase advance between the IPs is an odd multiple of $\pi/2$. In what follows we will assume this choice of phase advance.

β *wave within the triplets*

We split the region between the 1st and 4th triplets into 3 regions. Let s denote the point of observation, μ_{s1} be the phase advance from the 1st triplet to this point and define κ_α and κ_β as $\kappa_\alpha = \cos(2\mu_{s1} - \mu_o)$ and $\kappa_\beta = \cos(2\mu_{s1} + \mu_o)$.

Region I : between the 1st and 2nd triplets ($0 < \mu_{s1} < \pi$)

$$\frac{\Delta \beta_1^C(s)}{\beta_o(s)} = \frac{1}{2\sin\mu_o} \{Q_1\beta_1\kappa_\alpha + [Q_2\beta_2 - (Q_3\beta_3 + Q_4\beta_4)]\kappa_\beta\} \qquad (4.3)$$

Region II : between the 2nd and 3rd triplets ($\pi < \mu_{s1} < 8.5\pi$)

$$\frac{\Delta \beta_1^C(s)}{\beta_o(s)} = \frac{1}{2\sin\mu_o} \{(Q_1\beta_1 + Q_2\beta_2)\kappa_\alpha - (Q_3\beta_3 + Q_4\beta_4)\kappa_\beta\} \qquad (4.4)$$

Region III : between the 3rd and 4th triplets ($8.5\pi < \mu_{s1} < 9.5\pi$)

$$\frac{\Delta \beta_1^C(s)}{\beta_o(s)} = \frac{1}{2\sin\mu_o} \{[(Q_1\beta_1 + Q_2\beta_2) - Q_3\beta_3]\kappa_\alpha - Q_4\beta_4\kappa_\beta\} \qquad (4.5)$$

β *wave outside the triplets*

Let s be an arbitrary point outside the region bounded by the 4 triplets and μ_{1s} be the phase advance from this point to the 1st triplet. The change in β to first order in δ is

$$\frac{\Delta \beta_1^C(s)}{\beta_o(s)} = \frac{1}{2\sin\mu_o} [Q_1\beta_1 + Q_2\beta_2 - (Q_3\beta_3 + Q_4\beta_4)]\kappa_\alpha. \qquad (4.6)$$

Tune shift

Similarly we consider the tune shift due to the chromatic error of the 4 IR triplets only, ignoring the effect of other quadrupoles and sextupoles. Then to 2nd order in the momentum deviation δ, the phase shift due to these 4 triplets is

$$\Delta\mu = \Delta\mu_1^C\delta + \Delta\mu_2^C\delta^2 + O(\delta^3) \qquad (4.7)$$

where

$$\Delta\mu_1^C = -\frac{1}{2}\sum_{i=1}^{4} Q_i\beta_i ,$$

$$\Delta\mu_2^C = \frac{1}{4\sin\mu_o} \sum_{i=1}^{3} \sum_{j=i+1}^{4} Q_i\beta_i Q_j\beta_j [\cos\mu_o - \cos(2\mu_{ji} - \mu_o)] - \Delta\mu_1^C - \frac{1}{2}\cot\mu_o (\Delta\mu_1^C)^2; \tag{4.8}$$

μ_{ji} is the phase advance from the ith triplet to the jth triplet and $\nu_o = \mu_o/2\pi$ is the global tune of the ring. The first-order chromaticity is independent of phase advances between the triplets. However the second-order chromaticity depends crucially on the relative phase advances between the triplets. If the phase advance between the IPs is $(2n+1)\pi/2$, then the relative phase advances have the following values,

$\mu_{21} = \pi$ $\quad\quad\quad \mu_{31} = (2n+1)\pi/2 \quad\quad\quad \mu_{41} = (2n+3)\pi/2$
$\quad\quad\quad\quad\quad\quad \mu_{32} = (2n-1)\pi/2 \quad\quad\quad \mu_{42} = (2n+1)\pi/2$
$\quad\quad\quad\quad\quad\quad\quad\quad\quad\quad\quad\quad\quad\quad\quad\quad\quad\;\; \mu_{43} = \pi$

With these values, the second-order contribution reduces to

$$\Delta\mu_2^C = \Delta\mu_{2Q}^C + \frac{1}{2}(Q_1\beta_1 + Q_2\beta_2 + Q_3\beta_3 + Q_4\beta_4) \tag{4.9}$$

where $\Delta\mu_{2Q}^C$ is the contribution from terms second order in the quad strengths,

$$\Delta\mu_{2Q}^C = \frac{\cot\mu_o}{2}(Q_1\beta_1 + Q_2\beta_2)(Q_3\beta_3 + Q_4\beta_4) -$$

$$\frac{\cot\mu_o}{8}(Q_1\beta_1 + Q_2\beta_2 + Q_3\beta_3 + Q_4\beta_4)^2. \tag{4.10}$$

The large β functions in the triplets ensure that $\Delta\mu_{2Q}^C$ completely dominates the contribution to $\Delta\mu_2^C$.

4.1 Different configurations of IPs

A) Two IPs with equal β^*

In the chosen design we have repetitive symmetry across the two IRs. Here this symmetry implies

$$Q_3\beta_3 = Q_1\beta_1 \text{ and } Q_4\beta_4 = Q_2\beta_2.$$

In this configuration, the β wave in the three regions within the triplets is

$$\frac{\Delta\beta_1^C(s)}{\beta_o(s)} = Q_1\beta_1 \sin 2\mu_{s1} \quad\quad\quad \text{I}$$

$$= (Q_1\beta_1 + Q_2\beta_2) \sin 2\mu_{s1} \quad\quad\quad \text{II}$$

$$= Q_2\beta_2 \sin 2\mu_{s1} \quad\quad\quad \text{III} \tag{4.11}$$

and the beta wave outside the triplets is

$$\frac{\Delta\beta_1^C(s)}{\beta_o(s)} = 0. \tag{4.12}$$

Figure 3 shows the first-order chromatic beta wave from the triplets in this configuration. The different contributions to the phase shift are

$$\Delta\mu_1^C = -(Q_1\beta_1 + Q_2\beta_2), \qquad \Delta\mu_{2Q}^C = 0. \tag{4.13}$$

The vanishing of $\Delta\mu_{2Q}^C$ is a consequence of the fact that the β wave is zero outside the triplets. Hence the entire second-order phase shift arises from the first-order phase shift,

$$\Delta\mu_2^C = (Q_1\beta_1 + Q_2\beta_2). \tag{4.14}$$

For this case $\Delta\mu_2^C$ is independent of the global tune ν_o.

B) Two unequal IPs with different β^*

Consider the first IP at $\beta^*=0.25$ and the second IP at $\beta^*=0.50$. With repetitive symmetry, this implies

$$Q_3\beta_3 = \frac{1}{2}Q_1\beta_1 \qquad \text{and} \qquad Q_4\beta_4 = \frac{1}{2}Q_2\beta_2 .$$

Within the triplets, the chromatic β wave is

$$\frac{\Delta\beta_1^C(s)}{\beta_o(s)} = \frac{1}{2\sin\mu_o}\left[Q_1\beta_1\left(\kappa_\alpha - \frac{1}{2}\kappa_\beta\right) + \frac{Q_2\beta_2}{2}\kappa_\beta\right],$$

$$\frac{\Delta\beta_1^C(s)}{\beta_o(s)} = \frac{1}{2\sin\mu_o}(Q_1\beta_1 + Q_2\beta_2)\left(\kappa_\alpha - \frac{1}{2}\kappa_\beta\right),$$

$$\frac{\Delta\beta_1^C(s)}{\beta_o(s)} = \frac{1}{2\sin\mu_o}\left[\frac{Q_1\beta_1}{2}\kappa_\alpha + Q_2\beta_2\left(\kappa_\alpha - \frac{1}{2}\kappa_\beta\right)\right], \tag{4.15}$$

for regions I, II and III respectively. The first-order change in β at an arbitrary point s outside the triplets is

$$\frac{\Delta\beta_1^C(s)}{\beta_o(s)} = \frac{1}{4\sin\mu_o}(Q_1\beta_1 + Q_2\beta_2)\cos(2\mu_{1s} - \mu_o). \tag{4.16}$$

The propagation of the first-order chromatic beta wave from the triplets into the arcs is

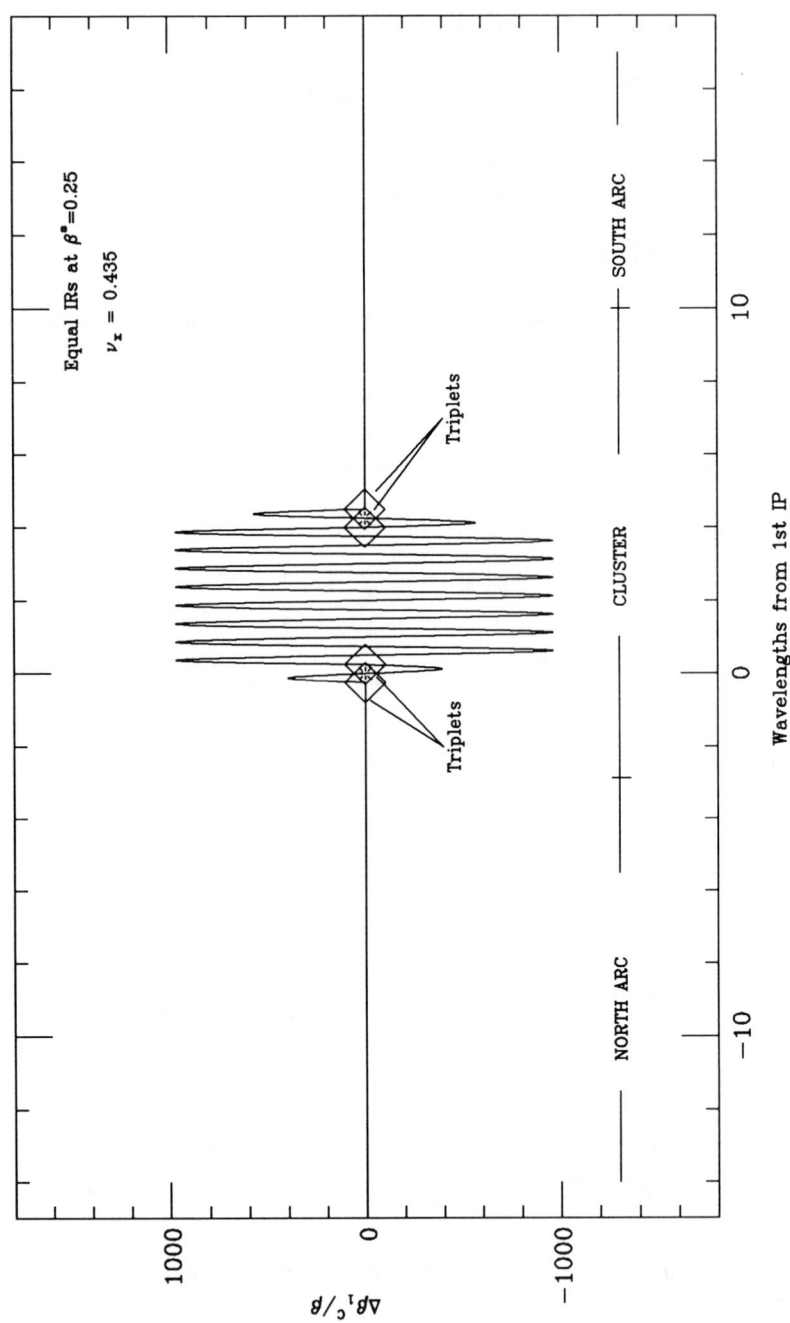

Figure 3 : 1st-order Chromatic β beat from the triplets (Case A)

shown in Figure 4. The phase shift due to the triplets is, up to second order,

$$\Delta\mu_1^C = -\frac{3}{4}(Q_1\beta_1 + Q_2\beta_2), \quad \Delta\mu_2^C = \frac{-\cot\mu_o}{32}(Q_1\beta_1 + Q_2\beta_2)^2 + \frac{3}{4}(Q_1\beta_1 + Q_2\beta_2). \tag{4.17}$$

C) Only one IP (effectively)

Let $\beta^* = 0.25$ m at the first IP and $\beta^* = 8$ m at the second IP. Then

$$Q_3\beta_3 = \frac{1}{32}Q_1\beta_1 \quad \text{and} \quad Q_4\beta_4 = \frac{1}{32}Q_2\beta_2.$$

Inside the region bounded by the triplets, the chromatic β wave (for regions I, II and III) is

$$\frac{\Delta\beta_1^C(s)}{\beta_o(s)} = \frac{1}{2\sin\mu_o}\left[Q_1\beta_1(\kappa_\alpha - \frac{1}{32}\kappa_\beta) + \frac{31}{32}Q_2\beta_2\kappa_\beta\right],$$

$$\frac{\Delta\beta_1^C(s)}{\beta_o(s)} = \frac{1}{2\sin\mu_o}(Q_1\beta_1 + Q_2\beta_2)(\kappa_\alpha - \frac{1}{32}\kappa_\beta),$$

$$\frac{\Delta\beta_1^C(s)}{\beta_o(s)} = \frac{1}{2\sin\mu_o}\left[\frac{31}{32}Q_1\beta_1\kappa_\alpha + Q_2\beta_2(\kappa_\alpha - \frac{1}{32}\kappa_\beta)\right], \tag{4.18}$$

and outside the triplets

$$\frac{\Delta\beta_1^C(s)}{\beta_o(s)} = \frac{31}{64\sin\mu_o}(Q_1\beta_1 + Q_2\beta_2)\kappa_\alpha, \tag{4.19}$$

which is nearly twice as large as the β wave at the same point with the two unequal IPs considered in Case B. This is evident in Figure 5.

The phase shifts are

$$\Delta\mu_1^C = -\frac{33}{64}(Q_1\beta_1 + Q_2\beta_2),$$

$$\Delta\mu_2^C = -\left[\frac{1}{2}(\frac{33}{64})^2 - \frac{1}{64}\right]\cot\mu_o(Q_1\beta_1 + Q_2\beta_2)^2 + \frac{33}{64}(Q_1\beta_1 + Q_2\beta_2). \tag{4.20}$$

The contribution $\Delta\mu_{2Q}^C$, of second order in the quad strengths, in this case is about 3.75

148 Chromaticity Correction for the Collider at the SSC

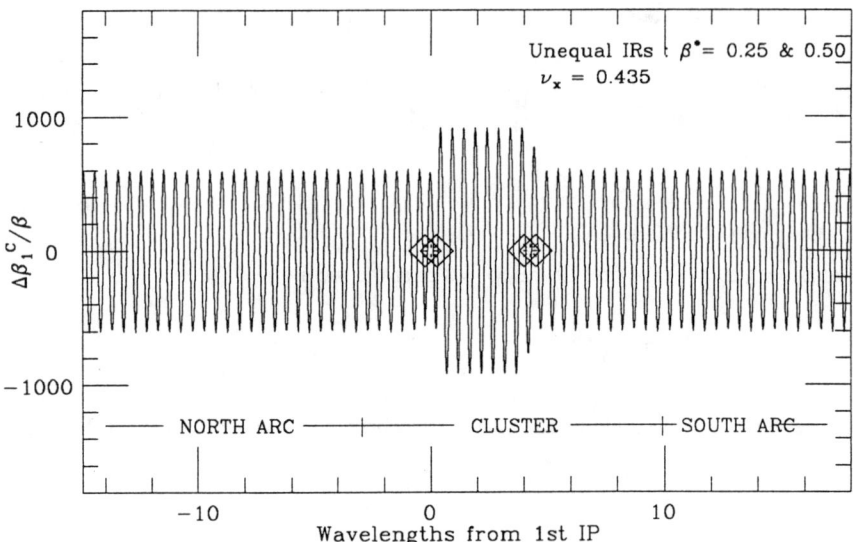

Figure 4 : 1st-order Chromatic β beat from the triplets (Case B)

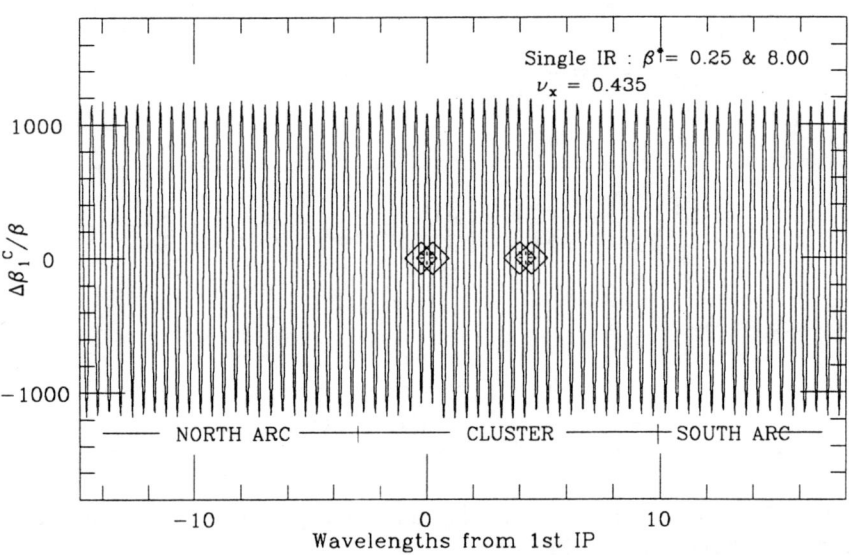

Figure 5 : 1st-order Chromatic β beat from the triplets (Case C)

times larger than in the case with two unequal IPs.

These three cases show that if the IPs are $(2n+1)\pi/2$ apart in phase advance, then the triplets at one IP wholly or partially cancel the contribution to the second-order chromaticity from the triplets at the other IP. For all cases, except the first with equally balanced IPs, the second-order chromaticity will be significantly amplified as $\nu_0 \rightarrow 0.5$.

In Table 2, we evaluate the tune shift due to the triplets in the three configurations and at two choices of tunes. The tune shift is

$$\Delta \nu = \xi_1 \delta + \xi_2 \delta^2 \equiv \frac{1}{2\pi} (\Delta \mu_1^C \delta + \Delta \mu_2^C \delta^2).$$

CASE	ξ_1	ξ_2	ξ_2	$\Delta\nu$ at $\delta=5*10^{-4}$	$\Delta\nu$ at $\delta=5*10^{-4}$
		$\nu_o=0.285$	$\nu_o=0.4$	$\nu_o=0.285$	$\nu_o=0.4$
A) equal IPs $\beta^*=0.25$ $\beta^*=0.50$	-154.0 -77.0	+154.0 +77.0	+154.0 +77.0	-0.07696 -0.03848	-0.07696 -0.03848
B) unequal IPs $\beta^*=0.25, \beta^*=0.5$	-115.5	+1156.4	+6524.8	-0.05746	-0.05612
C) one IP $\beta^*=0.25, \beta^*=8.0$	-79.4	+3977.5	+24132.6	-0.03871	-0.03367

Table 2: Tune shift due to the triplets

The total tune shift is dominated by the linear contribution in any configuration. Removing the linear tune shift is relatively simple, so our concern is with minimizing the higher-order contributions. Clearly the largest second-order chromaticity occurs for case C at $\nu_0=0.4$.

The total chromaticity of an IR includes contributions from the triplets, the quadrupoles in the $M=-I$ section and the variable strength quadrupoles in the tuning section. We concentrated on the triplets in this section because at collision the contribution of the triplets is about 76%, that of the $M=-I$ section about 19%, and the remainder is due to the tuning section. At injection, where higher-order chromaticity is not an issue, all three sections contribute about equally to the chromaticity, which is now dominated by the contributions from the arc quadrupoles.

5 THEORY OF 2nd-ORDER CHROMATICITY CORRECTION

Our aim is to correct the 2nd-order chromaticity of the Interaction Regions. We assume that the linear chromaticity is corrected for. We start with two facts derived in the previous sections.

a) The 2nd-order chromaticity is driven by the 1st-order chromatic beta wave, and
b) the beta wave propagates at twice the betatron frequency.

The 4 triplets at the two IPs contribute the most to the 2nd-order chromaticity.

Figure 6 : The triplets at the two IPs

Since the triplets on either side of an IP are π apart in phase (see Figure 6), the chromatic beta waves produced by them add in phase. Hence we can combine the two triplets into a composite lens at each IP. (See Figure 7.)

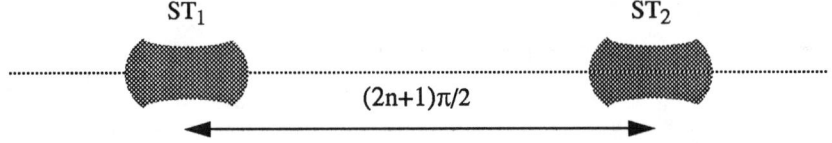

Figure 7 : Effective super triplets at each IP

The beta waves produced by the two super triplets are exactly out of phase in the region outside the triplets. If the two super triplets have the same strength i.e. $ST_1 = ST_2$, then no beta wave gushes out from the IRs and the 2nd-order chromaticity is a minimum. (See Figure 8.)

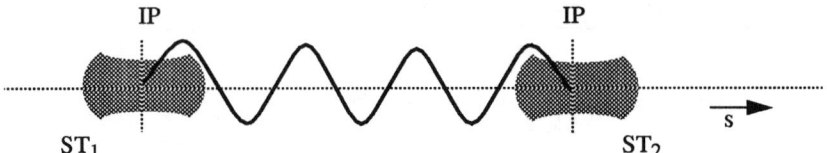

Figure 8 : Exact cancellation with equal IRs

Now turn off one of the IRs (with linear chromaticity still corrected). The chromatic beta wave from the remaining IR now flows uncorrected into the arcs. (See Figure 9.)

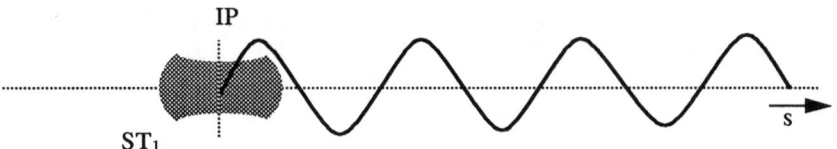

Figure 9 : Beta wave with 1 IR

Here we have drawn only the beta wave propagating out from the supertriplet ST_1 and not the periodic 1st-order beta wave in the ring. How do we stop this beta wave from ST_1 from going all around the ring?

Put another source of chromatic beta waves (e.g. a sextupole) $\pi/2$ from the IP to interfere destructively with the waves produced by ST_1 (i.e. to do the same as the missing ST_2). (See Figure 10.)

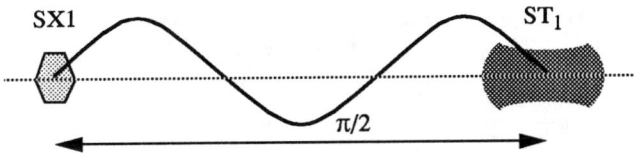

Figure 10 : Cancelling the beta wave from 1 IR

The SX1 sextupole produces a beta wave with the opposite phase. Hence the 2nd-order chromaticity should be a minimum. However there is an unwanted side effect. The SX1 sextupole introduces linear chromaticity into the ring.
Put another sextupole SX2 to cancel the linear chromaticity due to SX1.(See Figure 11.)

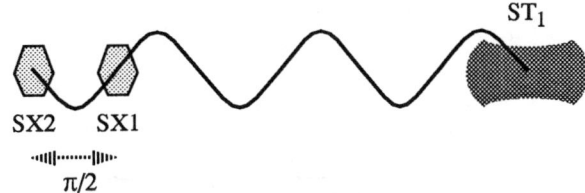

Figure 11 : Correction of 1st- and 2nd-order chromaticity with 2 sextupoles

The beta functions and the dispersion at the locations of SX1 and SX2 are the same.
Zero linear chromaticity $\rightarrow$ SX1 + SX2 = 0. Since these two sextupoles are $\pi/2$ apart and their strengths have opposite signs, their beta waves add in phase.
Other details: The beta wave has to be corrected in each plane. This requires 2 sextupoles, SX1F and SX2F, to correct for the 1st-and 2nd-order chromaticity in the horizontal plane (primarily) and sextupoles SX1D and SX2D to correct for the effects in the vertical plane (primarily). For each sextupole to be at the proper phase with respect to the IP requires a specific choice of phase advance across each IR.
Only 1 sextupole in each family would require too large a strength for each sextupole. With 24 members in a family, the required strength would be under the maximum allowed strength for each sextupole. 12 of these are placed at the edge of the North Arc adjacent to the cluster and the other 12 at the edge of the South Arc.
To have each member in a family produce a beta wave in phase with the other members of the family, there must be a phase advance of π between the members of a family. This has the additional advantage of removing the second-order geometrical aberrations[2]. This is easily achieved if the phase advance across each cell is 90°.
The sextupole distribution over 1 betatron wavelength is shown in Figure 12. The entire distribution spans 6 wavelengths on either side of the cluster.

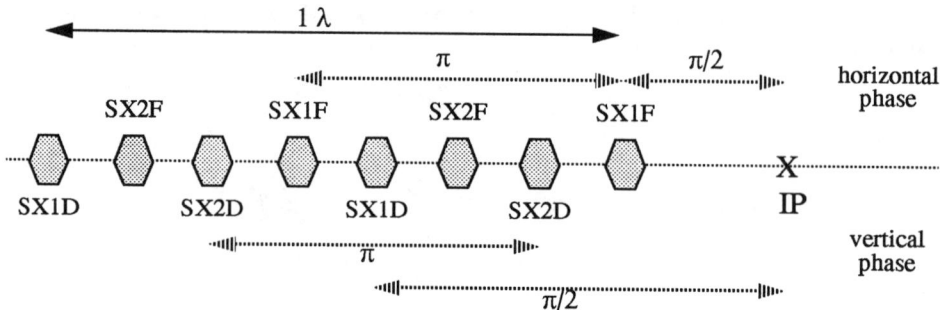

Figure 12 : Sextupole distribution

Imperfections: The contribution to the beta wave in the x plane from the sextupoles SX1D and SX2D is not cancelled and similarly for the y plane contributions from SX1F and SX2F. To see this, we represent the beta waves from each source in a phasor diagram. (See Figure 13.)

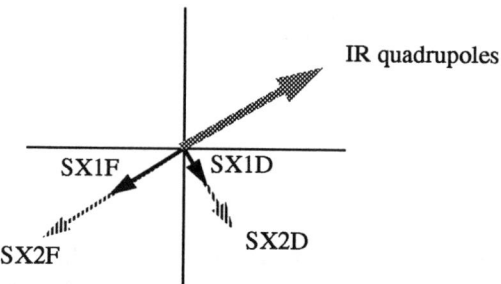

Figure 13 : 2ψ Phasor diagram for the beta waves

The angle between any two vectors in this diagram will be twice the phase advance between the corresponding sources. We will refer to this as a 2ψ diagram. Let the phase be measured from an arbitrary point in the lattice.

SX1F is at $\pi/2$ phase from ST_1 while SX2F is at $\pi/2$ phase from SX1F but has the opposite sign to SX1F. Hence the vectors representing 1F and 2F are parallel and combine to cancel the vector from the IR quads. However SX1D is at $\pi/4$ phase from SX1F, hence its vector is orthogonal to SX1F. SX2D is $\pi/2$ away from SX1D but with the opposite strength so its vector adds in phase to that of SX1D. The orthogonal contributions from SX1D and SX2D are not removed by any source and remain as a residual beta wave. The relative amplitude of the beta wave in the horizontal plane from the D sextupoles is $\beta_{min}/\beta_{max} \sim 1/6$ (for a FODO cell with 90 degrees phase advance per cell). With the scheme outlined above, the second-order chromaticity is thus reduced by a factor of 6.

6 PERFORMANCE OF THE CHROMATICITY CORRECTION SYSTEM

We correct for the chromaticity of the collider ring by placing sextupoles next to each quadrupole in the two arcs of the collider. This amounts to 392 D and 392 F sextupoles. Of these, 96 sextupoles of each type are placed in the 24 cells at both ends of each arc for correcting the nonlinear chromaticity of both the interaction regions. In all, eight families of sextupoles labelled SX1F, SX1D, ..., SX4F, SX4D are available for the nonlinear correction. These will be referred to as the *local sextupoles*. The families (SX1F, SX4F) are $\pi/2$ (mod 2π) in horizontal phase away from the North IP and (SX1D, SX4D) are $\pi/2$ (mod 2π) in vertical phase from this IP. The same statement can be made for the families (SX2F, SX3F) and (SX2D, SX3D) with respect to the South IP. The 296 sextupoles in each of two families SXF, SXD correct for the linear chromaticity of the total ring including the interaction regions. These will be called the *global sextupoles*.

The collider has two low-β (or high luminosity) interaction regions on the east side of the ring and two medium-β IRs on the west side. In what follows, the chromaticity of only the low-β IRs has been corrected for. Consequently, for this report, we use 2 x 48 local sextupoles and 2 x 344 global sextupoles. The optimization of the nonlinear correction included the minimization of the tune shift as a function of momentum to the 2nd and 3rd order. It was done using the module HARMON in MAD [3].

We have seen in earlier sections that the second-order chromaticity is largest when the fractional part of the global tune is close to 0.5 and one of the interaction regions in a cluster is tuned to collision optics while the other is tuned to injection optics. For brevity, we will present results for this "worst" case only. The tunes in the two planes are chosen to be (0.435, 0.415). An important advantage of correcting this configuration is that the phase advances between the IPs become irrelevant and we do not need to rely on the chromatic cancellation of one IR by another. This is important in practice since the detectors at the two IPs will possibly be operating at different luminosities and the phase advance between IPs may not be an odd multiple of $\pi/2$.

First we look at the chromatic behaviour with only the linear chromaticity corrected. This will show if nonlinear chromaticity correction is needed and serve as a benchmark by which to compare the improvement due to the nonlinear correction. All 392 sextupoles in each of the D and F families are used for the correction in this case. Figure 14 shows the variation of the tune shift with the relative momentum deviation δ and Figure 15 shows how the relative β at the IP varies with δ. The standard deviation σ_p for the relative momentum spread in the beam is approximately 6×10^{-5} at 20 TeV. For stable operation of the beam we require the tune shift $\Delta\nu < 0.002$. We find that the linear correction provides us with a momentum aperture of approximately 2.5 σ_p, which is inadequate. The relative variation in β^* is also large, reaching 10% at $1\sigma_p$. Clearly, higher-order chromaticity correction is needed under these conditions. Now we examine the performance of the nonlinear chromaticity correction system. The local sextupole strengths were limited to be less than 0.25 T-m at 1 cm. Tables 3 and 4 compare the strengths of the sextupoles in the two cases ($\beta^* = 0.25$ and 8 m, $\nu_x = 123.435$, $\nu_y = 122.415$).

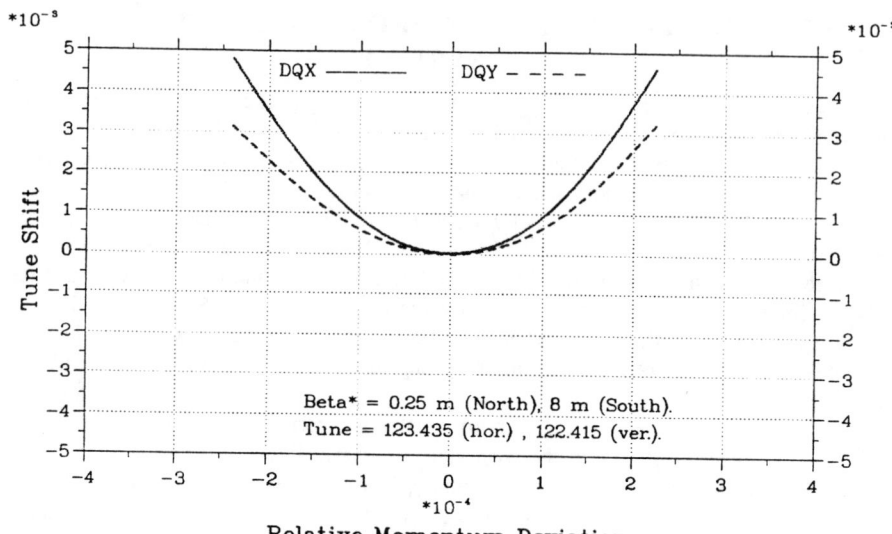

Figure 14: Tune Variation with Momentum: Global Linear Correction.

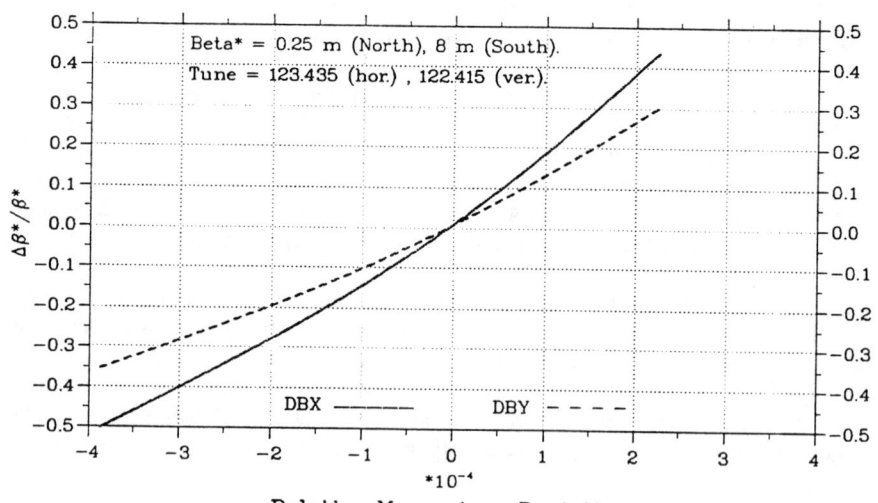

Figure 15: Variation of Beta* with Momentum: Global Linear Correction.

Sextupole name	Integrated Field Strength T-m at 1cm
Global F	0.068
Global D	-0.136

Table 3: Linear chromaticity correction only ("global")

Sextupole name	Integrated Field Strength T-m at 1cm
Global F	0.078
Global D	-0.155
Local F1, F4	0.168
Local D1, D4	-0.250
Local F2, F3	-0.164
Local D2, D3	0.247

Table 4: Linear and nonlinear chromaticity correction ("local")

We note that the strengths of F1 and F2 are approximately equal and opposite, and similarly for D1 and D2. This is required to cancel the net linear chromaticity of the local sextupoles. For the correction of a single IR, only 4 families (F1,D1,F2,D2) of sextupoles are required. When both IRs are tuned to collision optics, F1 and F4 are allowed to have different strengths, and so too for the other pairs of sextupoles which have been constrained to be equal above.

Figure 16 shows the variation of tune with δ in the presence of the local sextupoles. The tune variation is flat over 2 σ_p and the momentum aperture is increased to approximately $8\sigma_p$ for $\Delta\nu < 0.002$. The relative β^* variation (shown in Figure 17) is limited to 2.5% at $1\sigma_p$. These are definite improvements in the chromatic behaviour and may be adequate for the stable operation of the collider. The final check of the chromaticity correction system is to examine the effect of the local sextupoles on the dynamic aperture.

7 DYNAMIC AND MOMENTUM APERTURE

The dynamic and momentum aperture of the collider ring in the presence of the local sextupole scheme has been studied in detail in order to establish the performance as well as the feasibility of the scheme itself. Sextupoles, as nonlinear elements, can potentially reduce the dynamic aperture of the machine, so the behaviour of the collider lattice has been checked with a realistic simulation model. For every configuration of the optics and setting of the local sextupoles correction scheme we determined the dynamic aperture, identified with the largest amplitude surviving 1024 turns. The choice of the short-term dynamic aperture as a figure of merit is justified by the fact that the main aim is to compare different machine configurations and not to investigate its long-term stability.

The behaviour of different configurations of the Interaction Region optics and of the sextupole scheme has been studied first for the ideal lattice, i.e. the first-order lattice plus the sextupoles as the only source of nonlinearity. The ideal lattice has then been compared with a

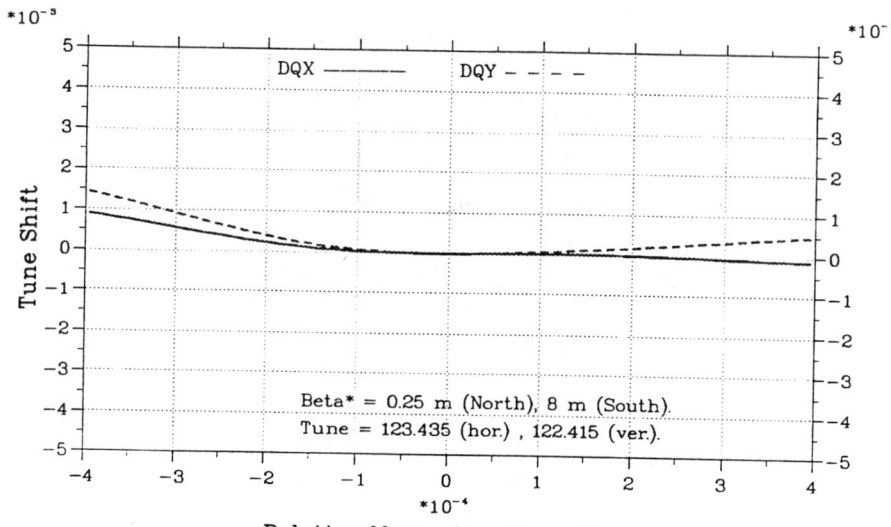

Figure 16: Tune Variation with Momentum: Local Nonlinear Correction.

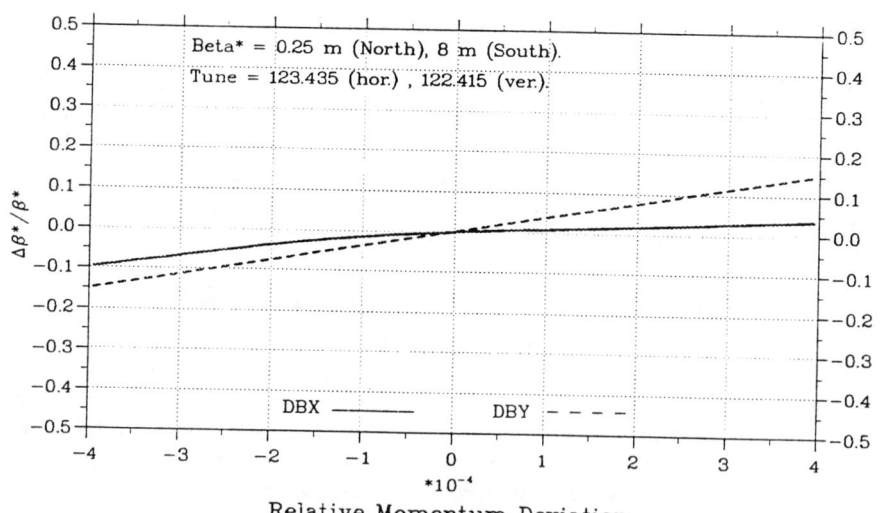

Figure 17: Variation of Beta* with Momentum: Local Nonlinear Correction.

realistic model of the lattice where the effect of errors and their operational corrections is taken into consideration.

The initial conditions for the increasing amplitudes of the particles tracked have been selected as follows:

$x_n = n\sigma_x$ where $\sigma_x = (\beta_x \varepsilon_x)^{1/2}$ and $\beta\gamma\varepsilon_x = 1$ mm mrad

$y_n = n\sigma_y$ where $\sigma_y = (\beta_y \varepsilon_y)^{1/2}$ and $\beta\gamma\varepsilon_y = 1$ mm mrad

At the beginning of the lattice $\beta_x \sim \beta_y \sim 460$ m, $\alpha_x \sim \alpha_y \sim 0$ and $D_x=0$, so that $\sigma_x \sim \sigma_y \sim 1.47 \times 10^{-4}$ m at 20 TeV.

7.1 Dynamic and momentum aperture of the ideal lattice

The β^* at the collision point of the collider low-beta interaction regions can be tuned between 0.25 m and 8 m, the latter value corresponding to the injection optics; the nominal value at collision is 0.5 m. Each interaction region, 2 in the East Cluster and 2 in the West Cluster, can be tuned independently to a value of β^* in this range. The optical configurations simulated have the injection optics in the West Cluster and several combinations of β^* in the East Cluster, i.e. in the north low-beta IR (ENLB) and in the south low-beta IR (ESLB). The configuration studied are labeled as follows:

N50-S50: baseline symmetric configuration, $\beta^*_{ENLB}=0.50$m, $\beta^*_{ESLB}=0.50$m
N25-S25: low β^*symmetric configuration, $\beta^*_{ENLB}=0.25$m, $\beta^*_{ESLB}=0.25$m
N25-S50: low β^*asymmetric configuration, $\beta^*_{ENLB}=0.25$m, $\beta^*_{ESLB}=0.50$m
N25-S800:asymmetric configuration, $\beta^*_{ENLB}=0.25$m, $\beta^*_{ESLB}=8$m

As previously discussed, the last configuration, where one east IR is tuned for maximum luminosity and the other is tuned to the injection optics, is expected to be the most sensitive to chromatic effects. The first configuration is the baseline optics for the Collider and it is the least sensitive to chromatic effects. In order to enhance the effect of higher-order chromaticity, the fractional tune of the lattice for this simulation has been chosen reasonably close to the half integer (ν_x=123.435, ν_y=122.415). Studies of beam-beam effects also suggest a working point close to 0.4.

Table 5 summarizes the results for the ideal lattice: for each configuration described above, the dynamic and momentum aperture for the lattice with the total linear chromaticity compensated by the sextupoles in the arcs only (global), is compared to one (local) where the linear and second-order chromatic effects arising from the IRs are corrected by the local sextupole scheme and the linear chromaticity from the rest of the machine is corrected by the arc sextupoles.

The local sextupoles cause a significant improvement of the momentum aperture, in particular for the asymmetric optics configurations. This confirms the improvement in machine performance expected, given the better tune versus amplitude and beta beat in the presence of the local scheme. The strong local sextupoles cause however a reduction of the dynamic aperture on momentum. This effect, together with the strength requirement on the local sextupoles, led us to limit the use of the local system only to compensate the higher-order chromaticity of the IRs and to correct the linear chromaticity, caused by both the arcs and the IRs, with the arc sextupoles.

Δp/p	N50-S50		N50-S50		N25-S25		N25-S25		N25-S50		N25-S50		N25-S800		N25-S800
	global		local		global		local		global		local		global		local
0.0000	100		70		100		40		100		50		100		50
0.0001	100		70		100		40		100		50		100		50
0.0002	100		70		100		40		100		50		100		50
0.0003	100		70		100		40		100		50		90		50
0.0004	100		70		100		40		100		40		50		50
0.0005	100		70		100		40		80		40		20		50
0.0006	100		70		80		30		70		40		0		50
0.0007	100		70		70		30		50		40		0		50
0.0008	100		70		50		30		40		40		0		50
0.0009	100		70		0		30		20		40		0		50
0.0010	100		70		0		30		0		40		0		50
0.0011	100		70		0		30		0		40		0		40
0.0012	100		70		0		30		0		40		0		30
0.0013	100		70		0		30		0		40		0		30
0.0014	100		70		0		30		0		40		0		30

Table 5: Dynamic and momentum aperture (σ) for the ideal lattice

7.2 Dynamic and momentum aperture of the lattice with errors

The investigation of performance of the local sextupole scheme done for the ideal lattice has been repeated and extended to a realistic model of the machine where the effect of errors and their corrections are accurately simulated. This study allows us to establish whether the benefits of the local scheme demonstrated for the ideal lattice still hold in the presence of errors that could potentially mask the effectiveness of sextupoles, and to investigate more thoroughly the issue of loss of dynamic aperture on momentum.

The model used for the simulation and implemented in the code TEAPOT [4] describes realistically the single-particle dynamics of the Collider as far as errors and corrections are concerned. Collective and beam-beam effects are not included in the model. Every relevant element in the lattice such as a bend, quadrupole, sextupole, beam position monitor, etc., is assigned random alignment errors and roll errors; main dipoles and quadrupoles also have systematic and random field errors associated with them, where normal and skew multipoles are specified up to the order 9. The issue of the error specifications for the Collider is a matter of continuing study and will not be discussed here in detail: except where otherwise specified, the assumptions for the alignment and field errors reflect the so-called Collider 3B specifications document [5]. We did not include alignment errors in the IR triplets: the triplets are extremely sensitive to these errors, and the correction of their effects on the collider dynamics is the topic of an ongoing independent study. Also, the effect of the crossing angle at the interaction point is not generally included in the results that follow. Preliminary results on the effect of the crossing angle on the baseline collider optics will however be discussed at the end.

The operational corrections necessary to operate the machine with imperfections are also accurately described in the model: the closed orbit is found by a steering algorithm, the lattice is retuned to the original fractional tune by means of trim quadrupole, and the local compensation of coupling is achieved by a set of 44 skew quadrupoles, 24 of them placed in the clusters and 20 in the arcs.

Several configurations have been studied with the above described set of errors and corrections: N25-S800, N50-S800 and the baseline collider optics N50-S50. We will limit the detailed discussion to the first one.

7.3 N25-S800

As already remarked, this optical setting has been studied in more detail since it represents a worst case scenario as far as chromatic effects from the IRs are concerned. The low-beta IR tuned at 0.25 m contributes about 100 units of chromaticity. We compared the following sextupole correction schemes:

global   Linear chromaticity ξ from arcs *and* IRs corrected with the arc sextupoles.

local_100 100 units of linear ξ corrected by the local system, the rest by the arc sextupoles. The local system minimizes the 2nd and 3rd-order tune shift with momentum.

local_50   50 units of linear ξ corrected by the local system, the rest by the arc sextupoles. The local system minimizes the 2nd- and 3rd-order tune shift with momentum.

local_0   All the linear chromaticity ξ is corrected with the arc sextupoles. The local system minimizes the 2nd- and 3rd-order tune shift with momentum.

For every correction scheme the dynamic aperture as a function of momentum has been determined for different error sets. The results are summarized in Figure 18.a-d.

Figure 18.a describes the ideal lattice; in Figure 18.b-d are summarized the results for the lattice with errors: they have the same set of alignment errors but differ in the assignment of field errors. When field errors are added to the arc dipoles and quadrupoles, the dynamic aperture obviously decreases, but there is a clear improvement in the momentum aperture with the local schemes compared to the global scheme.

The dynamic aperture of the global scheme on momentum is still larger than for the local schemes. The assignment of field errors to the IR quadrupoles (Figure 18.c) and successively to the IR triplets (Figure 18.d) further reduces the dynamic aperture of the machine as expected, but the increase in momentum aperture over the global scheme is verified.

Furthermore, the reduction of aperture for $\Delta p/p=0$ of the local schemes versus the global one is no longer present, because the effect of the field errors in the IR quadrupoles dominates the dynamics on momentum. Lattice performance in the presence of errors led us to select the local_0 scheme as the most effective way of correcting the chromatic effects of the IR. The local_0 scheme is also preferred because of the minimum strength of the local sextupoles.

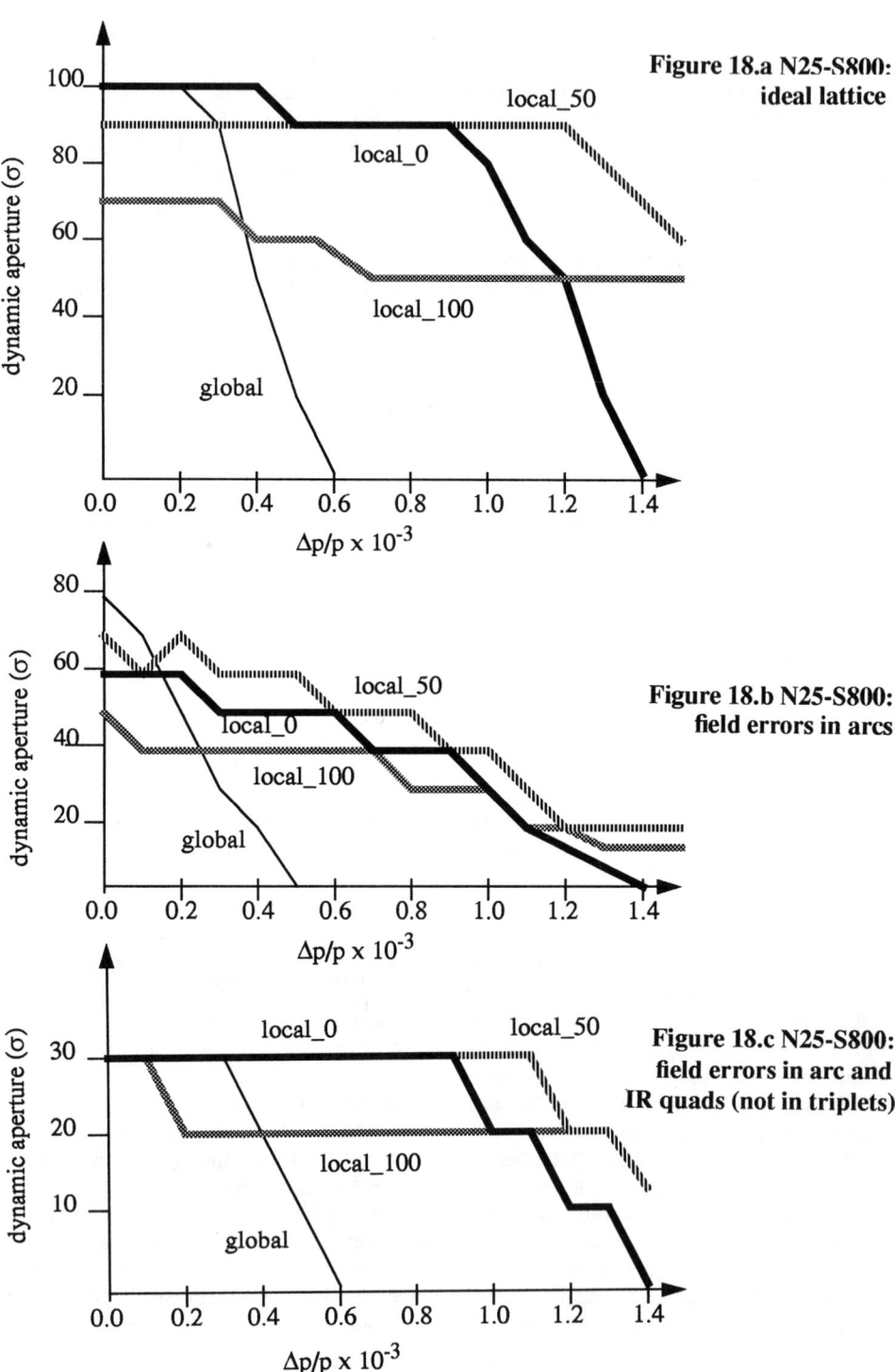

Figure 18.a N25-S800: ideal lattice

Figure 18.b N25-S800: field errors in arcs

Figure 18.c N25-S800: field errors in arc and IR quads (not in triplets)

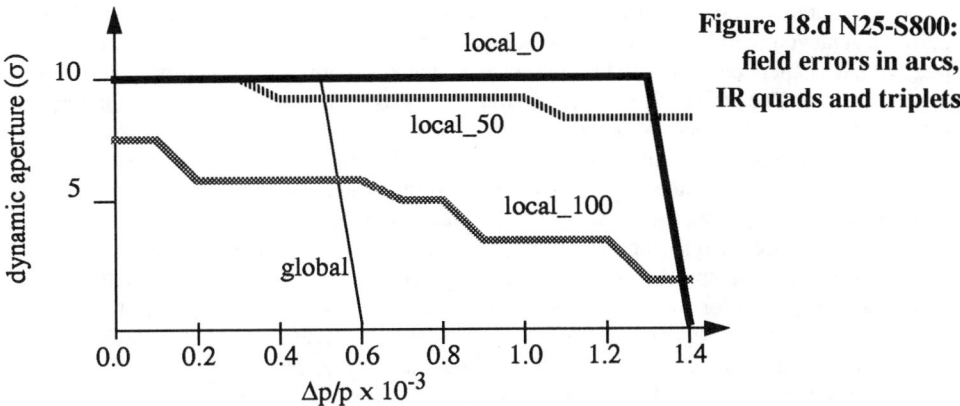

Figure 18.d N25-S800: field errors in arcs, IR quads and triplets

7.4 Effect of the crossing angle and field quality in the IR triplets

As previously remarked, the former results about the collider dynamic aperture at top energy do not take into consideration the effect of the crossing angle and assume the 3B specifications for the field quality in the IR quadrupoles. Both assumptions have important consequences as far as the effect of the IR triplet quadrupoles on the dynamics is concerned. Work is presently in progress that specifically addresses IR triplet issues: some preliminary results will be summarized here for the N50-S50 baseline optics configuration.

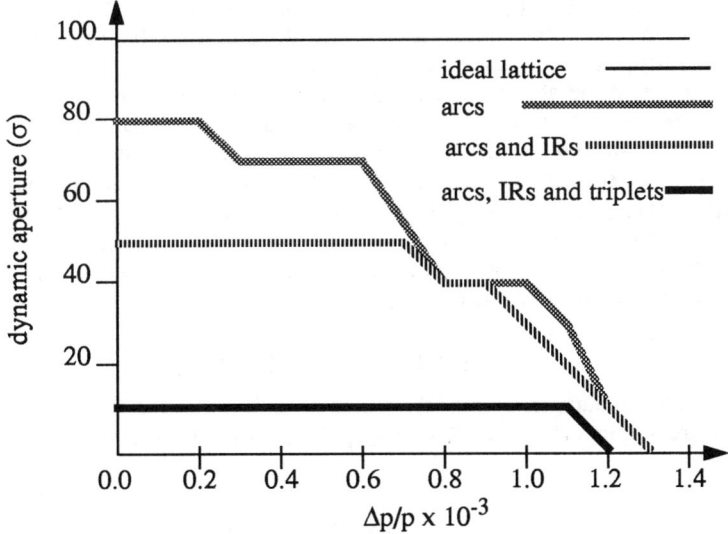

Figure 19 Baseline configuration N50-S50

The dynamic aperture for this optics, without crossing angle and assuming the standard 3B specifications for the IR quadrupoles, is described in Fig.19. The global sextupole correction scheme is used here since the optical symmetry makes this optics less sensitive to chromatic effects.

A horizontal (vertical) crossing angle of 135 μrad between the two beams at interaction points is achieved with a system of 4 horizontal (vertical) kickers per IP. The residual horizontal (vertical) dispersion produced by the system is matched with a set of 6 normal (skew) quadrupoles per IP. The effect of the crossing angle is to make the beam pass off axis through the triplets, increasing the effect of the higher-order multipoles in the quadrupoles. For a crossing angle of 135 μrad the maximum closed-orbit offset in the triplets is 5 mm: this effect has been simulated and the reduction of the aperture at collision found to be at the 1to 2 sigma level.

The multipoles assumed so far for the triplets have been derived from the specifications for the 40-mm-aperture arc quadrupoles by appropriately rescaling the values to an aperture of 50 mm in the IR quadrupoles. A study is now in progress towards the exact determination of the field quality required for the IR triplets, in particular the higher-order multipoles responsible for aperture reductions. Preliminary results show that the systematic b_5 multipole in the triplets, the first multipole allowed by symmetry in a quadrupole, has a significant effect on the aperture. Lowering b_5 from 0.534×10^{-4} (at 1 cm) to 0.1×10^{-4} increases the dynamic aperture by 3 to 4 sigma. A typical value for the dynamic aperture at collision, taking into consideration the crossing angle and the b_5 multipole, is 12 sigma.

8 Summary

Our scheme for correcting the nonlinear chromaticity of each IR consists of placing sextupoles in 4 families in the regular cells adjacent to the IRs and spread out over 6 betatron wavelengths into the arcs on each side of a cluster. These 'local' sextupoles correct primarily for the second- and to a lesser extent the third-order chromaticity of the IRs while contributing net zero linear chromaticity. The linear chromaticity of the entire collider ring is removed by two families of sextupoles in the remaining cells in the arcs.

We have tested the above scheme with different configurations of IRs. It improves the chromatic and dynamic behaviour for every configuration studied. Even for the worst case with one IP at $\beta^*=0.25$ m and the other at $\beta^*=8$ m, the data on tune shift with momentum show that the nonlinear correction scheme increases the momentum aperture by more than a factor of 3. This increased momentum aperture is obtained at the expense of a slight reduction in the dynamic aperture for particles on momentum, when no field errors in the magnets are included. When we add a realistic set of errors, specially the field errors in the IR triplets, the local sextupoles do not affect the dynamic aperture on momentum.

ACKNOWLEDGEMENTS

We thank R. Stiening for guiding the course of this work and D. Ritson for discussions and ideas. We are also indebted to K. Brown for pointing out the phasor diagram shown in Figure 13.

REFERENCES

[1] E.D. Courant and H.S.Snyder, *Annals of Physics* **3**, 1 (1958).
[2] K.L. Brown, SLAC-PUB-2257 (1979).
[3] H.Grote and F.C. Iselin, MAD, CERN/SL/90-13(AP), 1990.
[4] L. Schachinger and R.Talman, TEAPOT, *Part. Acc.* **22**, 35 (1987).
[5] Element Specification (Level 3B) Collider Accelerator Arc Sections, SSCL E10000027.

Longitudinal Phase-Space Measurements at IUCF*

D.D. Caussyn[a], M. Ball[a], B. Brabson[a], J. Budnick[a], A. W. Chao[b],
J. Collins[a], V. Derenchuk[a], S. Dutt[b], G. East[a], M. Ellison[a], T. Ellison[a],
D. Friesel[a], W. Gabella[e], B. Hamilton[a], H. Huang[a], W.P. Jones[a],
S.Y. Lee[a], D. Li[a], M.G. Minty[c], S. Nagaitsev[a], K.Y. Ng[e],
X. Pei[a], G. Rondeau[a], T. Sloan[a], M. Syphers[b],
S. Tepikian[d], Y. Wang[a], Y. Yan[b], P.L. Zhang[b]

Abstract

Synchrotron motion in the IUCF cooler ring was studied using turn-by-turn beam tracking on ten-turn intervals, where the beam phase relative to the rf and the closed-orbit position in a high-dispersion region were measured. The synchrotron tune shift with amplitude was measured and is compared with theory. The driven reponse of the system was also studied using the same techniques, and was found to share many of the same characteristics of other parametric resonant systems. The experimental results did not exhibit any effects of bunch decoherence expected from the tune shift with amplitude.

1 Introduction

The Indiana University Cyclotron Facility cooler storage ring and synchrotron accelerator ($B\rho = 3.6$ Tesla meter) was the first of the many similar accelerator storage rings designed specifically to employ electron cooling to produce and use high-quality medium- energy ion beams for nuclear research. The status and development of the IUCF cooler ring[1] and cooling system [2] have been reported in numerous accelerator conference proceedings. This machine has also been a near ideal laboratory for conducting accelerator physics experiments. The 95% emittance, or phase-space area, of the proton beam is electron-cooled to about 0.3π mm mrad in less than 3 s. The resulting relative momentum spread full width at half maximum, FWHM, of the beam is about ± 0.0001. Such a high-quality beam bunch can closely simulate single-particle motion. Several experiments studying transverse motion near resonances[3] have demonstrated this advantage.

Recently, we have applied many of the same techniques for studying transverse motion on a turn-by-turn basis to a study of longitudinal motion. Since its discovery in 1945 by McMillan and Veksler[4], synchrotron motion has been relatively well understood, sharing as it does many of the same characteristics of

---

*Work supported by the National Science Foundation under Grant NSF PHY 90-15957.
[a] IUCF, Indiana University, Bloomington, IN 47405; [b] The SSC Laboratory, 2550 Beckleymeade Avenue, Dallas, TX 75237-3946; [c] SLAC, MS26, Box 4349, Stanford, CA 94309; [d] Brookhaven National Laboratory, Upton, NY 11973; [e] Fermilab, P.O. Box 500, Batavia, IL 60510.

the dynamics of other parametric resonant systems. By tracking the beam in ten-turn intervals, we have easily measured the longitudinal beam motion and made comparisons with the theory. From forcing the motion, by modulating the phase of the rf, the response of the beam was also easily measured with this technique. From these data, the chaotic transition from one stable solution to another near the bifurcation point could be observed.

2 Procedure and Results

The experiment began with a beam bunch stored in the IUCF cooler ring having about 3×10^8 protons with kinetic energy of about 45 MeV. The IUCF cooler ring has a circumference of 86.82 m, and we were using an rf system with a frequency of 1.03148 MHz at a harmonic number h of one. The beam bunch was about 60 ns (or 5.4 m) FWHM, having a revolution period of about 969 ns. The beam was injected into the cooler ring in a 10-s cycle, with injection and electron cooling being completed within the first 5 s, leaving a period of about 5 s for making the measurements.

Since measurements of longitudinal motion were being made, the phase-lock feedback loop for the rf, which is normally on, was switched off. Phase shifting, and phase modulation, of the beam relative to the rf were achieved by phase shifting, or modulating, the rf control signal. The rf cavity had a Q value of about 40 at a frequency of 1 MHz, with a resulting half-power bandwidth of 25 kHz. Thus, the fastest phase shift we could obtain occurred in about 40 turns, which was relatively fast compared to the synchrotron period (for this experiment) of about 1920 turns.

The synchrotron motion was characterized by measuring the phase of the beam relative to the rf and its conjugate variable, the momentum deviation. The relative phase of the beam was measured by using a type II or a type III phase detector [5] to compare the relative phase between the signal from a pickup coil in the rf cavity, and the sum signal from a beam position monitor (BPM) after it had been passed through a 1-MHz low-pass filter. The type III phase detector has an effective range of 360 deg whereas the type II phase detector had a range of only 180 deg. However the type III phase detector had a flaw which made it unreliable for measurements in a range of about 5 deg around zero degrees, and for this reason it was not used when the beam response was measured (to be discussed later).

The momentum deviation of the beam was measured by finding the deviation of the closed orbit in a region of high dispersion. The normalized position signal from a BPM located in a region of the ring where the dispersion D_x was 3.9 m was used for this purpose. The position signal was passed through a 3-kHz low-pass filter to remove any oscillation which might be attributable to a coherent betatron oscillation, leaving only the slower closed-orbit oscillations. The measured variations in the closed-orbit Δx_{co} could then be used to determine the fractional momentum deviation δ, which is given by $\delta = \Delta x_{co}/D_x$.

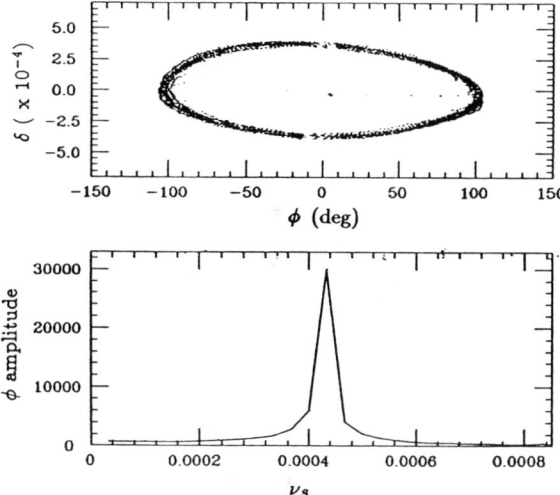

Figure 1: A longitudinal phase-space plot after inducing a synchrotron oscillation by making a sudden shift in the rf phase (top) and the FFT of the phase signal to find the synchrotron tune ν_s (bottom). The type III phase detector was used in this case.

A typical measurement of the longitudinal phase-space variables after inducing a synchrotron oscillation is shown in Fig. 1 as a Poincaré plot, with only every tenth turn plotted. Similar data were taken for induced synchrotron oscillations having phase amplitudes in the range from 10 to 150 deg. The synchrotron tune ν_s, which is the ratio of the synchrotron frequency and the revolution frequency, for each case was determined from a fast-Fourier transform (FFT), or from a measurement of the synchrotron period. The result of the measurement of the synchrotron tune with amplitude is shown in Fig. 2. Since the equation of motion for a synchrotron oscillation for a stored beam is the same as that for a pendulum, it shares the same functional dependence of the tune on phase amplitude as that of a pendulum, which is plotted in Fig. 2 as a solid line.

The longitudinal response of the beam to forced phase oscillations was also studied. Experimentally this was accomplished by modulating the phase of the rf sinsuoidally, and varying both the amplitude and the frequency of this phase modulation. Suspecting that the damping due to electron cooling would not be fast enough for the homogeneous solution to damp out (leaving the steady-state solution) before decoherence would become a significant complication, the transient response of the beam was measured. The response was characterized in two different ways; in terms of the amplitude of the response, and the frequencies of the response. The amplitude response A_m was found by measuring the maximum peak-to- peak amplitude and normalizing it with respect to the amplitude of the external phase modulation. The frequency response was characterized by measuring the beat period T_m of the phase signal. See Fig. 3 for an illustration of the phase signal which was produced when the rf phase modulation was begun.

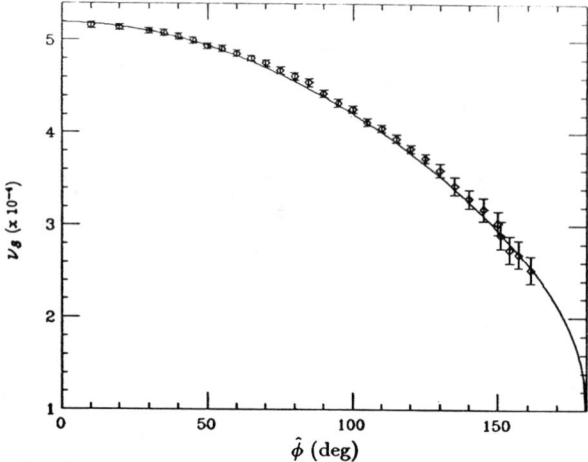

Figure 2: A plot of the measured synchrotron tunes ν_s versus synchrotron amplitude $\hat{\phi}$. The solid line is the theoretically expected result.

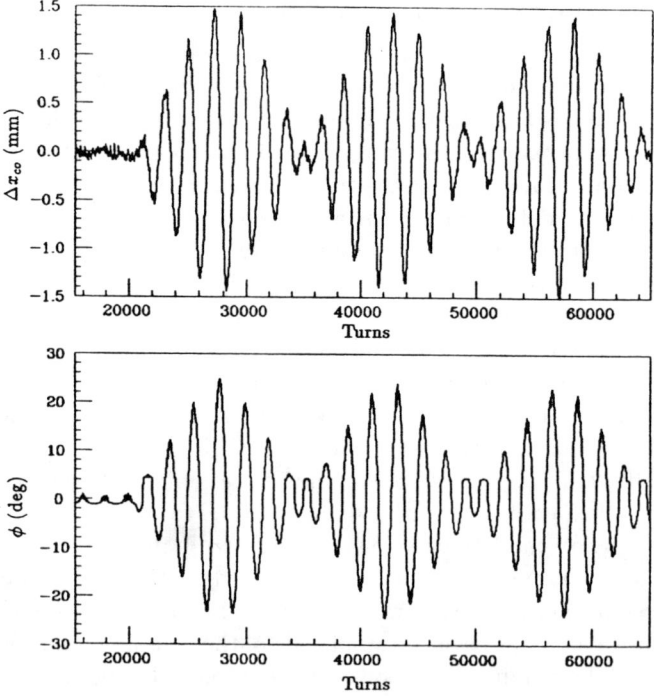

Figure 3: A plot of the measured phase when the rf phase modulation was started. Note the beat in the response, due prinicipally to the beat between the frequencies of the homogeneous and steady-state solutions.

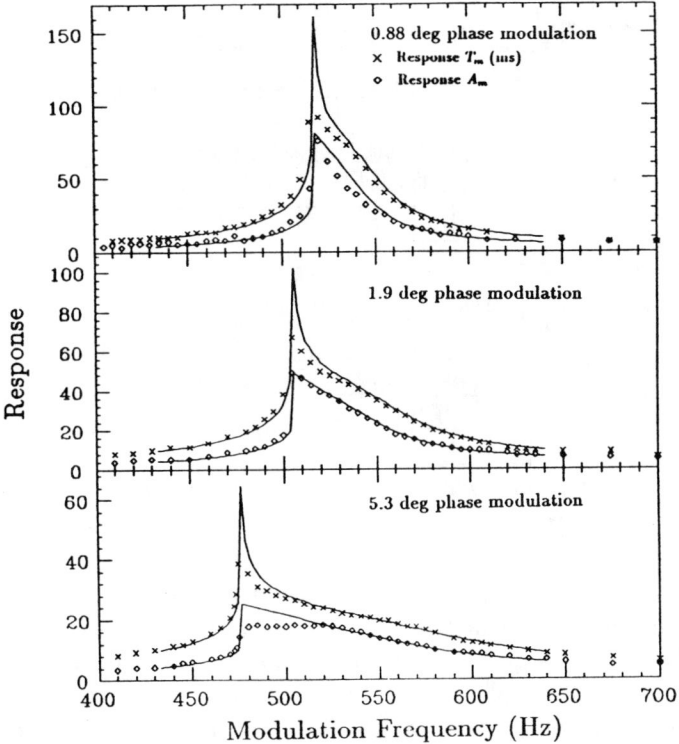

Figure 4: A plot of the measured beam response to phase modulation for three different phase modulation amplitudes. The solid lines are the results from the simulation. The type II phase detector was used for these measurements, whose limitation of ±90 deg resulted in the flat amplitude response observed in the 5.3 deg case (bottom).

In Fig. 4, the measured responses as a function of phase modulation frequency are plotted for three different phase modulation amplitudes. The sudden change in the beam response in the neighborhood of the resonant frequency followed by a slow decrease with increasing frequency is characteristic of a resonant parametric system near its bifurcation point[6,7].

This system was studied numerically using difference equations. For an undamped system in the absence of the external rf phase modulation, the difference equations describing the motion are

$$\delta_{n+1} = \delta_n + \frac{2\pi \nu_{s_0}^2}{h|\eta \cos\phi_0|}(\sin\phi_n - \sin\phi_0), \quad (1)$$

$$\phi_{n+1} = \phi_n + 2\pi\eta\delta_{n+1}, \quad (2)$$

where δ and ϕ are the variables described above with the subscripts indicating the turn number, η is the phase slip factor which is about -0.86 in this case, ϕ_0 is the

168 Longitudinal Phase-Space Measurements at IUCF

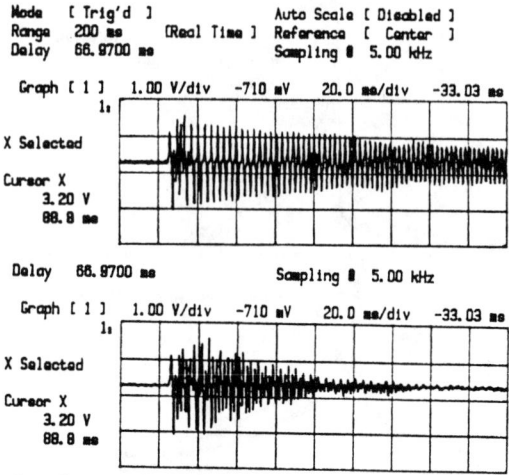

Figure 5: A plot showing the ϕ signal on digital oscilloscope. The upper trace is with the electron-cooling on, and the lower trace is with the cooling off. With the cooling off, there is rapid decoherence.

synchronous phase which for this stored beam is 0 deg, h is the harmonic number which was one, and ν_{s_0} is the small-amplitude synchrotron tune which was about 5.20×10^{-4}. When the rf phase is modulated, an additional term must be included in the equation for the beam phase, $\Delta\phi_{\rm rf}$, where $\Delta\phi_{\rm rf} = a(\sin\omega_m t_{n+1} - \sin\omega_m t_n)$ with a the phase modulation amplitude, ω_m the angular frequency of the rf phase modulation, and t_n and t_{n+1} the beam arrival times at the rf cavity, which are approximately multiples of the revolution period for a synchronous particle. Since it was the transient response which was measured, the effect of electron cooling, which would be included as a very small damping term, was neglected. The results of the simulation are shown as solid lines in Fig. 4 with the data for comparison. The agreement with this simple model is quite good.

Another intriguing aspect of these measurements is the relative unimportance of decoherence, and its near absence while cooling the beam. In Fig. 5, the phase amplitude as a function of time is shown for two cases where the applied phase shift was about 150 deg. In the first case the electron energy was optimized to maximize cooling. In the second case, the electron energy was deliberately shifted so that electron cooling would be neglible. Note that the damping of the coherent signal due to decoherence in the absence of electron cooling is relatively rapid, whereas the damping of the coherent signal due to actual damping by electron cooling, perhaps combined with decoherence, is relatively slow! In fact, the damping of the coherent signal we measure is not unreasonable when compared to the damping rates expected from the electron cooling, implying that with the electron cooling on, there is little decoherence. Furthermore, attempts to add damping from electron cooling to the simulation and to do the calculation for a collection of particles having a reasonable phase-space distribution has thus far failed to duplicate this result. The most realistic fit to the data for the beam response has come from the simplest model using a single particle with no damping.

III Conclusions

We have found that the experimental techniques used for turn-by-turn tracking in studying transverse motion are equally effective for studying longitudinal motion. The synchrotron tune shift versus the amplitude was measured easily and accurately using a coherent synchrotron oscillation, as was the response of the beam to an external phase modulation of the rf. The response of the beam is of interest as it was found to have features in common with other resonant parametric systems, the most interesting of these being the transition or bifurcation from one stable solution to another[6]. It is not inconceivable that aspects of this motion may prove to be useful, perhaps in a beam extraction scheme, and similar measurements can shed light on the effects of rf noise.

The multiparticle aspects of the motion we have studied remain somewhat a mystery. Further analysis, and perhaps experimental work, will be necessary in order to understand the relationship between electron cooling and the observed decoherence. It seems quite likely to a number of us that space charge, which we have not yet included in our model, will play a significant role in the explanation of the behavior we observed as shown in Fig. 5.

References

[1] R.E. Pollock, "IUCF cooler ring status 1989," *Proc. 1989 IEEE P.A.C., Acc. Eng. and Tech.*, Chicago, IL, pp. 17–21; D.L. Friesel, T.J. Ellison, and P. Schwandt, "Status report on the IUCF cooler-storage ring," *Nucl. Inst. Meth.* **B40/41**, 927–933 (1989).

[2] T. Ellison, D.L. Friesel, and R.J. Brown, "Status and performance of the IUCF 270 keV electron cooling system," ibid, pp. 633–635; T. Ellison, "Electron Cooling," Indiana University, Bloomington, Indiana, Ph.D. Dissertation, 1990 (unpublished).

[3] S.Y. Lee et al., *Phys. Rev. Lett.* **67**, 3768 (1991); D.D. Caussyn et al., *Phys. Rev. A* **46**, 7942 (1992).

[4] E.M. McMillan, *Phys. Rev.* **68**, 143 (1945); V.I. Veksler, *Compt. Rend. Acad. Sci. U.R.S.S.* **43**, 329 (1944); ibid. **44**, 365 (1944).

[5] R.E. Best, *Phase Locked Loops, Theory, Design, and Applications* (McGraw-Hill, 1984), pp. 7-9.

[6] M. Ellison et al., *Phys. Rev. Lett.*, to be published.

[7] D. D'Humieres, M.R. Beasley, B.A. Huberman, and Libchaber, *Phys. Rev. A* **26**, 3483 (1982).

Betatron Coupling Correction at the IUCF Cooler, Leading to Improved Determination of Fourth-Order Resonance Hamiltonian[*]

M. Ellison,[a] M. Ball,[a] B. Brabson,[a] J. Budnick,[a] D. D. Caussyn,[a]
J. Collins,[a] S. Curtis,[a] V. Derenchuk,[a] G. East,[a] T. Ellison,[a] D. Friesel,[a]
B. Hamilton,[a] H. Huang,[a] W. P. Jones,[a] W. Lamble,[a] S. Y. Lee,[a] D. Li,[a]
S. Nagaitsev,[a] X. Pei,[a] G. Rondeau,[a] T. Sloan,[a] A. W. Chao,[b] S. Dutt,[b]
M. Syphers,[b] Y. Yan,[b] M. G. Minty,[c] S. Tepikian,[d] K. Y. Ng,[e] W. Gabella[e]

Abstract

Recently the Hamiltonian of particle motion near a resonance condition was deduced at the Indiana University Cyclotron Facility Cooler Ring. It was found that linear betatron coupling complicated the analysis. A coupling correction scheme is described which reduced the coupling coefficient from 0.03 to 0.0012. When particles were kicked onto resonance islands with the coupling reduced, the island motion was stable for upwards of 1×10^6 turns. This stable island motion allowed for the determination of higher-order terms in the Hamiltonian.

1 Introduction

The study of nonlinear beam dynamics in synchrotrons has become increasingly popular in recent years. This interest has been motivated by the realization that higher-order multipoles in the superconducting magnets being designed for the Superconducting Super Collider (SSC) and the Relativistic Heavy Ion Collider (RHIC) will be relatively large. It is important to verify that these higher-order multipoles and the resulting nonlinear particle motion will not significantly reduce the stability of particle motion in these machines. While many excellent theoretical studies exist,[1] experimental studies of resonant behavior are essential to better understand the approximations which these studies use.

The Indiana University Cyclotron Facility (IUCF) cooler ring provides an excellent laboratory for nonlinear resonance studies. Electron cooling reduces the rms emittance of the beam to <0.3 (π mm mrad), which allows the proton bunch to act

---

[*]Work supported in part by the National Science Foundation and the U.S. Department of Energy.
[a] IUCF, Indiana University, Bloomington, IN 47405; [b] SSC Laboratory, 2550 Beckleymeade Avenue, Dallas, TX 75237-3946; [c] SLAC, MS26, Box 4349, Stanford, CA 94309; [d]Brookhaven National Laboratory, Upton, NY 11973; [e]Fermilab, P.O. Box 500, Batavia, IL 60510.

as a single macroparticle. The IUCF cooler ring is hexagonal with a circumference of 86.82 m. This experiment was performed with a stored 45-MeV proton beam; the cooler operated with a ten-second cycle time. The stored beam typically had 3×10^8 protons per bunch and a bunch length of about 3.6 m (or 40 ns) FWHM. The revolution frequency, f_{rev}, in the accelerator was 1.03168 MHz, and the beam was bunched on the first harmonic of the revolution frequency. The beam was kicked with various angular deflections, by a pulsed deflecting magnet with a pulse width of 500 ns and with rise and fall times of 100 ns. The subsequent beam-centroid displacement was measured by two beam position monitors (BPMs),[2] with an rms position resolution of about 0.1 mm. The transverse positions were recorded on a turn-by-turn basis for 4096 turns.

2 Coupling Correction

In a previous study[3] it was found that linear coupling between the horizontal and vertical planes of oscillation complicated the particle dynamics. Thus, a coupling correction scheme was devised and implemented prior to the nonlinear resonance study reported here. The observed coupling in the cooler ring originates from a combination of random tilts of the 36 quadrupoles in the ring, imperfect compensation of the solenoidal field employed in the electron cooling system, off-axis transit through sextupole fields, and stray fields from the Lambertson injection magnet. Since the magnitude and phase of the coupling constant were difficult to predict, we adopted an on-line coupling correction scheme so we could "tune" the coupling toward zero. Using a perturbative treatment in classical dynamics, it can be shown that the coupling coefficient for linear coupling, near the $v_x - v_z = p$ resonance, where v_x and v_z are the horizontal and vertical tunes respectively and p is an integer, is given by[4]

$$C = \frac{1}{2\pi} \oint \sqrt{\beta_z \beta_x} \left[K + \frac{M}{4} \left(\frac{\beta_z'}{\beta_x} - \frac{\beta_x'}{\beta_z} \right) - i\frac{M}{2} \left(\frac{1}{\beta_z} + \frac{1}{\beta_x} \right) \right] e^{i\Phi} \, ds, \quad (1)$$

where $\Phi = \Psi_x - \Psi_z - (v_x - v_z + 1)\theta$, $K = (\partial B_x/\partial x)/B\rho$, $M = B_{sol}/B\rho$, $\beta_{x/z}$ are the beta-functions, $B\rho = p/e$ is the magnetic rigidity, $\Psi_{x/z}$ are the betatron phase advances, s is the distance measured along the reference orbit in the ring and $\theta = s/R$ is the orbital angle around the ring. In Eq. (1), the first term arises from skew quadrupole fields, while the second and third terms result from the axial fields of solenoids. Note that the effects of axial fields and skew quadrupole fields are phase shifted by 90°. The magnitude of C is a measure of the minimum possible tune separation between the upper and lower normal modes of oscillation. Since the coupling coefficient is a complex number, generally two correction elements are needed to correct for it. In the IUCF cooler there are two skew quadrupoles in place, which, although located only 3.5 m apart, have phase factors Φ differing by about 60°. This relatively large difference in the phase factors results from a strong focus in the z-plane between the elements and hence a rapid betatron phase advance in that plane relative to the x-plane. These skew quadrupoles were connected to bipolar supplies which calculations showed would be able to correct any coupling coefficient up to $|C| \approx 0.07$, much larger than the typically observed value of 0.03.

The beat period between the two planes of oscillation is given by[5]

$$T = \frac{1}{f_{rev}\sqrt{\Delta^2 + |C_R|^2 + |C_I|^2}} \quad \text{where} \quad \Delta = \nu_x - \nu_z + 1 \quad (2)$$

and C_R and C_I are the real and imaginary components of the coupling coefficient respectively. In order to "tune" the coupling to as small a value as possible the beam was kicked horizontally while coincidentally monitoring the difference signal from a BPM. The difference signal was input to a spectrum analyzer operated in zero span mode. The center frequency was set to the first betatron sideband of the revolution frequency $f_{rev}(1+\Delta \nu_x)$, and the resolution bandwidth was set to 30 kHz. The spectrum analyzer was triggered in coincidence with the horizontal kick and the power in the sideband was monitored as a function of time. The beat period, as power was transferred into the vertical plane and then back into the horizontal plane, was clearly visible. An iterative process was then used where the betatron tune was adjusted to minimize Δ, and then the skew quadrupole strength adjusted to minimize the coupling coefficient $|C|$ (or maximize the beat period). The maximum observed beat periods were up to 1 ms in length which implies that the magnitude of $|C|$ was reduced to about 0.001.

As a test of this correction scheme, the strength of one skew quadrupole was varied independently while monitoring the beat period. The phase of the skew quadrupole being varied was designated as 0° and Eq. (1) was used to calculate C_R. The values of the horizontal and vertical beta-functions were obtained by running the program MAD. Assuming that we had succesfully adjusted the tunes to make $\Delta = 0$, C_I was then the only adjustable parameter in Eq. (2). The fit is shown in Fig. 1. The experimental beat period shown is the average beat period over 5 oscillations. For this case C_I was found to be 0.0012. The value of C_I was the value of the coupling coefficient when C_R was adjusted to zero and thus represents the minimum value of $|C|$ obtained.

3 Fourth-Order Resonance

For particle motion in a circular accelerator, the horizontal deviation from the closed orbit satisfies Hill's equation

$$\frac{d^2x}{ds^2} + K(s)x = \frac{\Delta B_z}{B\rho}, \quad (3)$$

where $K(s)$ is a function of the quadrupole strength, and s is the longitudinal particle coordinate which advances from 0 to C, the storage ring circumference. Normally the anharmonic term $\Delta B_z/B\rho$ is small. Neglecting the anharmonic term, Hill's

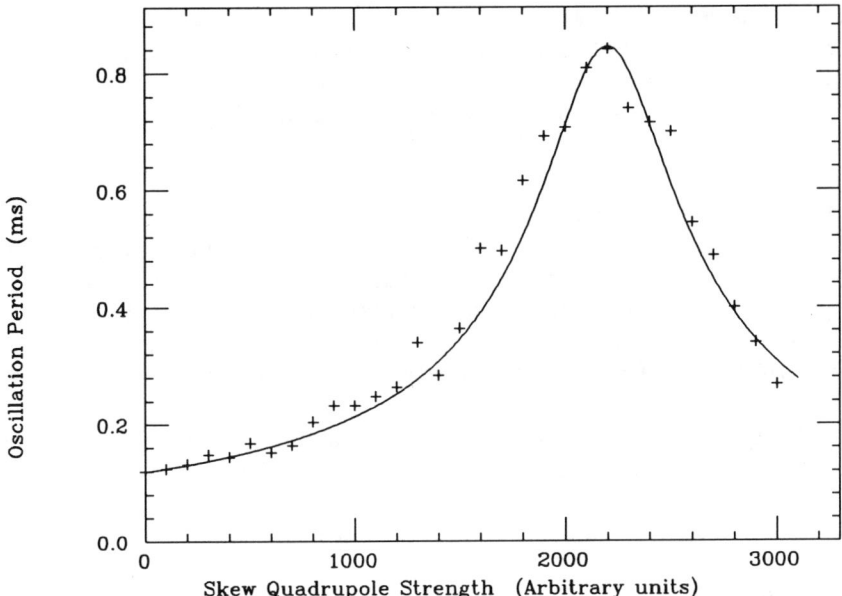

Figure 1. The beat period is shown as a function of skew quadrupole strength. The solid line is a fit using Eq. (1).

equation can be solved using the Floquet transformation to obtain the solution $x = (2J\beta_x)^{1/2}\cos\phi$ where J and ϕ are action-angle variables, β_x is the betatron amplitude function, and $2J$ is the phase-space area of the betatron motion.

The transverse displacements x_1 and x_2 were measured at two locations in the ring separated in betatron phase by ϕ_{12}. Considering the horizontal plane, the canonically conjugate variable to x is p_x, where $p_x = \beta_x x' + \alpha x = -(2J\beta_x)^{1/2}\sin\phi$, and $\alpha = -1/2 \, d\beta_x/ds$. For linear motion, the Poincaré map generated by plotting the turn-by-turn tracking motion in phase space (x, p_x) is a circle. The value of p_x in terms of x_1 and x_2 is given by $p_x = -x_1\cot\phi_{12} + x_2(\beta_1/\beta_2)^{1/2}/\sin\phi_{12}$. The values of ϕ_{12} and β_1/β_2, needed to transform the position variables to the normalized momentum, were determined by fitting the experimental data in both the x and z planes.

Normally the effects of the nonlinearities are small. When the betatron tunes are close to a resonant condition, which for one-dimensional motion occurs at $m\nu_x = n$, where m, n are integers, one must take into account the particular harmonics of the error distribution which act coherently with betatron motion. The Hamiltonian near the resonance can then be approximated by

$$H = H_o(J) + \epsilon J^2 \cos(m\Phi - n\theta - \chi) \qquad (4)$$

where (J, Φ) are the conjugate action-angle variables of the betatron motion, χ is a phase factor having little physical significance, and ϵ is the resonant strength, which is determined by the distribution of nonlinearities in the accelerator. The betatron tune is given by $\nu(J) = \partial H/\partial J \approx \nu_o + \alpha J$, where we have used a first-order Taylor series expansion in the action variable with ν_o the betatron tune at zero betatron amplitude, and α is now the coefficient of the first-order expansion. For the present study, $m=4$ and $n=15$.

The Hamiltonian in Eq. (4) is not a constant of the motion. A canonical transformation with generating function $F_2(\Phi,J_1) = [\Phi - (n/m)\theta]J_1$ can be performed which is a transformation to a rotating coordinate frame. This transformation has the effect of "freezing" the resonance in the new phase space and yields a new Hamiltonian $H_1 = H_o(J_1) - (n/m)J_1 + \epsilon J_1^2 \cos(m\Phi_1 - \chi)$ which *is* a constant of the motion. Here (J_1,Φ_1) are the new action-angle variables with $J_1 = J$ and $\Phi_1 = \Phi - (n/m)\theta$.

Finally, the new Hamiltonian H_1 can be expanded about the resonant action J_r, where $m\nu_x(J_r) = n$. This expansion yields

$$H_1 = \frac{1}{2}\alpha(J_1 - J_r)^2 + \epsilon J_1^2 \cos(m\Phi_1 - \chi) + \ldots , \qquad (5)$$

where H_1 is a constant of the motion. This Hamiltonian can be used to track particle trajectories around the stable fixed points, where both J_1 and Φ_1 are constant. A series of kicks were performed when we were close to the resonance condition and the data are shown in Fig. 2.

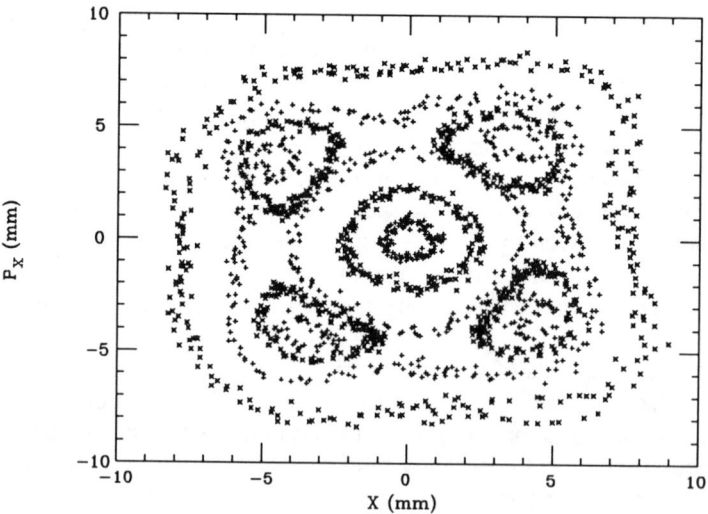

Figure 2. Poincaré maps for motion near the fourth integer resonance for seven different kicker amplitudes. Motion inside the separatrix, in the resonance islands, and outside the separatrix is evident.

The tune shift with amplitude α was determined from the measured betatron tunes across the resonance. These data are shown in Fig. 3, where α was found to be $6.5 \times 10^{-4} \pm 0.5 \times 10^{-4}$ (π mm mrad)$^{-1}$. The flat region resulted from kicking into resonance islands where the fractional tune was locked onto 0.2500. The values of J_r, ϵ and χ were determined by the fit to all the different kicker amplitudes. We found the resonant action J_r to be 1.2 ± 0.1 (π mm mrad), the phase factor χ to be 0.5 ± 0.05 radians, and the resonance strength to be $8.0 \times 10^{-5} \pm 1.0 \times 10^{-5}$ (π mm mrad)$^{-1}$. The x-p_x phase-space data shown in Fig. 2 were plotted in action-angle coordinates in Fig. 4. The solid lines are fits to the data using Eq. (5) with all parameters except H_1 held constant.

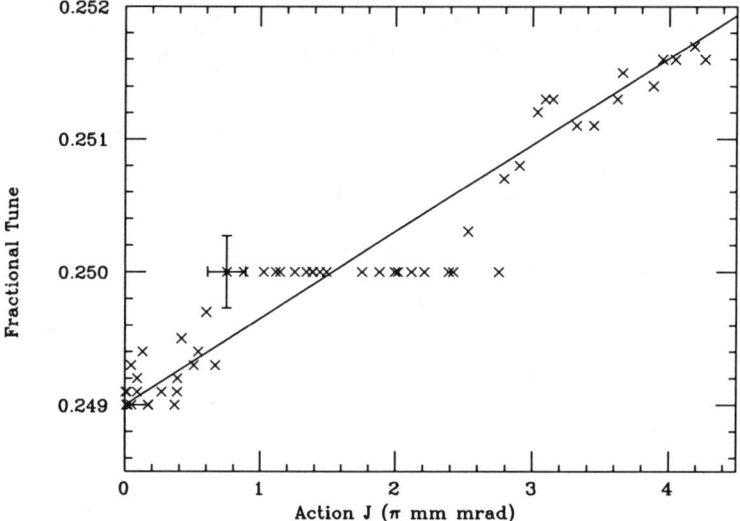

Figure 3. Plot of Δv_x vs. the average value of J produced by a variety of kicker amplitudes. The error bar shown is representative. The slope of the solid line was used to determine α.

4 Conclusion

A linear coupling correction scheme was tested at the IUCF cooler ring which reduced the magnitude of the coupling coefficient C to approximately 0.001. This removed a serious complication to both the analysis and dynamics of the fourth-order resonance. The particle tracking done after the coupling was removed could be described with the Hamiltonian of Eq. (5). The resonance strength determined in this study was about a factor of two smaller than that determined in the previous study.[3] The sextupole strengths were different in these two cases. Both strengths will be compared to predictions in a later publication.

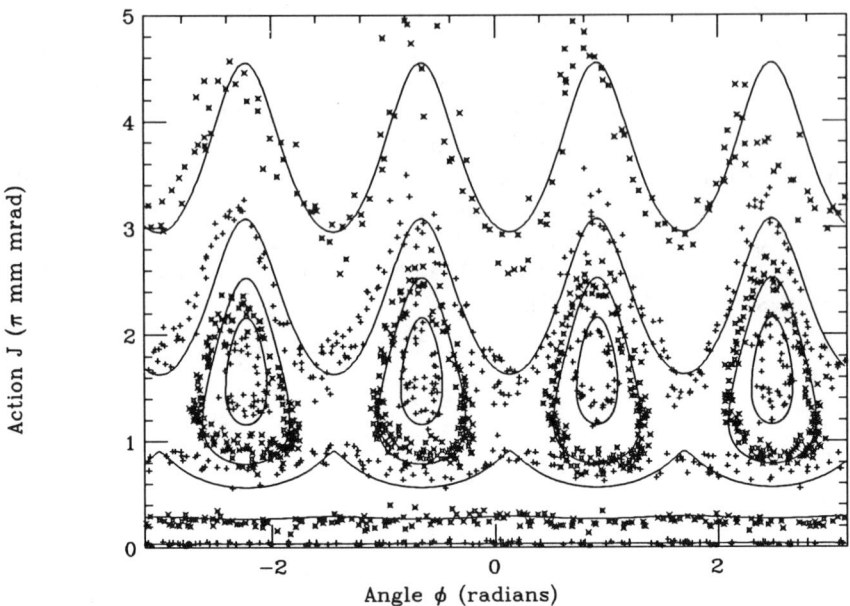

Figure 4. Plot of the same data shown in Fig. 2 except now shown in (J,Φ) coordinates. The solid lines are fits using Eq. (4) with the value of H_1 being the only adjustable parameter.

References

1. R. Ruth, in *Physics of Particle Accelerators*, AIP Conf. Proc. No. **153**, p. 150 (1987); R.L. Warnock, R. D. Ruth, and K. Ecklund, in *Proc. IEEE Part. Accel. Conf., Chicago*, p. 1325 (1989); A. J. Dragt, in *Physics of High Energy Particle Accelerators*, AIP Conf. Proc. No. **87**, p. 147 (1982); G. Guignard, CERN Report No. **78-11**, (1978) (unpublished).

2. B. J. Hamilton, M. Ball, and T. J. Ellison, Betatron 'Ping' Tune Measurement System for the IUCF Cooler Synchrotron/Storage Ring, in *Proc. IEEE 1993 Part. Accel. Conf., Washington*, to be published; T. J. Ellison, C. M. Fox, and S. W. Koch, NIM B24/25, 873 (1987).

3. S. Y. Lee *et al.*, Experimental Determination of a Nonlinear Hamiltonian in a Synchrotron, *Phys. Rev. Lett.* **67**, p. 3768 (1991).

4. G. Guignard, CERN Report No. **76-06** (1976).

5. E. Wilson, in *CERN Accelerator School General Accelerator Physics*, CERN No. **85-19**, p. 114 (1985).

SUCCESS IN ONE-TURN MAPS FOR DYNAMIC APERTURE STUDIES—A BRIEF REVIEW[1]

Yiton T. Yan
Superconducting Super Collider Laboratory, Dallas, TX 75237

Abstract

Progress in the use of one-turn maps for long-term tracking of particles—useful in dynamic aperture studies on large circular accelerators—is reviewed briefly. It is recommended that long-term tracking of particles for large circular accelerators such as the Superconducting Super Collider (SSC) or its high-energy booster be performed with one-turn maps. The advantages are twofold: not only is tracking speed dramatically enhanced (by about two orders of magnitude for the SSC), but one-turn-map tracking also offers an easier reliability check of the tracking results and easier order-by-order analysis.

Since my last comment on one-turn maps for long-term tracking,[1] there has been more progress in this field. Indeed, it is time to recommend and encourage our colleagues to consider using one-turn maps for particle orbit advancement, especially for long-term tracking of large circular accelerators. In this letter the author wishes to give a very brief review of the progress of the one-turn map concept for tracking.

Given the advantage that the Lie transformations can be concatenated (to form a one-turn map) via the Campbell-Baker-Hausdorf theorem,[2] I believe that the concept of using one-turn maps for particle tracking in circular accelerators should have been planted in Dragt's mind since he and his associates introduced the Lie-algebraic treatment of beam dynamics in the mid-1970s.[3] However, one-turn maps for particle tracking, particularly for long-term tracking, were not seriously tried—at least not for practical cases—until the late 1980s. It was then that Alex Chao headed the accelerator physics division of the Universities Research Association (URA)/Superconducting Super Collider (SSC) Central Design Group (CDG) at Berkeley. Thanks to Chao's encouragement and collaboration, several individuals were able to get together at Lawrence Berkeley Laboratory for a period of time (a review paper was written that summarize work done during that period).[4] Berz introduced the computational use of differential algebra, which allows computational extraction of one-turn maps more easily and in a more general way than before.[5] Forest, having been associated with Dragt, was able to see the advantage of using these differential algebras for computational Lie-algebraic formulations and to develop the Lie-algebraic normal form in collaboration with Berz and Irwin.[6] After performing the long-term dynamic aperture

---

[1]Operated by the Universities Research Association, Inc., for the U.S. Department of Energy under Contract No. DE-AC35-89ER40486.

studies for the SSC with the time-consuming (both computer time and human time) single-particle trackings, Irwin and I knew that in order to achieve a more complete study of the dynamic apertures of large circular accelerators such as the SSC, advancements in tracking techniques must be made. I went on to take advantage of the supercomputer vectorization and parallelization (multi-tasking) capabilities for multi-particle tracking and was able to create "Livingston dynamic aperture plots."[7] (The name was changed to "survival plots" at Richard Talman's suggestion.) Meanwhile, Irwin factorization was developed,[8] enabling one to convert a series of Lie transformations of different orders into a reasonable number of kicks and rotations for symplectic tracking. Forest, even though it was against his philosophy of not using one-turn maps for tracking, was kind enough to collaborate with me in programming the Irwin factorization, hoping to make the study of the SSC dynamic aperture more efficient. About the same time, some of the Zlib[9] vectorized subroutines were developed, making possible the programming of Zmaps[10] for extracting high-order, one-turn maps of the SSC.

Tracking with Irwin factorization for the SSC dynamic aperture study was tried during the transition period in which the CDG was merged with the SSC Laboratory; it was not quite successful. Part of the reason, it was suspected, was that the Irwin factorization bases were chosen randomly instead of by following Irwin's description that would have helped minimize the high-order spurious terms. However, for various reasons this effort was not continued.

In the spring of 1990, the issue of whether a one-turn map can be used for long-term tracking and thus for dynamic aperture study was raised at the Capri nonlinear-beam dynamics workshop.[11] This topic attracted many colleagues to a somewhat controversial discussion, but no conclusion was reached because of a lack of sufficient practice. After the workshop, we tested an 11th-order Taylor map of the SSC by advancing the phase-space orbits of particles turn-by-turn via direct evaluation of the truncated 11th-order Taylor map (not exactly symplectic due to truncation). We found that the survival plot was roughly the same as that of the previous results with element-by-element tracking.[12] This was the first time, to my knowledge, that the one-turn map showed some promise for dynamic aperture study, although there were still some concerns about the non-exact symplecticity. The same one-turn map was also tested with 10th-order Taylor-map tracking, resulting in somewhat different survival plots. However, the 10th-order Taylor map—after it had been Lie-transformed (by Dragt-Finn factorization)[3] and re-expanded into an 11th- or 12th-order Taylor map to gain a higher degree of symplecticity—showed correct dynamic aperture up to 10^6 turns.[13] What we have learned from these results is that a moderate-order (lower than 11th-order), one-turn differential algebraic Taylor map is usually accurate enough for dynamic aperture studies, but its degree of symplecticity is usually not enough for long-term tracking.[14] The wrong survival plots obtained with the direct Taylor-map tracking of 10th order are due not to inaccuracy of the maps but to artificial diffusion of the particle orbits because of the lack of sufficient symplecticity. How to symplectify the Taylor map without imposing large spurious errors in the map becomes the key to success when using one-turn maps for long-term tracking.

An accurate symplectification method for a truncated one-turn Taylor map is a series of order-by-order Lie transformations called Dragt-Finn factorization.[3] Unfortunately, nobody yet knows how to directly evaluate a Lie transformation in general. Therefore, re-formation of a series of order-by-order general Lie transformations into a special kind of Lie transformation that can be evaluated has attracted a great deal of effort. The methods will, in general, induce high-order spurious terms, thereby reducing the accuracy of the map. Irwin factorization, mentioned above, is one of these symplectification methods. Monomial factorization is another worth mentioning.[15] Since neither method guarantees no symmetry violation, the symplectified map will, in general, preserve a strong image imposed by how the method is used. Therefore the map, although symplectic, can be inaccurate for dynamic aperture study unless it is extended to an order higher than what one would consider adequate. Recently, we have been investigating a method for the evaluation of Lie transformations that will take care of the symmetry property. It is hoped that this method is to be more accurate than both the Irwin factorization and the monomial factorization at the same order.[16] However, before we can know how useful this method is for long-term tracking and thus for dynamic aperture study, it must be tested for practical cases. Such testing will be performed once a numerical program has been completed.

Before I go on to review another accurate symplectification method, I wish to acknowledge the efforts contributed by colleagues in Europe.[17] They have recently tested an interesting method of evaluating one-turn Taylor maps with dynamic rescaling, using linear corrections to suppress the artificial diffusion due to non-exact symplecticity of the truncated Taylor map.[18] Although not mathematically rigorous, this method, in my opinion, can be used for the study of certain cases, especially given the fact that accelerators are not exactly static.

Another symplectification method for "order-by-order symplectic" but truncated Taylor maps (thus not exactly symplectic) is the use of generating functions for converting the explicit truncated Taylor maps into implicit ones. Since symplecticity means canonical transformation, it is very likely that many colleagues may have considered the possibility of using generating functions for symplectification. For example, we know Dragt and his associates have considered such a method.[19]

Evaluation of these implicit maps can lead to numerical instabilities. Large spurious errors can also be generated if one imposes a certain form (not type) of generating functions. It is, perhaps, due to these drawbacks that such an implicit method has been overlooked in the past.

Recently, Yan, Channell, and Syphers have considered, tested, and successfully used an implicit Taylor-map tracking scheme.[20] First, the one-turn map is separated into two maps, a linear Courant-Snyder matrix followed by a nonlinear truncated Taylor map to enhance numerical stability. Then, differential algebras are used to convert the truncated nonlinear Taylor map into an implicit type of truncated Taylor map without imposing a certain form of generating function. (Of course, one of the four types of generating functions exists implicitly.) Because this method does not impose a pre-determined form (not type) of generating function, it provides the same degree of accuracy as the explicit Taylor map at the same order, and it has shown much success in practical use. It has been used to save enormous amounts of computer time for long-term tracking of the SSC and its high-energy booster.[21] There are two advantages to using such a one-turn-map tracking scheme over the traditional element-by-element tracking scheme: it not only provides faster tracking speed (about two orders of magnitude faster for the SSC), thus allowing lifetime tracking of large circular accelerators, but it also provides easier order-by-order analysis, thus allowing easier reliability checks of the global results such as the survival plots. Note that due to slow tracking speed, one of the drawbacks of element-by-element tracking is that one usually assigns nonlinear multipole errors up to certain orders without checking whether they are adequate. (Although one-turn-map tracking cannot be more accurate than element-by-element tracking, due to fast tracking speed, tracking with various orders can be easily performed to make sure that the order of the map and the order of multipole errors are indeed adequate.)

In summary, since one is interested only in phase-space regions where one-turn Taylor maps converge, it is fine to use one-turn maps not only for order-by-order analysis but also for turn-by-turn tracking. It is especially economical to use one-turn maps for long-term tracking of large circular accelerators. In certain cases, dynamical (time-dependent) one-turn maps are also workable.[22]

References

1. Y. Yan, "Brief Comment on One-turn Map for Long-term Tracking," *AIP Conf. Proc.* No. 255, p. 305, A. Chao, ed. (1992).
2. J.E. Campbell, *Proc. London Math. Soc.* **29**, 14 (1898); H.F. Baker, ibid. **34**, 347 (1902); F. Hausdorf, Ber. Verhandl. Akad. Wiss. Leipzig, *Math-naturwiss* **58**, 19 (1906).
3. A. Dragt and J. Finn, "Lie Series and Invariant Functions for Analytic Symplectic Maps," *J. Math Phys.* **17**, 2215 (1976); A. Dragt et al., "Lie Algebraic Treatment of Linear and Nonlinear Beam Dynamics," *Ann. Rev. Nucl. Part. Sci.* **38**, 455 (1988).
4. A. Chao, "Recent Non-Linear Dynamics Studies for the SSC," *AIP Conf. Proc.* No. 230, p. 203, Y.H. Ichikawa and T. Tajima, eds. (1990).
5. M. Berz, "Differential Algebraic Description of Beam Dynamics to Very High Orders," *Part. Accel.* **24**, 109 (1989).
6. E. Forest, M. Berz, and J. Irwin, "Normal Form Methods for Complicated Periodic Systems," *Part. Accel.*, **24**, 91 (1989).
7. Y. Yan, "Supercomputing for the Superconducting Super Collider," *Energy Sciences Supercomputing 1990*, pp. 9–13 (1990), published by DOE National Energy Research Supercomputer Center, A. Mivin and G. Kaiper, eds.

8. J. Irwin, "A Multi-kick Factorization Algorithm for Nonlinear Maps," SSCL-228 (1989); A. Dragt, "Methods for Symplectic Tracking," presented at *Workshop on Nonlinear Problems in Future Particle Accelerators*, Capri, Italy (1990).
9. Y. Yan and C. Yan, "Zlib—A Numerical Library for Differential Algebra," SSCL-300 (1990); Y.T. Yan, "Zlib 2.0—A Numerical Library for Differential Algebra and Lie Algebraic Treatment of Beam Dynamics," *Proc. 1991 IEEE Part. Accel. Conf.*, Vol. **1**, p. 333 (1991).
10. Y. Yan, "Zmap—A Differential Algebraic High-Order Map Extraction Program for Teapot Using Zlib," SSCL-299 (1990); L. Schachinger and R. Talman, "Teapot: A Thin-Element Accelerator Program for Optics and Tracking," *Part. Accel.* **22**, 35 (1987).
11. *Proc. Workshop on Nonlinear Problems in Future Particle Accelerators*, Capri, Italy (April 1990), W. Scandale and G. Turehetti, eds., published by World Scientific.
12. Y. Yan, T. Sen, A. Chao, G. Bourianoff, A. Dragt, and E. Forest, "Comment on Round-off Errors and on One-Turn Taylor Maps," SSCL-301 (1990); also in Ref. 11, p. 77.
13. T. Sen, Y. Yan, A. Chao, and E. Forest, "Taylor Maps for Long-term Tracking at the SSC," SSCL-497 (1991).
14. Y. Yan, "Applications of Differential Algebra to Single-Particle Dynamics in Storage Rings," SSCL-500 (1991); also in *Physics of Particle Accelerators*, M. Month and M. Dienes, eds., *AIP Conf. Proc.* No. 249, Vol. **1**, pp. 378–455 (1992).
15. I. Gjaja, "Monomial Factorization of Symplectic Maps," University of Maryland preprint 1-92 (1992).
16. J. Shi and Y. Yan, to be presented at *IEEE Part. Accel. Physics Conf.*, May 1993.
17. R. Kleiss, F. Schmidt, Y. Yan, and F. Zimmermann, "On the Feasibility of Tracking with Differential-Algebra Maps in Long-Term Stability Studies for Large Hadron Colliders," DESY HERA 92-01, CERN SL/92-02 (AP) and SSCL-564 (1992).
18. R. Kleiss, F. Schmidt, and F. Zimmermann, "Experience with a Simple Method to 'Symplectify' Differential Algebra Maps," DESY HERA 92-16 and CERN SL/92-31 (AP) (1992).
19. A. Dragt, Private communication.
20. Y. Yan, P. Channell, and M. Syphers, "An Algorithm for Symplectic Implicit Taylor-Map Tracking," SSCL-Preprint-157 (1992); submitted to *J. Comp. Phys.*
21. Y. Yan, P. Channell, M. Li, and M. Syphers, to be presented in *IEEE Part. Accel. Physics Conf.*, in May 1993.
22. Y. Yan *et al.*, work in progress.

TRACKING EXPERIENCE WITH PATRIS AT BNL[*]

J. Milutinovic and A.G. Ruggiero
Brookhaven National Laboratory, Upton, NY 11973, USA

ABSTRACT

We present some of the results of our tracking investigations using PATRIS. Both short-term and long-term tracking, using large numbers of particles, were performed and analyzed. When switching from short-term (1000 turns) to long-term (the maximum achieved was 10 million turns) tracking mode, we observed a small (about 14%) reduction of the dynamic aperture. We also empirically confirmed that the distinction between short-term and long-term tracking becomes meaningful for a number of turns of the order of 100,000 rather than 10,000, i.e. that 10^5 is the threshold which must be crossed to get any appreciable difference between a 1000-turn run and a longer one. We have also found that a high density of initial conditions employed in tracking is somewhat more important for accurate determination of the dynamic aperture than a large number of turns.

1 INTRODUCTION

One of the main purposes of particle tracking in an accelerator is to determine the dynamic aperture, i.e. the region of stability in the machine, containing the initial conditions for which the particle motion will be stable. "Stable" in the tracking context means survival of a certain specified number of turns. This number, chosen for the purposes of computer simulation, was selected in conformity with the available computational resources, rather than being close to the actual number of turns being operational in a well-designed machine. In the past, when computer resources were more limited, this number was limited to 400 to 500 turns, while at present it is typically 1000 for short-term tracking and significantly greater for long-term tracking, where the available resources usually dictate the magnitude of that "significantly greater" number. Currently, this number will be in the vicinity of 10^6 or even beyond it.

2 MOTIVATION FOR TRACKING STUDIES; TOOLS SELECTED FOR TRACKING

In our tracking studies, we were motivated by our desire to understand better the nature of the dynamic aperture, in particular its possible dependence on the number of tracking turns. We wanted to verify or disprove some claims that it shrinks as one starts probing a very large number of turns (vicinity of one million or more).

---

[*] Work performed under the auspices of the U.S. Department of Energy.

The tracking tool of our choice was the tracking code PATRIS,[1] run on a CRAY-2 computer. PATRIS was chosen for two reasons. First, it belongs to a class of codes which have been historically acceptable for tracking. Second, we know it very well and fully control its procedures and algorithms. The code is symplectic; it treats drifts, dipoles and quadrupoles to first order, by propagating the particles along trajectories which are solutions of linearized equations of motion. On the other hand, nonlinearities are simulated as thin lenses. The CRAY-2 was selected as the most powerful computer we had access to. It is fast and fairly accurate in the single precision mode, and it allows vectorization, which was crucial for simultaneous tracking of a large number of particles, up to 5000 per tracking run.

The lattice for tracking was RHICAGR.[2] The reason for its selection was that it is simpler than the "official" RHIC lattice, and unlike the "official" lattice it was not subject to constant changes. Moreover, the results are probably easier to understand, but the conclusions we derive should have a more general validity. Up to now, this lattice has been "ideal," i.e. without errors of any kind and with chromaticity correcting sextupoles as the only type of nonlinearities.

The program activities can be roughly classified into two categories: (a) short-term tracking activities, and (b) long-term tracking activities.

3 SUMMARY OF SHORT-TERM TRACKING ACTIVITIES

3.1 Short-Term Tracking; General Information

We started with short-term tracking activities, with the intention to move on later to work on long-term tracking. The number of turns we selected for short-term activities was $N = 1000$. Particles were launched from the middle of a defocusing quadrupole QD, with the initial conditions being always constrained by the

$$X' = Y' = 0 \tag{1}$$

condition, and in the eight different directions in the (ϵ_x, ϵ_y) plane. These directions are shown in Figure 1. Because of mid-plane symmetry of the ideal lattice, only directions 1 through 5 had to be examined.

In the initial phase of the investigations both physical and dynamic aperture were determined for all eight directions and for the momentum deviation values

$$\delta = -0.6\%, -0.4\%, -0.2\%, \quad 0.0\%, +0.2\%, +0.4\%, +0.6\%. \tag{2}$$

Subsequent studies were restricted to the on-momentum ($\delta = 0$) case.

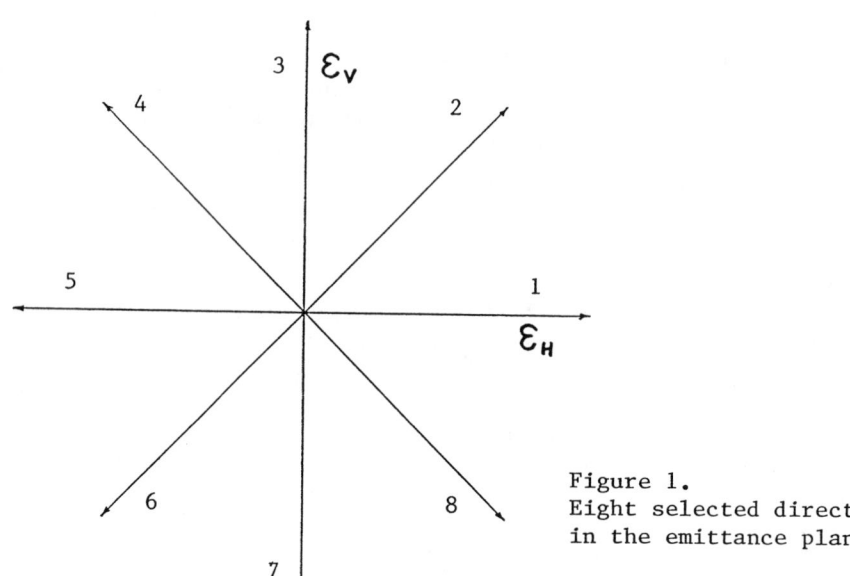

Figure 1.
Eight selected directions in the emittance plane

3.2 Dynamic Aperture and Particle Loss Criterion

Whereas in the search for physical aperture any particle was lost when it struck a magnet, in the search for dynamic aperture it was considered lost if its transverse coordinates (X, Y) violated the condition

$$\sqrt{X^2 + Y^2} \leq 1000 \text{ mm}. \tag{3}$$

This condition is being checked at every nonlinear element and also at the two aperture limitations which appeared to be in insertion dipoles.

3.3 Automated Searching Procedure

To determine the boundary of the dynamic aperture in a particular direction in the emittance plane, one has to launch several particles with different magnitudes of emittance all lying on the same direction. If these initial values are properly chosen, some particles will survive the tracking while others will be lost. One then concludes that the boundary of the dynamic aperture lies between the smallest emittance that gives rise to particle loss and the greatest emittance of all surviving particles being closer to the origin than any lost particle along this direction. Initial attempts to investigate this by adjusting the initial tracking emittances by hand, and then readjusting them subsequently for more accurate limits on the dynamic aperture, proved to be too time consuming, given the variety of directions in the emittance plane and the need to explore both dynamic and physical aperture (the latter with sextupoles both on and off) for on-momentum and several off-momentum values. To overcome this problem, we developed an automated searching procedure.

The automated procedure worked as follows. An interval $(\epsilon_0^{\min}, \epsilon_0^{\max})$ was chosen on the particular direction of interest. $\epsilon_0^{\min}$ was usually taken to be zero, while $\epsilon_0^{\max} = 15$ πmm·mrad was empirically found to be large enough to guarantee the loss of a particle launched from an $\epsilon_0^{\max}$ initial condition. This interval $(\epsilon_0^{\min}, \epsilon_0^{\max})$ was divided into N equal subintervals, and $(N+1)$ particles equally spaced along that direction were launched. Our choice was $N = 25$. The emittances of the launched particles were

$$\epsilon_0^{\min}, \ \epsilon_0^{\min} + \Delta\epsilon_0, \ \epsilon_0^{\min} + 2\Delta\epsilon_0, \ \ldots, \epsilon_0^{\min} + N\Delta\epsilon_0 = \epsilon_0^{\max} \tag{4}$$

where

$$\Delta\epsilon_0 = \left(\epsilon_0^{\max} - \epsilon_0^{\min}\right)/N. \tag{5}$$

Next we determined which particles were lost by violating the condition (3) or equivalent conditions in the case of searching for the physical aperture. Among them, we select the particle with the smallest emittance $\epsilon_{\text{LOST}}^{\min}$. All tracked particles with smaller emittances, of course, survived. Among these surviving particles we selected the survivor with the largest emittance $\epsilon_{\text{SURV}}^{\max}$. (See Figure 2.) Of course, $\epsilon_{\text{LOST}}^{\min} > \epsilon_{\text{SURV}}^{\max}$ and they satisfy the relation

$$\epsilon_{\text{LOST}}^{\min} - \epsilon_{\text{SURV}}^{\max} = \Delta\epsilon_0 = \left(\epsilon_0^{\max} - \epsilon_0^{\min}\right)/N. \tag{6}$$

Now we select new bounds

$$\epsilon_1^{\min} = \epsilon_{\text{SURV}}^{\max}, \quad \epsilon_1^{\max} = \epsilon_{\text{LOST}}^{\min}, \tag{7}$$

subdivide this new interval into N subintervals of the size

$$\Delta\epsilon_1 = \left(\epsilon_1^{\max} - \epsilon_1^{\min}\right)/N, \tag{8}$$

and repeat the tracking with a new set of $N+1$ initial conditions, corresponding to the emittances that would result from a subdivision of the new interval $(\epsilon_1^{\min}, \epsilon_1^{\max})$ in a manner analogous to (4) and (5). This is then the second iteration of our automated searching procedure, which continues with iterating until the subinterval between the lost particle with the smallest emittance and its closest surviving neighbor with even smaller emittance becomes smaller than some predetermined ϵ. We then conclude that the boundary of the dynamic aperture lies in this final subinterval. In this manner, the dynamic and physical apertures were determined for all directions and all values of momentum deviation.

At this stage tune dependence on the amplitude was also determined from the tracking data by using FFT. One example of the results is shown in Figure 3. The solid line shows the results obtained by MAD[3] and SYNCH[4] on the basis of linear theory.[5] The groups of hollow circles are the values obtained from the tracking data supplied by PATRIS. For small values of amplitude, where the linear theory is a good approximation, the agreement is very good.

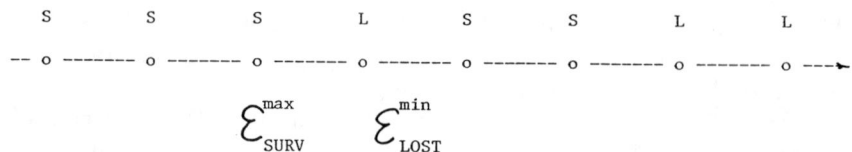

Figure 2.
Equidistant particles along a direction in the emittance plane. All are tracked with these initial conditions. Some are lost (L) while others survive (S) the tracking.

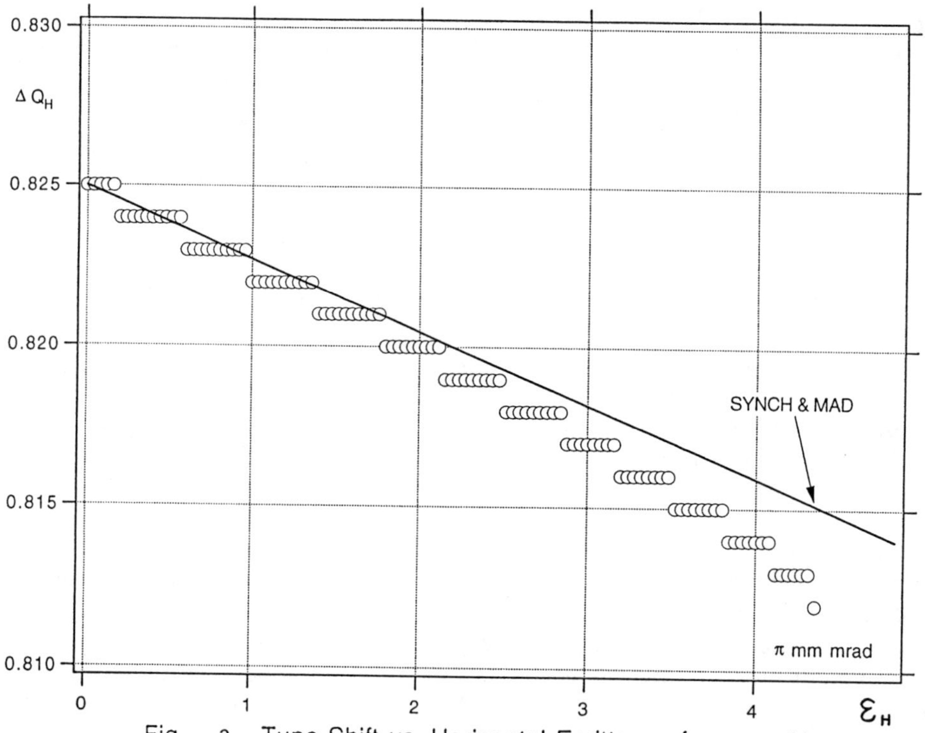

Fig. 3 Tune-Shift vs. Horizontal Emittance for case #1

Phase-space plots were also generated for future use. They revealed the presence of third- and fifth-integer resonances, as obvious from Figures 4 and 5. Details are reported in Ref. 6, 7, and 8.

3.4 Problems with the Automated Searching Procedure; Elusiveness of the Dynamic Aperture

The previously described method of searching for the dynamic aperture by the automated iterative procedure also presented some problems. The searching procedure was rapidly converging but in doing so it constricted the searched interval on the emittance direction so quickly that it also missed some lost particles. The reason became obvious. The initial subinterval $\Delta\epsilon_0$ was big enough to allow the existence of some unstable but not tracked particles with emittances smaller than ϵ_{SURV}^{max}, the emittance of the last acceptable survivor. After that, the algorithm iterated in the interval (ϵ_{SURV}^{max}, ϵ_{LOST}^{min}) which was already above the instability threshold. This threshold was not discovered since no particles were tracked in that region of instability. This is shown in Figure 6. This resulted in a systematic overestimation of the dynamic aperture, in particular in directions 2 and 4 (Figure 1).

The remedy for this problem was to start tracking with a much larger number of particles and then in the next iteration to choose a less constricted interval, which extends further toward the origin in the (ϵ_x, ϵ_y) plane from the detected lost particle with the smallest emittance ϵ_{LOST}^{min}. This new interval is now big enough to contain presumably all unstable particles with emittances smaller than ϵ_{LOST}^{min}, not tracked in the first iteration. The new iteration, with the higher density of particles per emittance interval, will now catch most of them. In this way, each iteration is catching unstable particles without grossly overshooting those with emittances where instabilities start. This method is slower but safer. To determine the dynamic aperture with an accuracy of 0.01 πmm·mrad, we selected the number of particles $N = 250$ for subdivision of a selected interval in any iteration, and after that iteration we selected as the interval for the next iteration, the part of this selected interval that contained the detected lost particle with the smallest emittance ϵ_{LOST}^{min} together with its first 25 adjacent survivors in the direction of the origin. One extra iteration was needed compared with the simpler scheme, but this time no overestimation of the dynamic aperture occurred, as confirmed by the investigations described in the next section.[8]

This improved procedure revealed the presence of many holes within the initially determined (and usually overestimated) dynamic aperture. We suspect that this problem arose for many researchers who relied on determining the dynamic aperture by tracking a few particles, and we caution against this practice. Some awareness of the problem was privately conveyed to us by our colleagues.[9,10] All these features were sufficiently intriguing to justify a more detailed investigation, which is the subject of the next section.

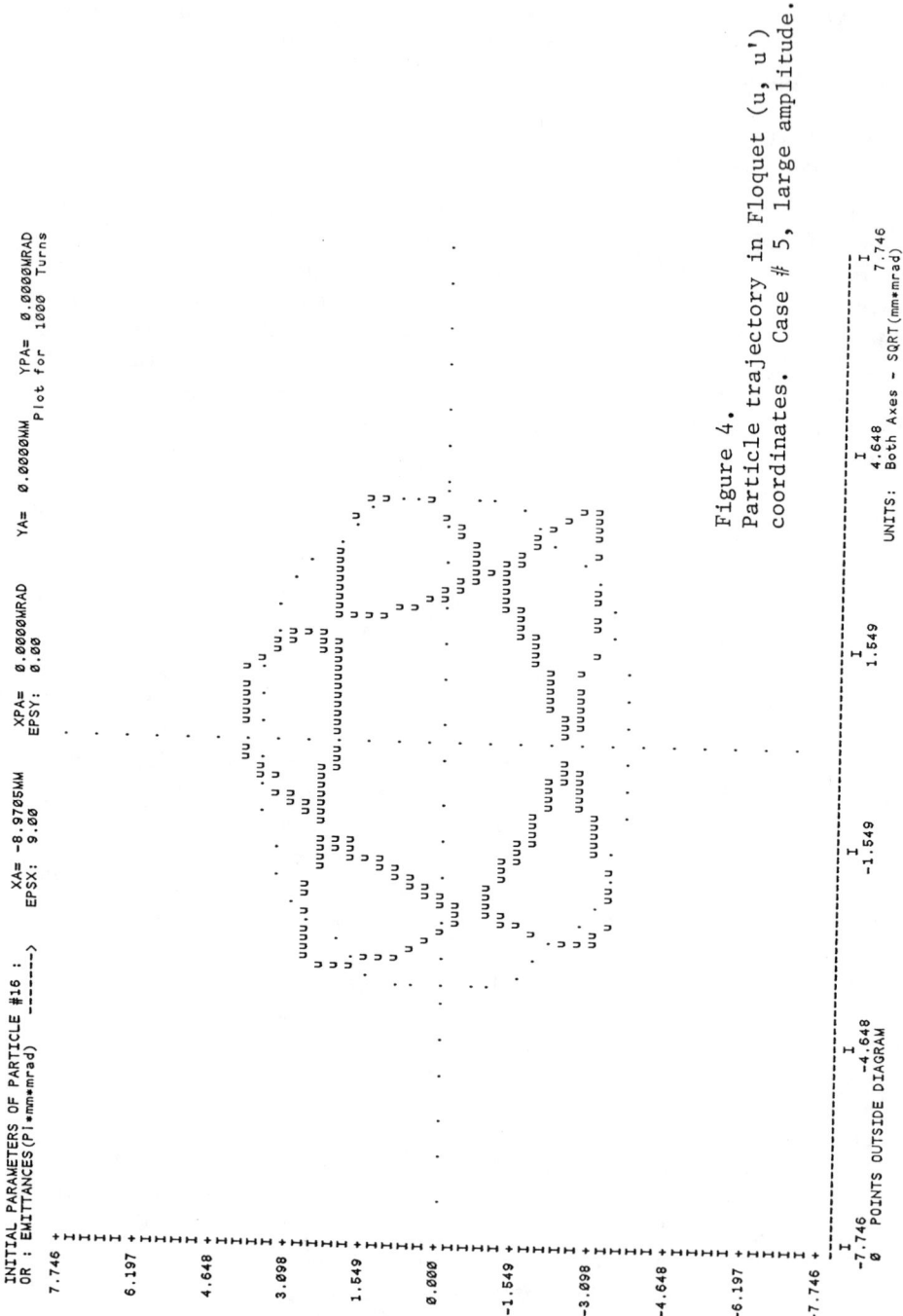

Figure 4.
Particle trajectory in Floquet (u, u') coordinates. Case # 5, large amplitude.

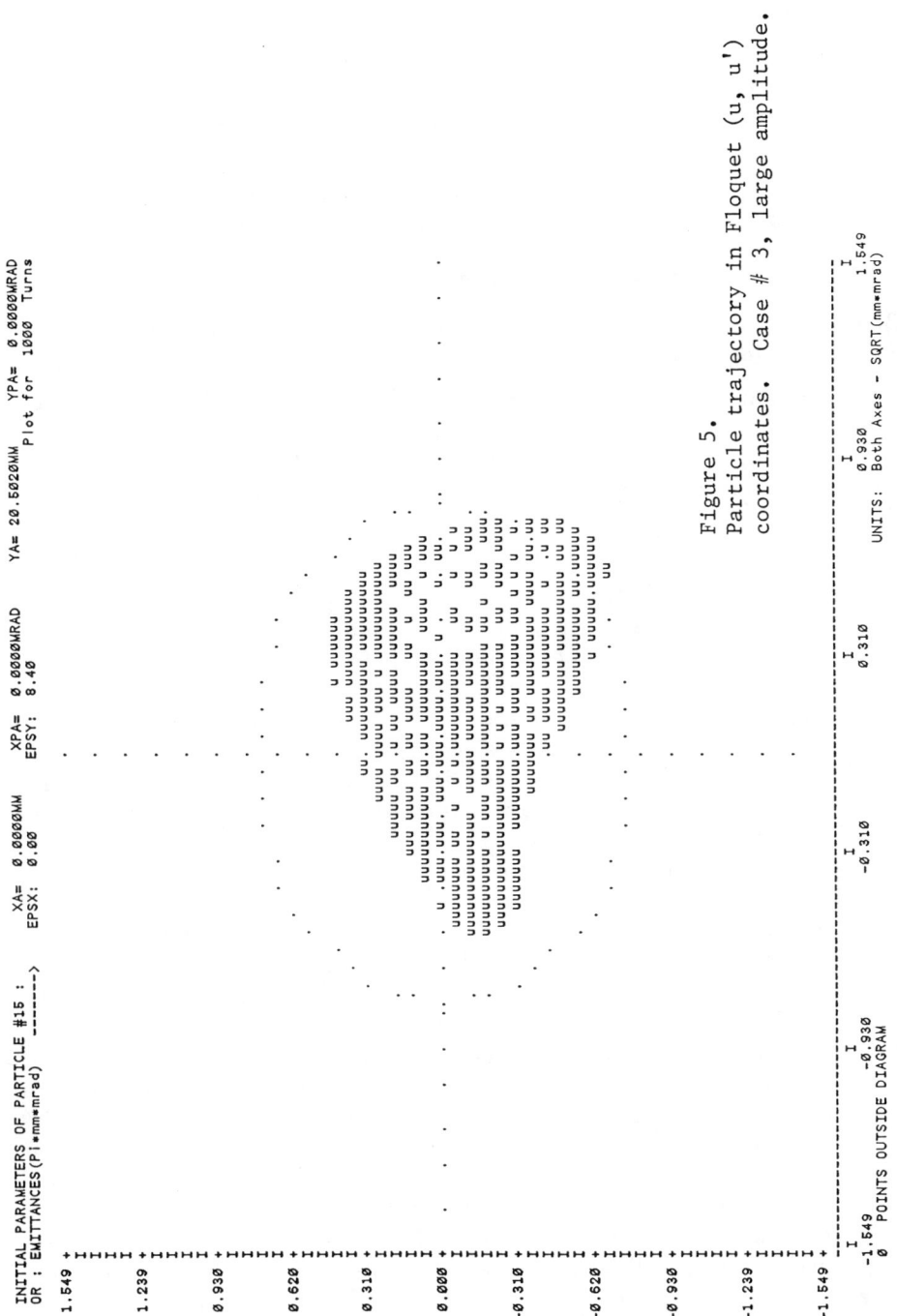

Figure 5. Particle trajectory in Floquet (u, u') coordinates. Case # 3, large amplitude.

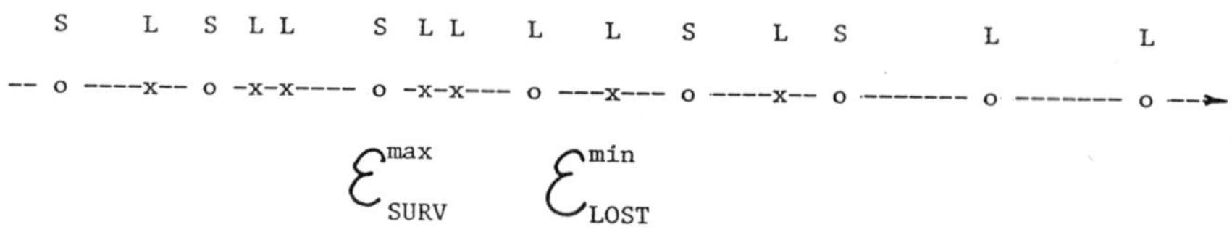

```
   S     L   S  L L      S  L L     L      L    S      L     S          L           L
-- o ----x-- o -x-x---- o -x-x--- o ---x--- o ----x-- o --------- o --------- o -->
```
$$\mathcal{E}^{max}_{SURV} \qquad \mathcal{E}^{min}_{LOST}$$

Figure 6.
Same equidistant particles as those tracked and represented in Figure 2. As before, they are represented by "o." Particles which were not tracked but which would have been lost had they been tracked are now represented by "x."

3.5 Probing the Dynamic Aperture with a Large Number of Particles; External Regions of Stability

In the subsequent investigations all five directions in the (ϵ_x, ϵ_y) plane were examined, in the emittance intervals $(0.0, 25.0)$ πmm·mrad. A total of 25,000 equally spaced particles were launched in each direction and tracked for 1000 turns, as usual. This means a subdivision into **intervals of the size**

$$\Delta\epsilon = (25/25000) = 0.001 \; \pi\text{mm·mrad}. \tag{9}$$

The number of turns that each particle made was recorded. Regions of stability interspersed by regions of instability were found. This is shown in Tables 1 through 6, which appear at the end of this paper.

Each table represents the number of turns a tracked particle successfully completed. There are four sections on each page. The first column represents a particle number in the set of 5000 particles tracked in one run. The second column represents the number of turns the particle completed. The third column represents the horizontal initial emittance. Vertical emittance is not shown since it is zero in this run. The next three sections are continuations of the first, placed beside it to save space.

Table 1 starts with particle number 3701 with the emittance 12.4500 πmm·mrad. The run started from a particle (number 1 in the sequence, not shown here) with the emittance 10.0 πmm·mrad. No lost particles were detected in two other runs that covered the emittance intervals below 10.0 πmm·mrad, and no particle in this run was lost in the sequence number 1 through 3700. For this reason the table starts at particle number 3701. From the table it is obvious that very few particles were lost in the emittance interval (12.450, 12.649) πmm·mrad, covered by this table. These lost particles look more like a set of "holes" at larger emittances than the edge of the dynamic aperture. The next 200 particles (numbers 3901 through 4100) all survived, hence we omitted the corresponding table. Table 2 covers particles number 4101 through 4300. It reveals more holes than Table 1 but still a vast majority of particles survive 1000 turns. Table 3 suddenly shows a different picture. Instead of holes, continuous emittance intervals where particles are lost appear. Table 4, on the other hand, looks again like Table 2, even though emittances are greater here than in Table 3. We now have continuous emittance intervals where particles are stable and some holes separating these intervals. This picture continues into Table 5, with the exception of its last section, where again we have regions of instability interspersed by regions of stability. Finally, Table 6 reveals unstable particles, with a couple of isolated particles which did survive 1000 turns.

From this set of tables it is obvious that what we call "dynamic aperture as the region of stability of particle motion" is somewhat tricky and also to some extent arbitrary. Figure 7 shows a rough qualitative sketch of the number of successfully completed turns as a function of the initial emittance for case 1 (see Fig. 1 for classification). Although every hole could not be represented in this sketch, it should convey the qualitative picture of regions of stability interspersed by regions of instability. We can conditionally use the name "dynamic aperture" for the compact central region of stability which in this case covers the emittances from 0.0 up to 12.4700 πmm·mrad, and we can call the more distant finite intervals, where particles survive the tracking, the "external regions of stability."

At this point we would like to offer a clear explanation of why it is so easy to overestimate the dynamic aperture. By using a small number of particles one has a good chance of missing not only isolated holes but also finite intervals of instability, e.g. the interval spanned by particles number 4324 and 4333 in Table 3, because of the presence of adjacent finite regions of stability such as the region spanned by particles 4341 and 4346. Or even worse, if one starts tracking from large emittances with few particles, which are mainly lost, and

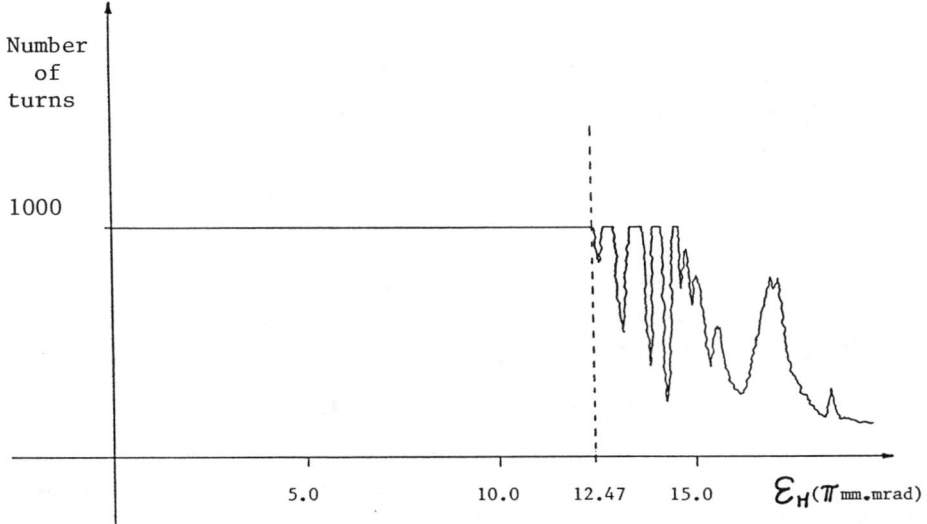

Figure 7.
Sketch of the number of completed turns as a function of the emittance. Direction # 1.

then reduces the amplitudes until all start surviving the tracking, then one is risking a significant overestimation of the dynamic aperture since even the innermost particle may end up in a distant external region of stability such as the thick region spanned by particles number 4766 and 4841 in Table 5. For this reason, our initially designed automated searching procedure malfunctioned. On the other hand, our improved version of the searching procedure pinpointed the real edge of the dynamic aperture, precisely at the value of the emittance 12.4700 πmm·mrad, revealed in our unusually dense probing of the stability regions under tracking.[8]

Three fundamental questions arise at this point:

1. Do these external regions of stability, lying beyond the central compact region, which we conditionally call the "dynamic aperture," survive if we increase the number of tracking turns beyond 1000?
2. What happens with the central region if the number of turns is much greater than 1000? Does it shrink and, if so, how much?
3. What happens if we significantly increase the density of particles used to probe the stability regions? Will holes start appearing in the external regions of stability or even in the central region?

The third question is the toughest, since it must be tackled in conjunction with the first two. Or, to be precise, given the potential dependence of a particle's stability verification on the number of tracking turns, one would have to increase the number of tracking turns and the density of initial conditions simultaneously. Before even contemplating this kind of research, one has to come up with some answer to the first two questions. This was indeed the subject of our long-term tracking investigations.

4 SUMMARY OF LONG-TERM TRACKING ACTIVITIES

4.1 Long-Term Tracking; General Information

Before embarking upon a long-term tracking study, one has to achieve the following two objectives: (a) an assessment of the possible contributions due to computer roundoff error, and (b) vectorization of the tracking code, if one intends to track a large number of particles per computer run.

Having in mind long-term tracking with a large number of particles per run, we made sure that both objectives had been achieved before we launched a real long-term tracking run.

4.2 Roundoff Error; An Assessment of Its Impact

To assess the impact of the roundoff error, tracking runs were undertaken with sextupoles present in the lattice, but with zero strength. In this case, the following two quantities are separately invariant:

$$u^2 + u'^2 = \epsilon_x, \tag{10}$$

$$v^2 + v'^2 = \epsilon_y, \tag{11}$$

where

$$u = X/\sqrt{\beta_x}, \quad u' = \alpha_x X/\sqrt{\beta_x} - \sqrt{\beta_x} X', \quad (12)$$

$$v = Y/\sqrt{\beta_y}, \quad v' = \alpha_y Y/\sqrt{\beta_y} - \sqrt{\beta_y} Y'. \quad (13)$$

Particles with several different initial emittances were tracked for 10^4 and 10^6 turns and these two invariants were computed and recorded as functions of the number of turns. The results are shown in Tables 7 and 8, at the end of this paper.

The entries in the tables show that these two invariants exhibit a linear change with the number of turns. Since the quadratic invariants Eqs. (10) and (11) grow (or diminish) proportionally to the number of turns N, it follows that the quantities (u, u', v, v') and subsequently (X, X', Y, Y') suffer from an error proportional to $\sqrt{N}$. This proves that what we are detecting in this test is indeed a result of the computer roundoff, which is proportional to $\sqrt{N}$.

Furthermore, the tables shows that the rate of error accumulation is closer to 10^{-12} than to 10^{-11} per turn. This means that even after 10^7 turns the total accumulation will affect the emittance by an amount lying between 10^{-5} and 10^{-4}. Therefore, the roundoff error will not affect the evaluation of the dynamic aperture by more than several parts in 10^{-5}, for 10 million turns!

4.3 Vectorization

This was simply a must, if a large number of particles per computer run were to be tracked. The action of each magnet contains a loop with a counter running over the tracked particles and this loop was a natural candidate for vectorization. This was tricky, since the condition (3), which verifies a particle's survival, inhibits vectorization. However, these obstacles were overcome by using some ideas from Ref. 11 and 12. After that, the CPU time for 5000 particles and 1000 turns dropped from 1 hour to 256 seconds.

4.4 Tracking Results

After successful vectorization and roundoff error analysis, a pilot run was made with 100 particles for 10^6 turns. The immediate consequence noted was that the dynamic aperture shrinks, but the external regions of stability do survive one million turns. The results are shown in Table 9, at the end of this paper. The first two sections show the results of the pilot run, if only 1000 turns are requested. As before, the first hole is for the initial emittance 12.4700 πmm·mrad. Even though the particle density is much smaller than in the run represented partially in Tables 1 through 6, one sees the presence of holes (particles number 37, 59, 66, 67) and an external region of stability, spanned by particles number 77 through 93. The remaining two sections show the results of the same run if the full number of 10^6 turns are requested. A brief look at these two sections reveals several interesting features. The dynamic aperture is smaller than the same quantity determined for 1000 turns. Its exact value unfortunately was not determined in this run since our selection of the innermost particle overshot the edge of the dynamic aperture. However, an external region of stability is revealed in the region that was previously part of the compact central region called the "dynamic aperture at 1000 turns." This region is now spanned by particles number 28 through 33. The next external region of stability is now in the area that was previously between the first two holes, but this time the stability region is narrower (particles number 47 through 56). The previous external region of stability (particles number 77 through 93, for 1000 turns) is not completely lost at 10^6 turns. It gets narrower, spanned by particles number 83 through 87, and is accompanied by two isolated stable particles, numbers 91 and 92.

Finally, encouraged by these interesting results, we made a big run with 250 particles for 10 million turns. It consumed about 25 hours of CRAY-2 CPU time. In the slow batch queue, it cost us about 9 hours of CRAY-2 CRU units. The results are presented in Tables 10 and 11, at the end of this paper.

Table 10 shows the results of this long tracking run, if only 1000 turns are being requested. It is presented for a better comparison with the results of the long run, which are displayed in Table 11. Both tables start with particle number 51 since all among the first 50 particles did survive the requested total of 10 million turns. The following remarkable features appear:

1. The dynamic aperture, rigorously determined for 10 million turns, shrinks to about 86% of the dynamic aperture determined with the same rigor for 1000 turns.

2. The external stable regions become narrower but do survive the requested total of 10 million turns. Furthermore, among the particles that do not survive 10 million turns, a relatively small number survive even 1 million turns.

3. There are no holes, i.e. isolated unstable particles imbedded in otherwise stable regions, under a 10-million-turn stability test. Analogs of particles number 187 or 209 from the 1000-turn test do not exist any more. Isolated stable particles such as numbers 119, 145 or 215 still exist. The last case is stable under a 10-million-turn test even though it has highly unstable particles such as 219 and 220 in its proximity. This absence of holes for a high number of turns may be a consequence of the limited density of the initial conditions, and it is quite possible that the holes would reappear under a much higher density of initial conditions. In this run, computer resources forced us to choose an initial emittance spacing of 0.02 πmm·mrad, which is 20 times coarser than the previously used finest spacing of 0.001 πmm·mrad.

4. Going from 1000 turns to 10,000 turns leaves the dynamic aperture practically unaffected (just a bare 2% decrease). Much higher thresholds must be crossed to see "real things happening."

5. For those particles that survive all 10 million turns, a phase-space diagram during the last 1000 turns is almost identical to a diagram taken during the first 1000 turns. The two are so similar that they were barely distinguishable, or even indistinguishable, under visual inspection. This similarity was present both in the central compact region that we call the dynamic aperture and in the external regions of stability.

6. By failing to probe the dynamic aperture with a sufficient density of initial conditions, one can miss some unstable particles and conclude that the edge of the dynamic aperture is what is actually the outer edge of one of the external regions of stability. Exactly this happened to us during the initial phases of our tracking investigations. This can cause overestimation of the dynamic aperture by much more than 14%, which was the amount of overestimation we discovered after comparing a run for 1000 turns with a run for 10 million turns started from the same initial conditions. Our grossest overestimation, in tracking direction 4 in the emittance plane, with the presence of coupling provided by the sextupoles, was close to 100% for tracking with a low density of initial conditions. Therefore, probing with a higher density of initial conditions offers a better limit on the dynamic aperture, per computer resources involved, than probing with fewer particles but a higher number of turns. This is especially true if the tracking code is properly vectorized. We even dare say that going to a very high number of turns is not very meaningful in the context of searching for the dynamic aperture, unless it is accompanied by a sufficient density of initial conditions!

5 CONCLUSION

We initiated our tracking studies with a clear motivation to understand better the somewhat fuzzy nature of the dynamic aperture. Even before this study we knew that holes in stability regions, i.e. initial conditions of some unstable particles, do exist near the edge of the dynamic aperture. We also knew that some of our colleagues were aware of the presence of these holes.[9,10] Various claims had also been made about the reduction of the dynamic aperture for a high number of tracking turns. We wanted either to verify or to disprove them. Our attitude was to examine more initial conditions than done by most of our colleagues, i.e. to examine eight directions in the emittance plane (Fig. 1), and for each direction to proceed in small steps. Under the influence of our preliminary results, this attitude eventually resulted in probing the dynamic aperture with a high density of initial conditions.

The results we obtained were interesting and at least partially unexpected. The most significant observation is the existence of external stability regions and their survival under a ten-million-turn test. They have an important practical consequence on the tracking procedures, namely the need for a relatively high density of initial conditions to avoid possible overestimation of the magnitude of the dynamic aperture. Going to large number of tracking turns also appears to be important. Unfortunately, thresholds that must be crossed are closer to 10^5 turns than to 10^4 turns, so that a significant increase in the number of tracking turns is needed to see real changes. Tracking for 1000 turns looks more like a crude estimate than like a meaningful determination of the dynamic aperture. It also appears that in the absence of unlimited resources one first has to probe the dynamic aperture with a high density of initial conditions to prevent its gross overestimation, and only after that should one try to examine the edge of the dynamic aperture and the adjacent regions by higher numbers of turns. In a practical sense, the dynamic aperture appears to be a fuzzy and elusive concept, and tracking with a few particles for 1000 turns provides only a very rough estimation of the extent of the stability region.

6 REFERENCES

1. J. Milutinovic and A.G. Ruggiero, Comparison of Accelerator Codes for a RHIC Lattice, AD/AP-9, BNL, 1988.
2. A.G. Ruggiero, RHICAGR. A Most Simplified RHIC Lattice, AD/AP-26, BNL, August 1991.
3. H. Grote and F.C. Iselin, The MAD Program, Ver. 8.4, CERN/SL/90-13 (AP) (Rev. 2).
4. A.A. Garren et al., A User's Guide to SYNCH, FN-420, Fermilab, June 1985.
5. T.L. Collins, Distortion Functions, Fermilab-84/114, Fermilab, October 1984.
6. J. Milutinovic and A.G. Ruggiero, Long-Term Tracking with PATRIS. Part 1. Search of Physical and Dynamic Aperture with Sextupoles, AD/AP-34, BNL, January 1992.
7. J. Milutinovic and A.G. Ruggiero, Long-Term Tracking with PATRIS. Part 2. Aberration Effects, AD/AP-40, BNL, April 1992.
8. J. Milutinovic and A.G. Ruggiero, Long-Term Tracking with PATRIS. Part 3. Dynamic Aperture – How Stringent Is This Concept? AD/AP-44, BNL, June 1992.
9. E. Forest, private communications (1987 and later).
10. G.F. Dell, private communication (1992).
11. M. Metcalf, *Fortran Optimization*, Academic Press, London (1985).
12. C.G. Lazou, *Supercomputers and Their Use*, Clarendon Press, Oxford (1988).

TABLE 1. Number of completed turns in the emittance interval (12.450, 12.649) Pi.mm.mrad. Direction # 1.

Part. No.	Turn No. Completed	EPSX	Part. No.	Turn No. Completed	EPSX	Part. No.	Turn No. Completed	EPSX	Part. No.	Turn No. Completed	EPSX
3701	1000	12.4500	3751	1000	12.5000	3801	1000	12.5500	3851	1000	12.6000
3702	1000	12.4510	3752	1000	12.5010	3802	1000	12.5510	3852	1000	12.6010
3703	1000	12.4520	3753	1000	12.5020	3803	1000	12.5520	3853	830	12.6020
3704	1000	12.4530	3754	1000	12.5030	3804	866	12.5530	3854	1000	12.6030
3705	1000	12.4540	3755	1000	12.5040	3805	1000	12.5540	3855	1000	12.6040
3706	1000	12.4550	3756	1000	12.5050	3806	1000	12.5550	3856	1000	12.6050
3707	1000	12.4560	3757	1000	12.5060	3807	999	12.5560	3857	1000	12.6060
3708	1000	12.4570	3758	1000	12.5070	3808	1000	12.5570	3858	1000	12.6070
3709	1000	12.4580	3759	1000	12.5080	3809	1000	12.5580	3859	505	12.6080
3710	1000	12.4590	3760	1000	12.5090	3810	1000	12.5590	3860	1000	12.6090
3711	1000	12.4600	3761	1000	12.5100	3811	1000	12.5600	3861	1000	12.6100
3712	1000	12.4610	3762	1000	12.5110	3812	1000	12.5610	3862	1000	12.6110
3713	1000	12.4620	3763	1000	12.5120	3813	1000	12.5620	3863	1000	12.6120
3714	1000	12.4630	3764	1000	12.5130	3814	1000	12.5630	3864	1000	12.6130
3715	1000	12.4640	3765	1000	12.5140	3815	1000	12.5640	3865	1000	12.6140
3716	1000	12.4650	3766	1000	12.5150	3816	1000	12.5650	3866	1000	12.6150
3717	1000	12.4660	3767	1000	12.5160	3817	1000	12.5660	3867	1000	12.6160
3718	1000	12.4670	3768	1000	12.5170	3818	1000	12.5670	3868	1000	12.6170
3719	1000	12.4680	3769	1000	12.5180	3819	1000	12.5680	3869	1000	12.6180
3720	1000	12.4690	3770	1000	12.5190	3820	992	12.5690	3870	1000	12.6190
3721	875	12.4700	3771	1000	12.5200	3821	1000	12.5700	3871	1000	12.6200
3722	1000	12.4710	3772	1000	12.5210	3822	721	12.5710	3872	1000	12.6210
3723	1000	12.4720	3773	1000	12.5220	3823	1000	12.5720	3873	696	12.6220
3724	1000	12.4730	3774	1000	12.5230	3824	1000	12.5730	3874	1000	12.6230
3725	1000	12.4740	3775	1000	12.5240	3825	1000	12.5740	3875	1000	12.6240
3726	1000	12.4750	3776	1000	12.5250	3826	1000	12.5750	3876	1000	12.6250
3727	1000	12.4760	3777	1000	12.5260	3827	1000	12.5760	3877	615	12.6260
3728	1000	12.4770	3778	1000	12.5270	3828	1000	12.5770	3878	1000	12.6270
3729	1000	12.4780	3779	1000	12.5280	3829	1000	12.5780	3879	1000	12.6280
3730	1000	12.4790	3780	1000	12.5290	3830	1000	12.5790	3880	1000	12.6290
3731	1000	12.4800	3781	1000	12.5300	3831	1000	12.5800	3881	1000	12.6300
3732	1000	12.4810	3782	1000	12.5310	3832	1000	12.5810	3882	1000	12.6310
3733	1000	12.4820	3783	1000	12.5320	3833	1000	12.5820	3883	1000	12.6320
3734	1000	12.4830	3784	1000	12.5330	3834	1000	12.5830	3884	1000	12.6330
3735	1000	12.4840	3785	1000	12.5340	3835	1000	12.5840	3885	1000	12.6340
3736	1000	12.4850	3786	1000	12.5350	3836	1000	12.5850	3886	1000	12.6350
3737	1000	12.4860	3787	1000	12.5360	3837	1000	12.5860	3887	1000	12.6360
3738	1000	12.4870	3788	1000	12.5370	3838	483	12.5870	3888	1000	12.6370
3739	1000	12.4880	3789	1000	12.5380	3839	1000	12.5880	3889	1000	12.6380
3740	1000	12.4890	3790	1000	12.5390	3840	349	12.5890	3890	1000	12.6390
3741	1000	12.4900	3791	1000	12.5400	3841	1000	12.5900	3891	1000	12.6400
3742	1000	12.4910	3792	1000	12.5410	3842	1000	12.5910	3892	1000	12.6410
3743	1000	12.4920	3793	1000	12.5420	3843	1000	12.5920	3893	1000	12.6420
3744	1000	12.4930	3794	1000	12.5430	3844	1000	12.5930	3894	1000	12.6430
3745	1000	12.4940	3795	1000	12.5440	3845	1000	12.5940	3895	1000	12.6440
3746	1000	12.4950	3796	1000	12.5450	3846	1000	12.5950	3896	1000	12.6450
3747	1000	12.4960	3797	1000	12.5460	3847	1000	12.5960	3897	1000	12.6460
3748	1000	12.4970	3798	1000	12.5470	3848	1000	12.5970	3898	1000	12.6470
3749	1000	12.4980	3799	1000	12.5480	3849	1000	12.5980	3899	1000	12.6480
3750	1000	12.4990	3800	1000	12.5490	3850	1000	12.5990	3900	1000	12.6490

TABLE 2. Number of completed turns in the emittance interval (12.850, 13.049) Pi.mm.mrad. Direction #1.

Part. No.	Turn No. Completed	EPSX	Part. No.	Turn No. Completed	EPSX	Part. No.	Turn No. Completed	EPSX	Part. No.	Turn No. Completed	EPSX
4101	1000	12.8500	4151	1000	12.9000	4201	1000	12.9500	4251	1000	13.0000
4102	1000	12.8510	4152	1000	12.9010	4202	390	12.9510	4252	1000	13.0010
4103	1000	12.8520	4153	1000	12.9020	4203	186	12.9520	4253	1000	13.0020
4104	1000	12.8530	4154	1000	12.9030	4204	924	12.9530	4254	404	13.0030
4105	1000	12.8540	4155	1000	12.9040	4205	317	12.9540	4255	1000	13.0040
4106	1000	12.8550	4156	1000	12.9050	4206	1000	12.9550	4256	1000	13.0050
4107	1000	12.8560	4157	1000	12.9060	4207	1000	12.9560	4257	1000	13.0060
4108	1000	12.8570	4158	1000	12.9070	4208	1000	12.9570	4258	1000	13.0070
4109	1000	12.8580	4159	1000	12.9080	4209	1000	12.9580	4259	1000	13.0080
4110	1000	12.8590	4160	569	12.9090	4210	1000	12.9590	4260	1000	13.0090
4111	1000	12.8600	4161	419	12.9100	4211	651	12.9600	4261	1000	13.0100
4112	1000	12.8610	4162	1000	12.9110	4212	878	12.9610	4262	1000	13.0110
4113	1000	12.8620	4163	1000	12.9120	4213	1000	12.9620	4263	1000	13.0120
4114	1000	12.8630	4164	1000	12.9130	4214	1000	12.9630	4264	1000	13.0130
4115	1000	12.8640	4165	1000	12.9140	4215	625	12.9640	4265	1000	13.0140
4116	1000	12.8650	4166	1000	12.9150	4216	972	12.9650	4266	1000	13.0150
4117	1000	12.8660	4167	1000	12.9160	4217	1000	12.9660	4267	1000	13.0160
4118	1000	12.8670	4168	1000	12.9170	4218	1000	12.9670	4268	1000	13.0170
4119	1000	12.8680	4169	1000	12.9180	4219	272	12.9680	4269	1000	13.0180
4120	1000	12.8690	4170	1000	12.9190	4220	591	12.9690	4270	1000	13.0190
4121	1000	12.8700	4171	1000	12.9200	4221	1000	12.9700	4271	1000	13.0200
4122	1000	12.8710	4172	944	12.9210	4222	1000	12.9710	4272	1000	13.0210
4123	1000	12.8720	4173	397	12.9220	4223	1000	12.9720	4273	1000	13.0220
4124	1000	12.8730	4174	1000	12.9230	4224	1000	12.9730	4274	1000	13.0230
4125	1000	12.8740	4175	1000	12.9240	4225	326	12.9740	4275	1000	13.0240
4126	1000	12.8750	4176	514	12.9250	4226	1000	12.9750	4276	1000	13.0250
4127	1000	12.8760	4177	325	12.9260	4227	1000	12.9760	4277	1000	13.0260
4128	1000	12.8770	4178	1000	12.9270	4228	569	12.9770	4278	1000	13.0270
4129	810	12.8780	4179	1000	12.9280	4229	1000	12.9780	4279	1000	13.0280
4130	1000	12.8790	4180	1000	12.9290	4230	341	12.9790	4280	1000	13.0290
4131	1000	12.8800	4181	1000	12.9300	4231	924	12.9800	4281	1000	13.0300
4132	696	12.8810	4182	1000	12.9310	4232	1000	12.9810	4282	1000	13.0310
4133	1000	12.8820	4183	1000	12.9320	4233	496	12.9820	4283	1000	13.0320
4134	1000	12.8830	4184	1000	12.9330	4234	1000	12.9830	4284	1000	13.0330
4135	1000	12.8840	4185	1000	12.9340	4235	1000	12.9840	4285	1000	13.0340
4136	1000	12.8850	4186	490	12.9350	4236	1000	12.9850	4286	1000	13.0350
4137	1000	12.8860	4187	1000	12.9360	4237	1000	12.9860	4287	1000	13.0360
4138	1000	12.8870	4188	480	12.9370	4238	1000	12.9870	4288	1000	13.0370
4139	1000	12.8880	4189	399	12.9380	4239	1000	12.9880	4289	1000	13.0380
4140	1000	12.8890	4190	1000	12.9390	4240	317	12.9890	4290	1000	13.0390
4141	1000	12.8900	4191	1000	12.9400	4241	1000	12.9900	4291	1000	13.0400
4142	1000	12.8910	4192	1000	12.9410	4242	1000	12.9910	4292	1000	13.0410
4143	1000	12.8920	4193	1000	12.9420	4243	1000	12.9920	4293	1000	13.0420
4144	1000	12.8930	4194	1000	12.9430	4244	1000	12.9930	4294	1000	13.0430
4145	1000	12.8940	4195	1000	12.9440	4245	1000	12.9940	4295	1000	13.0440
4146	1000	12.8950	4196	486	12.9450	4246	876	12.9950	4296	1000	13.0450
4147	1000	12.8960	4197	1000	12.9460	4247	1000	12.9960	4297	463	13.0460
4148	1000	12.8970	4198	713	12.9470	4248	1000	12.9970	4298	1000	13.0470
4149	1000	12.8980	4199	1000	12.9480	4249	1000	12.9980	4299	1000	13.0480
4150	1000	12.8990	4200	1000	12.9490	4250	1000	12.9990	4300	1000	13.0490

TABLE 3. Number of completed turns in the emittance interval (13.050, 13.249) Pi.mm.mrad. Direction # 1.

Part. No.	Turn No. Completed	EPSX	Part. No.	Turn No. Completed	EPSX	Part. No.	Turn No. Completed	EPSX	Part. No.	Turn No. Completed	EPSX
4301	581	13.0500	4351	132	13.1000	4401	301	13.1500	4451	207	13.2000
4302	1000	13.0510	4352	321	13.1010	4402	1000	13.1510	4452	180	13.2010
4303	955	13.0520	4353	1000	13.1020	4403	542	13.1520	4453	628	13.2020
4304	1000	13.0530	4354	330	13.1030	4404	739	13.1530	4454	303	13.2030
4305	1000	13.0540	4355	1000	13.1040	4405	795	13.1540	4455	476	13.2040
4306	1000	13.0550	4356	979	13.1050	4406	1000	13.1550	4456	695	13.2050
4307	540	13.0560	4357	581	13.1060	4407	243	13.1560	4457	331	13.2060
4308	1000	13.0570	4358	126	13.1070	4408	879	13.1570	4458	564	13.2070
4309	1000	13.0580	4359	275	13.1080	4409	1000	13.1580	4459	320	13.2080
4310	1000	13.0590	4360	924	13.1090	4410	855	13.1590	4460	1000	13.2090
4311	1000	13.0600	4361	96	13.1100	4411	798	13.1600	4461	397	13.2100
4312	711	13.0610	4362	118	13.1110	4412	278	13.1610	4462	164	13.2110
4313	350	13.0620	4363	109	13.1120	4413	1000	13.1620	4463	239	13.2120
4314	284	13.0630	4364	288	13.1130	4414	1000	13.1630	4464	791	13.2130
4315	1000	13.0640	4365	394	13.1140	4415	422	13.1640	4465	1000	13.2140
4316	600	13.0650	4366	777	13.1150	4416	471	13.1650	4466	606	13.2150
4317	1000	13.0660	4367	1000	13.1160	4417	662	13.1660	4467	210	13.2160
4318	649	13.0670	4368	143	13.1170	4418	145	13.1670	4468	706	13.2170
4319	1000	13.0680	4369	159	13.1180	4419	232	13.1680	4469	220	13.2180
4320	508	13.0690	4370	423	13.1190	4420	643	13.1690	4470	258	13.2190
4321	240	13.0700	4371	1000	13.1200	4421	398	13.1700	4471	1000	13.2200
4322	362	13.0710	4372	157	13.1210	4422	1000	13.1710	4472	425	13.2210
4323	1000	13.0720	4373	364	13.1220	4423	346	13.1720	4473	891	13.2220
4324	310	13.0730	4374	82	13.1230	4424	1000	13.1730	4474	1000	13.2230
4325	777	13.0740	4375	86	13.1240	4425	501	13.1740	4475	748	13.2240
4326	122	13.0750	4376	84	13.1250	4426	231	13.1750	4476	781	13.2250
4327	114	13.0760	4377	91	13.1260	4427	180	13.1760	4477	1000	13.2260
4328	105	13.0770	4378	134	13.1270	4428	185	13.1770	4478	1000	13.2270
4329	114	13.0780	4379	1280	13.1280	4429	297	13.1780	4479	451	13.2280
4330	303	13.0790	4380	80	13.1290	4430	704	13.1790	4480	557	13.2290
4331	134	13.0800	4381	78	13.1300	4431	164	13.1800	4481	395	13.2300
4332	196	13.0810	4382	1000	13.1310	4432	144	13.1810	4482	954	13.2310
4333	114	13.0820	4383	175	13.1320	4433	573	13.1820	4483	416	13.2320
4334	1000	13.0830	4384	1000	13.1330	4434	452	13.1830	4484	1000	13.2330
4335	1000	13.0840	4385	130	13.1340	4435	418	13.1840	4485	1000	13.2340
4336	1000	13.0850	4386	251	13.1350	4436	189	13.1850	4486	1000	13.2350
4337	704	13.0860	4387	814	13.1360	4437	1000	13.1860	4487	416	13.2360
4338	918	13.0870	4388	657	13.1370	4438	820	13.1870	4488	1000	13.2370
4339	1000	13.0880	4389	680	13.1380	4439	187	13.1880	4489	1000	13.2380
4340	498	13.0890	4390	325	13.1390	4440	248	13.1890	4490	891	13.2390
4341	1000	13.0900	4391	564	13.1400	4441	1000	13.1900	4491	1000	13.2400
4342	1000	13.0910	4392	1000	13.1410	4442	1000	13.1910	4492	1000	13.2410
4343	1000	13.0920	4393	1000	13.1420	4443	885	13.1920	4493	1000	13.2420
4344	1000	13.0930	4394	568	13.1430	4444	183	13.1930	4494	1000	13.2430
4345	1000	13.0940	4395	588	13.1440	4445	268	13.1940	4495	471	13.2440
4346	1000	13.0950	4396	587	13.1450	4446	532	13.1950	4496	506	13.2450
4347	359	13.0960	4397	316	13.1460	4447	817	13.1960	4497	625	13.2460
4348	197	13.0970	4398	1000	13.1470	4448	758	13.1970	4498	1000	13.2470
4349	563	13.0980	4399	194	13.1480	4449	292	13.1980	4499	1000	13.2480
4350	421	13.0990	4400	194	13.1490	4450	215	13.1990	4500	998	13.2490

TABLE 4. Number of completed turns in the emittance interval (13.250, 13.449) Pi.mm.mrad. Direction # 1.

Part. No.	Turn No. Completed	EPSX	Part. No.	Turn No. Completed	EPSX	Part. No.	Turn No. Completed	EPSX	Part. No.	Turn No. Completed	EPSX
4501	295	13.2500	4551	1000	13.3000	4601	1000	13.3500	4651	1000	13.4000
4502	304	13.2510	4552	1000	13.3010	4602	1000	13.3510	4652	1000	13.4010
4503	1000	13.2520	4553	1000	13.3020	4603	1000	13.3520	4653	1000	13.4020
4504	1000	13.2530	4554	1000	13.3030	4604	1000	13.3530	4654	1000	13.4030
4505	1000	13.2540	4555	1000	13.3040	4605	1000	13.3540	4655	1000	13.4040
4506	1000	13.2550	4556	1000	13.3050	4606	1000	13.3550	4656	1000	13.4050
4507	321	13.2560	4557	1000	13.3060	4607	1000	13.3560	4657	1000	13.4060
4508	404	13.2570	4558	1000	13.3070	4608	1000	13.3570	4658	1000	13.4070
4509	781	13.2580	4559	1000	13.3080	4609	1000	13.3580	4659	1000	13.4080
4510	944	13.2590	4560	1000	13.3090	4610	1000	13.3590	4660	1000	13.4090
4511	1000	13.2600	4561	1000	13.3100	4611	1000	13.3600	4661	1000	13.4100
4512	966	13.2610	4562	1000	13.3110	4612	1000	13.3610	4662	1000	13.4110
4513	1000	13.2620	4563	1000	13.3120	4613	1000	13.3620	4663	1000	13.4120
4514	1000	13.2630	4564	1000	13.3130	4614	1000	13.3630	4664	1000	13.4130
4515	375	13.2640	4565	1000	13.3140	4615	1000	13.3640	4665	1000	13.4140
4516	774	13.2650	4566	1000	13.3150	4616	654	13.3650	4666	1000	13.4150
4517	1000	13.2660	4567	1000	13.3160	4617	802	13.3660	4667	1000	13.4160
4518	487	13.2670	4568	1000	13.3170	4618	1000	13.3670	4668	1000	13.4170
4519	1000	13.2680	4569	1000	13.3180	4619	1000	13.3680	4669	1000	13.4180
4520	1000	13.2690	4570	1000	13.3190	4620	1000	13.3690	4670	1000	13.4190
4521	1000	13.2700	4571	1000	13.3200	4621	1000	13.3700	4671	1000	13.4200
4522	1000	13.2710	4572	1000	13.3210	4622	1000	13.3710	4672	1000	13.4210
4523	1000	13.2720	4573	1000	13.3220	4623	1000	13.3720	4673	1000	13.4220
4524	898	13.2730	4574	1000	13.3230	4624	1000	13.3730	4674	1000	13.4230
4525	674	13.2740	4575	1000	13.3240	4625	1000	13.3740	4675	1000	13.4240
4526	1000	13.2750	4576	1000	13.3250	4626	1000	13.3750	4676	1000	13.4250
4527	1000	13.2760	4577	1000	13.3260	4627	720	13.3760	4677	1000	13.4260
4528	1000	13.2770	4578	1000	13.3270	4628	1000	13.3770	4678	1000	13.4270
4529	1000	13.2780	4579	1000	13.3280	4629	1000	13.3780	4679	1000	13.4280
4530	1000	13.2790	4580	1000	13.3290	4630	1000	13.3790	4680	1000	13.4290
4531	1000	13.2800	4581	1000	13.3300	4631	1000	13.3800	4681	1000	13.4300
4532	1000	13.2810	4582	1000	13.3310	4632	1000	13.3810	4682	1000	13.4310
4533	808	13.2820	4583	1000	13.3320	4633	1000	13.3820	4683	1000	13.4320
4534	1000	13.2830	4584	1000	13.3330	4634	1000	13.3830	4684	1000	13.4330
4535	1000	13.2840	4585	1000	13.3340	4635	1000	13.3840	4685	1000	13.4340
4536	1000	13.2850	4586	1000	13.3350	4636	1000	13.3850	4686	1000	13.4350
4537	1000	13.2860	4587	1000	13.3360	4637	1000	13.3860	4687	1000	13.4360
4538	1000	13.2870	4588	1000	13.3370	4638	1000	13.3870	4688	1000	13.4370
4539	1000	13.2880	4589	1000	13.3380	4639	1000	13.3880	4689	1000	13.4380
4540	1000	13.2890	4590	1000	13.3390	4640	1000	13.3890	4690	1000	13.4390
4541	1000	13.2900	4591	1000	13.3400	4641	1000	13.3900	4691	1000	13.4400
4542	1000	13.2910	4592	1000	13.3410	4642	1000	13.3910	4692	1000	13.4410
4543	1000	13.2920	4593	1000	13.3420	4643	1000	13.3920	4693	1000	13.4420
4544	1000	13.2930	4594	1000	13.3430	4644	1000	13.3930	4694	1000	13.4430
4545	1000	13.2940	4595	1000	13.3440	4645	1000	13.3940	4695	1000	13.4440
4546	923	13.2950	4596	1000	13.3450	4646	1000	13.3950	4696	1000	13.4450
4547	1000	13.2960	4597	1000	13.3460	4647	1000	13.3960	4697	1000	13.4460
4548	1000	13.2970	4598	1000	13.3470	4648	1000	13.3970	4698	1000	13.4470
4549	1000	13.2980	4599	1000	13.3480	4649	1000	13.3980	4699	1000	13.4480
4550	1000	13.2990	4600	1000	13.3490	4650	1000	13.3990	4700	1000	13.4490

TABLE 5. Number of completed turns in the emittance interval (13.450, 13.649) Pi.mm.mrad. Direction #1.

Part. No.	Turn No. Completed	EPSX	Part. No.	Turn No. Completed	EPSX	Part. No.	Turn No. Completed	EPSX	Part. No.	Turn No. Completed	EPSX
4701	1000	13.4500	4751	1000	13.5000	4801	1000	13.5500	4851	454	13.6000
4702	1000	13.4510	4752	1000	13.5010	4802	1000	13.5510	4852	1000	13.6010
4703	1000	13.4520	4753	1000	13.5020	4803	1000	13.5520	4853	1000	13.6020
4704	1000	13.4530	4754	1000	13.5030	4804	1000	13.5530	4854	928	13.6030
4705	1000	13.4540	4755	1000	13.5040	4805	1000	13.5540	4855	755	13.6040
4706	1000	13.4550	4756	1000	13.5050	4806	1000	13.5550	4856	1000	13.6050
4707	1000	13.4560	4757	1000	13.5060	4807	1000	13.5560	4857	1000	13.6060
4708	1000	13.4570	4758	1000	13.5070	4808	1000	13.5570	4858	457	13.6070
4709	1000	13.4580	4759	1000	13.5080	4809	1000	13.5580	4859	1000	13.6080
4710	1000	13.4590	4760	1000	13.5090	4810	1000	13.5590	4860	631	13.6090
4711	1000	13.4600	4761	1000	13.5100	4811	1000	13.5600	4861	374	13.6100
4712	1000	13.4610	4762	1000	13.5110	4812	1000	13.5610	4862	242	13.6110
4713	1000	13.4620	4763	1000	13.5120	4813	1000	13.5620	4863	396	13.6120
4714	1000	13.4630	4764	1000	13.5130	4814	1000	13.5630	4864	746	13.6130
4715	1000	13.4640	4765	881	13.5140	4815	1000	13.5640	4865	576	13.6140
4716	1000	13.4650	4766	1000	13.5150	4816	1000	13.5650	4866	1000	13.6150
4717	1000	13.4660	4767	1000	13.5160	4817	1000	13.5660	4867	678	13.6160
4718	1000	13.4670	4768	1000	13.5170	4818	1000	13.5670	4868	1000	13.6170
4719	1000	13.4680	4769	1000	13.5180	4819	1000	13.5680	4869	1000	13.6180
4720	1000	13.4690	4770	1000	13.5190	4820	1000	13.5690	4870	1000	13.6190
4721	1000	13.4700	4771	1000	13.5200	4821	1000	13.5700	4871	1000	13.6200
4722	1000	13.4710	4772	1000	13.5210	4822	1000	13.5710	4872	453	13.6210
4723	1000	13.4720	4773	1000	13.5220	4823	1000	13.5720	4873	583	13.6220
4724	1000	13.4730	4774	1000	13.5230	4824	1000	13.5730	4874	559	13.6230
4725	1000	13.4740	4775	1000	13.5240	4825	1000	13.5740	4875	404	13.6240
4726	1000	13.4750	4776	1000	13.5250	4826	1000	13.5750	4876	546	13.6250
4727	1000	13.4760	4777	1000	13.5260	4827	1000	13.5760	4877	174	13.6260
4728	1000	13.4770	4778	1000	13.5270	4828	1000	13.5770	4878	1000	13.6270
4729	1000	13.4780	4779	1000	13.5280	4829	1000	13.5780	4879	1000	13.6280
4730	875	13.4790	4780	1000	13.5290	4830	1000	13.5790	4880	1000	13.6290
4731	1000	13.4800	4781	1000	13.5300	4831	1000	13.5800	4881	237	13.6300
4732	893	13.4810	4782	1000	13.5310	4832	1000	13.5810	4882	1000	13.6310
4733	1000	13.4820	4783	1000	13.5320	4833	876	13.5820	4883	299	13.6320
4734	1000	13.4830	4784	1000	13.5330	4834	1000	13.5830	4884	1000	13.6330
4735	1000	13.4840	4785	1000	13.5340	4835	1000	13.5840	4885	1000	13.6340
4736	1000	13.4850	4786	1000	13.5350	4836	1000	13.5850	4886	425	13.6350
4737	1000	13.4860	4787	1000	13.5360	4837	1000	13.5860	4887	505	13.6360
4738	1000	13.4870	4788	1000	13.5370	4838	1000	13.5870	4888	245	13.6370
4739	1000	13.4880	4789	1000	13.5380	4839	1000	13.5880	4889	1000	13.6380
4740	1000	13.4890	4790	1000	13.5390	4840	1000	13.5890	4890	371	13.6390
4741	1000	13.4900	4791	1000	13.5400	4841	1000	13.5900	4891	1000	13.6400
4742	1000	13.4910	4792	1000	13.5410	4842	876	13.5910	4892	393	13.6410
4743	1000	13.4920	4793	1000	13.5420	4843	1000	13.5920	4893	367	13.6420
4744	1000	13.4930	4794	1000	13.5430	4844	1000	13.5930	4894	104	13.6430
4745	1000	13.4940	4795	1000	13.5440	4845	1000	13.5940	4895	1000	13.6440
4746	1000	13.4950	4796	1000	13.5450	4846	1000	13.5950	4896	289	13.6450
4747	560	13.4960	4797	1000	13.5460	4847	1000	13.5960	4897	215	13.6460
4748	560	13.4970	4798	1000	13.5470	4848	1000	13.5970	4898	153	13.6470
4749	1000	13.4980	4799	1000	13.5480	4849	1000	13.5980	4899	146	13.6480
4750	1000	13.4990	4800	1000	13.5490	4850	410	13.5990	4900	219	13.6490

TABLE 6. Number of completed turns in the emittance interval (13.650, 13.750) Pi.mm.mrad. Direction # 1.

Part. No.	Turn No. Completed	EPSX	Part. No.	Turn No. Completed	EPSX	Part. No.	Turn No. Completed	EPSX	Part. No.	Turn No. Completed	EPSX
4901	664	13.6500	4951	390	13.7000	5001	205	13.7500			
4902	1000	13.6510	4952	150	13.7010						
4903	1000	13.6520	4953	131	13.7020						
4904	244	13.6530	4954	165	13.7030						
4905	1000	13.6540	4955	510	13.7040						
4906	355	13.6550	4956	80	13.7050						
4907	1000	13.6560	4957	274	13.7060						
4908	132	13.6570	4958	172	13.7070						
4909	122	13.6580	4959	1000	13.7080						
4910	452	13.6590	4960	196	13.7090						
4911	775	13.6600	4961	184	13.7100						
4912	320	13.6610	4962	350	13.7110						
4913	1000	13.6620	4963	515	13.7120						
4914	436	13.6630	4964	545	13.7130						
4915	186	13.6640	4965	178	13.7140						
4916	276	13.6650	4966	159	13.7150						
4917	756	13.6660	4967	1000	13.7160						
4918	375	13.6670	4968	295	13.7170						
4919	1000	13.6680	4969	188	13.7180						
4920	218	13.6690	4970	379	13.7190						
4921	782	13.6700	4971	235	13.7200						
4922	486	13.6710	4972	1000	13.7210						
4923	424	13.6720	4973	282	13.7220						
4924	265	13.6730	4974	100	13.7230						
4925	676	13.6740	4975	172	13.7240						
4926	316	13.6750	4976	525	13.7250						
4927	102	13.6760	4977	580	13.7260						
4928	460	13.6770	4978	1000	13.7270						
4929	250	13.6780	4979	103	13.7280						
4930	834	13.6790	4980	162	13.7290						
4931	167	13.6800	4981	105	13.7300						
4932	334	13.6810	4982	72	13.7310						
4933	82	13.6820	4983	81	13.7320						
4934	540	13.6830	4984	93	13.7330						
4935	1000	13.6840	4985	82	13.7340						
4936	195	13.6850	4986	504	13.7350						
4937	1000	13.6860	4987	108	13.7360						
4938	278	13.6870	4988	232	13.7370						
4939	170	13.6880	4989	140	13.7380						
4940	1000	13.6890	4990	330	13.7390						
4941	168	13.6900	4991	401	13.7400						
4942	160	13.6910	4992	130	13.7410						
4943	73	13.6920	4993	282	13.7420						
4944	112	13.6930	4994	305	13.7430						
4945	186	13.6940	4995	81	13.7440						
4946	1000	13.6950	4996	51	13.7450						
4947	1000	13.6960	4997	55	13.7460						
4948	536	13.6970	4998	91	13.7470						
4949	117	13.6980	4999	81	13.7480						
4950	1000	13.6990	5000	255	13.7490						

TABLE 7. Invariants EH=(u**2+u'**2) & EV=(v**2+v'**2) vs. Number of Turns

Direction # 2	Epsx: 1.052000000	Epsy: 1.052000000
Turn #	EH(Turn #)	EV(Turn #)
(IC)	1.051999999999957	1.051999999999985
100	1.052000000219628	1.051999999884494
200	1.052000000438760	1.051999999767702
300	1.052000000658936	1.051999999651585
400	1.052000000878230	1.051999999536122
500	1.052000001098470	1.051999999421142
600	1.052000001319215	1.051999999306496
700	1.052000001538168	1.051999999191430
800	1.052000001758287	1.051999999075491
900	1.052000001978151	1.051999998959033
1000	1.052000002197452	1.051999998842440
1100	1.052000002417813	1.051999998726792
1200	1.052000002637897	1.051999998611514
1300	1.052000002857177	1.051999998497514
1400	1.052000003077268	1.051999998383742
1500	1.052000003297564	1.051999998270070
1600	1.052000003516198	1.051999998153910
1700	1.052000003736403	1.051999998037978
1800	1.052000003957389	1.051999997923453
1900	1.052000004176570	1.051999997808501
2000	1.052000004395886	1.051999997694764
2100	1.052000004615657	1.051999997580609
2200	1.052000004836231	1.051999997465565
2300	1.052000005057842	1.051999997349505
2400	1.052000005277577	1.051999997233168
2500	1.052000005498577	1.051999997117974
2600	1.052000005719783	1.051999997002490
2700	1.052000005940030	1.051999996887879
2800	1.052000006160455	1.051999996774526
2900	1.052000006380183	1.051999996660470
3000	1.052000006599428	1.051999996545540
3100	1.052000006820712	1.051999996429380
3200	1.052000007040050	1.051999996314265
3300	1.052000007260332	1.051999996199243
3400	1.052000007478327	1.051999996084845
3500	1.052000007699327	1.051999995970874
3600	1.052000007919638	1.051999995856143
3700	1.052000008140020	1.051999995740303
3800	1.052000008358377	1.051999995625053
3900	1.052000008577714	1.051999995509171
4000	1.052000008797087	1.051999995394695
4100	1.052000009018478	1.051999995278926
4200	1.052000009239578	1.051999995164387
4300	1.052000009459697	1.051999995049883
4400	1.052000009680043	1.051999994934071
4500	1.052000009899729	1.051999994819184
4600	1.052000010119635	1.051999994704282
4700	1.052000010339192	1.051999994589082
4800	1.052000010559837	1.051999994474350
4900	1.052000010781043	1.051999994360010
5000	1.052000011000437	1.051999994246025

TABLE 8. Invariants EH=(u**2+u'**2) & EV=(v**2+v'**2) vs. Number of Turns

Direction # 2   Epsx: 1.052000000   Epsy: 1.052000000

Turn #	EH(Turn #)	EV(Turn #)
(IC)	1.051999999999957	1.051999999999985
49500	1.052000108881757	1.051999942967129
99500	1.052000218854772	1.051999885366563
149500	1.052000328824143	1.051999827763709
199500	1.052000438773248	1.051999770133946
249500	1.052000548739919	1.051999712533004
299500	1.052000658707705	1.051999654943629
349500	1.052000768697667	1.051999597336930
399500	1.052000878679223	1.051999539747207
449500	1.052000988646270	1.051999482110766
499500	1.052001098608685	1.051999424535474
549500	1.052001208571397	1.051999366932961
599500	1.052001318562318	1.051999309306339
649500	1.052001428516220	1.051999251699669
699500	1.052001538493833	1.051999194090953
749500	1.052001648485870	1.051999136474464
799500	1.052001758436433	1.051999078853235
849500	1.052001868380316	1.051999021253039
899500	1.052001978350489	1.051998963637736
949500	1.052002088302366	1.051998906025780
999500	1.052002198268774	1.051998848435758
1000000	1.052002199370214	1.051998847858670

TABLE 9. Number of completed turns in the emittance interval (11.750, 13.730) Pi.mm.mrad. Direction # 1.

Part. No.	Turn No. Completed	EPSX	Part. No.	Turn No. Completed	EPSX	Part. No.	Turn No. Completed	EPSX	Part. No.	Turn No. Completed	EPSX
1	1000	11.7500	1	1000	12.7500	1	37568	11.7500	51	1000000	12.7500
2	1000	11.7700	2	1000	12.7700	2	290907	11.7700	52	1000000	12.7700
3	1000	11.7900	3	1000	12.7900	3	382607	11.7900	53	1000000	12.7900
4	1000	11.8100	4	1000	12.8100	4	662217	11.8100	54	1000000	12.8100
5	1000	11.8300	5	1000	12.8300	5	91464	11.8300	55	1000000	12.8300
6	1000	11.8500	6	1000	12.8500	6	67606	11.8500	56	1000000	12.8500
7	1000	11.8700	7	1000	12.8700	7	477270	11.8700	57	8818	12.8700
8	1000	11.8900	8	1000	12.8900	8	430425	11.8900	58	2021	12.8900
9	1000	11.9100	9	419	12.9100	9	36020	11.9100	59	419	12.9100
10	1000	11.9300	10	1000	12.9300	10	13838	11.9300	60	101337	12.9300
11	1000	11.9500	11	1000	12.9500	11	73706	11.9500	61	5919	12.9500
12	1000	11.9700	12	1000	12.9700	12	15416	11.9700	62	12310	12.9700
13	1000	11.9900	13	1000	12.9900	13	194063	11.9900	63	1677	12.9900
14	1000	12.0100	14	1000	13.0100	14	1000000	12.0100	64	2613	13.0100
15	1000	12.0300	15	1000	13.0300	15	339146	12.0300	65	1000000	13.0300
16	1000	12.0500	16	581	13.0500	16	463594	12.0500	66	581	13.0500
17	1000	12.0700	17	240	13.0700	17	113851	12.0700	67	240	13.0700
18	1000	12.0900	18	1000	13.0900	18	32797	12.0900	68	1283	13.0900
19	1000	12.1100	19	96	13.1100	19	110649	12.1100	69	96	13.1100
20	1000	12.1300	20	78	13.1300	20	367118	12.1300	70	78	13.1300
21	1000	12.1500	21	301	13.1500	21	19272	12.1500	71	301	13.1500
22	1000	12.1700	22	398	13.1700	22	157461	12.1700	72	398	13.1700
23	1000	12.1900	23	613	13.1900	23	49723	12.1900	73	613	13.1900
24	1000	12.2100	24	397	13.2100	24	6369	12.2100	74	397	13.2100
25	1000	12.2300	25	395	13.2300	25	10853	12.2300	75	395	13.2300
26	1000	12.2500	26	295	13.2500	26	40901	12.2500	76	295	13.2500
27	1000	12.2700	27	1000	13.2700	27	11271	12.2700	77	3680	13.2700
28	1000	12.2900	28	1000	13.2900	28	1000000	12.2900	78	2072	13.2900
29	1000	12.3100	29	1000	13.3100	29	1000000	12.3100	79	1678	13.3100
30	1000	12.3300	30	1000	13.3300	30	1000000	12.3300	80	2897	13.3300
31	1000	12.3500	31	1000	13.3500	31	1000000	12.3500	81	9893	13.3500
32	1000	12.3700	32	1000	13.3700	32	1000000	12.3700	82	1630	13.3700
33	1000	12.3900	33	1000	13.3900	33	1000000	12.3900	83	1000000	13.3900
34	1000	12.4100	34	1000	13.4100	34	8397	12.4100	84	1000000	13.4100
35	1000	12.4300	35	1000	13.4300	35	6415	12.4300	85	1000000	13.4300
36	1000	12.4500	36	1000	13.4500	36	17358	12.4500	86	1000000	13.4500
37	875	12.4700	37	1000	13.4700	37	875	12.4700	87	1000000	13.4700
38	1000	12.4900	38	1000	13.4900	38	1489	12.4900	88	4323	13.4900
39	1000	12.5100	39	1000	13.5100	39	3978	12.5100	89	1643	13.5100
40	1000	12.5300	40	1000	13.5300	40	29463	12.5300	90	106943	13.5300
41	1000	12.5500	41	1000	13.5500	41	2821	12.5500	91	1000000	13.5500
42	1000	12.5700	42	1000	13.5700	42	2198	12.5700	92	1000000	13.5700
43	1000	12.5900	43	1000	13.5900	43	1219	12.5900	93	1180	13.5900
44	1000	12.6100	44	374	13.6100	44	1924	12.6100	94	374	13.6100
45	1000	12.6300	45	237	13.6300	45	7116	12.6300	95	237	13.6300
46	1000	12.6500	46	664	13.6500	46	830119	12.6500	96	664	13.6500
47	1000	12.6700	47	782	13.6700	47	1000000	12.6700	97	782	13.6700
48	1000	12.6900	48	168	13.6900	48	1000000	12.6900	98	168	13.6900
49	1000	12.7100	49	184	13.7100	49	1000000	12.7100	99	184	13.7100
50	1000	12.7300	50	105	13.7300	50	1000000	12.7300	100	105	13.7300

206 Tracking Experience with PATRIS at BNL

TABLE 10. Number of completed turns in the emittance interval (9.750, 13.730) Pi.mm.mrad. Direction # 1.

Part. No.	Turn No. Completed	EPSX	Part. No.	Turn No. Completed	EPSX	Part. No.	Turn No. Completed	EPSX	Part. No.	Turn No. Completed	EPSX
51	1000	9.7500	101	1000	10.7500	151	1000	11.7500	201	1000	12.7500
52	1000	9.7700	102	1000	10.7700	152	1000	11.7700	202	1000	12.7700
53	1000	9.7900	103	1000	10.7900	153	1000	11.7900	203	1000	12.7900
54	1000	9.8100	104	1000	10.8100	154	1000	11.8100	204	1000	12.8100
55	1000	9.8300	105	1000	10.8300	155	1000	11.8300	205	1000	12.8300
56	1000	9.8500	106	1000	10.8500	156	1000	11.8500	206	1000	12.8500
57	1000	9.8700	107	1000	10.8700	157	1000	11.8700	207	1000	12.8700
58	1000	9.8900	108	1000	10.8900	158	1000	11.8900	208	1000	12.8900
59	1000	9.9100	109	1000	10.9100	159	1000	11.9100	209	419	12.9100
60	1000	9.9300	110	1000	10.9300	160	1000	11.9300	210	1000	12.9300
61	1000	9.9500	111	1000	10.9500	161	1000	11.9500	211	1000	12.9500
62	1000	9.9700	112	1000	10.9700	162	1000	11.9700	212	1000	12.9700
63	1000	9.9900	113	1000	10.9900	163	1000	11.9900	213	1000	12.9900
64	1000	10.0100	114	1000	11.0100	164	1000	12.0100	214	1000	13.0100
65	1000	10.0300	115	1000	11.0300	165	1000	12.0300	215	1000	13.0300
66	1000	10.0500	116	1000	11.0500	166	1000	12.0500	216	581	13.0500
67	1000	10.0700	117	1000	11.0700	167	1000	12.0700	217	240	13.0700
68	1000	10.0900	118	1000	11.0900	168	1000	12.0900	218	1000	13.0900
69	1000	10.1100	119	1000	11.1100	169	1000	12.1100	219	96	13.1100
70	1000	10.1300	120	1000	11.1300	170	1000	12.1300	220	78	13.1300
71	1000	10.1500	121	1000	11.1500	171	1000	12.1500	221	301	13.1500
72	1000	10.1700	122	1000	11.1700	172	1000	12.1700	222	398	13.1700
73	1000	10.1900	123	1000	11.1900	173	1000	12.1900	223	613	13.1900
74	1000	10.2100	124	1000	11.2100	174	1000	12.2100	224	397	13.2100
75	1000	10.2300	125	1000	11.2300	175	1000	12.2300	225	395	13.2300
76	1000	10.2500	126	1000	11.2500	176	1000	12.2500	226	295	13.2500
77	1000	10.2700	127	1000	11.2700	177	1000	12.2700	227	1000	13.2700
78	1000	10.2900	128	1000	11.2900	178	1000	12.2900	228	1000	13.2900
79	1000	10.3100	129	1000	11.3100	179	1000	12.3100	229	1000	13.3100
80	1000	10.3300	130	1000	11.3300	180	1000	12.3300	230	1000	13.3300
81	1000	10.3500	131	1000	11.3500	181	1000	12.3500	231	1000	13.3500
82	1000	10.3700	132	1000	11.3700	182	1000	12.3700	232	1000	13.3700
83	1000	10.3900	133	1000	11.3900	183	1000	12.3900	233	1000	13.3900
84	1000	10.4100	134	1000	11.4100	184	1000	12.4100	234	1000	13.4100
85	1000	10.4300	135	1000	11.4300	185	1000	12.4300	235	1000	13.4300
86	1000	10.4500	136	1000	11.4500	186	1000	12.4500	236	1000	13.4500
87	1000	10.4700	137	1000	11.4700	187	875	12.4700	237	1000	13.4700
88	1000	10.4900	138	1000	11.4900	188	1000	12.4900	238	1000	13.4900
89	1000	10.5100	139	1000	11.5100	189	1000	12.5100	239	1000	13.5100
90	1000	10.5300	140	1000	11.5300	190	1000	12.5300	240	1000	13.5300
91	1000	10.5500	141	1000	11.5500	191	1000	12.5500	241	1000	13.5500
92	1000	10.5700	142	1000	11.5700	192	1000	12.5700	242	1000	13.5700
93	1000	10.5900	143	1000	11.5900	193	1000	12.5900	243	1000	13.5900
94	1000	10.6100	144	1000	11.6100	194	1000	12.6100	244	374	13.6100
95	1000	10.6300	145	1000	11.6300	195	1000	12.6300	245	237	13.6300
96	1000	10.6500	146	1000	11.6500	196	1000	12.6500	246	664	13.6500
97	1000	10.6700	147	1000	11.6700	197	1000	12.6700	247	782	13.6700
98	1000	10.6900	148	1000	11.6900	198	1000	12.6900	248	168	13.6900
99	1000	10.7100	149	1000	11.7100	199	1000	12.7100	249	184	13.7100
100	1000	10.7300	150	1000	11.7300	200	1000	12.7300	250	105	13.7300

TABLE 11. Number of completed turns in the emittance interval (9.750, 13.730) Pi.mm.mrad. Direction # 1.

Part. No.	Turn No. Completed	EPSX	Part. No.	Turn No. Completed	EPSX	Part. No.	Turn No. Completed	EPSX	Part. No.	Turn No. Completed	EPSX
51	10000000	9.7500	101	10000000	10.7500	151	37568	11.7500	201	10000000	12.7500
52	10000000	9.7700	102	10000000	10.7700	152	290907	11.7700	202	10000000	12.7700
53	10000000	9.7900	103	10000000	10.7900	153	382607	11.7900	203	10000000	12.7900
54	10000000	9.8100	104	10000000	10.8100	154	662217	11.8100	204	10000000	12.8100
55	10000000	9.8300	105	10000000	10.8300	155	446683	11.8300	205	10000000	12.8300
56	10000000	9.8500	106	10000000	10.8500	156	67606	11.8500	206	10000000	12.8500
57	10000000	9.8700	107	747728	10.8700	157	477270	11.8700	207	8818	12.8700
58	10000000	9.8900	108	940747	10.8900	158	430425	11.8900	208	2021	12.8900
59	10000000	9.9100	109	632786	10.9100	159	36020	11.9100	209	419	12.9100
60	10000000	9.9300	110	721092	10.9300	160	13838	11.9300	210	101337	12.9300
61	10000000	9.9500	111	545183	10.9500	161	73706	11.9500	211	5919	12.9500
62	10000000	9.9700	112	1012562	10.9700	162	15416	11.9700	212	12310	12.9700
63	10000000	9.9900	113	1631016	10.9900	163	194063	11.9900	213	1677	12.9900
64	10000000	10.0100	114	510895	11.0100	164	2881408	12.0100	214	2613	13.0100
65	10000000	10.0300	115	786077	11.0300	165	339146	12.0300	215	10000000	13.0300
66	10000000	10.0500	116	150013	11.0500	166	463594	12.0500	216	581	13.0500
67	10000000	10.0700	117	179461	11.0700	167	113851	12.0700	217	240	13.0700
68	10000000	10.0900	118	10000000	11.0900	168	32797	12.0900	218	1283	13.0900
69	10000000	10.1100	119	344702	11.1100	169	110649	12.1100	219	96	13.1100
70	10000000	10.1300	120	632501	11.1300	170	367118	12.1300	220	78	13.1300
71	10000000	10.1500	121	660123	11.1500	171	19272	12.1500	221	301	13.1500
72	10000000	10.1700	122	1089453	11.1700	172	157461	12.1700	222	398	13.1700
73	10000000	10.1900	123	1183279	11.1900	173	49723	12.1900	223	613	13.1900
74	10000000	10.2100	124	516210	11.2100	174	6369	12.2100	224	397	13.2100
75	10000000	10.2300	125	270377	11.2300	175	10853	12.2300	225	395	13.2300
76	10000000	10.2500	126	922212	11.2500	176	40901	12.2500	226	295	13.2500
77	10000000	10.2700	127	122583	11.2700	177	11271	12.2700	227	3680	13.2700
78	10000000	10.2900	128	219116	11.2900	178	10000000	12.2900	228	2072	13.2900
79	10000000	10.3100	129	265061	11.3100	179	10000000	12.3100	229	1678	13.3100
80	10000000	10.3300	130	426372	11.3300	180	10000000	12.3300	230	2897	13.3300
81	10000000	10.3500	131	1138604	11.3500	181	29463	12.3500	231	9893	13.3500
82	10000000	10.3700	132	1117658	11.3700	182	10000000	12.3700	232	1630	13.3700
83	10000000	10.3900	133	806675	11.3900	183	2821	12.3900	233	10000000	13.3900
84	10000000	10.4100	134	2015107	11.4100	184	2198	12.4100	234	10000000	13.4100
85	10000000	10.4300	135	2204965	11.4300	185	8397	12.4300	235	10000000	13.4300
86	10000000	10.4500	136	314954	11.4500	186	6415	12.4500	236	10000000	13.4500
87	10000000	10.4700	137	826148	11.4700	187	3660	12.4700	237	10000000	13.4700
88	10000000	10.4900	138	665675	11.4900	188	875	12.4900	238	4323	13.4900
89	10000000	10.5100	139	953619	11.5100	189	1489	12.5100	239	1643	13.5100
90	10000000	10.5300	140	33497	11.5300	190	3978	12.5300	240	106943	13.5300
91	10000000	10.5500	141	53599	11.5500	191	1219	12.5500	241	2821	13.5500
92	10000000	10.5700	142	682700	11.5700	192	2198	12.5700	242	10000000	13.5700
93	10000000	10.5900	143	480733	11.5900	193	1924	12.5900	243	1180	13.5900
94	10000000	10.6100	144	10000000	11.6100	194	7116	12.6100	244	374	13.6100
95	10000000	10.6300	145	132346	11.6300	195	830119	12.6300	245	237	13.6300
96	10000000	10.6500	146	319832	11.6500	196	10000000	12.6500	246	664	13.6500
97	10000000	10.6700	147	117664	11.6700	197	10000000	12.6700	247	782	13.6700
98	10000000	10.6900	148	602083	11.6900	198	10000000	12.6900	248	168	13.6900
99	10000000	10.7100	149	628656	11.7100	199	10000000	12.7100	249	184	13.7100
100	10000000	10.7300	150	628656	11.7300	200	10000000	12.7300	250	105	13.7300

Study of Stability of Beam in the Fermilab Main Injector

C.S. Mishra and F. Harfoush
Fermi National Accelerator Laboratory, Batavia, IL 60510

ABSTRACT

The Fermilab Main Injector is a new 150-GeV proton synchrotron, designed to replace the Main Ring and improve the high energy physics potential of Fermilab. The status of the Fermilab accelerator complex upgrade will be discussed. Study of the stability of the beam in the Main injector will be discussed. Detuning and corrector schemes to improve the dynamic aperture of the Main Injector will be presented. Tune modulation caused by octupolar detuning will be discussed.

1   Introduction

The Fermilab Tevatron is the highest energy proton-antiproton collider in the world today and it will remain so until either the Superconducting Super Collider (SSC) or the Large Hadron Collider (LHC) is operational. At present the Fermilab accelerator complex is in the middle of an upgrade plan, FERMILAB III. This upgrade plan will slowly increase the luminosity in the Tevatron by at least a factor of thirty. Major components of this upgrade plan are installation of electrostatic separators in the Tevatron, installation of low-β systems at the two collider detectors located at B0 and D0, antiproton source improvements, upgrading the linac energy from 200 MeV to 400 MeV, installation of cold compressors and fast kickers, and the construction of a new accelerator, the Fermilab Main Injector.

This upgrade plan is designed to extend the discovery potential of the U.S. high energy physics program. Some of the physics goals of Fermilab III are to discover and study the properties of the top quark, the last unobserved fundamental particle; to provide a factor of two increase in the mass scales characterizing possible extensions to the Standard Model; high rate b-quark hadron production and decay experiments, perhaps leading to the study of CP violation in b-quark hadron; and support of new initiatives in the neutral kaon physics and neutrino oscillation investigations.

2   Status of Upgrades

At present, Fermilab is under collider operation Run 1a. This run is underway after the successful installation of electrostatic separators in the Tevatron. These separators create helically separated orbits in the Tevatron and keep proton and antiproton bunches separated everywhere except at the two collider detectors. Significant improvements have been made to the antiproton source to increase the accumulation rate and reduce the emittance of a given stack size. During this run Fermilab has been making a record in antiproton stacking. The new low-β system is operational and has allowed the implementation of a second high luminosity interaction region. At present the Tevatron is running at a luminosity larger than 4.5×10^{30} and is slowly heading towards the goal of 6×10^{30}.

After this run, the linac upgrade will be completed by replacing the second half of the existing drift tube linac with a side coupled structure generating about 300 MeV in the same length. This increase in energy will improve the injection into the 8-GeV

Booster due to the reduction in space-charge forces. This will also increase the proton transverse beam densities and will benefit antiproton production by increasing the proton flux through the Main Ring. Before the Main Injector, cold compressors and fast kickers will be installed in the Tevatron to increase the beam energy from 900 GeV to 1000 GeV and the number of bunches from 6 to 36. These upgrades will provide at least a factor of two increase over current luminosity.

3 The Main Injector

The Fermilab Main Injector (FMI) is a new 150-GeV proton synchrotron designed to remove the limitations of the Main Ring in the delivery of high intensity proton and antiproton beams to the Tevatron and to increase the antiproton production rate. The Main Ring aperture (12π mm-mr) is about half the size of the current booster aperture. After the 400-MeV linac upgrade the booster aperture will increase to about 30π mm-mr. The FMI is designed to have a transverse aperture of 40π mm-mr. The FMI will increase the number of protons targeted for antiproton production from 5.0×10^{15}/hour to 1.2×10^{16}/hour and will be capable of efficiently accelerating antiprotons in larger stacks containing 2×10^{12} antiprotons for injection into the Tevatron collider. It will also increase the total number of protons that can be delivered to the Tevatron to 6×10^{13}, with proton bunches containing up to 3×10^{11} protons. The FMI will be capable of supporting a luminosity of 5×10^{31} in the existing collider. A new added feature due to FMI will be intense slow extracted beams, 3×10^{13} protons every 2.9 sec with 33% duty factor, for use in the studies of CP violation and rare kaon decays, and for experiments designed to search for neutrino oscillations. In a similar amount of running time with FMI, a state of the art kaon experiment will improve the upper limits of rare decays by two orders of magnitude.

The FMI will be constructed using newly designed conventional dipole magnets. The new dipole magnets are being built based on considerations of field quality, aperture, and reliability. The FMI lattice has two different types of cells, the normal FODO cells in the arcs and straight sections and the dispersion-suppressor FODO cells adjacent to the straight sections to reduce the dispersion to zero in the straight sections. There are eight straight sections, at present four are being used for beam transfer and one for radio-frequency (rf) cavities.

Two full-scale prototypes of the FMI dipoles have been built and have undergone extensive measurements of their field quality. [1] Measurements show that these magnets meet the design specifications and are well described by the computer models. In the FMI we have two different length dipole magnets, 6 m and 4 m. Several iterations of the dipole end design have been made to reduce the change in the effective length of the magnet and the sextupole component of the dipole ends. The change in the effective length introduces closed-orbit error. [2] We have also initiated a dipole power supply R&D program. The power supply and magnet systems are designed to allow a significant increase in the number of 120-GeV acceleration cycles for simultaneous operation of antiproton production and 120-GeV slow spill beam. Besides the newly constructed dipoles, FMI will use existing components including quadrupoles, and 18 rf systems from the Main Ring.

4 Accelerator Physics

Accelerator physics studies are underway to understand the FMI better and also to improve its performance. These studies include incorporation of magnetic measurements into tracking studies, transition crossing studies, impedance budgeting, beam-line design and study of slow extraction. Using the measurements of the two prototype dipoles, Main Ring dipoles and quadrupoles, and PE2D static field calculations we have made a database for the systematic and random errors of FMI magnets. [3] This database is used in simulations of the dynamical performance of the FMI [4] and other studies such as power supply requirements, corrector strength, etc.

Simulations to study the performance of the Main Injector at the injection energy of 8.9 GeV are described in this paper. We present a detailed study of the Main Injector lattice including the closed-orbit errors, betatron function errors, tune versus amplitude, and dynamic aperture. The tracking calculations include the magnetic field errors, both systematic and random, and misalignment errors. In this paper we will briefly describe these errors along with the tracking conditions. A detailed description of these errors and their calculation can be found in MI66.[1] A thin-element tracking program TEAPOT [5] has been used for these simulations.

4.1 Tracking Conditions and Errors

The Main Injector lattice has two different sized dipole magnets, their magnetic lengths are 6.096 and 4.064 m at 120 GeV. The magnetic length of these dipoles decreases with energy due to the saturation of ends, and at 8.9 GeV their length is 2.5 mm larger than the nominal at 120 GeV. This change in length introduces a nonzero dipole multipole at each end of the magnet, and is represented in TEAPOT by a horizontal kick given by

$$H_{kick} = (\Delta L/2L_{ref}) \times (2\pi/904/3) \text{ radian}$$

where ΔL is 2.5 mm. This additional bending of the particle is corrected by decreasing the dipole excitation, calculated by Eq. (14) of Ref. 3.

The ends of the magnet have different magnetic multipoles than the body of the magnet. Furthermore the two ends of the dipole are slightly different due to the presence or absence of nearby bus work, leading to the labels "BUS END" or "NO BUS END." For the tracking calculation the two ends and the body are treated as separate magnets. The end multipoles, both normal and skew, are calculated by using the method described in Section 2.2 of Ref. 3. The multipoles used for these calculations were calculated by using the measurements where the 80-in. rotating coil was placed 50 in. inside the magnet rather than 30 in. as described in Ref. 3. This change gives us more consistent results at all the energies. These values are also in better agreement with the fit to the flat coil data between −1 in. and +1 in. in x. Multipole error values quoted for the dipole ends in Table 1 are obtained by dividing the integrated multipole moments by eight (the length of a long dipole magnet, 240", divided by 30")so that their values can be directly compared with the dipole body multipole errors. The random errors of the body multipoles are calculated by using the measurements of the B2 dipoles at 210 A.

The values of the systematic and random errors of the quadrupoles are calculated using the Main Ring quadrupole measurements. A very limited number of measurements are available for MR quads. Normal multipoles are calculated by using the 195-A measurements, whereas the skew multipoles are calculated using the measurements at 1575 A. The variation of the octupole strength and random errors with current are small.

All skew quadrupole field errors are turned off, for the convenience of the simulation. By using a coupling compensation scheme, any linear coupling effects due to the presence of skew quadrupole can be removed.

Table 1 summarizes all of the multipoles as used in the input file to TEAPOT. Multipole field errors are quoted in units of 10^{-4} at a displacement of one inch.

Table 1 Magnetic errors used in the 8.9-GeV simulation

	Multipole Order	Normal $<b_n>$	Normal σb_n	Skew $<a_n>$	Skew σa_n
Dipole body	dipole	−4.68	10.0	--	--
	quadrupole	−0.13	0.45	--	--
	sextupole	0.43	0.61	−0.04	0.22
	8	0.09	0.13	0.00	0.41
	10	0.18	0.32	0.03	0.15
	12	−0.03	0.10	0.00	0.19
	14	−0.01	0.23	−0.05	0.08
Dipole end BUS	dipole	2.05	--	0.0	--
	quadrupole	0.03	--	--	--
	sextupole	0.92	--	0.03	--
	8	−0.02	--	0.02	--
	10	−0.09	--	0.04	--
	12	0.04	--	−0.03	--
	14	−0.07	--	0.00	--
Dipole end NO BUS	dipole	2.05	--	0.0	--
	quadrupole	0.03	--	--	--
	sextupole	0.99	--	−0.07	--
	8	−0.08	--	−0.02	--
	10	−0.11	--	−0.05	--
	12	−0.06	--	0.03	--
	14	−0.09	--	0.00	--
Recycled new Main Ring quadrupole	quadrupole	--	24.0	--	--
	sextupole	0.50	2.73	0.12	1.85
	8	5.85	1.02	−1.16	2.38
	10	−0.10	1.12	0.42	0.47
	12	−1.82	0.63	0.40	0.70
	14	0.21	0.64	−0.55	0.44
	16	1.41	0.64	--	--
	18	−0.03	0.12	0.14	0.16
	20	−0.80	0.06	0.02	0.07
Newly built Main Injector quads	quadrupole	--	24.0	--	--
	sextupole	--	2.73	--	--
	8	−0.39	1.02	--	--
	10	--	1.12	--	--
	12	−1.39	0.63	--	--
	14	--	0.64	--	--
	16	1.29	0.64	--	--
	18	--	0.12	--	--
	20	−0.73	0.06	--	--

The misalignment of all the magnetic elements and beam position monitors has been included in this calculation. The sigma of the alignment error with respect to the closed orbit is 0.25 mm in both horizontal and vertical directions. In addition dipole magnets have a roll angle of 0.5 mrad sigma.

Base tunes of $(Q_x, Q_y) = (26.425, 25.415)$ were used in all the simulations. This tune is different from (26.407, 25.409) which was used for the Main Injector calculations before. This change in tune was necessary to increase the dynamic aperture, with all magnetic and misalignment errors turned on, the presence of an rf, and chromaticity adjusted to −5,−5. In the lattice there are 18 rf cavities, each operating at V_{rf} = 0.0218 MV. The rf frequency is set to 54 MHz, corresponding to a harmonic number of 588.

4.2 Closed-Orbit Errors and Corrector Strength

In the Main Injector lattice there are 208 quadrupoles. Located inside these quadrupoles are the beam position monitors. The vertical and horizontal beam positions are measured at the focusing and defocusing quadrupoles respectively. The vertical and horizontal displacements of the particles are corrected by applying corresponding kicks just after these position monitors.

A typical uncorrected closed orbit in both the horizontal and vertical planes is shown in Fig. 1. The average RMS closed-orbit deviation before correction is 7.2 mm horizontal and 5.2 mm vertical for the selected seed. After three iterations of the orbit corrections the average RMS closed-orbit deviation is reduced to 0.12 mm (H) and 8×10^{-3} mm (V).

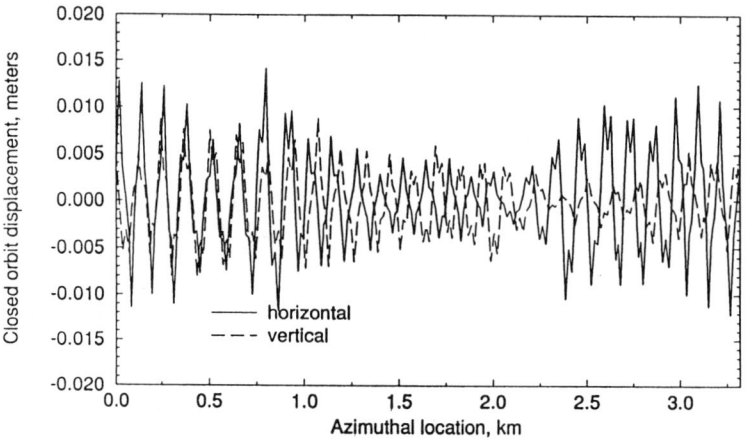

Fig. 1 Closed-orbit errors before correction.

We have studied the contribution of each magnetic error and the displacement error to the average RMS closed-orbit deviation for this seed. The total error is not a simple combination of all of these errors. There are some cancellations between errors. The result is summarzed in Table 2. Most of the orbit deviation is due to random errors. Fig. 2 shows the distribution of uncorrected horizontal and vertical RMS closed-orbit errors for 20 different seeds. The average RMS deviation of each seed is 7 mm and 6 mm in the horizontal and vertical planes respectively. The maximum corrector strength required to correct these orbit deviation is 150 μradians in both planes. In the Main Injector we plan to use recycled Main Ring dipole correctors and also newly built dipole correctors. At 8.9 GeV the Main Ring dipole correctors can provide 750 μradian and 350 μradian of horizontal and vertical corrections respectively. The new correctors will be stronger, which will help correct the orbit at higher energy.

Table 2 Closed-orbit errors for one seed

Errors	RMS deviation, mm	
	Horizontal	Vertical
All	6.3	4.3
Dipole systematic (including ΔL)	1.1	0.0
Dipole random	**5.1**	0
Quad systematic	0	0
Quad random	0	0
Displacement and rotation error	2.6	4.2
Change of effective length	1.1	0

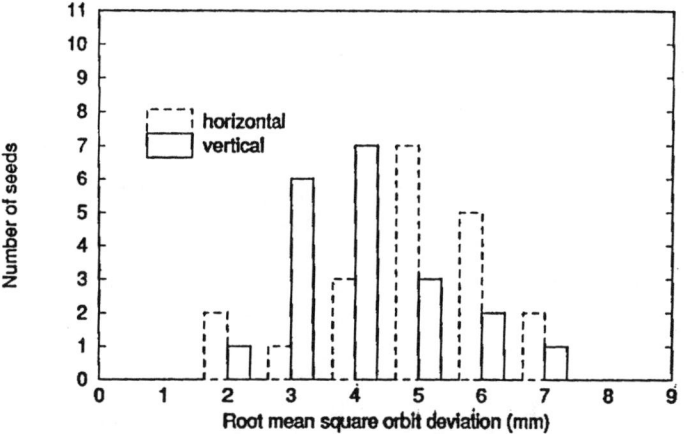

Fig. 2 Histogram of closed-orbit errors before correction.

4.3 Tune Versus Amplitude and Dynamical Aperture Results

We have studied the survival of particles launched at different amplitudes in the Main Injector at the injection energy. A single particle will go around 35,000 turns at the injection energy of 8.9 GeV during any operation that involves filling the ring with six Booster bunches. A particle is launched with a maximum horizontal displacement of A at a location where the horizontal beta function is at its maximum of 80 meters. The maximum vertical displacement of the same particle is $0.4A$ ($x/y = 2.5$) also at a beta of 80 meters. Synchrotron oscillation were included in the simulation by launching all particles with an amplitude of $\delta_{max} = (\Delta P/P)_{max} = 2.0 \times 10^{-3}$.

Figure 3 shows the variation of horizontal and vertical tunes as the amplitude of the motion was increased. The numbers on the tune plot correspond to the initial amplitude A of a test particle, in millimeters. Points on the plot lie on a straight line up to an amplitude of about 17 mm, with the spacing between points increasing linearly. Both the horizontal and vertical tunes depend quadratically on amplitude, for moderate amplitudes. This octupolar detuning is dominated by a combination of the systematic octupole error in the recycled Main Ring quadrupole and second-order sextupole effects.

Particles were launched from 1 mm to 25 mm amplitude. Particles with an amplitude above 19 mm did not survive for the full 35,000 turns of the simulation, for this particular seed. Similar simulations were performed for five different seeds. Fig. 4 is a survival plot, displaying how many turns a particle survives the 35k turns in the

214 Stability of Beam in the Fermilab Main Injector

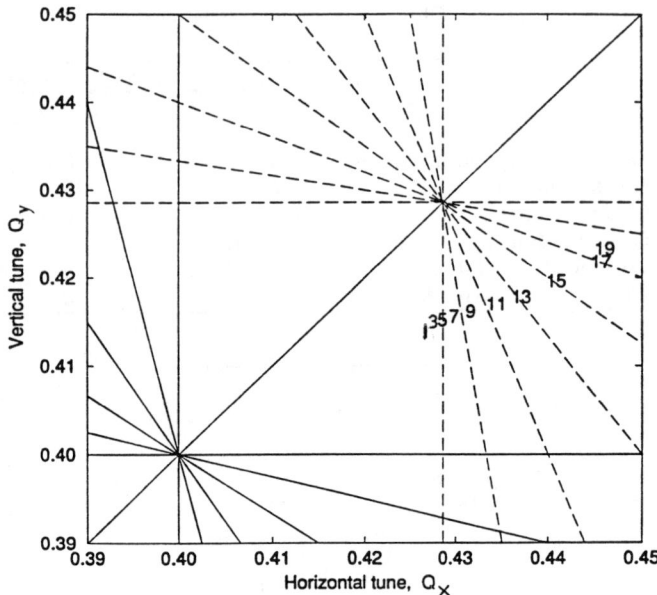

Fig. 3 Tune-tune plot. Numbers 1 to 19 on the plot indicate the initial amplitude of the launched particle.

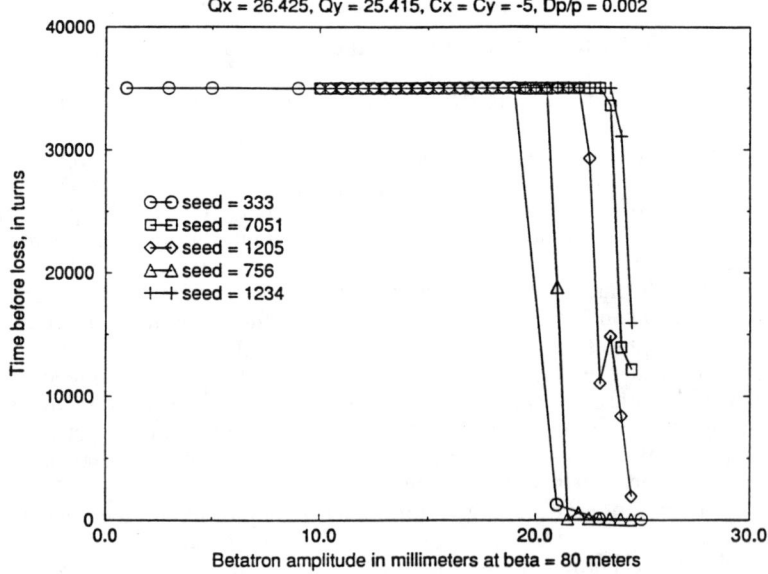

Fig. 4 Injection survival plot for five seeds.

Main Injector, as a function of initial amplitude. If the dynamical aperture of the machine is defined as the smallest amplitude particle that did not survive for 35,000 turns, then the dynamical aperture for the Main Injector at the injection energy is predicted to be 22 ± 1.4 mm, corresponding to a normalized emittance of 59.2 ± 10.2 πmm-mradian.

To study the detuning effects discussed above, we have varied the octupole strength of the Main Ring quadrupole and the sextupole strength of the Main Injector dipole ends. Reducing the end sextupole to half the nominal value has no significant effect on the quadratic detuning. Also there was no change to the dynamic aperture of the machine. When we set the octupole (b_3) component of the quadrupole to zero, the detuning is very small. Fig. 5 is a tune-tune plot for different initial amplitudes with half nominal sextupole strength of the dipole ends and $b_3 = 0$ for the MR quads. This study was done for only one seed. For this seed with nominal MR quads, octupole particles with amplitude larger than 19 mm did not survive. Particles with amplitude larger than 24 mm did not survive when we set the $b_3 = 0$ for the MR quads. We are in process of developing a correction scheme, using the octupole correctors placed in the ring for 120-GeV slow extraction, to cancel or reduce the total octupole of the ring at 8.9 GeV. This will help us improve the dynamic aperture of the MI. We can improve this further by utilizing a quadrupole shuffling scheme, which will help to reduce the effect of quadrupole random error. Study of the FMI dynamical performance at 120 GeV and simulation of slow extraction is in progress.

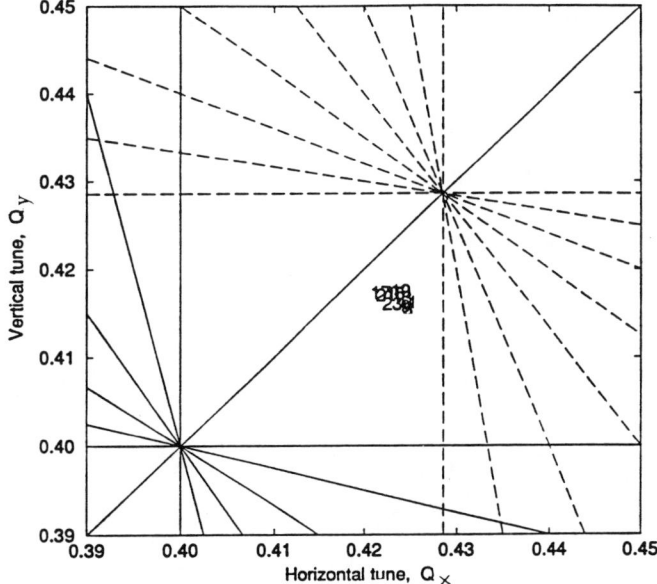

Fig. 5 Same as Fig. 3, but for half the nominal dipole end sextupole and no MR quads octupole.

The FMI has approval to begin the Title II work, below and above grade construction at MI-60 straight section, where rf will be located. We are requesting approvals for copper coil, steel lamination, and start of general site preparation. The scheduled completion date and commencement of operations is 1997.

References

1. D.J. Harding et al., Design considerations and prototype performance of the Fermilab Main Injector dipole, *Proc. 1991 IEEE Part. Accel. Conf.*
2. C.S. Mishra, H.D. Glass and F.A. Harfous, Effective length of the Main Injector Dipole and its effect on the Main Injector, *FMI Internal Report 0072*, 1992.
3. F.A. Harfoush and C.S. Mishra, Systematic and random errors for the Main Injector tracking, *FMI Internal Report 0066*, 1992.
4. C.S. Mishra and F.A. Harfoush, Simulation of the dynamical performance of the Main Injector at 8.9 GeV, *FMI Internal Report 0070*, 1992.
5. L. Schachinger and R. Talman, *Part. Accel.* 22, 35 (1987).

AN OPTIMAL PROCEDURE FOR MAGNET SORTING*

Rabinder Kumar Koul

APS, Argonne National Laboratory, Argonne, IL 60439

ABSTRACT

A new magnet sorting method for accelerators is developed. It is based on the linearized analysis of the effects of errors on accelerators. It is implementable in two steps. The first step is completely analytical in character while the second step involves the comparison of computed values with the measured error values. The whole process is repeated at most n times, where n is the number of magnets to be chosen from at a time. Simulations of the method, using Mathematica®, have been implemented for sorting the APS injector synchrotron dipoles and quadrupoles with excellent results.

1 Introduction

Magnetic elements are among the most important component of accelerators. The stability (both short-term and long-term) and other characteristics of the beam are crucially dependent upon the quality of the magnetic field in the magnets. It is only too well known that the designed magnetic fields are never precisely achieved in actual construction of the magnetic elements. Such deviations of actual magnetic fields from the designed magnetic field are called errors. In principle these errors can be either systematic or random in origin. We will treat random errors here, but the results could be used for systematic errors as well.

In the design of accelerators a significant portion of the effort is spent to control the effect of errors on the beam. One of the methods used for controlling the effects of the errors is to sort the magnets in terms of the measured errors and place them around the accelerator ring so that different errors tend to cancel each other.

---

* This manuscript has been authored under contract number W31-109-ENG-38 with the U.S. Department of Energy. Accordingly, the U.S. Government retains a non-exclusive, royalty-free license to publish or reproduce the published form of this contribution, or allow others to do so, for U.S. Government purposes.

2 The Definition of the Problem

Suppose we are given a set of magnets whose field error distribution has been measured. Then the question arises whether there is some optimal way of placing them around the accelerator ring (i.e relating the errors through phase of the particle) to minimize the effect of the field errors.

Of course the problem is not new and several methods have been employed in the past to find solutions to this problem. For example, one of the simplest ways is to try all possible permutations of the known errors on the magnet locations and choose the combination that produces the minimum effect. However, the number of combinations grows as $n!$, where n is the number of magnets to be sorted.

In the past a different method has been used which, for lack of better name, we will call the "Main Harmonic Cancellation Method." (For the APS injector synchrotron this method was suggested by Frederick Mills and implemented by Frederick Lopez[1] for the simulated errors. It is my understanding that a similar method was used for sorting at Fermilab.) The idea of this method is as follows. Let us say we want to sort dipole magnets on dipole errors. The magnets of equal strength are selected and placed π radians apart in phase of the particle trajectory so that the equal strength errors make equal but opposite contributions to the particle trajectory.

Sometimes the method of simulated annealing may be used for sorting. (This method was also used in simulation for sorting the injector synchrotron dipoles over simulated errors. The method gave a slight improvement over the previous method when used in simulation for the APS injector synchrotron, see Sampson.[2]) For details see Kirkpatrick[3] or Flannery.[4]

3 Optimal Procedure

The idea behind the optimal procedure is as follows. It is well known that the errors in a magnetic element are much smaller than the main field component. Hence their effects on the beam motion are given by the linearized equations of motion. For details see Koul[5] or Courant.[6] The effect of errors is characterized by some function of the error value, the phase of the particle trajectory at which the error occurs, and certain coefficients characterizing the motion of the particle at that phase. The function characterizing the error effect is often some positive definite invariant and will be called the amplitude function. The form of the function is

$$I = C \sum_{i=1}^{i=n} \sum_{j=1}^{j=n} H(\beta_i, L_i) H(\beta_j, L_j) f_i f_j \cos \psi_{ij}. \qquad (1)$$

$H(\beta_i, L_i)$, is some function of β_i and L_i, where β_i and L_i are the beta function of the beam at the ith element and the length of the ith element respectively; f_i is the measured error at the ith location; and $\psi_{ij} \approx (\phi_i - \phi_j)$ is the phase difference of the trajectory between the ith and the jth locations phases ϕ_i and ϕ_j.

The appropriate thing to do would be to minimize this amplitude function with respect to the errors, giving phases at which the errors occur. However the straightforward variation of the function with errors gives an obvious solution that the value of all the errors must vanish. In that case the amplitude function vanishes.†

The optimal procedure has two steps. In the first step a certain linear set of equations has to be solved, and in the second step some matching of the calculated and the measured values has to be performed. The first step can be carried out in either of two ways. Both ways are outlined below under the headings "Procedure 1" and "Alternate Procedure 1."

3.1 Step 1 (Procedure 1)

Since the straightforward minimization fails, let us ask a different question. Suppose that instead of minimizing the invariant amplitude function I with respect to the errors, we minimize it with respect to the phase differences ψ_{ij}, thereby considering the ψ_{ij} as variables. Once we obtain the equations from the first functional derivative of I with respect to ψ_{ij}, we ask another question: What values should f_i (i.e errors) assume, such that the given ψ_{ij} satisfy the variational equations? If we answer this question, then we have found that for the calculated values of f_i, given ψ_{ij} minimize the amplitude function I.

---

† (1) The vanishing of the amplitude function shows that the lower bound of I given in Eq. (1) is zero. (2) It may be pointed out at this stage that if some function characterizing the solution of the perturbed equations is not positive definite, then the above procedure will have to be modified appropriately. (3) However, the question of whether in the process one is finding the global minimum or the local minimum cannot be decided.

However, before we carry out this variation it is important to point out the following.

Note that for $i \neq j$, $\psi_{ij} = -\psi_{ji}$ and $\psi_{ii} = 0$. We are left with only $n(n-1)/2$ variables, but all of these $n(n-1)/2$, ψ_{ij} are not linearly independent. Since we are dealing with the phase differences, we can fix a position, say position number 1, and compute ψ_{1j} for all $j \neq 1$, out of the total of $n(n-1)/2$, ψ_{ij}. Then it is easy to see that for all i and j, $\psi_{ij} = \psi_{1j} - \psi_{1i}$. Hence for n positions corresponding to the n magnets, there are only $(n-1)$ linearly independent ψ_{1j}. Therefore, to minimize I with respect to the phase differences we will give variations only with respect to ψ_{1j}. Finding the first derivative of the I in Eq. (1), we get

$$\frac{\delta I}{\delta \psi_{1k}} = H(\beta_k, L_k) f_k \sum_{i=1}^{i=n} H(\beta_i, L_i) f_i \sin \psi_{ki} \qquad (2)$$

which is true for all k, $(k = 2, \ldots, n)$. Since $\delta I/\delta \psi_{1k} = 0$ we get

$$H(\beta_k, L_k) f_k \sum_{i=1}^{i=n} H(\beta_i, L_i) f_i \sin \psi_{ki} = 0. \qquad (3)$$

As in the case of Eq. (2), Eq. (3) is true for all k, where $k = (2, \ldots, n)$. If the set of f_k are non-zero, the above equation further simplifies to the following:

$$\sum_{i=1}^{i=n} H(\beta_i, L_i) f_i \sin \psi_{ki} = 0 \qquad (4)$$

for all k for which $f_k \neq 0$.

Hence in Eq. (3) we have $(n-1)$ equations with n unknowns. If we specify, for instance, the value of the first error from the measured errors, we have $(n-1)$ linear inhomogeneous equations in $(n-1)$ unknowns. If we pick the error at the first magnet, for example from the measured errors, then we can write the inhomogeneous equation in the following form:

$$\sum_{i=2}^{i=n} H(\beta_i, L_i) f_i \sin \psi_{ki} = H(\beta_1, L_1) f_1 \sin \psi_{1k}, \qquad (5)$$

which can be formally written as a matrix equation:

$$\sum_{i=2}^{i=n} M_{ki} f_i = H(\beta_1, L_1) f_1 \sin \psi_{1k}. \qquad (6)$$

Again, Eq. (5)† is true for all k, $k = (2,\ldots,n)$ and $M_{ki} = H(\beta_i, L_i)\sin\psi_{ki}$. It is important to remember that the calculated values $(f_2,\ldots,f_n)$ will be different from the actual measured values in the magnets. However we will delay the discussion of this point.

Note that in carrying out the above procedure we had to choose at least one of the measured errors to start with and calculate the remaining errors in order to cancel the effect of that particular error. This shows that if we want to give variations with respect to the errors in order to obtain a non-trivial solution to the sorting, based on a minimization procedure, we must assume that one or more of the errors are known and are different from zero. That is the basis of our "Alternate Procedure 1."

3.2 Step 1 (Alternate Procedure 1)

In view of the above comments, we will consider the f_i as variables and take the derivative of the amplitude function I with respect to them. Let us assume that we know the value of the first error. Then $\delta I/\delta f_k = 0$ is a set of linear equations whose solution, in terms of the known phases and the known errors, gives the set of unknown errors whose placement at the corresponding phases would minimize I. (In fact, for the calculated values, I takes its lower bound.) We get

$$\frac{\delta I}{\delta f_k} = 2cH(\beta_k, L_k)\sum_{j=1}^{j=n} H(\beta_j, L_j) f_j \cos\psi_{k,j} \qquad (7)$$

which is true for all k, $k = (2,\ldots,n)$. Again, since $H(\beta_k, L_k) \neq 0$ for any k, and $\delta I/\delta f_k = 0$ for all k, we get

$$\sum_{j=1}^{j=n} H(\beta_j, L_j) f_j \cos\psi_{k,j} = 0 \qquad (8)$$

---

† It may be pointed out that it is not necessary to specify only one of the unknowns to start. We can specify more than one measured value at different positions and solve for the remaining ones in terms of the known quantities. Even though we formally have $(n-1)$ linearly independent equations in Eq. (5), it may still be that a set of equations is not linearly independent by virtue of the coefficient of the matrix $M_{k,i}$ vanishing for certain values.

for all k, $k = (2, ..., n)$. These are again a set of $(n-1)$ inhomogeneous equations with $(n-1)$ unknowns. Since the error f_1 is known, we can write

$$\sum_{j=2}^{j=n} H(\beta_j, L_j) f_j \cos \psi_{k,j} = -H(\beta_1, L_1) f_1 \cos \psi_{k,1} \qquad (9)$$

for all k, $k = (2, \ldots, n)$. The above equations basically form the alternate first step in the two-step procedure.

Comments similar to ones made about Procedure 1 are valid for this alternate procedure. We will have more to say about it later; see general discussion.

3.3 Step 2

Having used one of the above procedures to solve the set of linear equations and having obtained the calculated values of the errors, at a given $\psi_{i,j}$, needed to cancel the effect of the known measured error, we ask how this information gets translated into the placement of the measured error values. It is noted at the outset that this question could have several different answers; however, the one which was adopted for the APS was the simplest one. We ordered the measured error values and compared the largest error value to the calculated values. It was decided that the measured error would be placed in the position of its closest calculated value. Starting from the largest absolute error value we worked our way through towards the smallest error†. In this way we associate the measured errors with a position in phase.

3.4 Implementation of the optimal procedure

Note that in the process of matching, most measured values will not match the calculated values. Consequently for every choice of measured errors, in terms of which we solve the linear set of equations for the remaining errors and finally choose the position of the remaining measured errors, the eventual

---

† Note that as we go from the largest value to the smaller values the number of choices for measured error placement keeps decreasing. In particular the last measured error has only one place to go to. For these reasons, this may not be an optimal way to relate the measured values to the calculated values. The only comment in favour of the procedure is that it worked extremely well for the simulations run with simulated random errors used for the APS injector synchrotron dipole and quadrupole errors.

cancellation of contributions to I will not be exact. The eventual value of the function I will be different from zero. Therefore, instead of carrying out the above procedure once by taking some particular measured error as f_1 in the first step, we carry out this procedure n times, each time picking up a different measured error as f_1. We then choose a solution, out of n different solutions, corresponding to n different choices of f_1, which gives the minimum value for the function I.

4 Some General Discussion of the Optimal Procedure and Its Implementation

As stated earlier the two procedures in step one are not identical. For example, from Eq. (7) it is easy to see that if the number of dipole magnets to be sorted is two (i.e $n = 2$) then the matrix in Eq. (7) is identically zero. It does not tell us anything about the possible placement of the two magnets. On the other hand Eq. (9) tells us that $f_2 = -H(\beta_1, L_1) f_1 \cos \psi_{2,1} / f_2 H(\beta_2, L_2)$ or $\cos \psi_{2,1} = f_2 H(\beta_2, L_2) / - H(\beta_1, L_1) f_1$. In particular if $f_1 = f_2$ (i.e two measured errors are of same strength), and $H(\beta_1, L_1) \sim H(\beta_2, L_1)$, then Eq. (9) says that the phase difference between the two magnets must be π radians, which is common wisdom.

The simulation of this optimal procedure was implemented at APS using Mathematica®. The Mathematica® routine "Lusolve" was used for solving the linear set of equations. During simulation the following points were noted. Whenever the determinant of the matrix was not very small, the two methods essentially gave the same result. However, if the determinant of the linear matrix was small then the two solutions for the calculated values were different. The difference of the two solutions was verified to be a member of the null space of both matrices.

The following bar diagrams and tables are the result of the computer simulation studies carried out for the dipole and quadrupole placement in the injector synchrotron. The error distributions used were simulated to be a Gaussian random normal distribution with three-sigma and two-sigma cut-off. The functional form of the $H(\beta_i, L_i)$, the form of the coefficient C, and the form of $\psi_{i,j}$ for the dipole errors in dipoles and the quadrupole errors in quadrupoles are as follows:

$$H(\beta_i, L_i) = (\beta_i)^{1/2} L_i, \qquad C = \frac{1}{2\sin \pi \nu} \qquad \psi_{i,j} = \nu(\phi_i - \phi_j), \qquad (10)$$

$$H(\beta_i, L_i) = \beta_i L_i, \qquad C = \frac{1}{2 \sin 2\pi \nu} \qquad \psi_{i,j} = 2\nu(\phi_i - \phi_j). \qquad (11)$$

An Optimal Procedure for Magnet Sorting

The injector synchrotron in the APS has 68 dipole magnets and 80 quadrupole magnets. The orbit parameters used in the computation were obtained from the "Mad" run of the perfectly designed machine. The bar chart in Fig. 1 is the sample error distribution with three-sigma cut-off used in the dipole sorting simulations, with dipole errors. The simulation for the dipole sorting, with dipole errors, was carried out by sorting different numbers of magnets at a time. The maximum number of dipoles sorted at a time was 68. Recall that for any given set of random dipole errors generated (which we will call a single machine), if we sorted all the 68 dipoles, there would be 68 possible solutions each corresponding to a possible choice for f_1. We ran 45 such machines for the dipole sorting, using all 68 magnets as a single batch. The next bar chart (see Fig. 2) shows the frequency with which the amplification factor I was improved in each machine, over the randomly placed dipoles, out of 68 possible solutions. For every machine sorted, in the mean, 93 percent of the solutions gave a smaller I than the $I_{(random)}$ obtained for randomly placed magnets. In the case of the quadrupole sorting we simulated 27 machines and sorted all 80 quadrupoles at the same time. Therefore, for simulation of a single machine we had 80 possible solutions for sorting. In this case, 95 percent of the solutions decreased the value of I over the I obtained for randomly placed magnets. This is shown in the bar graph of Fig. 3.

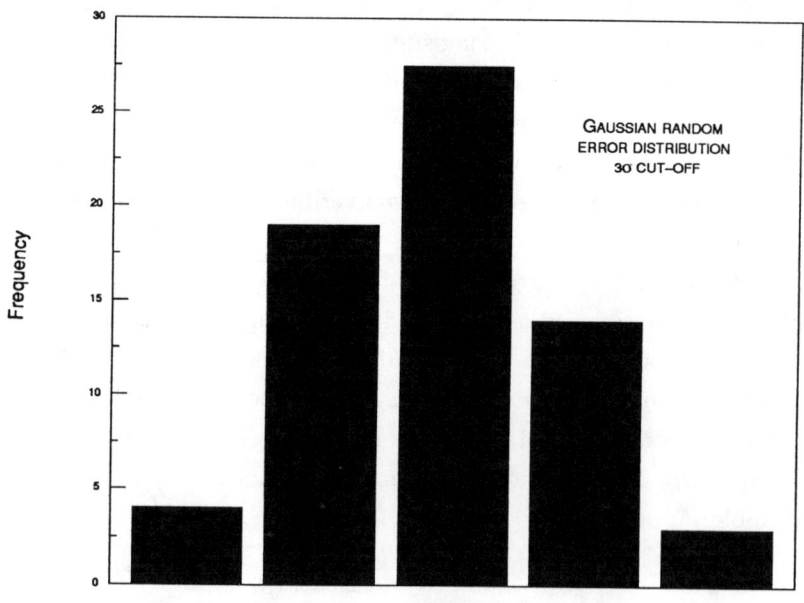

Fig. 1

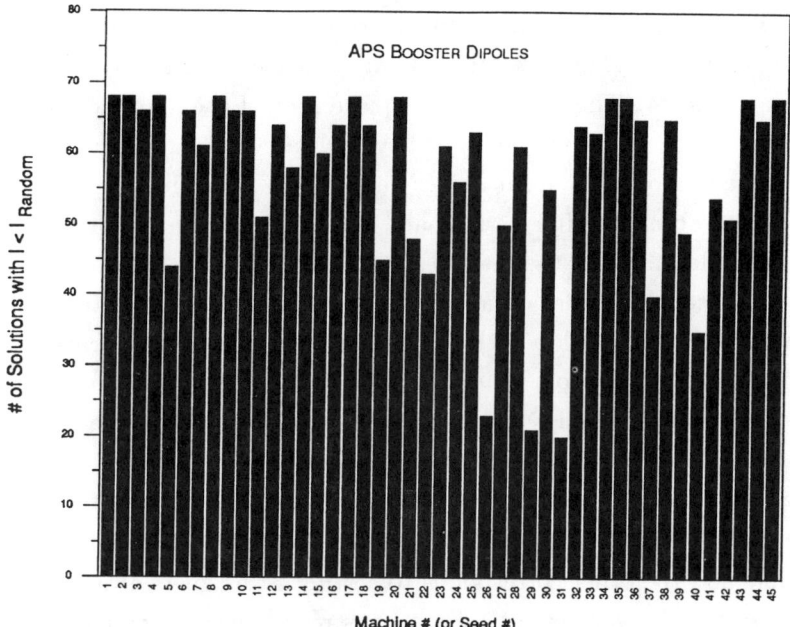

Fig. 2

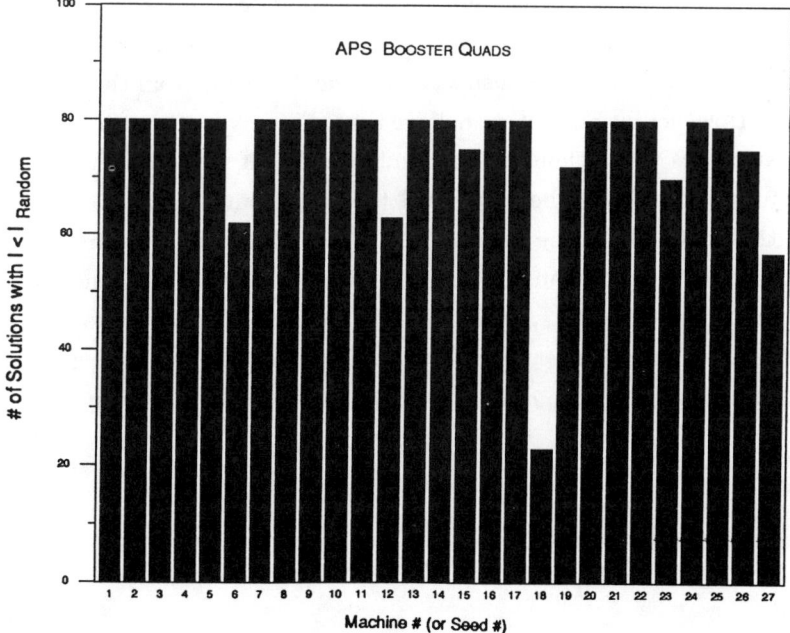

Fig. 3

As stated, we simulated 45 machines for the dipole errors in dipoles. The frequency distribution of the amplification factor for sorted and random machines is shown in Fig. 4. The x-axis is the square root of the amplification factor I, and the y-axis is the frequency with which a certain amplification factor was obtained among the 45 machines simulated. Both the minimum amplification factor obtained after sorting and the amplification factors obtained for random placement of the magnets are shown on the plot. One can see that the amplification factors obtained after sorting fall neatly into the first three bins and do not overlap with the distribution of the amplification factors obtained from the random placement of the magnets. The amplification factors for sorted machines also fall in a narrow region, whereas the amplification factors obtained for randomly placed machines have a wide distribution. Fig. 5 is a similar plot, but here both sets in the plot refer to the amplification factors obtained after sorting. One corresponds to sorting 68 dipoles at a time and the other corresponds to sorting 34 magnets at a time. Also note that sorting 34 magnets at a time gives results nearly comparable to sorting 68 magnets at a time.

How does the result of sorting depend on the number of magnets sorted at a time? The answer to this question is shown in Fig. 6, where $\sqrt{I_{mean}}$ is plotted against the number of magnets sorted at a time. One can easily see that the larger the number of magnets sorted at a time the lower the amplification factor obtained. There seems to be some deviation from this rule for $n = 12$ cases. However, at present we believe that this may be the reflection of the smaller statistics (even though this number is about two wavelengths in phase). Figure 6 also shows the mean and the minimum $\sqrt{I_{random}}$. One can easily see that not only is the mean $\sqrt{I_{random}}$ much larger than any of the sorted $\sqrt{I_{mean}}$, but the minimum $\sqrt{I_{random}}$ out of 45 random machines is much larger than the sorted $\sqrt{I_{mean}}$ as long as the number of magnets sorted is at least 18. Figure 7 shows the same result, but here the plot is for the gain factor, defined as $\sqrt{I_{random}/I_{sorted}}$, as a function of the number of dipoles sorted.

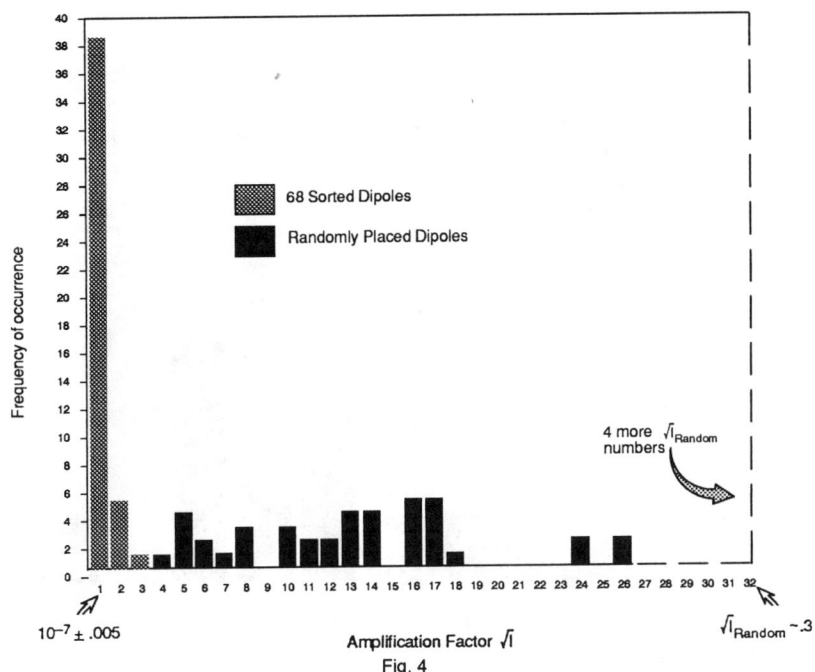

Fig. 4

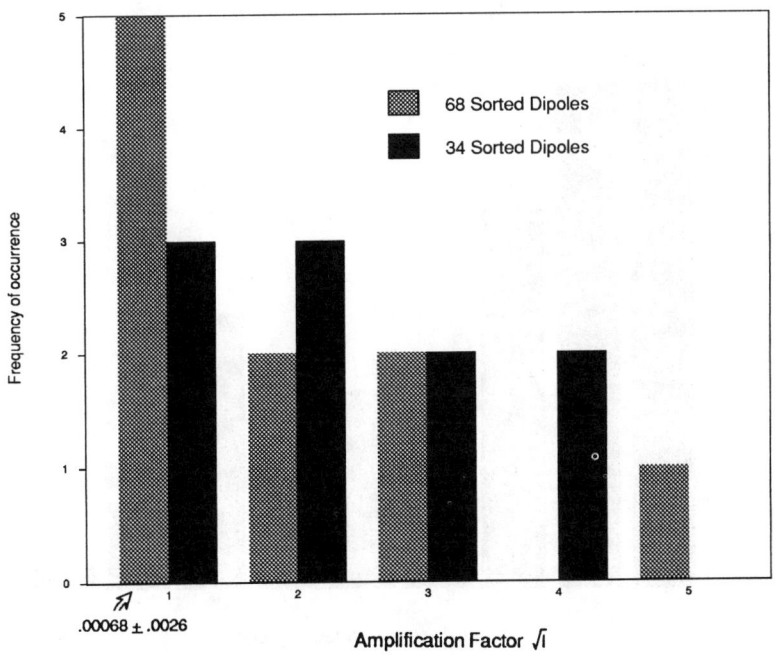

Fig. 5

An Optimal Procedure for Magnet Sorting

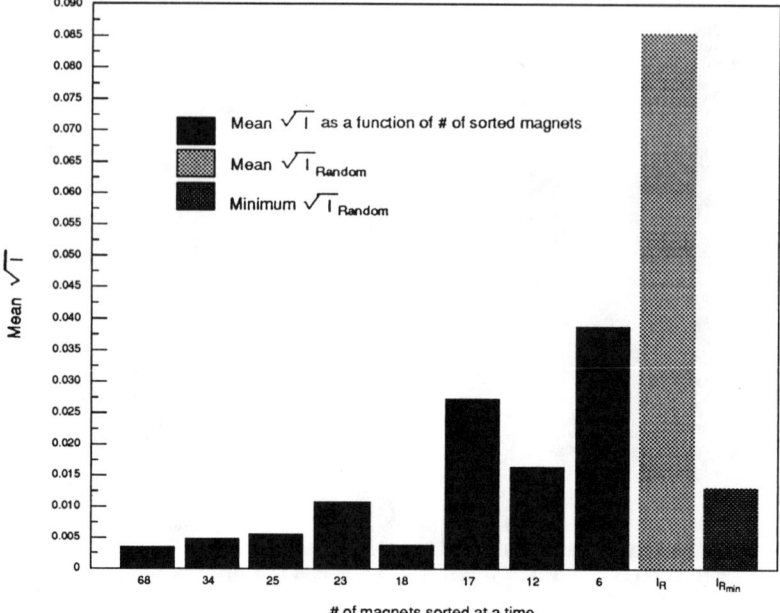

Fig. 6

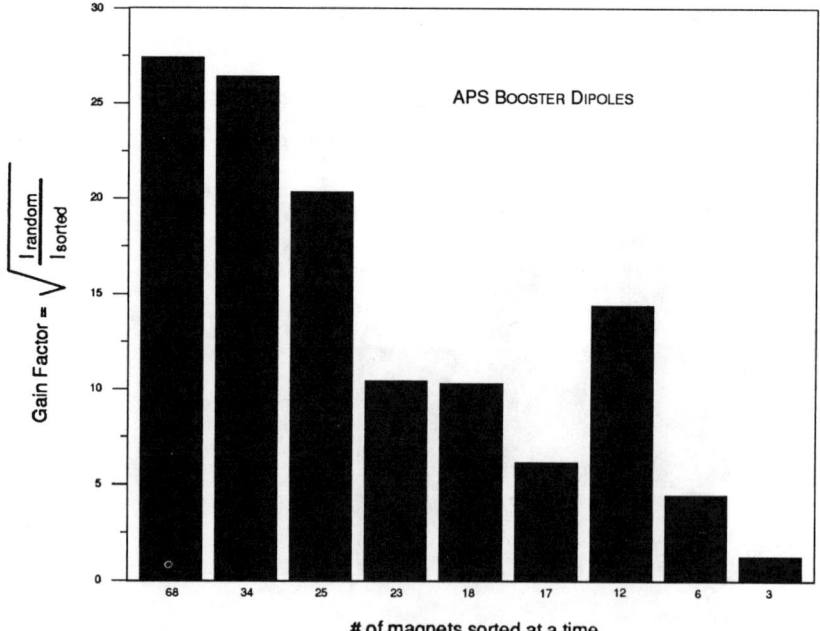

Fig. 7

Finally, two tables summarize the results of the optimal procedure (referred to as "Min. Method") and compare these results with those obtained from the two known methods described at the beginning of this paper. Table 1 shows the results for the APS injector synchrotron dipoles and Table 2 lists similar results for the APS injector synchrotron quadrupoles. These tables clearly indicate that the optimal procedure is an improvement over the known methods. At present the utility of this method is limited by the accuracy of the magnet measurements.

Table 1: SUMMARY TABLE — BOOSTER DIPOLES

No. of Dipoles at a Time	Mean Gain Factor Aneal.Meth.	Mean Gain Factor Reson.Meth	Mean Gain Factor Min. Meth.	Max Gain Factor Min. Meth.	Min Gain Factor Min Meth.	RMS Gain Factor Min. Meth.
68	4.96	3.82	27.33	84.4	4.5	34.98
34		2.63	26.44	80.05	3.8	36.17
25			20.34	57.19	4.63	27.78
23			10.48	23.68	2.60	13.62
18			10.32	22.95	4.32	12.21
17			8.19	13.63	2.90	8.23
12			14.45	55.65	2.65	23.99
06			4.48	9.59	1.65	5.70

Table 2: SUMMARY TABLE — BOOSTER QUADRUPOLES

Gain Factor $= \sqrt{I_{random} / I_{choice}}$

Number of Quadrupoles at a Time	Mean Gain Factor Min. Meth.	Max Gain Factor Min. Meth.	Min Gain Factor Min. Meth.
80	17.12	62.69	4.5
25			
20	15.29	52.2	4.09
14	12.73	27.82	2.49
10	41.39	172.04	6.54
3	3.24	3.46	3.03

References

1. Frederick Lopez, Private Communication.
2. S. Sampson, Reduction of Closed Orbit Distortion at the APS Synchrotron by Method of Simulated Annealing, Internal Summer Student Report, APS, Argonne, 1992.
3. S. Kirkpatrick, Journal of Statistical Physics 34, 975 (1984).
4. B. P. Flannery, W.H.Press, S.A.Teukolsky, and W.T.Vetterling, in Numerical Recipes in C: The Art of Scientific Computing, Cambridge University Press, Cambridge, 1989.
5. R. K. Koul and Frederick Mills, "Linearized Error Analysis for an Accelerator" (to be published).
6. E.D. Courant and H.S.Snyder, "Theory of Alternating Gradient Synchrotron," Ann. Phys. 3, 1 (1958).

Taylor Series Maps and Their Domain of Convergence*

Dan T. Abell and Alex J. Dragt
Department of Physics, University of Maryland
College Park, MD 20742-4111

1 Introduction

Researchers have found the concept of a transfer map useful in both theoretical and numerical studies of a wide variety of dynamical systems. In the field of accelerator physics, mapping techniques—from simple linear maps to more sophisticated nonlinear ones—have become essential tools in the analysis, characterization, and design of accelerators. While maps come in different flavors, and may be cooked according to different recipes, almost all start with the same basic ingredient: a Taylor series map, which represents the condition of a dynamical system after some definite interval as a Taylor series in the initial conditions. For example, the techniques based on Lie algebra attempt to find a set of Lie polynomials that generates a symplectic transformation that agrees with the Taylor series through some desired order. We would therefore like to understand thoroughly just what limits the validity of a Taylor series map, and how.

Talman [1] investigated the suitability of Taylor series maps (or differential maps) for tracking purposes. In particular, he used a modest-order Taylor series map with a time-step of 1.6π to track the motion of a simple pendulum and compared those results with the exact solution, which is known in terms of elliptic functions. His results showed that one can obtain accurate tracking results with a modest-order Taylor series map for only a limited number of iterations, and furthermore that the number of accurate iterations decreases with increasing amplitude.

In a subsequent report that attempted to explain Talman's results, Hagel and Zotter [2] examined the motion of the simple pendulum by expanding its solution as a Taylor series in the time variable. They pointed out—quite correctly—that the exact solution of the simple pendulum has poles in the complex time domain, and that therefore a solution based on a Taylor series in the time t will diverge for t greater than the distance to the nearest singularity in the plane of complex t. Since this distance is about 0.84π, nearly half the step-size used by Talman, it was suggested by Hagel and Zotter that the complex singularities in t were responsible for limiting the number of iterations for which Talman obtained accurate results.

---

*Work supported in part by the U. S. Department of Energy under Contract No. DEFG05-92ER40748.

While we may need to know the locations of the complex time singularities in order to address certain questions concerning the simple pendulum, such knowledge does not, in fact, directly answer the specific questions raised by Talman's data. The issue is more subtle. Because the state of the simple pendulum at any instant depends on both the initial conditions and the elapsed time, one may expand the current state either as a Taylor series in the initial conditions with the time real and held fixed, as Talman did, or as a Taylor series in the elapsed time with the initial conditions real and held fixed, as Hagel and Zotter did. When using the first method, with t real and held fixed, one discovers that singularities in the complex planes of the initial conditions limit the domain of convergence of the Taylor series. On the other hand, when using the latter method, with the initial conditions strictly real and held fixed, one discovers that a limited domain of convergence in t results from the singularities in the complex plane of t. We aim in this paper to make these points more clear.

In the next section we describe the concept of a transfer map and quote some theorems that justify not only their existence but also their advantages. Then in Sec. 3 we describe the Taylor series representation for transfer maps. Following that, we attempt in Secs. 4 and 5 to elucidate some of the basic theorems from the theory of functions of one and several complex variables. This material forms the core of our understanding of what limits the domain of convergence of Taylor series maps. In the last section we use the concrete example of a simple anharmonic oscillator to illustrate how the theorems from several complex variable theory affect the domain of convergence of Taylor series maps. There we describe the singularities of the anharmonic oscillator in the complex planes of the initial conditions, show how they constrain our use of a Taylor series map, and then discuss our findings.

2 The Concept of a Transfer Map

In the most general terms a *transfer map* is a function, or mapping, that reveals how a system changes during a fixed interval of time. To be more specific, we start by giving a mathematical description of the physical system of interest—or, more precisely, of an idealized image of the physical system. In particular, a large and very general class of systems can be modelled by finite sets of coupled first-order ordinary differential equations (ODEs) having the form

$$\dot{z} = f(t, z; c). \tag{1}$$

Here $z = (z_1, \ldots, z_m)$ is a vector describing the state of the system as a point in an abstract m-dimensional space called the *phase space* of the system; $c = (c_1, \ldots, c_k)$ is a vector of *control parameters*, generally held fixed, whose values affect the global behavior of the system; and the dot, as usual, denotes differentiation with respect to the time, or some *time-like* variable, t. Using this model, we can translate any question about the time-evolution of the physical system into a question about the motion of points in the corresponding phase space.

In a simple analysis we might start from some initial state z^i at time t^i and describe the time-evolution of the system as a path traced in phase space by the moving point $z(t)$. On the other hand we often wish to know how the system evolves not from a *single* initial condition, but rather from a *range* of initial conditions. We envision, for example, wanting to know the behavior of a distribution of particles, or wanting to know the sensitivity of a given trajectory to its initial state. To answer such questions, we can imagine using a small set of initial conditions as markers and then taking snapshots of the phase space at a sequence of equally spaced times. By observing our markers, we can determine how the motion transforms the phase space.

In essence a transfer map describes how a system changes between successive snapshots. Suppose z^i represents the initial state of our dynamical system at time t^i, and that we wish to know the final state z^f at some later time t^f. We may then express the transformation from the phase space at time t^i to the phase space at time t^f as

$$z^f = \mathcal{M}^{t^f \leftarrow t^i} z^i. \tag{2}$$

Here Eq. 2 describes the final state as the result of the transfer map $\mathcal{M}^{t^f \leftarrow t^i}$ acting on the initial state. In other words, $\mathcal{M}^{t^f \leftarrow t^i}$ maps any given point in phase space to a new point in phase space. If only the time interval $\tau = t^f - t^i$ is relevant, then we may write Eq. 2 in the simpler form

$$z^f = \mathcal{M}^\tau z^i. \tag{3}$$

While this form certainly applies to autonomous (*i.e.*, time-independent) systems, we want to emphasize that it also applies to systems for which f is periodic in t with period τ.

The form shown in Eq. 3, if it applies, is especially useful because the map $\mathcal{M}^\tau$ contains all of the dynamical information about the system. In the case of periodic systems, $\mathcal{M}^\tau$ may also depend on where t^i occurs within one period, but it remains true that $\mathcal{M}^\tau$ contains all of the dynamical information. For example, suppose that our map has a fixed point z^o, so that $z^o = \mathcal{M}^\tau z^o$. The existence of such a fixed point implies that the system has a periodic orbit. Then we can easily ascertain the stability of the system in the neighborhood of the periodic orbit by examining the linear part of the map $\mathcal{M}^\tau$. In particular, one may represent the linear part of the map about the point z^o by the *Jacobian matrix* M with elements

$$M_{ab} = \left.\frac{\partial z_a^f}{\partial z_b^i}\right|_{z^o}.$$

The eigenvalues of M reveal the nature of the system in the neighborhood of the periodic orbit. As another example, one may follow, or "track," the motion of the system simply by iterating the map.

We now ask the natural question: "Under what circumstances can we express a solution to Eq. 1 in the form given in Eq. 2?" The process described above—of extracting the essence of a physical system and exhibiting it in the form shown in Eq. 1—works so well that we often forget the extraordinary intellectual difficulties that hindered its conception. Since the remarkable birth of classical physics some three hundred years ago, mathematicians have learned and then taught us a great deal about differential equations. For our purposes here, their most important teachings comprise the famous existence and uniqueness theorems about solutions to Eq. 1. We may summarize these as follows(*e.g.*, see Brauer and Nohel [3]).

Theorem 2.1 *Consider any set of m first-order ordinary differential equations given in the form of Eq. 1 with c held fixed. Assume that f and its partial derivatives $\partial f_i/\partial z_j$ all exist and are continuous and bounded on some domain $\mathcal{D}$ in the $(m+1)$-dimensional space of (t, z). Then there exists a unique solution or integral curve of Eq. 1 passing through each point of $\mathcal{D}$. In other words, given any point (t^i, z^i) in $\mathcal{D}$, there exists a unique*

$$z(t) = \phi(t; t^i, z^i; c) \tag{4}$$

satisfying Eq. 1 and having the property

$$z(t^i) = \phi(t^i; t^i, z^i; c) = z^i. \tag{5}$$

The solution (4) can be extended both forward and backward in time as long as $(t, z) = (t, \phi(t))$ remains in $\mathcal{D}$. Furthermore, this solution is continuous in the variables t, t^i, and z^i.[1]

On writing the equations of motion (1) in terms of a time-derivative, we appear to give special attention to the time (or the time-like variable). Expressing the solution (4) as $z(t)$ further emphasizes the role time plays. But Thrm. 2.1 informs us that we may equally well view the solution (4) as a function of the initial conditions. When we adopt this point of view, we see that Thrm. 2.1 guarantees us that any system described by Eq. 1 and satisfying the appropriate conditions has a transfer map $\mathcal{M}^{t^f \leftarrow t^i}$, and furthermore that transfer map is uniquely defined.

With the existence of transfer maps established, we now address two additional questions: "How do the control parameters affect the map?" and "What can we say about the map if f is analytic?"

In any physical system that we model, there always exist parameters whose values we must measure. We then include these parameters by making them part of the vector of control parameters, c, in Eq. 1. But since errors affect all measurements, we would like some assurance that small differences between the

---

[1]Other versions of the existence-uniqueness theorem (*e.g.*, see Petrovski [4]) place less stringent requirements on the function f, but this simpler rendition meets our needs.

measured and true values of the c_j will not prove too harmful. This is the subject of

Theorem 2.2 *Suppose the partial derivatives of f with respect to the control parameters, $\partial f_i/\partial c_j$, all exist and are continuous and bounded on some domain of control-space. Then the solution (4) will be continuous in each of the parameters c_j* [3]*.*

In the language of transfer maps, Thrm. 2.2 assures us that sufficiently small variations in c produce only small changes in the transfer map $\mathcal{M}$. Therefore, if our model contains an accurate representation of the system, then we have a good chance of making successful predictions. The mathematician, however, offers little insight concerning the construction of good models, so we must always look to experiment for assurance about the quality of our models.

For a host of reasons, including computational convenience, we often choose to use analytic functions for the right-hand side of Eq. 1. Since the class of analytic functions has very special properties, we might expect the corresponding solution (4) to have similarly special properties. The following theorem addresses just this point.

Theorem 2.3 (Poincaré) *Suppose f in Eq. 1 is a set of analytic functions on some domain of the variables t, z, and c. Then the solution (4) will be analytic in the variables t, t^i, z^i, and c* [4]*.*

This theorem derives its value from the fundamental fact that one may express any analytic function as a convergent power series. (We shall say more about this later.) A common application of Thrm. 2.3 is to the computation of perturbative power series expansions. In this paper we shall take advantage of this theorem to express the final conditions z^f as a power series in the initial conditions z^i.

We close this section by commenting that satisfying the analyticity requirements of Thrm. 2.3 often (at least in physics) proves less onerous than it appears. For example, imagine that we wish to model the motion of a charged particle in a system of electric and magnetic fields. It can be shown that even though the charges and currents that generate the electric and magnetic fields may not be analytic distributions, the fields themselves are analytic in any region free of sources [5]. This remarkable fact derives from the mathematical properties of Maxwell's equations and allows us to apply Thrm. 2.3 to many models of charged- particle dynamics.

3 The Taylor Series Representation

We now confine our attention to systems covered by Thrm. 2.3. In other words, we consider systems described by an equation of motion having the form given in Eq. 1 for which the right-hand side, $f(t, z; c)$, is analytic in all of its variables on some domain. With these restrictions, as we shall see in Secs. 4 and 5, we may write the final conditions as a convergent power series (on some domain) expanded in the initial conditions. In terms of the individual components, such an expansion has the general form

$$z_a^f = K_a + \sum_b R_{ab} z_b^i + \sum_{bc} T_{abc} z_b^i z_c^i + \sum_{bcd} U_{abcd} z_b^i z_c^i z_d^i + \cdots. \quad (6)$$

We call this a *Taylor series map* because it conveys, in Taylor series form, exactly the same information contained in Eq. 2 (or, if it applies, Eq. 3): the Taylor series map acts on an initial point in phase space and returns a final point in phase space. The coefficients R_{ab} define the (first-order) transfer matrix associated with the map. The coefficients T_{abc}, U_{abcd}, etc. define generalized higher-order transfer matrices. In passing, we note that the Taylor series coefficients depend on the times t^i and t^f (or, for autonomous systems, on the time interval $\tau = t^f - t^i$). Furthermore, the coefficients may not be independent. In Hamiltonian systems, for example, the transfer matrix R must satisfy the symplectic condition [6], so the elements R_{ab} must be interrelated. In such systems similar restrictions apply to the higher-order coefficients.

The principal virtue of the representation in Eq. 6 for a mapping $\mathcal{M}$ is computational speed. After calculating a suitable number of coefficients (the hard part), we can compute z^f from z^i with great speed. Furthermore, we can apply the Taylor series map to any number of initial conditions—a valuable feature. We may, for example, want a very finely grained image of the phase-space transformation effected by some map. We can obtain this information by applying the corresponding Taylor series map once to a large number of closely spaced initial conditions. On the other hand, we might want to track just a few initial conditions through many cycles of a periodic system. In this example we can choose to represent one cycle either as a single map or as the composition of several maps. Whether we use one or several maps does not matter: we still recycle the code. Thus in both of these examples we can use the same map or maps repeatedly.

The principal drawback of the Taylor series representation is that for most systems Eq. 6 does not terminate. This unfortunate fact has a number of important consequences. To begin with, a *truncated* Taylor series map may differ significantly from the original map. For example, even if the original map corresponds to a conservative Hamiltonian system, the truncated Taylor series map generally violates both the symplectic condition and the conservation of energy. Carrying the series to higher order may alleviate the problem but does not cure the disease. For a system in an N-dimensional phase space, the order d terms in the Taylor series map require $N(N + d - 1)!/[d!(N - 1)!]$ coefficients. The use of a high-order Taylor series map therefore requires the computation and storage of a large number of coefficients. Furthermore, and most important, the series may not converge. As the title of our paper suggests, we shall focus our attention on illuminating the principal features of the domain of convergence: what does it look like, and what factors affect its size and shape?

As a simple example of a Taylor series map, consider the dynamical system described by the Hamiltonian

$$H = \frac{1}{2}(p^2 + q^2) - \frac{1}{4}q^4, \tag{7}$$

where q is the coördinate, and p is its conjugate momentum. This Hamiltonian represents a particle of unit mass in the potential of an anharmonic oscillator—a harmonic oscillator with a quartic "correction." As H is polynomial in q and p, the corresponding equations of motion,

$$\begin{aligned} \dot{q} &= \partial H/\partial p = p, \\ \dot{p} &= -\partial H/\partial q = -q + q^3, \end{aligned}$$

have everywhere analytic right-hand sides, and Thrm. 2.3 therefore applies. Using the techniques of Lie algebra [6] or automatic differentiation [7-9], we can compute the coefficients of the Taylor series map to very high order. Here we present the first few terms of the map expanded about the origin, $z = (q, p) = (0, 0)$, using a time-step $\tau = 7$:

$$\begin{aligned} q^f &= 0.7539\, q^i + 0.6570\, p^i + 1.765\, q^{i3} - 1.626\, q^{i2}\, p^i + 1.603\, q^i\, p^{i2} - 1.768\, p^{i3} + \cdots, \\ p^f &= -0.6570\, q^i + 0.7539\, p^i + 2.283\, q^{i3} + 2.091\, q^{i2}\, p^i + 2.052\, q^i\, p^{i2} + 1.603\, p^{i3} + \cdots. \end{aligned} \tag{8}$$

The coefficients of the linear terms turn out to be $\cos(\tau)$ and $\pm\sin(\tau)$ and arise from the quadratic terms in the Hamiltonian (7). The presence of the quartic term in H generates the infinite series of higher-order terms.

4 Theory of Functions of a Single Complex Variable

As a prelude to the next section, we briefly recall a couple of the salient definitions and results from the theory of functions of a single complex variable. We shall then recognize the theorems for functions in several variables as natural extensions of more well-known results.

By a *domain* we shall mean a non-empty open connected set in the complex plane. By *connected* we mean that one may join any two points using a curve that lies entirely within the domain. We call such a domain *simply connected* if any closed curve in the domain can be deformed to a point without it passing out of the domain. In other words, a simply connected domain contains no interior boundaries, or "holes." A function $w = f(z)$ is called *analytic* on a domain $\mathcal{D}$ if it possesses a derivative everywhere in $\mathcal{D}$. Then we have the following two important theorems. (For proofs of these theorems, consult any moderately complete text on complex analysis.)

Theorem 4.1 (Cauchy's Integral Formula) *Suppose $f(z)$ is analytic on a simply connected domain $\mathcal{D}$, and suppose also that Γ is a simple closed curve contained in $\mathcal{D}$. Then for any z_0 inside Γ*

$$f(z_0) = \frac{1}{2\pi i} \oint_\Gamma \frac{f(z)}{z - z_0} \, dz. \qquad (9)$$

Theorem 4.2 (Taylor Series) *Suppose $f(z)$ is analytic on a domain $\mathcal{D}$ containing the point z_0. Define R as the radius of the largest circle which is centered at z_0 and which has its interior entirely within $\mathcal{D}$. One may then write $f(z)$ as a power series in z:*

$$f(z) = \sum_{n=0}^{\infty} c_n (z - z_0)^n. \qquad (10)$$

This sum is the Taylor series for f about the point z_0, and it converges absolutely inside the circle $|z - z_0| < R$. Furthermore, the coefficients c_n are given by

$$c_n = \frac{1}{n!} f^{(n)}(z_0) = \frac{1}{2\pi i} \oint_\Gamma \frac{f(z)}{(z - z_0)^{n+1}} \, dz, \qquad (11)$$

where Γ has the same properties as in Thrm. 4.1.

When we examine functions of several complex variables, we shall learn that Thrms. 4.1 and 4.2 possess natural analogues. What will be different, and what we shall pay especial attention to, is the *shape* of the domain of convergence. Whereas the natural domain for a Taylor series in a single complex variable has a circular boundary, the natural domain for a Taylor series in several complex variables has a much more dynamic shape.

5 Theory of Functions of Several Complex Variables

In this section we describe specifically the theory of functions of *two* complex variables; but, as the discerning reader will note, all the results we state have natural generalizations [10]. The discussion we give here follows that of Kaplan [11].

Given two complex numbers z_1 and z_2, we define $\mathcal{C}^2$ as the two-dimensional complex space of points (z_1, z_2). The natural metric for $\mathcal{C}^2$ gives the distance between points (z_1', z_2') and (z_1'', z_2'') as

$$d = (|z_1' - z_1''|^2 + |z_2' - z_2''|^2)^{1/2}.$$

By identifying $z_k = x_k + i\, y_k$, we note that $\mathcal{C}^2$ is equivalent to the four-dimensional Euclidean space $\mathcal{R}^4$ having points labelled by (x_1, y_1, x_2, y_2) together with the usual metric. Then one may define a domain in $\mathcal{C}^2$ exactly as in the one-dimensional case: a non-empty open connected set in the space $\mathcal{C}^2$ (or its equivalent $\mathcal{R}^4$).

For functions in $\mathcal{C}^2$, a natural domain is the *polycylinder*, a set of points (z_1, z_2) for which

$$|z_1 - a_1| < r_1 \quad \text{and} \quad |z_2 - a_2| < r_2,$$

where both r_1 and r_2 are positive real numbers, and a_1 and a_2 are complex. (Said another way, a polycylinder is the Cartesian product of two circular domains in $\mathcal{C}^1$.) The *boundary* of such a polycylinder is the set of points (z_1, z_2) satisfying

$$|z_1 - a_1| = r_1 \quad \text{and} \quad |z_2 - a_2| \le r_2,$$

together with those points satisfying

$$|z_1 - a_1| \le r_1 \quad \text{and} \quad |z_2 - a_2| = r_2.$$

A *generalized polycylindrical domain* in $\mathcal{C}^2$ is the set of points (z_1, z_2) for which z_j belongs to $\mathcal{D}_j$, where $\mathcal{D}_j$ is some domain in the complex z_j plane.

A function of two complex variables is a function $F(z_1, z_2)$ whose arguments z_1 and z_2 each vary over a complex plane. Such a function is called *holomorphic*, or *analytic*, on a domain $\mathcal{D}$ if it is analytic everywhere in $\mathcal{D}$ with respect to each of its variables. In particular, F is holomorphic in $\mathcal{D}$ if the partial derivatives $\partial F/\partial z_1$ and $\partial F/\partial z_2$ both exist in $\mathcal{D}$.

For functions in $\mathcal{C}^2$, the analogue to Thrm. 4.1 is

Theorem 5.1 *Suppose F is holomorphic in the generalized polycylinder $\mathcal{D}$: z_1 in $\mathcal{D}_1$ and z_2 in $\mathcal{D}_2$. Suppose also that Γ_1 and Γ_2 are piecewise smooth simple closed curves in $\mathcal{D}_1$ and $\mathcal{D}_2$ respectively, and that the interiors of Γ_1 and Γ_2 are contained in $\mathcal{D}_1$ and $\mathcal{D}_2$ respectively. Then, if z_1^0 lies inside Γ_1, and z_2^0 lies inside Γ_2,*

$$F(z_1^0, z_2^0) = \frac{1}{(2\pi i)^2} \oint_{\Gamma_1} \oint_{\Gamma_2} \frac{F(z_1, z_2)}{(z_1 - z_1^0)(z_2 - z_2^0)}\, dz_1\, dz_2. \tag{12}$$

Note the remarkably restrictive constraint this theorem places on holomorphic functions of two complex variables: the values of F on the *two*-dimensional set $\Gamma_1 \times \Gamma_2$ determine the value of F at any other point in a *four*-dimensional space. Here we are counting the number of real dimensions.

In analogy with Thrm. 4.2 we also have the following two theorems for functions in $\mathcal{C}^2$.

Theorem 5.2 *Under the same hypotheses as Thrm. 5.1, the function F has partial derivatives of all orders, and they are given by*

$$\frac{\partial^{m+n} F(z_1^0, z_2^0)}{\partial z_1^m \, \partial z_2^n} = \frac{m!\, n!}{(2\pi i)^2} \oint_{\Gamma_1} \oint_{\Gamma_2} \frac{F(z_1, z_2)}{(z_1 - z_1^0)^{m+1}(z_2 - z_2^0)^{n+1}} \, dz_1 \, dz_2. \tag{13}$$

Theorem 5.3 *Suppose $F(z_1, z_2)$ is holomorphic on a domain $\mathcal{D}$ containing the point (z_1^0, z_2^0). Then in a (perhaps small) neighborhood of (z_1^0, z_2^0), one may expand $F(z_1, z_2)$ as a convergent power series in z_1 and z_2:*

$$F(z_1, z_2) = \sum_{m,n=0}^{\infty} c_{mn} (z_1 - z_1^0)^m (z_2 - z_2^0)^n. \tag{14}$$

This double sum is the Taylor series for F and has coefficients given by

$$c_{mn} = \frac{1}{m!\, n!} \frac{\partial^{m+n} F(z_1^0, z_2^0)}{\partial z_1^m \, \partial z_2^n} = \frac{1}{(2\pi i)^2} \oint_{\Gamma_1} \oint_{\Gamma_2} \frac{F(z_1, z_2)}{(z_1 - z_1^0)^{m+1}(z_2 - z_2^0)^{n+1}} \, dz_1 \, dz_2. \tag{15}$$

Furthermore, the double sum converges absolutely in any polycylinder $|z_1 - z_1^0| < r_1$ and $|z_2 - z_2^0| < r_2$ contained in $\mathcal{D}$.

After an examination of Eq. 14 one might suspect that the r_1 and r_2 of Thrm. 5.3 must depend upon one another: as r_1 increases, the maximum possible value for r_2 decreases (or at least never increases). The following theorem supports this conclusion, where now we specialize to Taylor series expansions taken about the origin $(z_1^0, z_2^0) = (0, 0)$.

Theorem 5.4 (Domain of Absolute Convergence) *Suppose that for some $z_1 = a_1 \neq 0$ and $z_2 = a_2 \neq 0$ the series*

$$\sum_{m,n=0}^{\infty} c_{mn} z_1^m z_2^n \tag{16}$$

converges. Then the series converges absolutely for all points (z_1, z_2) in the polycylinder

$$|z_1| < |a_1| = r_1 \quad \text{and} \quad |z_2| < |a_2| = r_2. \tag{17}$$

This theorem places certain restrictions on the shape of the domain of convergence for a two-dimensional Taylor series expanded about the origin. In particular, since only absolute values matter, we may describe the real four-dimensional domain of convergence using a two-dimensional plot that shows all values of $|z_1|$ and $|z_2|$ for which the series converges. Such a plot is called the *absolute convergence diagram*, and we show a typical example in Fig. 1.

240 Taylor Series Maps

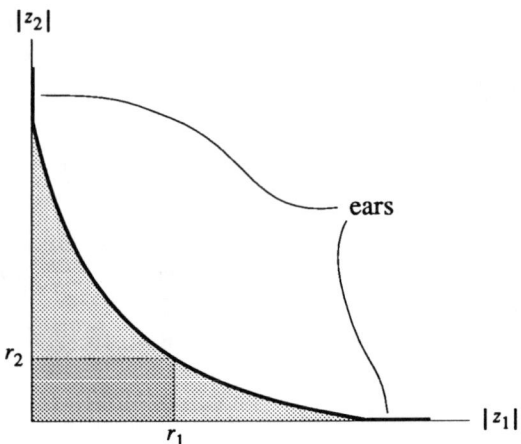

Figure 1: Example of an absolute convergence diagram. Theorem 5.4 shows that if a series converges for some point (r_1, r_2), then it converges for all pairs (z_1, z_2) having moduli $(|z_1|, |z_2|)$ inside the rectangle defined by the origin and the point (r_1, r_2).

The last theorem suggests the following recipe for determining the absolute convergence diagram of the function $F(z_1, z_2)$ (for Taylor series expansions about the origin). Begin by holding z_2 fixed at the value $r_2 e^{i\phi_2}$ and writing $f_{z_2}(z_1)$ for the resulting function of a single complex variable z_1. Then locate in the z_1-plane the singularities of f_{z_2} that lie nearest the origin $z_1 = 0$. Figure 2(a) shows a possible arrangement for a function whose only singularities are poles. Now alter the function f_{z_2} by varying the phase—but not the modulus— of z_2. As z_2 sweeps through a full circle of radius r_2, the singularities of f_{z_2} will trace out (possibly quite complicated) closed curves. Figure 2(b) illustrates a possible scenario based on the arrangement shown in Fig. 2(a). Now measure the distance r_1 from the origin of the z_1-plane to the nearest point on any of the curves traced out by the singularities of f_{z_2}. Then, according to Thrm. 5.3, the point (r_1, r_2) lies on the boundary of the absolute convergence diagram for the Taylor series expansion of the function $F(z_1, z_2)$ in the neighborhood of the origin. By repeating the process just described for a range of values of the modulus $r_2 = |z_2|$, one can trace out the entire boundary of the absolute convergence diagram.

As the discussion of the previous paragraph implies, we shall concentrate much of our attention on the singularities of the maps we wish to Taylor expand. We therefore want to comment on how the singularity structure of functions of several complex variables differs from that of the more familiar functions of one complex variable. In the study of analytic functions of a single complex variable one frequently encounters non-removable isolated singularities such as poles and branch points. (One may also encounter an infinite number of non-isolated singularities that form a natural boundary.) Indeed, such isolated singularities play an important role in the theory. On the other hand, functions holomorphic in $\mathcal{C}^n$ for $n \geq 2$ behave very differently: non-removable singularities are *never* isolated. This means that by making appropriate adjustments to z_2, one may follow continuously a singularity of $F(z_1, z_2)$ in the argument z_1 as the other argument z_2 changes. Hence there exists a singularity curve of the form $z_1(z_2)$. Furthermore, there exists the possibility that "one" singularity may trace out part (or perhaps all) of the boundary of the absolute convergence diagram.

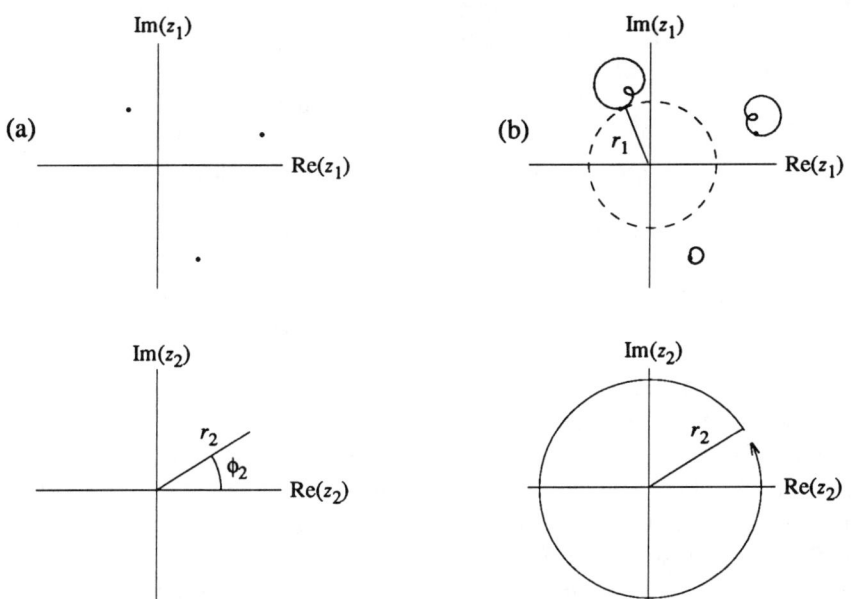

Figure 2: Finding the absolute convergence diagram. We sketch here a possible scenario for a function $F(z_1, z_2)$ whose only singularities are poles. The upper pair of graphs represents the complex z_1-plane, while the lower pair represents the complex z_2-plane. For a fixed value of $z_2 = r_2 e^{i\phi_2}$, the function $F(z_1, z_2)$ might have singularities near the origin as shown on the left in (a). As the phase ϕ_2 of z_2 sweeps through an interval of 2π, the singularities in the z_1-plane trace out closed curves, as shown on the right in (b).

In a similar vein we note that functions holomorphic in $\mathcal{C}^n$ for $n \geq 2$ have no isolated zeroes. Unless such a holomorphic function F is identically zero, every neighborhood of a zero z^0 contains points other than z^0 at which $F = 0$ as well as points at which $F \neq 0$.

Theorem 5.4 places straightforward limitations on the possible shapes for absolute convergence domains, but it turns out that even more restrictive conditions exist. Before we describe these, however, we must introduce the concepts of a Reinhardt domain and its logarithmic image.

A *Reinhardt domain with center* (z_1^0, z_2^0) is a domain $\mathcal{D}$ in $\mathcal{C}^2$ having the property that for each (z_1', z_2') in $\mathcal{D}$, all (z_1, z_2) for which

$$|z_1 - z_1^0| = |z_1' - z_1^0| \quad \text{and} \quad |z_2 - z_2^0| = |z_2' - z_2^0|$$

are also in $\mathcal{D}$. One may view a Reinhardt domain as a generalization to $\mathcal{C}^2$ of the annular domain in $\mathcal{C}^1$. However, a Reinhardt domain has a more dynamic structure than a simple Cartesian product between two annular domains in $\mathcal{C}^1$: the size (*i.e.*, both the radius and the width) of the annulus in the z_1 plane can vary with the size of the annulus in the z_2 plane.

A *complete Reinhardt domain with center* (z_1^0, z_2^0) is a domain $\mathcal{D}$ in $\mathcal{C}^2$ having the property that for each (z_1', z_2') in $\mathcal{D}$, all (z_1, z_2) for which

$$|z_1 - z_1^0| \leq |z_1' - z_1^0| \quad \text{and} \quad |z_2 - z_2^0| \leq |z_2' - z_2^0|$$

are also in $\mathcal{D}$. In other words, one completes a Reinhardt domain by adding the centers of the annuli to the domain. Like the Reinhardt domain, a complete Reinhardt domain represents more than a simple Cartesian product between two disks: the size of the disk in the z_1 plane can vary with the size of the disk in the z_2 plane.

Now suppose $\mathcal{D}$ is a complete Reinhardt domain with center $(0,0)$. For each point (z_1, z_2) in $\mathcal{D}$ form the point $(\xi_1, \xi_2) = (\ln|z_1|, \ln|z_2|)$. Then the points (ξ_1, ξ_2) form a (two-dimensional) set denoted $\mathcal{D}_{log}$ and called the *logarithmic image of* $\mathcal{D}$. We describe the domain $\mathcal{D}$ as *logarithmically convex* if $\mathcal{D}_{log}$ is convex. (The set $\mathcal{D}_{log}$ is called *convex* if one can join any two points P and Q in $\mathcal{D}_{log}$ by a straight line in $\mathcal{D}_{log}$). We illustrate this concept by showing in Fig. 3 the logarithmic image of the domain shown earlier in Fig. 1. The reader may note from this example that convexity of the logarithmic image does not imply convexity of the domain itself. Observe also that given the domain of convergence for the power series (16), one may easily form the logarithmic image of that domain by logarithmically scaling the axes of the absolute convergence diagram.

We can now state the following theorem:

Theorem 5.5 *Suppose the series (16) converges for all points (z_1, z_2) in some set $\mathcal{A}$ and that $\mathcal{A}$ has the non-empty interior $\mathcal{A}^{int}$. Then $\mathcal{A}^{int}$ is a complete Reinhardt domain with center $(0,0)$. The series converges absolutely in $\mathcal{A}^{int}$ and uniformly in any closed polycylinder contained in $\mathcal{A}^{int}$. And, furthermore, $\mathcal{A}^{int}$ is a logarithmically convex complete Reinhardt domain.*

The last statement of this theorem—that $\mathcal{A}^{int}$ is logarithmically convex—represents a significant constraint on the shape of the absolute convergence domain. Suppose, for example, that one seeks a rough picture of the absolute convergence diagram for a power series having the form given in Eq. 16. After establishing a few points on the boundary of the absolute convergence domain, one may transform to the logarithmic image, form the smallest convex set having the known locations on the boundary, and then re-transform back to the original $(|z_1|, |z_2|)$-plane. Then Thrm. 5.5 guarantees that the resulting domain belongs to the true domain of absolute convergence.

We note in passing that Thrms. 5.4 and 5.5 do allow for the possibility that the set of points for which the power series (16) converges may include sections in the coördinate planes (*i.e.*, $z_1 = 0$ or $z_2 = 0$) that extend beyond where one would expect them based on the interior of the absolute convergence diagram. This possibility explains the "ears" that stick out along the axes in Fig. 1.

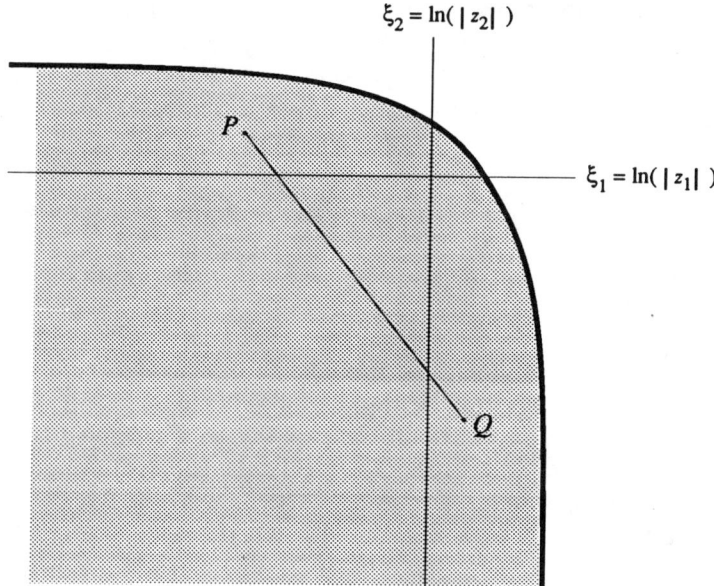

Figure 3: Example of a logarithmic image $\mathcal{D}_{log}$. We show here the logarithmic image of the domain $\mathcal{D}$ shown earlier in Fig. 1.

6 Example: The Anharmonic Oscillator

We shall use the anharmonic oscillator as a concrete example to illustrate the theorems discussed in Sec. 5. We examine first the singularity structure of this system to determine the domain of convergence of its Taylor series map. We then inspect the coefficients of the Taylor series map to see what information they contain concerning the domain of convergence. After that we examine the impact of the convergence domain on the accuracy of the Taylor series map. At the end we discuss our results.

6.1 The Singularity Structure

In the last part of Sec. 3 we very briefly introduced the (autonomous) Hamiltonian for the anharmonic oscillator,

$$H = \frac{1}{2}(p^2 + q^2) - \frac{1}{4}q^4, \tag{18}$$

and the corresponding equations of motion,

$$\begin{aligned} \dot{q} &= \partial H/\partial p = p, \\ \dot{p} &= -\partial H/\partial q = -q + q^3. \end{aligned} \tag{19}$$

We express the solution to these equations in the form of a transfer map, writing $z^f = \mathcal{M}^\tau z^i$ where $z = (q, p)$ and $\mathcal{M}^\tau$ is the map that advances the system through a time τ. As we observed earlier, the right-hand sides of the equations of motion are polynomial in q and p, and hence analytic everywhere except at infinity. Then by Thrm. 2.3 the solution is analytic in the time and the initial conditions as long as neither q nor p becomes infinite. According to Thrm. 5.3, we may therefore write the solutions, $q^f(\tau; q^i, p^i)$ and $p^f(\tau; q^i, p^i)$, as a pair of Taylor series in q^i and p^i. This pair of series is the Taylor series representation of the map $\mathcal{M}^\tau$. (We displayed in Eq. 8 the leading terms of the Taylor series map for time $\tau = 7$ expanded about the origin.)

Now for the crucial point: the *singularities* of the map, i.e., the locations z^i for which $z^f = \mathcal{M}^\tau z^i = \infty$, determine the domain of convergence of the Taylor series map for time τ. For example, the Taylor series map in Eq. 8 converges for any q^i and p^i lying inside a domain of (q^i, p^i)-space determined by locations where $q^f(7; q^i, p^i) = p^f(7; q^i, p^i) = \infty$. To calculate this domain of convergence, we can use the recipe described in Sec. 5. In particular, we shall find the domain of convergence for the Taylor series expanded about the origin $z^i = (0, 0)$.

We begin by noting that the Hamiltonian H in Eq. 18 is a constant of the motion,

$$H = E. \tag{20}$$

In other words, H takes on a constant value E along any trajectory in (q^i, p^i)-space *including* complex values of q and p. Also, we note from the explicit form of H that if either q or p becomes infinite on a trajectory, they must both become infinite in order to keep H constant. Consequently the series for q^f and the series for p^f have identical convergence domains, and we therefore need examine only one of them. We concentrate on determining the convergence domain for q^f.

Let us first consider only those trajectories for which both q and p are real. We define the potential energy $V(q)$ by the relation

$$V(q) = \frac{1}{2}q^2 - \frac{1}{4}q^4, \tag{21}$$

and show a graph of $V(q)$ in Fig. 4. An examination of this graph makes clear the following points:

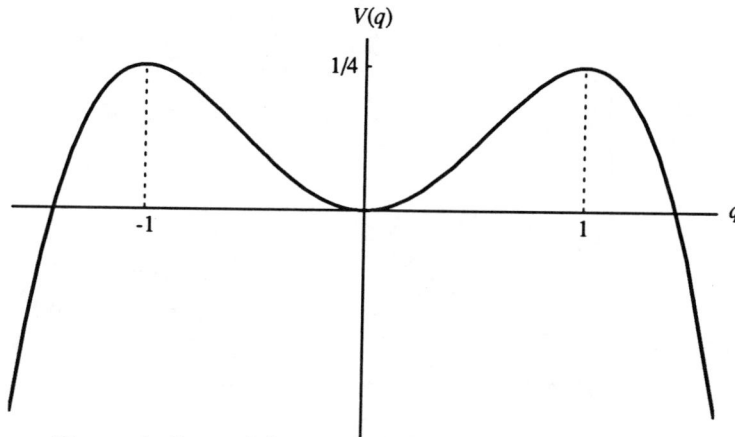

Figure 4: Potential energy of the anharmonic oscillator.

- Oscillatory motion occurs when $E < 1/4$ and the initial coördinate, q^i, lies inside the well i.e. $q^i \in (-1, 1)$.

- At small amplitudes, the motion is essentially simple harmonic with period 2π. Furthermore, since the restoring force, $-dV/dq$, weakens with amplitude, the period lengthens as amplitude increases.

- At $E = 1/4$, the system has unstable fixed points at $p^i = 0$, $q^i = \pm 1$. There is a separatrix that separates the phase-space regions of bounded and unbounded motion, and these points lie on the separatrix.

- When $E > 1/4$ or q^i lies outside the well, the system moves rapidly out to infinity and, as a simple calculation shows, reaches infinity in a *finite* amount of time. At the energy $E = 1/4$, for example, the time to reach infinity from a point $q^i > 1$ is given by the relation

$$\tau_\infty = \int_{q^i}^\infty dq/p = \sqrt{2} \int_{q^i}^\infty dq/(q^2-1) = \sqrt{2}\coth^{-1}(q^i) = \sqrt{2}\tanh^{-1}(1/q^i). \quad (22)$$

The phase-space portrait shown in Fig. 5 presents an elegant distillation of these points.

For our purposes, the last point is of particular interest. We can obtain an outer limit on the domain of convergence in the following manner. We begin by choosing τ and then setting $p^i = 0$. In this case we can always find some $q^i > 1$ for which $\mathcal{M}^\tau(q^i, 0) = \infty$. Now increase p^i to some small positive value. Then the point q^i that $\mathcal{M}^\tau$ carries to infinity moves back to a slightly smaller value. As we continue to increase p^i, the corresponding q^i will continue to decrease until we find some value of p^i for which $\mathcal{M}^\tau(0, p^i) = \infty$. A plot of these (q^i, p^i) pairs will produce a curve that bounds the domain of convergence of the Taylor series map for the given time τ. In Fig. 6 we show this curve for the case $\tau = 7$.

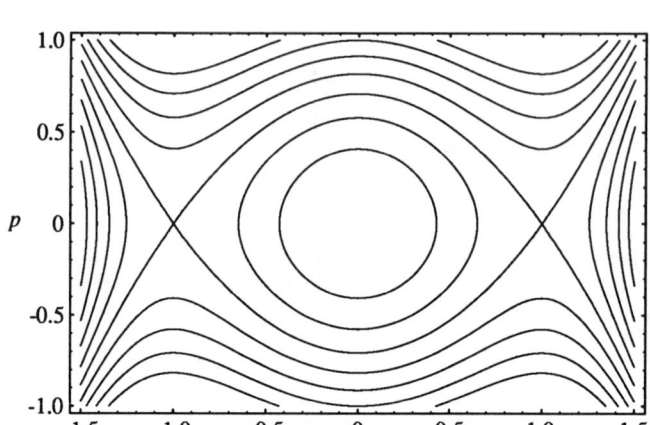

Figure 5: Phase-space portrait of the anharmonic oscillator.

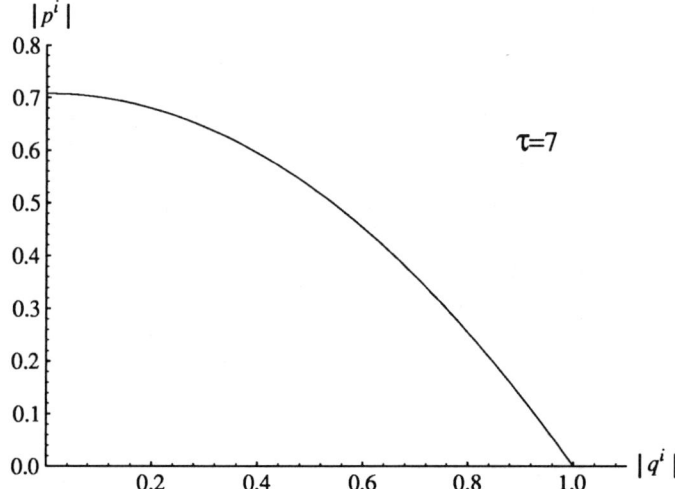

Figure 6: Real singularities of the time-seven map for the anharmonic oscillator. The curve shown here represents a portion of the locus of points (q^i, p^i) for which $\mathcal{M}^{\tau=7}(q^i, p^i) = \infty$. The reader may note that this curve lies very near the separatrix. This happens because the time-step $\tau = 7$ is relatively long in this example.

If only real values of q^i and p^i were relevant, the picture described in the previous paragraph would show us all we want to know. To view the entire landscape, however, we must survey the complex planes of the initial conditions q^i and p^i. On doing so, we discover that singularities of the map $\mathcal{M}^{\tau}$ can exist for values of q^i and p^i whose absolute values lie inside the curve determined by the singularities on the real axes.

The anharmonic oscillator described by the Hamiltonian in Eq. 18 can be solved exactly in terms of elliptic functions. In particular, the solution expresses the motion as a rational fraction involving elliptic functions in both the numerator and the denominator. A careful analysis of this solution shows that all of the singularities of the finite-time transfer map occur as simple poles, and, furthermore, the zeroes of the denominator determine the locations of those singularities. Therefore, to locate singularities of the transfer map for the anharmonic oscillator, we searched for zeroes of the denominator. We shall present a detailed description of this portion of our work in a separate publication. For the moment we just mention that we used *Mathematica* to perform all calculations involving elliptic functions at complex arguments [12].

Using our recipe from Sec. 5 for determining the domain of absolute convergence, we obtained the results shown in Fig. 7. There we plot the paths traced in the $(|q^i|, |p^i|)$ plane by three different singularities. The shaded region represents the domain of absolute convergence: the Taylor series map for time-seven converges for any initial point $z^i = (q^i, p^i)$ inside this domain. Furthermore, the Taylor series map diverges for any initial point outside this domain. Note that two different complex singularities cut inside the region we showed earlier in Fig. 6.

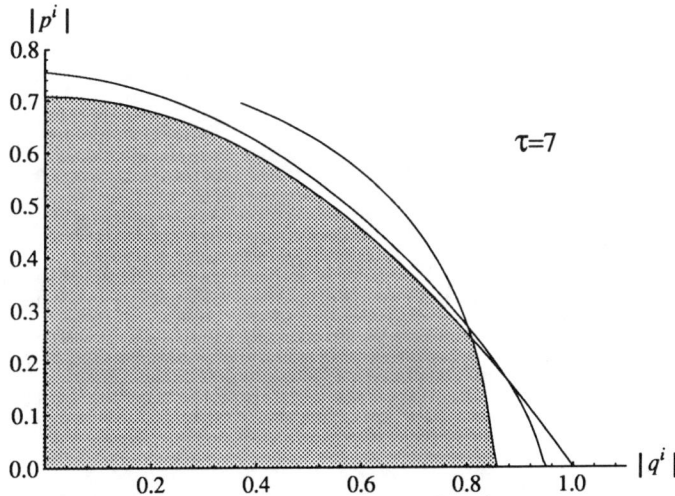

Figure 7: Absolute convergence diagram for the time-seven map. This graph shows the path in the $(|q^i|, |p^i|)$ plane of three singularities of the time-seven map for the anharmonic oscillator. The inner envelope of these curves bounds the domain of absolute convergence (the shaded region) for the corresponding Taylor series map.

Now consider how the situation changes for different size time-steps. If we choose a *larger* time-step τ, then we must, in some sense, start farther from infinity in order to reach infinity in exactly one step. Hence, the singularities of the map $\mathcal{M}^\tau$ move inwards as τ increases. We illustrate this behavior in Fig. 8, which shows the boundaries of the domains of absolute convergence for three different time-steps: $\tau = 1$, 7, and 15. For a time-step of one, the boundary of the domain of convergence lies well outside the separatrix. Then, as the time-step increases, the convergence domain shrinks. At a time-step of 15 the boundary of the convergence domain lies well inside the separatrix.

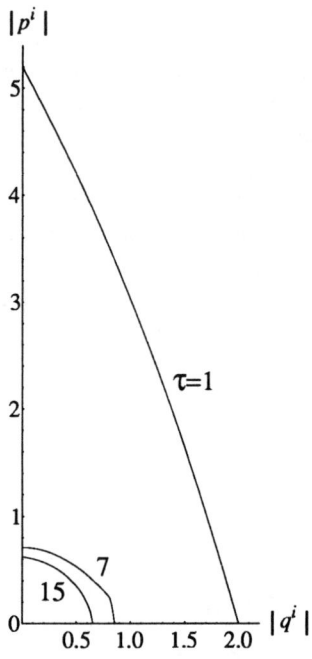

Figure 8: The absolute convergence diagram for the anharmonic oscillator at several different time-steps. This graph shows the boundaries of the convergence domains for the Taylor series map at three different time-steps: $\tau = 1$, 7, and 15. The boundary for $\tau = 1$ lies well outside the separatrix. For $\tau = 7$, a relatively long time-step, part of the boundary lies right next to the separatrix, while the rest of the boundary cuts inside the separatrix. And for $\tau = 15$ all of the boundary lies inside the separatrix.

6.2 Estimating the Convergence Domain from a Taylor Series Map

To emphasize the theoretical results of the previous section, we calculate some Taylor series maps for the anharmonic oscillator and then examine them for evidence of their domains of convergence. As mentioned earlier, we can use the techniques of Lie algebra or automatic differentiation to compute the coefficients of a Taylor series map to essentially arbitrary order. By examining the Taylor series coefficients, we can then estimate the corresponding domain of convergence.

In this section we begin by describing how we estimate the domain of convergence, and then we present our results.

Consider first an analytic function f of a single complex variable. According to Thrm. 4.2, we may Taylor expand f in the form $\sum_n a_n z^n$ within some neighborhood of the origin, where the coefficients a_n are given by the relation

$$a_n = \frac{1}{2\pi i}\oint_\Gamma \frac{f(\xi)}{\xi^{n+1}}\,d\xi = \frac{1}{2\pi R^n}\int_0^{2\pi} \frac{f(Re^{i\theta})}{e^{in\theta}}\,d\theta. \tag{23}$$

Here we integrate around the circle of radius $R = |\xi|$ centered on the origin. We can use Eq. 23 to determine an upper bound on the size of these Taylor series coefficients. One simple limit takes the form

$$|a_n| \leq \frac{1}{2\pi R^n}\int_0^{2\pi}|f(Re^{i\theta})|\,d\theta \equiv \frac{\tilde{f}(R)}{R^n}, \tag{24}$$

where $\tilde{f}(R)$ represents the average value of $|f|$ over the circle about which we integrate. If we fix the value of R, Eq. 24 tells us that for all n

$$|a_n|R^n \leq \text{constant},$$

or, equivalently,

$$\ln(|a_n|R^n) \leq \text{constant}. \tag{25}$$

This result implies that for R inside the radius of convergence, the values of $\ln(|a_n|R^n)$ will, on average, fall as n increases. Then as R approaches the radius of convergence, the values of $\ln(|a_n|R^n)$ will fall less rapidly. And when R reaches the radius of convergence, the values of $\ln(|a_n|R^n)$ will, on average, remain constant with n. Consequently, we concoct the following recipe for estimating the radius of convergence of a Taylor series $\sum_n a_n z^n$. Select a value for R and then plot $\ln(|a_n|R^n)$ versus n. Fit a straight line through the points and determine the slope $m(R)$. Call the best estimate of the radius of convergence that value R_e for which $m(R_e) = 0$. Using this approach, we obtain a simple formula for estimating the radius of convergence:

$$R_e = \exp\left(\frac{(1/N)\sum_n n \sum_n \ln|a_n| - \sum_n n \ln|a_n|}{\sum_n n^2 - (1/N)(\sum_n n)^2}\right). \tag{26}$$

Here each sum includes only those terms that correspond to non-zero coefficients in the Taylor series; and N is the number of non-zero coefficients.

To estimate the domain of convergence of the Taylor series map for the anharmonic oscillator, we continue to concentrate on q^f. Because here we want to find the convergence domain of a Taylor series in two variables, q^i and p^i, we look for some way to reduce it to a one-variable Taylor series—to which we can apply the formula in Eq. 26. We do this by writing $q^i = r\cos\theta$ and $p^i = r\sin\theta$

and substituting these expressions into the Taylor series for q^f. If we give θ a definite value, say $\pi/6$, we then obtain a Taylor series for q^f in the single variable r. Applying Eq. 26 to this series gives us an estimate $R_e(\pi/6)$ of where the line $(r\cos\pi/6, r\sin\pi/6)$ crosses the boundary of the domain of absolute convergence (see Fig. 9).

To obtain a more accurate estimate of the domain of convergence, we must add an important refinement. Because q^i and p^i can have complex values, the single point $(|q^i|, |p^i|)$ in the absolute convergence diagram includes all points of the form $(q^i, p^i e^{i\alpha})$. Hence, we write $q^i = r\cos\theta$ and $p^i = r e^{i\alpha}\sin\theta$ and insert these expressions into the Taylor series for q^f to obtain a Taylor series in the single variable r. Now when we apply Eq. 26, we determine an estimate $R_e(\theta, \alpha)$. We therefore define our best estimate of the radius of convergence as

$$R_e(\theta) = \min_{\alpha} R_e(\theta, \alpha). \tag{27}$$

Using a modified version of the program Tlie [13], we determined the coefficients of the Taylor series map for the anharmonic oscillator at the time-steps $\tau = 1, 7$, and 15. Because this problem involves just two phase-space variables, we could obtain extremely high-order coefficients using only a few minutes of computer time. For no particular reason we chose to compute the series through order forty-seven. To verify that the coefficients we obtained were accurate, we determined a few of the high-order coefficients by evaluating numerically the contour integral in Eq. 15. Doing so using even a crude method of numerical integration gave agreement to better than five significant figures. With Taylor series coefficients thus in hand, we then estimated the domains of convergence for these Taylor series maps using the procedure described in the last two paragraphs. We show our results in Figs. 9 and 10.

6.3 Accuracy of the Taylor Series Map

A finite domain of convergence has a very real effect on the accuracy of a Taylor series map. For the anharmonic oscillator—which we can solve exactly—one easy way to see this effect is to use the Taylor series map to track a few initial conditions for many iterations. We can then compare the approximate phase-space portrait generated by the Taylor series map with the true phase-space portrait shown in Fig. 5. In a different scheme for evaluating the effect of the domain of convergence on the accuracy of a Taylor series map, we simply measure the distance in phase space between the results of applying the exact and Taylor series maps to the same initial conditions. In this section we present both of these approaches.

Suppose we start the anharmonic oscillator in some initial state $z^i = (q^i, p^i)$. Because our map $\mathcal{M}^\tau$ carries this system forward in time by an amount τ, we can determine the future state of the anharmonic oscillator at any time that is an

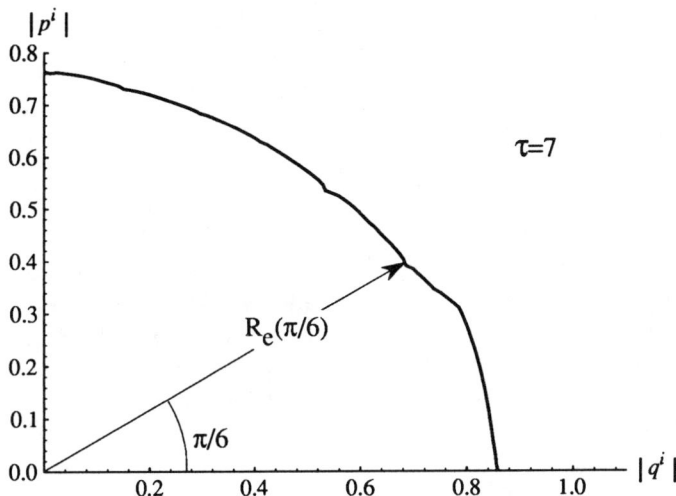

Figure 9: Estimate of the convergence domain for the time-seven map. We used the values of the Taylor series coefficients through order forty-seven to estimate the domain of convergence of the time-seven Taylor series map for the anharmonic oscillator.

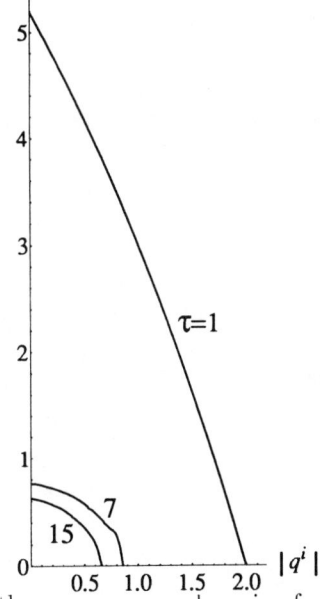

Figure 10: Estimates of the convergence domains for several different time-steps. In this graph we show our estimates of the domains of convergence of the Taylor series maps for time-steps $\tau = 1$, 7, and 15. These estimates are based on the values of the Taylor series coefficients of the different maps.

integer multiple of the time-step τ simply by iterating the equation

$$z^{n+1} = \mathcal{M}^\tau z^n.$$

Here we use the initial condition $z^0 = z^i$. In Fig. 11 we show the results of 500 iterations on several initial conditions using the order-47 Taylor series maps for time-steps $\tau = 1, 7$, and 15. We chose most of the initial conditions on the q^i axis ($p^i = 0$). For $\tau = 15$ the phase-space orbits generated by the Taylor series map agree well with the true phase-space orbits up to about $q^i = 0.50$. Similarly, for $\tau = 7$ the orbits agree well up to about $q^i = 0.65$. And for $\tau = 1$ the Taylor series map generates an accurate phase-space portrait throughout the region enclosed by the separatrix. We also show that the Taylor series map for $\tau = 1$ renders accurately even some orbits (the large dots) that lie outside the separatrix.

Because almost all points outside the separatrix travel rapidly to infinity, we cannot really use iterations of the map to see how well the Taylor series map works in this region. Instead we measure the distance in phase space between the results of applying the exact and Taylor series maps to the same initial conditions. In other words, we evaluate

$$\Delta z = \sqrt{(q^f_{ex} - q^f_{tay})^2 + (p^f_{ex} - p^f_{tay})^2}, \tag{28}$$

where (q^f_{ex}, p^f_{ex}) is the result of applying the exact map to some initial condition, and (q^f_{tay}, p^f_{tay}) is the result of applying the order-47 Taylor series map to the same initial condition. In Figs. 12, 13, and 14, we have plotted contour levels of Δz for our Taylor series maps with time-steps $\tau = 1, 7$, and 15. In all three figures the phase-space error Δz changes by a factor of ten between each of the contour levels, with $\Delta z = 10^{-4}$ on the inner-most contour and $\Delta z = 10^{-1}$ on the outer-most contour.

The Hamiltonian (18) for the anharmonic oscillator contains only even-order terms in the phase-space variables. As a consequence it turns out that the Taylor series map contains only odd-order terms. Since we truncated the Taylor series at order forty-seven, the first terms neglected in the series are those of order forty-nine. We therefore expect the phase-space error Δz to increase at a rate proportional to r^{49}. The approximately uniform spacing between the contour levels of Δz reflects this power-law behavior; and a rough calculation confirms that the exponent is about forty-nine.

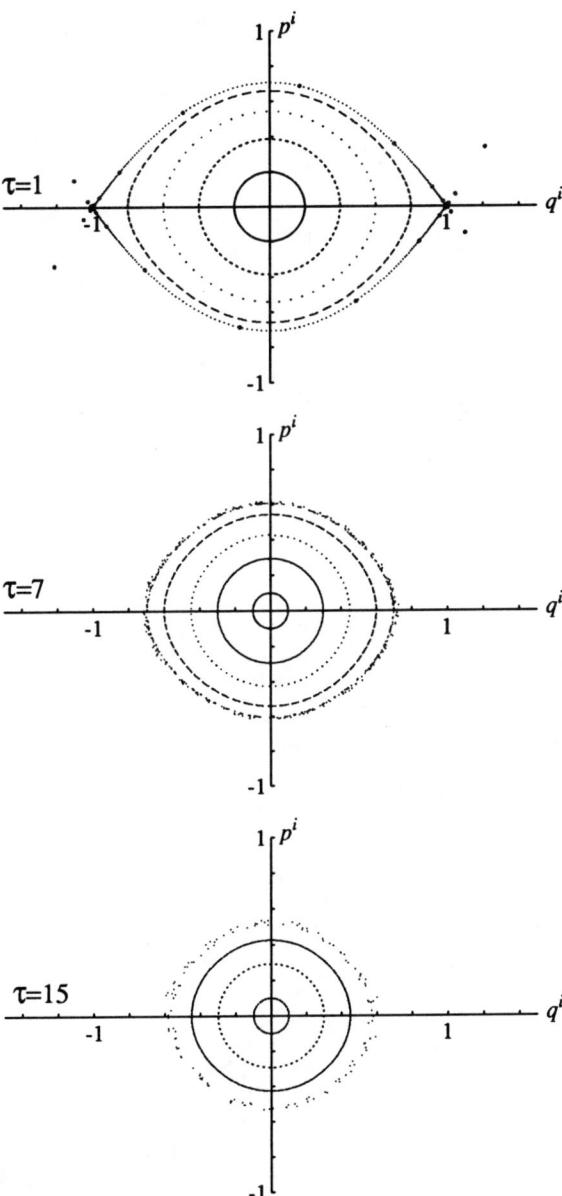

Figure 11: The results of tracking the motion of the anharmonic oscillator using Taylor series maps with three different time-steps: from top to bottom we show $\tau = 1$, 7, and 15. The large points in the $\tau = 1$ plot indicate orbits whose initial conditions lie outside, but close to, the separatrix.

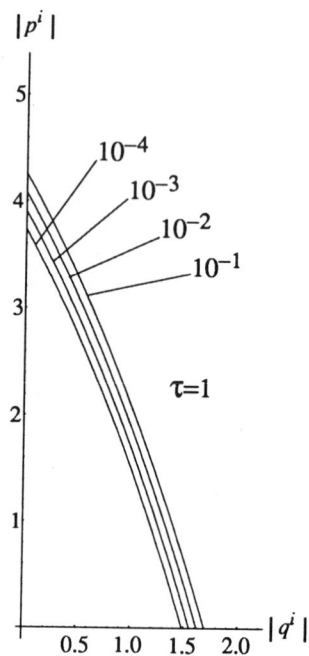

Figure 12: Contour levels of the phase-space error Δz for the order-47 Taylor series map for the anharmonic oscillator using a time-step $\tau = 1$.

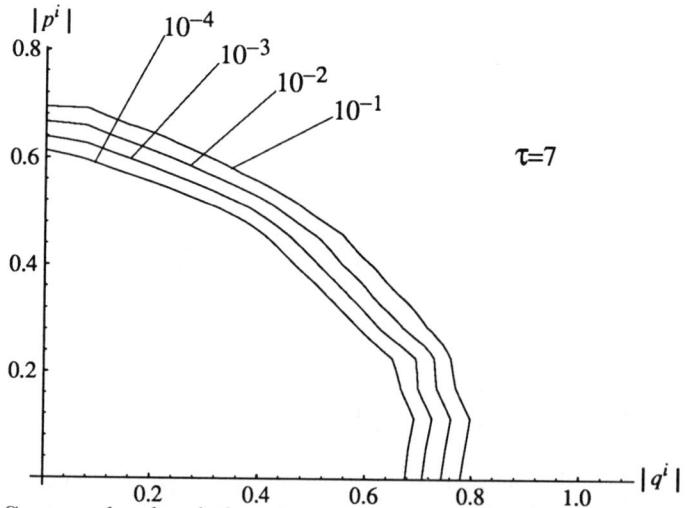

Figure 13: Contour levels of the phase-space error Δz for the order-47 Taylor series map for the anharmonic oscillator using a time-step $\tau = 7$.

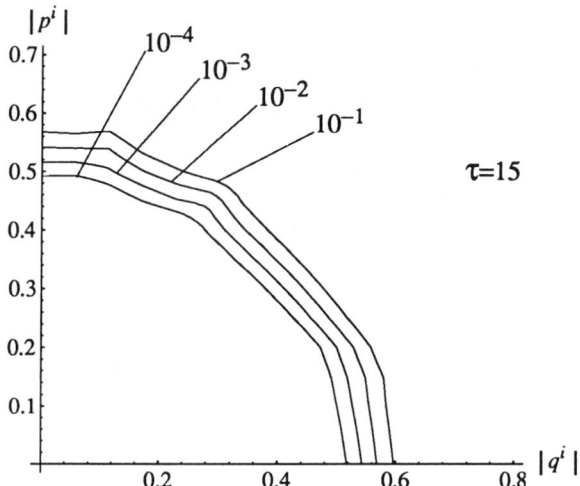

Figure 14: Contour levels of the phase-space error Δz for the order-47 Taylor series map for the anharmonic oscillator using a time-step $\tau = 15$.

6.4 Discussion

We have shown that the theory of functions of several complex variables tells us that the domain of convergence of a Taylor series map is determined by the singularities—both real and complex—of the corresponding transfer map. The striking agreement between our estimates of the convergence domains (in Fig. 10) and the true convergence domains (in Fig. 8) emphasizes the fact that the Taylor series maps see the singularities.

For time-steps $\tau = 1$ and 15, the curves in Figs. 8 and 10 agree extremely well with one another. In fact, for $\tau = 1$ we can see no visible difference between the two graphs. For $\tau = 7$, however, a small discrepancy exists, which we illustrate in Fig. 15. There we see that the estimate of the convergence domain agrees very well with the true domain along only part of the boundary. A comparison of our estimated domain with all three of the singularities that we tracked suggests why this happens. In Fig. 16 we show that, along the section where the estimated boundary does not see the true boundary, it instead sees one of the other singularities. Indeed, the estimated boundary follows that other singularity very closely. We computed numerically the residues of these two singularities and found that the one seen by our estimated boundary (i.e., the one seen by the Taylor series coefficients) has a residue that is about sixty-three times as large as the other singularity, which bounds the domain of convergence. We therefore expect that one must keep many more terms in the Taylor series before the values of the coefficients reflect the presence of the pole with the smaller residue.

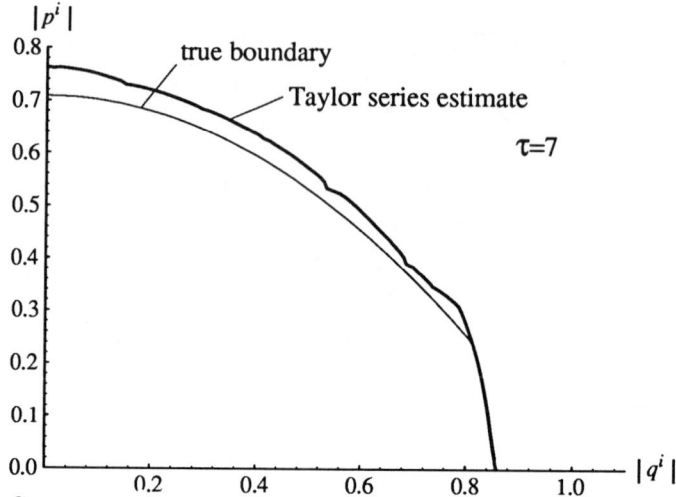

Figure 15: Comparison between the domain of absolute convergence for the time-seven Taylor series map and the estimate based on the Taylor series coefficients through order forty-seven.

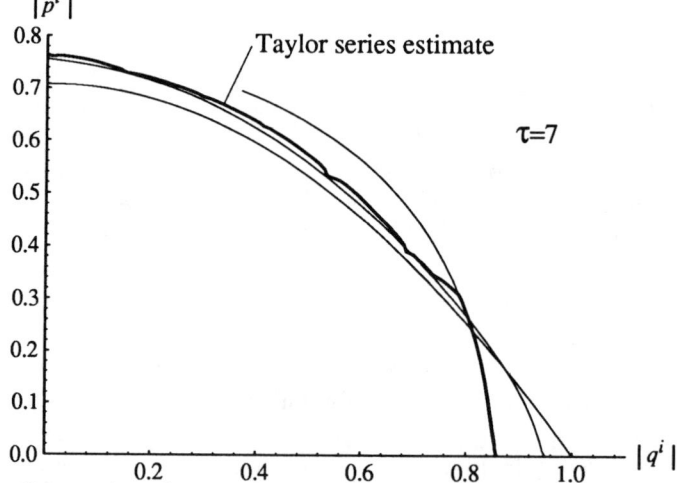

Figure 16: Comparison between the singularities of the time-seven map and the estimate of the convergence domain based on the Taylor series coefficients through order forty-seven.

We have just seen that the coefficients of the Taylor series maps contain information about the singularities of the corresponding transfer map. This is also made evident in our results from iterating the Taylor series maps. For example, we note from Fig. 11 that the Taylor series map for $\tau = 15$ gives an accurate portrait of the phase space for amplitudes smaller than about 0.5—somewhat inside the domain of convergence we found for $\tau = 15$ (see Fig. 8). For the smaller time-steps, the domain of convergence is larger, and the result of iterating the Taylor series map produces an accurate portrait of the phase space over a correspondingly larger range of amplitudes.

We can make a similar comparison between the domains of convergence for the Taylor series maps (Fig. 8) and the contour levels of the phase-space error Δz (Figs. 12-14). In particular, we note from Fig. 12 that the Taylor series map for $\tau = 1$ gives accurate results even in regions well outside the separatrix. In those regions we cannot iterate the Taylor series map more than a few times (or perhaps only once), but as long as the result of one iteration remains well inside the domain of convergence, we can go on to perform the next iteration. The contour levels of Δz give us some measure of what constitutes "well inside the domain of convergence."

One may very naturally ask, "What happens to the domain of convergence as the time-step τ continues to increase?" First note that the point $z^i = 0$ is a fixed point of the anharmonic oscillator—i.e., for any time-step τ, $\mathcal{M}^\tau 0 = 0$. Then according to Thrm. 2.3, for any *finite* time-step τ there exists a domain of z^i, surrounding the origin, within which $\mathcal{M}^\tau z^i$ remains finite. Therefore, as the time-step τ approaches infinity, the size of the convergence domain shrinks to zero; but *for any $\tau < \infty$ the domain of convergence has a finite non-zero size.*

A principal lesson to learn here concerns the relationship between the dynamics of a system under study and the domain of convergence of a corresponding Taylor series map. Poincaré's theorem (Thrm. 2.3) tells us that the fundamental constraints on the Taylor series map come from those locations where the right-hand side of the equations of motion (1) fail to be analytic. As illustrated by the $\tau = 1$ curve in Fig. 8 for the anharmonic oscillator, the presence of a separatrix does not pose any fundamental problem. A finite size for the domain of convergence results when the system can arrive in a finite time at a non-analytic location. We emphasize again that here the initial conditions may be either real or complex.

We underscore the point made in the previous paragraph by examining the following modification of our anharmonic oscillator. Suppose we change the sign of the quartic term in the Hamiltonian (18) for our anharmonic oscillator. In other words, consider the Hamiltonian

$$H = \frac{1}{2}(p^2 + q^2) + \frac{1}{4}q^4, \tag{29}$$

which has the corresponding equations of motion,

$$\begin{aligned} \dot{q} &= \partial H/\partial p = p, \\ \dot{p} &= -\partial H/\partial q = -q - q^3. \end{aligned} \tag{30}$$

This system executes bounded periodic motion for all real values of the initial conditions. Consequently, in this case there exist no singularities of $\mathcal{M}^\tau$ in the real (q^i, p^i) plane. However, we must not be lulled into a false sense of security. As the reader can easily show, replacing q by iq and p by ip transforms the equations of motion (30) for the anharmonic oscillator with the *positive* quartic term into the equations of motion (19) for the anharmonic oscillator with the *negative* quartic term. Therefore, the oscillator with the positive quartic term has complex initial conditions that map to infinity in a finite time. In fact, because the transformation between these two systems corresponds to simple rotations in the complex planes of q^i and p^i, the domain of convergence for the anharmonic oscillator does not depend on whether the quartic term is positive or negative! In other words, the same convergence diagrams (Figs. 7 and 8) apply to the transfer maps $\mathcal{M}^\tau$ generated by the two different Hamiltonians (18) and (29). This state of affairs resembles that surrounding, for example, the function $1/(1+x^2)$. Although this function is perfectly well-behaved for all real values of x, we know that its Taylor series expansion about the origin converges only inside the unit circle because of the singularities at $x = \pm i$.

Experience teaches us to exercise caution when using large steps in time (or the independent variable) for a Taylor series map. We have tried in this paper to spell out the principles that underlie this intuition. In particular, it is the presence of singularities in the equations of motion that lead to finite domains of convergence. Indeed, there exists no inherent limit to the size of the time-step τ that we may employ. On the other hand, the choice of τ does affect the size of the region within which a Taylor series map gives good results.

References

[1] R. Talman, *Part. Accel.* **34**, 1 (1990).

[2] J. Hagel and B. Zotter, *Part. Accel.* **34**, 67 (1990).

[3] F. Brauer and J.A. Nohel, *The Qualitative Theory of Ordinary Differential Equations* (W. A. Benjamin, New York, NY, 1969).

[4] I.G. Petrovski, *Ordinary Differential Equations* (Prentice-Hall, Englewood Cliffs, NJ, 1966).

[5] Avner Friedman, *Partial Differential Equations* (Holt, Rinehart and Winston, New York, NY, 1969).

[6] A.J. Dragt, "Lectures on Non-linear Orbit Dynamics." In *Physics of High-Energy Particle Accelerators*, AIP Conf. Proc., No. 87, pp.147-313, ed. R. A. Carrigan, et. al. (AIP, New York, NY, 1982).

[7] M. Berz and H. Wollnik, *Nucl. Instr. Meth.* **A258**, 364 (1987).

[8] M. Berz, *Part. Accel.* **24**, 109 (1989).

[9] G. Corliss and L. B. Rall, *Transactions of the First Army Conference on Applied Mathematics and Computing*, vol. ARO Report 84-1 (1984).

[10] W. Kaplan, *Functions of Several Complex Variables* (Ann Arbor Publishers, Ann Arbor, MI, 1964).

[11] W. Kaplan, *Introduction to Analytic Functions* (Addison-Wesley, Reading, MA, 1966).

[12] S. Wolfram, *Mathematica: A System for Doing Mathematics by Computer*, 2nd ed. (Addison- Wesley, Redwood City, CA, 1991).

[13] Johannes van Zeijts, Private communication.

Is the Momentum Space Optimally Used with the FODO Lattices?

D. Trbojevic, K.Y. Ng* and S.Y. Lee[†]

Brookhaven National Laboratory, Upton, NY 11973

ABSTRACT

The available momentum space of a FODO lattice is determined by the maximum value of the dispersion function ($\delta x = D_x \, \partial p/p$). In a regular FODO lattice the dispersion function oscillates between its maximum and minimum values, which are always positive. The maximum value of the dispersion function in a FODO cell of a fixed length depends on the cell phase difference. An example of a new lattice, in which the dispersion function is lowered to half its value in the same FODO cell, is presented. The available momentum space in the new lattice is raised to twice that in the FODO lattice by allowing the dispersion function to oscillate between the same positive and negative values. The maxima of the dispersion function in the new lattice have half the value of those within the regular 90° cells.

1 INTRODUCTION

The optimum value of the dispersion function, in a lattice that consists of a row of FODO cells, depends primarily on the phase difference within the cell. A phase difference of $\Phi=90°$ for a fixed-length cell represents an optimum [1] value with respect to the dispersion function as well as the betatron functions. An example of the FODO cell used in the present Relativistic Heavy Ion Collider (RHIC) lattice at Brookhaven is used for comparison. A new lattice with better momentum dependence is designed by using a combination of the same FODO cells and additional insertions [2] with a π-phase difference.

2 TWOFOLD IMPROVEMENT IN MOMENTUM DEPENDENCE

The available momentum space of the lattice depends on the maximum value of the dispersion function. For the same value of the admittance-available beam offset, the maximum momentum ($\sigma_{p\,max}$) is defined from the definition of the beam size σ_{tot}:

$$\sigma^2_{tot} = \sigma^2_{Twiss} + (D_x \, \sigma_p)^2, \quad \text{with} \quad \sigma^2_{Twiss} = \varepsilon_N \, \beta_{Twiss} / 6\pi \, \gamma\beta, \tag{1}$$

where $\gamma\beta$ are the relativistic factors, ε_N is the normalized emittance in mm mrad, D_x is the horizontal dispersion function, and β_{Twiss} is the betatron amplitude. Thus the available momentum aperture is

$$\sigma_p = [\sqrt{(\sigma^2_{tot} - \sigma^2_{Twiss})}] / D_x. \tag{2}$$

---

*Permanent address: Fermilab, P.O.Box 500, Batavia, IL 60510
[†]Permanent address: Indiana U., Dept. of Physics, Bloomington, IN 47405

Equation (2) shows that the momentum aperture is indirectly proportional to the value of the dispersion function.

The dispersion function could oscillate between positive and negative values instead of always having positive values as in the lattice made of FODO cells. This was accomplished [3-6] in previous designs of the imaginary γ_t lattices. A lattice in which the dispersion function is both positive and negative could avoid the transition energy, which is an additional advantage over regular FODO-cell-based lattices. A method of designing lattices without transition was reported earlier.[4, 7]

The FODO cell used for comparison is like a cell from the present RHIC lattice design. The RHIC lattice within the arcs consists of a row of FODO cells. The magnets are 9.45 meters long, and the focussing and defocussing quadrupoles are both 1.11 meters long. If the phase difference in the FODO cell is 90°, the maxima of the beta functions are $\beta_{x\,max} = 49.98$ m, $\beta_{y\,max} = 50.02$ m; and $D_{x\,max} = 1.554$ m, $D_{x\,min} = 0.751$ m. The phase difference of the FODO cells in the two rings of the present RHIC lattice is slightly different from 90° ($\phi_x=80.43°$, $\phi_y=85.30°$ in the inner arc cells). The maximum of the dispersion function is $D_{x\,max} = 1.841$ m.

Figure 1 shows the betatron functions within a lattice made of three FODO cell in a row, with a total length L= 88.761 m. The betatron functions are also presented in Table 1. The emittance is assumed to be $\varepsilon=20\ \pi$ mm mrad, and the acceptance depends on the available aperture.

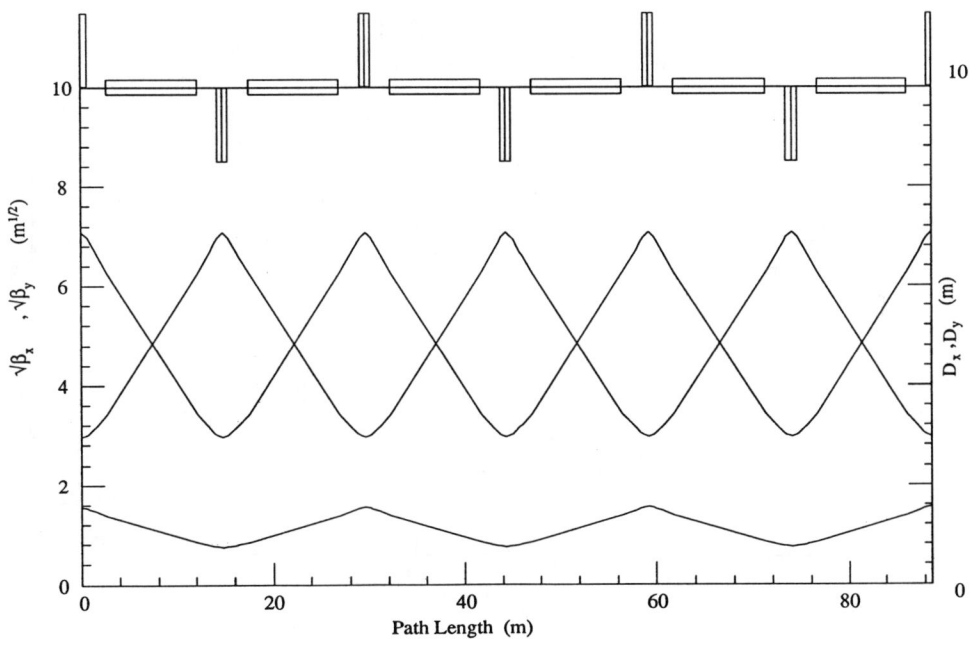

Figure 1 Betatron functions within three FODO cells in a row.

TABLE 1

	$\beta_{x\,max}$	$\beta_{x\,min}$	$\beta_{y\,max}$	$D_{x\,max}$	$D_{x\,min}$	v_x	v_y	ξ_x	ξ_y	γ_t
FODO(90°)	49.98	8.75	50.02	1.554	0.751	.750	.750	−.897	−.900	18.79
RHIC	49.71	10.55	48.55	1.841	0.939	.670	.711	−.777	−.809	17.01
New	55.17	1.35	49.49	0.887	−.880	.727	.411	−1.69	−.589	i176.0

A new block of three cells is constructed by using two of the FODO cells described above plus a π-insertion constructed from two doublet quadrupoles with a 9.45-meter-long dipole between. The first quadrupole of the doublet replaces the focussing quadrupole of the FODO cell in front of the insertion. The total length of this block is L= 73.451 m. Figure 2 shows the elements and the betatron functions within the new block. Analytical formulae and other details of the construction of such blocks are presented elsewhere.[7] The excellent properties and good stability of this kind of lattice were also studied earlier.[7]

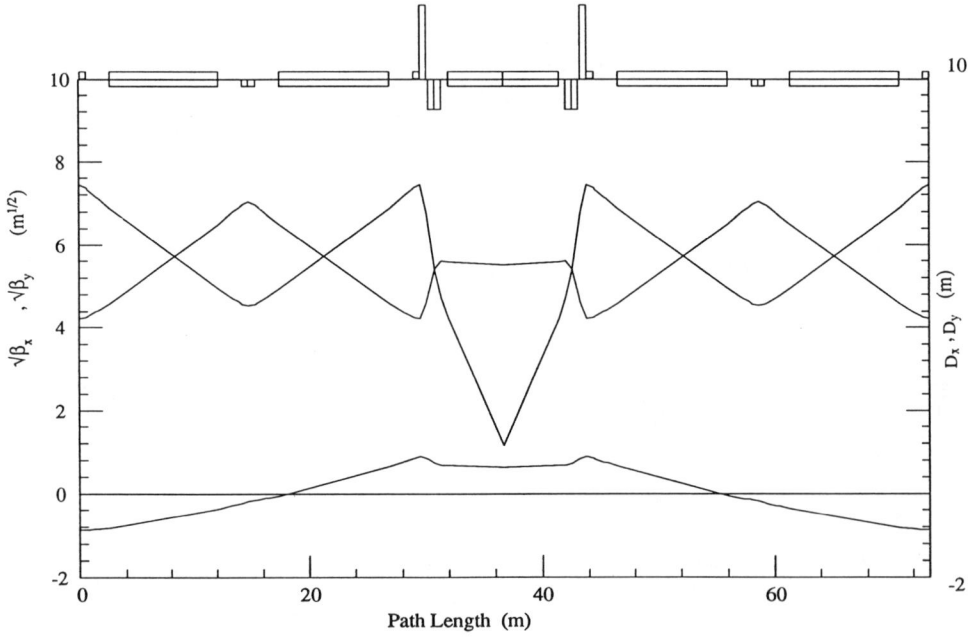

Figure 2   Betatron function within the basic block of the new lattice. The block consists of two FODO cells and a π-insertion.

3 CONCLUSION

A new type of lattice could be constructed, by using a combination of FODO cells and π-insertions, which provides double the momentum space of a regular FODO cell without losing much in the geometrical compaction factor. The new lattice not only improves the momentum dependence over that of the FODO design but could even be optically very stable, with smaller sextupole-induced second-order tune shifts.

REFERENCES

[1] Tom Collins, The Beta Theory, Technical Memo, Fermilab, 1988.
[2] D. Trbojevic, D. Finley, R. Gerig, and S. Holmes, Design Method for High Energy Accelerators without Transition Energy, in Proc. 2nd European Part. Accel. Conf., Nice, June 1990, pp.1536-1538.
[3] L.C Teng, Infinite Transition Energy Synchrotron Lattice Using π-straight Section, Part. Accel. 4, 81-85 (1972).
[4] K.Y. Ng, D. Trbojevic, and S.Y. Lee, A Transitionless Lattice for the Fermilab Main Injector, in Proc.1991 IEEE Part. Accel. Conf., San Francisco, May 1991, pp.159-161.
[5] V.V. Vladimirski and E.K. Tarasov, Theoretical Problems of the Ring Accelerators, USSR Academy of Sciences, Moscow (1955).
[6] G. Guinard, A Lattice with No Transition and Largest Aperture, in Proc. 1989 IEEE Part. Accel. Conf., Chicago, March 1989, pp. 915-917.
[7] S.Y. Lee, K.Y. Ng, and D. Trbojevic, Minimizing Dispersion in Flexible Momentum Compaction Lattices, to be published in Phys. Rev.

IV. Contributions Working Group B

Long-Term Stability

QUASI-CLOSED ORBIT IN A HARMONICALLY PERTURBED MAGNETIC FIELD

G.V. Stupakov

Superconducting Super Collider Laboratory, Dallas, TX 75237

ABSTRACT

The paper generalizes a notion of the closed orbit for the case when the accelerator lattice is perturbed by a time-dependent harmonic dipole field. The problem is motivated by effects of current ripple in a proton accelerator. Our result allows to estimate the amplitude of the beam excursions as a function of the amplitude and the frequency of the perturbation. It predicts that the deviation of the beam increases as the frequency of the ripple approaches the sideband betatron frequency.

1 INTRODUCTION

Perturbation of the dipole magnetic field of an accelerator ring causes a distortion of the original closed orbit. If the perturbation is time independent, a new closed orbit is given by the well-known equation

$$y_{c.o.} = \frac{\delta B l \sqrt{\beta \beta_0}}{2 B \rho \sin \pi \nu} \cos(\zeta - \pi \nu - \zeta_0), \quad \zeta_0 \leq \zeta \leq \zeta_0 + 2\pi \nu, \qquad (1)$$

where y is the particle displacement, ζ denotes the betatron phase, β is the beta-function, ρ is the bending radius, B is the bending magnetic field, ν is the tune. In Eq. (1) we have assumed that the magnetic field is perturbed by δB in a short dipole magnet whose length is equal of l; ζ_0 and β_0 refer to the position of the magnet. The most general motion of a particle in the perturbed lattice consists of free betatron oscillations around the new closed orbit.

In the case when δB is a function of time, Eq. (1) is no longer valid. Generally speaking, the notion of the closed orbit itself loses sense and one has to compute a particle trajectory using a given function $\delta B(t)$ and proper initial conditions. In applications such as studies of current ripple in dipoles or quadrupole magnet vibrations it is often necessary to analyze the particle motion for harmonic dependence δB on t.

In this paper, we show that for a sinusoidal function δB one can define a particular trajectory that retains many useful features of the time-independent closed orbit. This orbit closes on itself after a few revolutions if the ratio of the frequency of the perturbation to the revolution frequency is a rational number. An arbitrary motion of a particle consists of free betatron oscillations around this orbit. To emphasize uniqueness of this trajectory we give it a special name – *quasi closed orbit* (QCO).

2 A CONCEPT OF A QUASI CLOSED ORBIT

A linear equation that describes the particle motion in the presence of a time-dependent perturbation of the magnetic field is given by

$$\frac{d^2\eta}{d\zeta^2} + \eta = \beta_0^{1/2}\epsilon(t)\sum_{m=-\infty}^{\infty}\delta\left(\zeta - m\mu - \zeta_0\right), \qquad (2)$$

where $\eta = yB^{-1/2}$, $\mu = 2\pi\nu$ and

$$\epsilon(t) = \delta B(t)\, l/B\rho. \qquad (3)$$

In Eq. (2) we have assumed that the length l of the dipole magnet is much shorter than the betatron wavelength: this allows to model the additional force acting on the particle from the perturbed magnetic field by a series of kicks described by the periodic delta function on the right of the equation. The betatron phase ζ plays the role of a time variable and changes from $-\infty$ to $+\infty$, each μ interval corresponding to a complete turn. Let us assume that

$$\epsilon(t) = \epsilon_0\cos(\omega t + \phi), \qquad (4)$$

where ω and ϕ are the frequency and the initial phase of the perturbation.

Before writing down a formal solution of Eq. (2) we should express the variable t in Eq. (4) through ζ. Since after each period T the betatron phase increases by μ, we have

$$t = \frac{\zeta}{\nu\Omega} + r(\zeta), \qquad (5)$$

where $r(\zeta)$ is a periodic function of ζ with period μ, and Ω is the revolution frequency, $\Omega = 2\pi/T$. Putting Eq. (5) into Eq. (4) and using the periodicity of $r(\zeta)$, one can expand the cosine in a Fourier series:

$$\epsilon(t) = \epsilon_0\sum_{m=-\infty}^{\infty} a_m\cos\left[\frac{1}{\nu}\zeta\left(\frac{\omega}{\Omega} + m\right) + \phi_m\right], \qquad (6)$$

where a_m and ϕ_m are the amplitude and the phase of different harmonics.

For the sake of simplicity we consider below the effect of only one harmonic in the series (6) corresponding to $m = 0$. This is a rather good approximation for the Supercollider because due to the large value of ν ($\nu = 123.28$) the function $r(\zeta)$ is small compared with 2π and can be neglected in Eq. (5). In what follows we assume that ϵ as a function of ζ can be given by

$$\epsilon(\zeta) = \epsilon_0\cos\left(\frac{\omega}{\nu\Omega}\zeta + \phi\right). \qquad (7)$$

A general solution of Eq. (2) is

$$\eta(\zeta) = \beta_0\int_{-\infty}^{\infty} d\xi\epsilon(\xi)\sin(\zeta - \xi)\sum_{m=-\infty}^{\infty}\delta(\xi - \zeta_0 - m\mu)$$

$$= \beta_0\sum_{m=-\infty}^{N}\epsilon(m\mu + \zeta_0)\sin(\zeta - \zeta_0 - m\mu), \qquad (8)$$

where N denotes an integer part of the ratio $(\zeta - \zeta_0)/\mu$. An integer N numbers complete turns performed by the particle. Putting Eq. (7) into Eq. (8) and trying to perform the summation we find that the sum does not converge in the limit $m \to -\infty$. This is related to the fact that we did not specify an initial condition at $t \to -\infty$. A particular trajectory we are looking for here is characterized by the perturbation of the magnetic field needing to be adiabatically slowly "turned on." That means that, strictly speaking, instead of a harmonic function given by Eq. (4) one has to consider a function that starts from zero value, slowly increases its amplitude, and after a transition period arrives at the sinusoid with a constant amplitude. If the transition from zero to the finite amplitude occurs slowly enough, the result will not depend on the specific form of the transition profile. The easiest way to perform the calculation is to assume that

$$\epsilon_0 = \hat{\epsilon} e^\gamma, \qquad (9)$$

where γ is a positive infinitesimally small number. Eq. (9) guarantees that the amplitude of the perturbation vanishes at $t \to -\infty$. Performing the summation and putting $\gamma \to 0$ in the final result one finds

$$\eta = \eta_{QCO} \equiv \frac{\epsilon_0 \beta_0 / 2}{\sin^2 \pi\nu \cos^2 \pi(\omega/\Omega) - \sin^2 \pi(\omega/\Omega) \cos^2 \pi\nu}$$

$$\times \left\{ \sin \pi\nu \cos \pi \frac{\omega}{\Omega} \cos\left(\zeta - \mu N - \frac{1}{2}\mu - \zeta_0\right) \cos\left[\frac{\omega}{\Omega}\left(2\pi N + \frac{\pi}{2} + \frac{\zeta_0}{\nu}\right) - \phi\right] - \right.$$

$$\left. \sin \pi \frac{\omega}{\Omega} \cos \pi\nu \sin\left(\zeta - \mu N - \frac{1}{2}\mu - \zeta_0\right) \sin\left[\frac{\omega}{\Omega}\left(2\pi N + \frac{\pi}{2} + \frac{\zeta_0}{\nu}\right) - \phi\right]\right\}. \qquad (10)$$

Eq. (10) gives a *quasi closed orbit* (QCO) that generalizes the notion of the closed orbit for a time-dependent case. One can easily check that in the limit $\omega \to 0$ Eq. (10) reduces to Eq. (1).

Since Eq. (10) gives a particular solution of the inhomogeneous equation (2), a general solution to Eq. (2) is the sum of η_{QCO} and a general solution of the homogeneous equation, that is free betatron oscillations,

$$\eta = \eta_{QCO} + A \cos(\zeta + \phi_0), \qquad (11)$$

where A and ϕ_0 are arbitrary amplitude and phase of the betatron oscillations.

Now we list the most important features of the QCO.

1. If the ratio ω/Ω is a rational number, $\omega/\Omega = k/l$, where k and l are integers, the trajectory given by Eq. (10) closes after l revolutions. This is illustrated by Fig. 1, which shows QCO for the following parameters: $\nu = 1.28$, $\omega/\Omega = 1/3$, $\phi = 0$, $\zeta_0 = 0$. Due to periodicity over ζ a particle that moves along the trajectory *1* and reaches the end right point at $\zeta = 2\pi\nu = 8.04$ jumps to the curve *2* at $\zeta = 0$ and continues the motion to the end right point on this curve and then again jumps to the curve *3*. After three revolutions the particle returns to the original curve *1*. Note that the three curves in Fig. 1 continuously match at their end points, however they have discontinuous first derivatives there (because the magnetic perturbation is located at $\zeta = 0$).

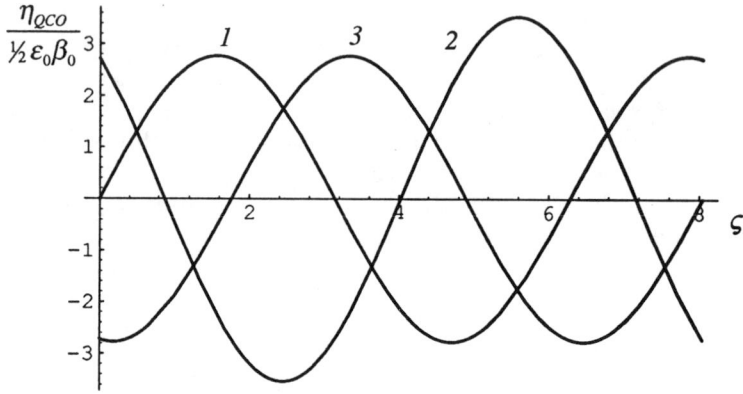

Fig. 1

2. The form of QCO is sensitive to the phase ϕ of the perturbation. Fig. 2 shows QCO for the same parameters as for Fig. 1 except that $\phi = \pi/2$. Typically, the phase ϕ is not as controllable as the frequency ω; it can easily fluctuate in the external perturbation even when ω is a well-defined quantity. Fluctuation of the phase would cause "breathing" of QCO and result in a picture similar to the case of the irrational ω/Ω (see below).

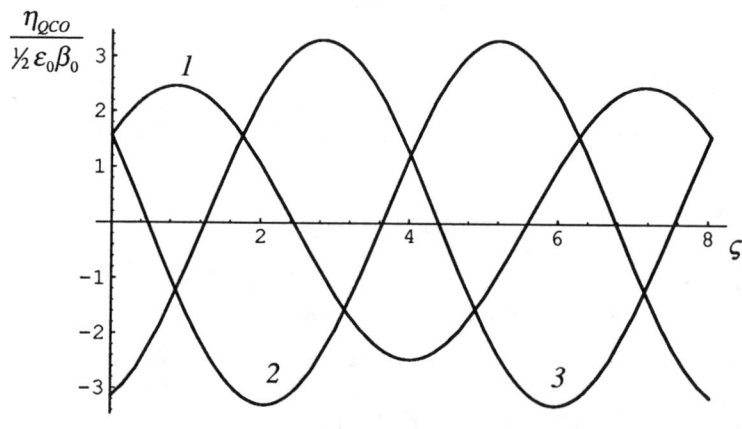

Fig. 2

3. For irrational ω/Ω the QCO does not close. Fig. 3 shows 20 turns of QCO for $\nu = 1.28$, $\omega/\Omega = 1/\pi$, $\phi = 0$, $\zeta_0 = 0$.

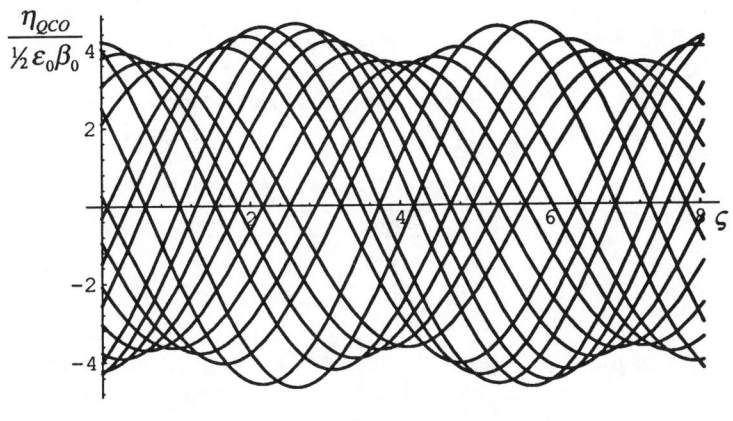

Fig. 3

4. As seen in Fig. 3 an important feature of the irrational case is the existence of the envelope of QCO. The equation of this envelope is given by

$$\eta_{\text{envelope}} = \frac{\pm \epsilon_0 \beta_0 / 2}{\sin^2 \pi \nu \cos^2 \pi (\omega/\Omega) - \sin^2 \pi (\omega/\Omega) \cos^2 \pi \nu}$$

$$\times \left\{ \sin^2 \pi \nu \cos^2 \pi \frac{\omega}{\Omega} \cos^2 \left(\zeta - \frac{1}{2}\mu - \zeta_0 \right) + \sin^2 \pi \frac{\omega}{\Omega} \cos^2 \pi \nu \sin^2 \left(\zeta - \frac{1}{2}\mu - \zeta_0 \right) \right\}^{1/2}.$$

(12)

For the parameters of Fig. 3 the envelope is shown in Fig. 4. As follows from Eq. (12), the envelope of QCO does not depend on the phase ϕ of the

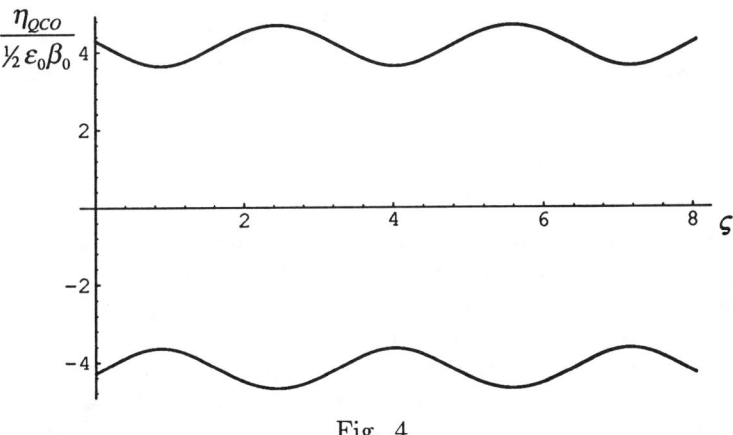

Fig. 4

perturbation. It can also be applied for the case of a rational ω/Ω when ϕ fluctuates in the range from 0 to 2π so that individual trajectories for a given ϕ (such as shown in Fig. 2 and 3) overlap. Using Eq. (12) one loses information about an exact offset of a particle at a given position of the ring but one can still predict the limits of the possible particle deviation at this position.

5. A more crude characteristic of the particle deflection on QCO is given by maximum and minimum value of η_{envelope}. A simple analysis yields

$$\max \eta_{\text{envelope}} = \frac{\epsilon_0 \beta_0 |\sin \pi \nu \cos \pi (\omega/\Omega)|/2}{|\sin^2 \pi \nu \cos^2 \pi (\omega/\Omega) - \sin^2 \pi (\omega/\Omega) \cos^2 \pi \nu|},$$
$$\min \eta_{\text{envelope}} = \frac{\epsilon_0 \beta_0 |\sin \pi (\omega/\Omega) \cos \pi \nu|/2}{|\sin^2 \pi \nu \cos^2 \pi (\omega/\Omega) - \sin^2 \pi (\omega/\Omega) \cos^2 \pi \nu|}. \quad (13)$$

The dependencies of max η_{envelope}(curve 1) and min η_{envelope} (curve 2) on ω/Ω for $\nu = 1.28$ are shown in Fig. 5. This figure also clearly demonstrates a resonant effect when the ratio ω/Ω approaches the fractional part of ν (equal to 0.28 in our case). For this frequency the denominator in Eqs. (13) vanishes causing unlimited increase of minimum and maximum η.

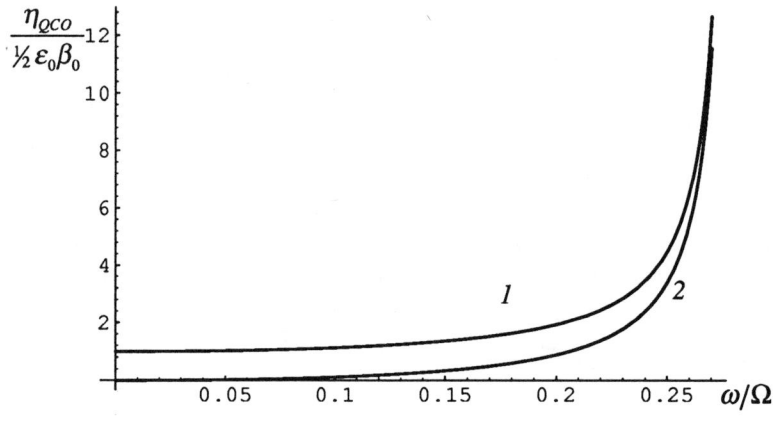

Fig. 5

3 CONCLUSION

A quasi closed orbit studied in this paper generalizes the notion of a perturbed closed orbit in a static magnetic field. Because of its dependence on the phase of the perturbation, the QCO itself is not particularly useful for applications, but its envelope allows one to determine the range of particle deviations at any position of the ring for a given amplitude of the fluctuation.

Due to linearity of basic equations, our consideration can be easily extended to include effects of many fluctuating magnets as well as several frequencies of the perturbation.

Dynamic Aperture and Emittance Growth Rates in the HERA Proton Ring

Frank Zimmermann
Deutsches Elektronen-Synchrotron DESY
Notkestr. 85 2000 Hamburg 52 Germany

Abstract

The dynamic aperture of the HERA proton ring is limited by persistent current field errors in the superconducting magnets. Its value as predicted by tracking studies is compared to preliminary aperture measurements in the machine. Since inclusion of tune modulation leads to a significant reduction of the aperture in the simulation, a theory is developed, which quantifies the impact of tune modulation on high-order resonances. For a typical working point, parameters of resonances up to order 11 are calculated for a model of HERA using differential algebra and normal-form algorithms. Based on these resonance parameters, estimate values are derived of amplitude-dependent emittance growth rates caused by the interplay of tune modulation and nonlinear field errors.

1 Introduction

At injection energy persistent current field errors can strongly limit the dynamic aperture of the HERA proton ring. In this report the effect of the measured nonlinear field errors is determined both by tracking studies and by an analytical treatment. Particular emphasis is put on the effect of an additional tune modulation.

2 Dynamic Aperture and Tracking Studies

2.1 Definition

The dynamic aperture is defined as the border in phase space, inside which particles are stable for a sufficiently large number of turns. A typical number is $6 \cdot 10^7$ turns, which corresponds to the time required to inject 200 proton bunches into HERA. Because such a number is beyond the reach of present computers, additional criteria are needed, which allow one to detect potentially unstable trajectories using tracking data for a much smaller number of turns.

2.2 Lyapunov Exponent

A very promising method of this kind determines the rate of divergence of two initially close-by trajectories in phase space, which is described by the Lyapunov exponent [1]. A trajectory in phase space is either regular or chaotic. Regular motion is characterized by a linear divergence of nearby trajectories, whereas chaotic motion shows an exponential divergence. The formal definition of the Lyapunov exponent is

$$\lambda \equiv \lim_{N\to\infty} \lim_{d(0)\to 0} \frac{1}{N} \ln \frac{d(N)}{d(0)}, \tag{1}$$

where $d(N)$ denotes the distance between the two trajectories after N turns. The limit of vanishing initial distance $d(0) \to 0$ in (1) can be performed exactly. This leads to an alternative definition of the Lyapunov exponent, which uses the largest eigenvalue $EV_{N,\max}$, of the N-turn Jacobian matrix J_N,

$$\lambda \equiv \lim_{N \to \infty} \lambda_N \equiv \lim_{N \to \infty} \frac{1}{N} \ln EV_{N,\max}, \tag{2}$$

where the Jacobian J_N expresses the dependence of the final phase-space coordinates in terms of the initial ones in a linear approximation. In the simulation studies of HERA always a sharp amplitude threshold for the onset of chaotic motion has been found. Even though the Lyapunov exponent primarily reflects the angular divergence of two trajectories, chaotic particles generally show an emittance growth on a longer time scale. Thus an exponential angular divergence indicates a long-term amplitude increase and final particle loss. This supposition is the main reason for the use of Lyapunov exponents in tracking studies and has been confirmed for all cases under consideration. An example is presented in Fig. 1.

2.3 The Pure Impact of Persistent Current Field Errors

Comprehensive tracking studies have been performed with modified versions of RACETRACK [2] and SIXTRACK [3], to simulate the effect of the nonlinear field errors of all superconducting dipoles and quadrupoles. The simulation is based on the individual measured multipole errors of each magnet. Due to a faulty database the 20-pole coefficient of the quadrupole magnets is ten times larger than the measured value. As a consequence of this the results presented for HERA might be too pessimistic. The strengths of the 6-pole, 10-pole and 12-pole correction coils are close to the actual ones. In the model of HERA the nonlinear fields have been taken into account by five thin lenses per half cell.

The dynamic aperture expected from the field errors is depicted in Fig. 2. Also indicated in the figure is a rough estimate of the actual dynamic aperture, as deduced from beam profile measurements [4] for poorly injected beams. The prediction differs from the actual value by about a factor of 2.

2.4 Tracking Studies including Tune Modulation

A better agreement between tracking and measurement is obtained, if a tune modulation is included in the simulation. Fig. 3 illustrates a typical emittance evolution in the presence of an external tune modulation. Clearly visible are step-like increases, which occur in intervals equal to the modulation period. The start coordinates for the shown trajectory lie far inside a region, which is regular in the absence of the modulation.

The addition of tune modulation causes a substantial reduction of the dynamic aperture in the simulation and thus may explain the remaining difference to the actual aperture. As an example, a 10-Hz tune modulation with an amplitude of 10^{-3} reduces the acceptance from $4.2\,\pi$ mm mrad to $1.5\,\pi$ mm mrad. The strong impact of tune modulation observed in the tracking calls for an analytical treatment.

Our further strategy is as follows.

- We outline a theory, which describes the effect of tune modulation on a single resonance.

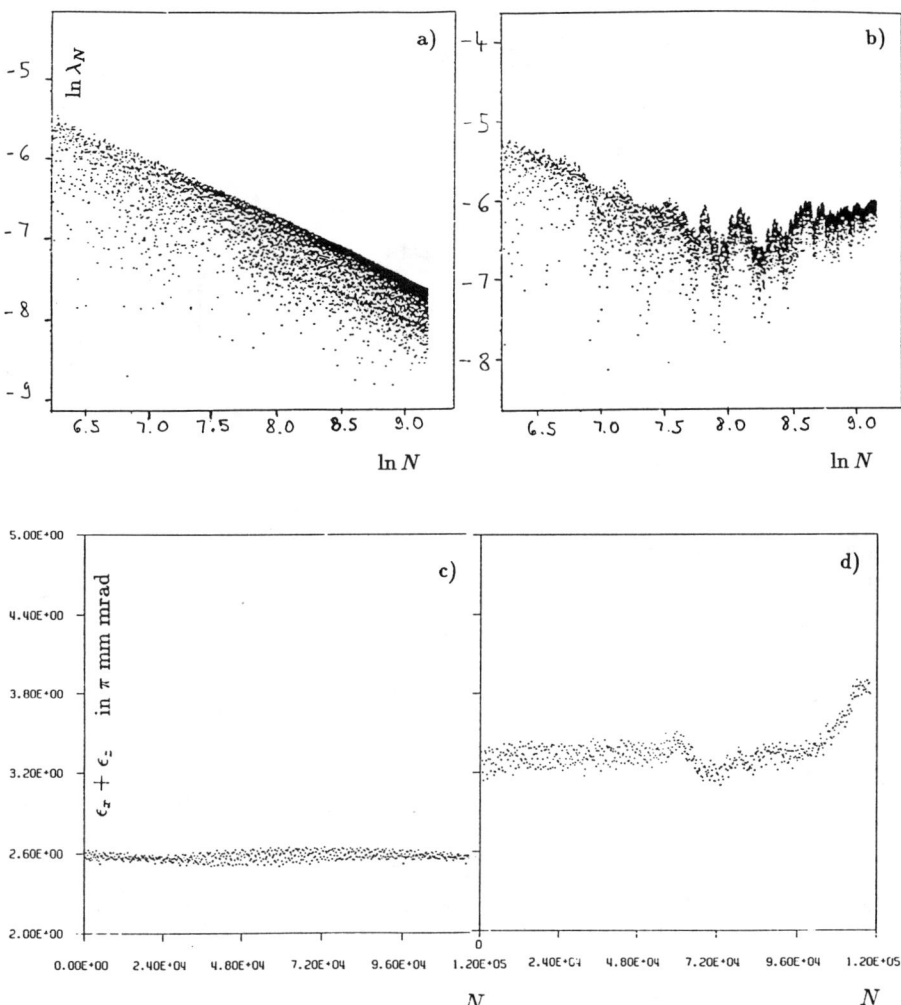

Figure 1: Lyapunaov exponent for a regular and a chaotic trajectory in HERA and the corresponding emittance evolution: a) and b) $\ln \lambda_N$ as a function of $\ln N$ for 10,000 turns at a start amplitude of 11.9 mm and 13.5 mm respectively ($\beta = 76$ m) with $\Delta p/p = 10^{-3}$, c) and d) transverse emittance $\epsilon_x + \epsilon_z$ over 120,000 revolutions for the same two trajectories; each dot represents the maximum value during 200 turns.

Dynamic Aperture and Emittance Growth Rates

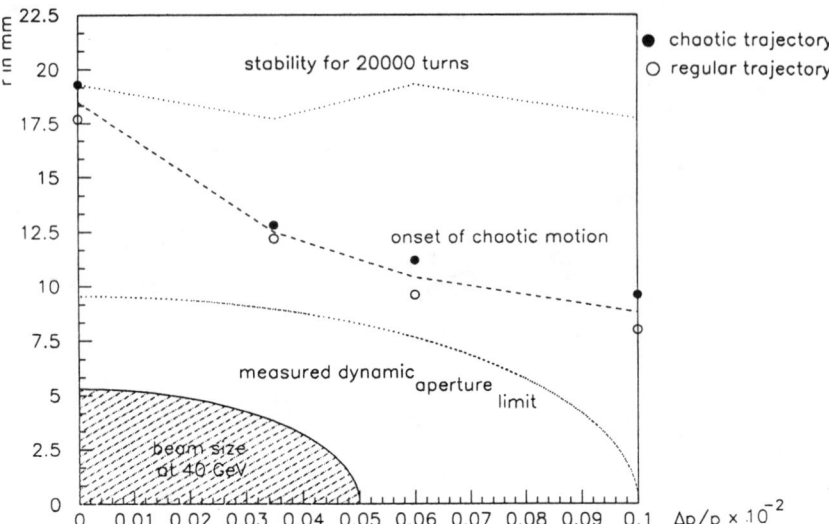

Figure 2: Predicted dynamic aperture in the arcs of the HERA proton ring for the working point $Q_x = 31.27$, $Q_z = 32.30$ and a circumferential rf voltage of 70 kV. Shown are amplitude values r in mm, defined by the equation $r = \sqrt{\beta(\epsilon_x + \epsilon_z)}$ with $\beta = 76$ m, as a function of the initial relative momentum deviation.

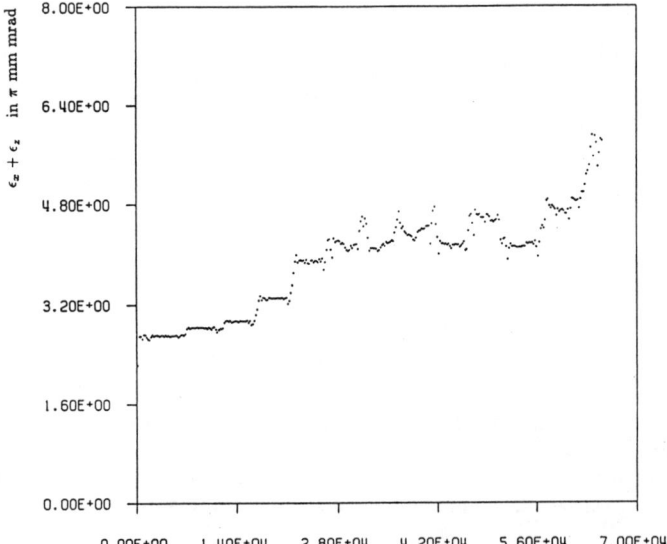

Figure 3: Transverse emittance $\epsilon_x + \epsilon_z$ for 70,000 revolutions in HERA at a start amplitude of 11.0 mm ($\beta - 76$ m). Each dot represents the maximum value during 200 turns. The step-like changes occur with the tune modulation period of 5,000 turns, which corresponds to a frequency of 10 Hz. The modulation amplitude is $q = 0.001$.

- Parameters of high-order resonances in HERA are extracted by means of differential algebra and normal-form transformations.
- Amplitude-dependent emittance growth rates are predicted and an analytical estimate of the dynamic aperture is derived.

3 Tune Modulation in a Nonlinear Hamiltonian System

3.1 Hamiltonian

The general form of a nonlinear Hamiltonian describing the single resonance $kQ_x + lQ_z \approx p$ is

$$H(I_x, I_z, \phi_x, \phi_z, \theta) = I_xQ_x + I_zQ_z + g(I_x, I_z) + h(I_x, I_z)\cos(k\phi_x + l\phi_z - p\theta) + q \cdot (k+l) \cdot (I_x + I_z) \cdot \cos(Q_m\theta + \alpha) \quad (3)$$

where the last term represents an additional tune modulation of amplitude q and frequency Q_m in both transverse planes. The modulation frequency is given in units of the revolution frequency, which is 47.3 kHz for HERA. Instead of a detailed theory, which can be found in [5], we only want to briefly list a few remarks here. It is well known [6] that the above Hamiltonian can be further approximated close to the resonance by a nonlinear pendulum Hamiltonian. An important parameter is then the island tune Q_I, that is the frequency at which small-amplitude particles oscillate around the elliptical fixed point. The island tune is given by

$$Q_I = \sqrt{\left[k^2 \frac{\partial^2 g(I_{x,r}, I_{z,r})}{\partial I_x^2} + 2kl\frac{\partial^2 g(I_{x,r}, I_{z,r})}{\partial I_x I_z} + l^2 \frac{\partial^2 g(I_{x,r}, I_{z,r})}{\partial I_z^2}\right] h(I_{x,r}, I_{z,r})}, \quad (4)$$

where the subindex r denotes the value at the resonance, and the total island width $\Delta I_{\text{tot}} = \sqrt{\Delta I_x^2 + \Delta I_z^2}$ of the resonance is written as

$$\Delta I_{\text{tot}} = \frac{4Q_I\sqrt{l^2 + k^2}}{|l^2 \frac{\partial^2 g}{\partial I_z^2} + 2kl\frac{\partial^2 g}{\partial I_x \partial I_z} + k^2 \frac{\partial^2 g}{\partial I_x^2}|} \quad (5)$$

A tune modulation causes new resonances in phase space. The i-th sideband resonance is defined by

$$kQ_x + lQ_z + iQ_m + p = 0. \quad (6)$$

3.2 Phase Diagram

In the parameter space of tune modulation, areas with qualitatively different motion can be graphically represented in a 'phase diagram' as proposed in [7]. The borderlines in the (q, Q_m)-plane, given by the equalities

$$\frac{Q_m^{\frac{3}{4}}(k+l)^{\frac{1}{4}}q^{\frac{1}{4}}}{Q_I} = \frac{4}{\pi^{\frac{1}{4}}},$$

$$\left|\frac{qQ_m}{Q_I^2 - Q_m^2}\right| = \frac{1}{k+l}, \quad (7)$$

separate regions with distinct dynamical behavior. The first equation specifies the onset of chaos due to an overlap of sidebands. The second gives the small-angle boundary, which corresponds either to the adiabatic limit or to a vanishing size of the first sideband for small and large modulation frequencies respectively [7]. The characteristic properties of the nonlinear system enter into the phase diagram only in the form of two parameters, namely the sum $k+l$ specifying the resonance and the island tune Q_I. Global chaos and particle losses occur in the left upper corner of the (q, Q_m)-plane.

3.3 Width of the Stochastic Layer

An alternative phase diagram is derived by an argumentation developed in [6]. The motion near the pendulum separatrix can be approximated by a two-dimensional map of the form

$$\bar{w} = w + \frac{(k+l)q}{2Q_I}\left(\mathcal{A}_2(\frac{Q_m}{Q_I}) - \mathcal{A}_2(-\frac{Q_m}{Q_I})\right)\sin\alpha; \tag{8}$$

$$\bar{\alpha} \approx \alpha + \pi\frac{Q_m}{Q(\bar{w})} \approx \alpha + \frac{Q_m}{Q_I}\ln\frac{32}{|\bar{w}|}. \tag{9}$$

Here w denotes the relative energy deviation from the separatrix, $w \equiv (H_0 - h)/h$, and the Melnikov-Arnold integral $\mathcal{A}_2$ is given by

$$\mathcal{A}_2\left(\frac{Q_m}{Q_I}\right) = 4\pi\frac{Q_m}{Q_I}\frac{e^{\frac{\pi Q_m}{2Q_I}}}{\sinh\left(\frac{\pi Q_m}{Q_I}\right)}. \tag{10}$$

Either by examining the range of stability for the fixed points of this map or by linearizing the map in w, the width of the stochastic layer close to the separatrix is obtained:

$$w_{sl} = \pi(k+l)q\frac{Q_m^2}{Q_I^3}\frac{\sinh\left(\frac{\pi Q_m}{2Q_I}\right)}{\sinh\left(\frac{\pi Q_m}{Q_I}\right)}. \tag{11}$$

The contour line $w = w_0$, for a properly chosen constant w_0, predicts a border between chaotic and regular regions alternatively to (7).

Figure 4 shows the width of the stochastic layer w as a function of the modulation frequency Q_m for a typical 10th-order resonance in HERA and a constant modulation amplitude $q = 10^{-3}$. For large modulation frequencies the width decreases exponentially, which indicates that a high-frequency tune modulation does not affect the particle dynamics.

The two alternative phase diagrams are compared in Fig. 5. The chaotic regions agree approximately for both approaches, but there are also a few important differences. According to the two equations (7) an arbitrarily small modulation amplitude can cause global chaos if the modulation frequency is close to the island tune Q_I, whereas the single equation (11) predicts a maximum stochastic width at $Q_M \approx 1.35 Q_I$ but no resonance-like behavior.

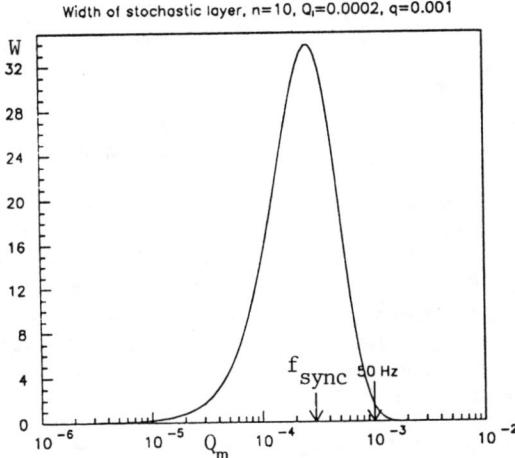

Figure 4: Width of the stochastic layer as a function of the modulation frequency for a typical 10th-order resonance in HERA according to (11).

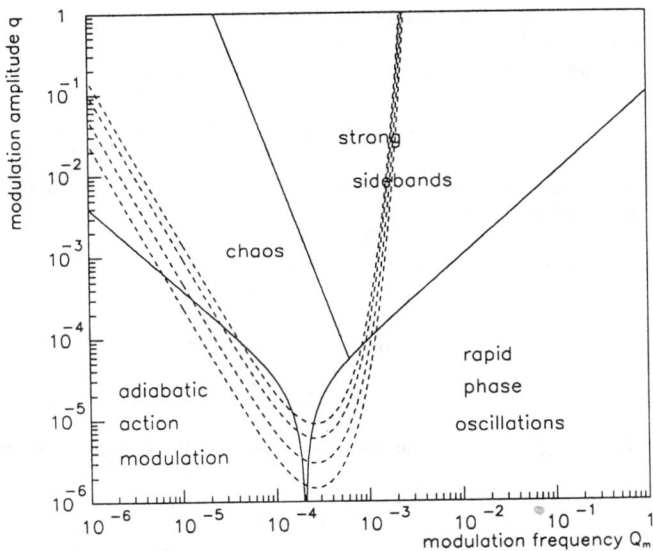

Figure 5: Phase diagram for the island tune $Q_I = 2 \cdot 10^{-4}$ and the resonance order $n = 10$ according to (7) (solid curves) and contour lines (11) with $w = 0.05, 0.1, 0.2, 0.3$ (dotted curves). This diagram represents a typical high-order resonance at injection energy in HERA.

3.4 Octupole Kick Map

It is worthwhile to compare the predictions of both schemes for a simple octupole kick map of the following form:

$$\begin{pmatrix} x \\ p \end{pmatrix}_f = \begin{pmatrix} \cos 2\pi Q_0 & \sin 2\pi Q_0 \\ -\sin 2\pi Q_0 & \cos 2\pi Q_0 \end{pmatrix} \begin{pmatrix} x \\ p - x^3 \end{pmatrix}_i. \qquad (12)$$

For our study we choose $Q_0 = 0.23$ and thus stay close to the resonance $4Q_0 = 1$. The island tune for this resonance is about $Q_I \approx 0.049$ and we have $k = 4$ and $l = 0$. The modulation amplitude is kept fixed at $q = 2 \cdot 10^{-3}$. Figure 6 presents the phase-space diagrams for ten different modulation frequencies. Each diagram shows six trajectories plotted once every modulation period. While according to (7) global chaos is expected only in the narrow range $0.044 < Q_m < 0.054$, (11) predicts large phase-space areas to be covered with chaotic trajectories roughly in the range $0.012 < Q_m < 0.2$, if $w_0 \approx 0.01$. It is evident from the figure that the prediction of (7) is much too restrictive.

3.5 Diffusion Rate in the Chaotic Region

Following earlier work [8–10] the diffusion rate in the chaotic regions of phase space can be estimated in a simple manner. The change in emittance is calculated for a single crossing of the resonance by integrating the equations of motion. This typically leads to the evaluation of Fresnel integrals. If one assumes that between successive resonance crossings the phase correlation of different particles is lost, the mean-squared change of the emittance can be calculated as an incoherent sum of single resonance crossings. Then the expression for the local emittance growth becomes

$$\frac{\overline{(\Delta(\epsilon_x + \epsilon_z))^2}}{\Delta t} = 32 \frac{f_{rev}(k+l)h^2(I_{x,r}, I_{z,r})}{q}. \qquad (13)$$

To examine this formula, the average emittance spread is determined numerically for a two-dimensional map similar to (12):

$$\begin{pmatrix} x \\ p \end{pmatrix}_f = \begin{pmatrix} \cos 2\pi Q & \sin 2\pi Q \\ -\sin 2\pi Q & \cos 2\pi Q \end{pmatrix} \begin{pmatrix} x \\ p - x^3 \end{pmatrix}_i, \qquad (14)$$

where in this case Q depends on the amplitude in the form

$$Q = Q_0 + \frac{\alpha}{4}(x_i^2 + p_i^2)^2. \qquad (15)$$

If we choose $Q_0 = 0.23$ and $\alpha = 1000$ the driving term of the fourth-integer resonance

$$h(I) = \frac{1}{16\pi} I^2 \qquad (16)$$

can be varied independently of the detuning. A tune modulation with amplitude $q \approx 0.1$ and frequency $Q_m \approx 6.25 \cdot 10^{-3}$, thus not too far from the island tune $Q_I = 7.6 \cdot 10^{-3}$, gives rise to global chaos in agreement with (7). The rms-emittance-spread of an ensemble of 200 particles, launched inside the chaotic region with a tiny initial spread in position of 10^{-7}, is shown in Fig. 7 as a function of the turn number. After a transient period of about 2000 turns, in which the initially very close particles lose their phase correlation, the evolution is well described by (13).

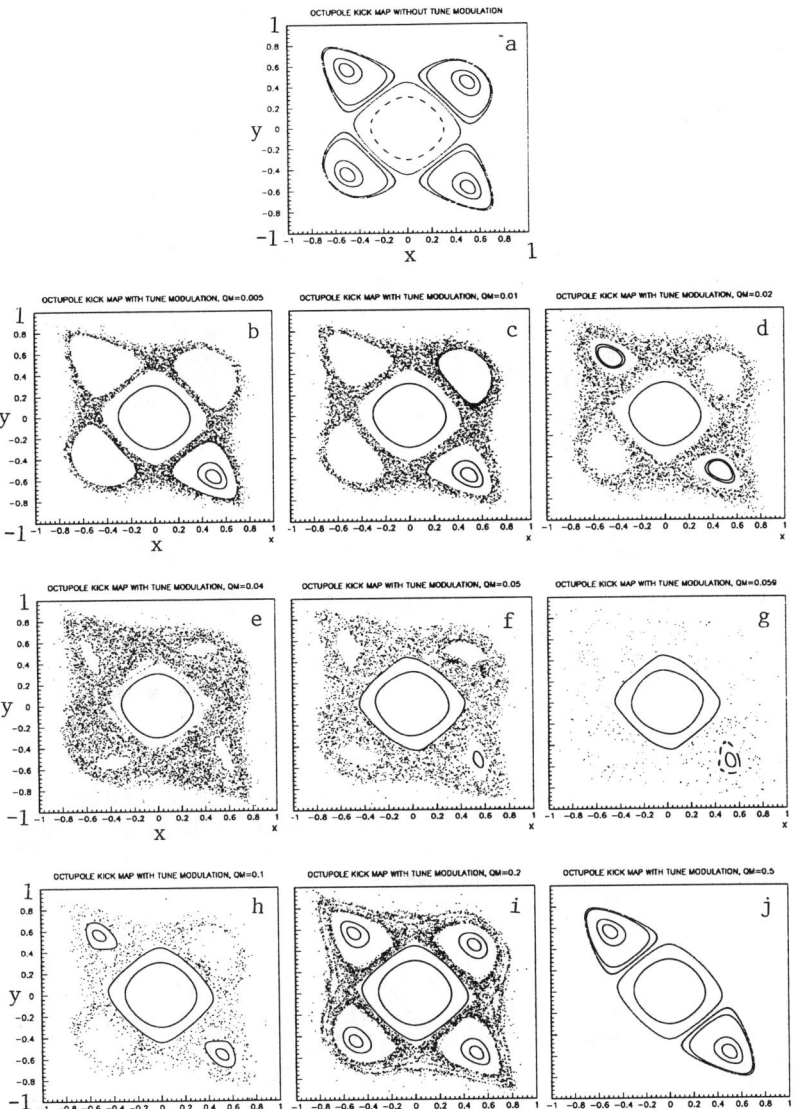

Figure 6: Phase-space diagrams of six trajectories for the octupole kick map (12) with a tune modulation amplitude $q = 0.002$; each diagram corresponds to a different modulation frequency Q_m, namely a) no modulation, b) $Q_m = 0.005$, c) $Q_m = 0.01 \approx \frac{1}{5}Q_I$, d) $Q_m = 0.02$, e) $Q_m = 0.04$, f) $Q_m = 0.05 \approx Q_I$, g) $Q_m = 0.1$, h) $Q_m = 0.1$, i) $Q_m = 0.2 \approx 4Q_I$, j) $Q_m = 0.5$.

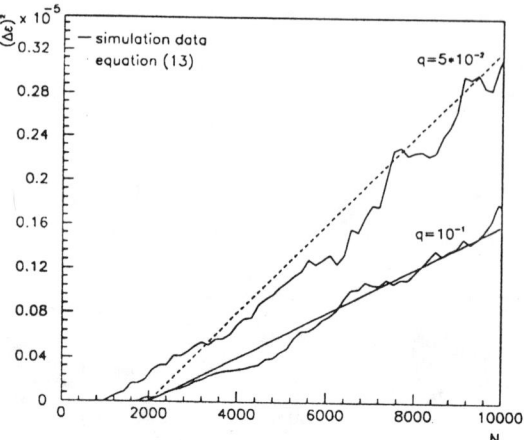

Figure 7: Emittance spread for 200 chaotic trajectories of the two-dimensional map (14)–(15) as a function of the turn number.

4 Parameters of High-Order Resonances in HERA

In the last section the effect of tune modulation on a single resonance was studied leading to expressions for the size of the resonance island ΔI_{tot}, the chaotic fraction of the island w and the emittance growth rate inside the stochastic layer $[\Delta(\epsilon_x+\epsilon_z)]^2/\Delta t$. To apply the developed theory to the HERA proton ring, typical values of $(k+l)$, $\partial g(I_x,I_z)/\partial I_{x,z}$, $\partial^2 g(I_x,I_z)/\partial I_{x,z}^2$ and $h(I_x,I_z)$ for high-order resonances have to be evaluated. This can be done by means of differential algebra [11] and normal-form procedures [12].

In general a normal-form transformation diverges above some order due to small resonance denominators. A verification of its well-behaving is therefore necessary to get reliable results. Appropriate methods to confirm the convergence are to apply the normal-form transformation to tracking data, which in case of no divergence should be transformed into almost a circle, or to compare the tunes predicted by the normal-form analysis with the results from tracking. According to Fig. 8 tunes from an 8th-order normal-form analysis for HERA do not agree with the tracking data over the full amplitude range of interest. An 11th-order normalization diverges strongly at about half the dynamic aperture. A way to overcome this difficulty and to determine resonances up to 11th order, is to first perform an 8th-order normal-form transformation and thereafter to rewrite the remainder as a Dragt-Finn factorization [13]. The original map M is then cast into the following form:

$$M = A^{-1} e^{:-2\pi Q I + t_3(I)+...+t_8(I):} e^{:f_9(I,\phi):} \ldots e^{:f_{11}(I,\phi):} A + \mathcal{O}(12), \quad (17)$$

where the t_n and the f_n are polynomials of degree n in $x_k = \sqrt{2I_k}\cos(\phi_k)$ and $p_k = -\sqrt{2I_k}\sin(\phi_k)$. A denotes the 8th-order normal-form transformation. The detuning is deduced from the approximate Hamiltonian

$$H_{approx} = A^{-1}\left[QI - \frac{1}{2\pi}\{t_3(I)+\ldots+t_8(I)+ < f_9(I,\phi)+\ldots+f_{11}(I,\phi) >_\phi\}\right]. \quad (18)$$

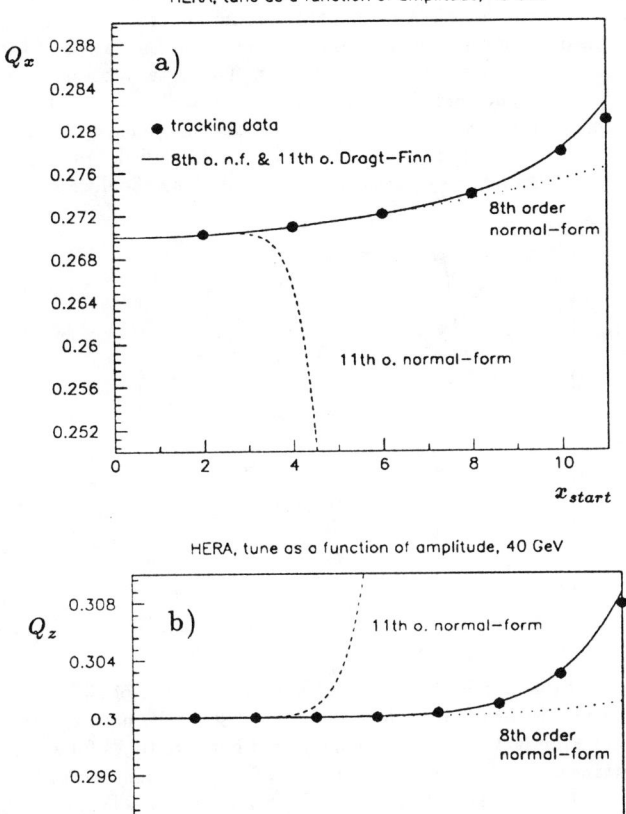

Figure 8: a) Horizontal tune Q_x as a function of the start amplitude x_{start} ($\beta_x = 28.5$ m, $z_{start} = 0$), b) vertical tune Q_z as a function of the start amplitude z_{start} ($\beta_x = 3.8$ m, $x_{start} = 0$).

Figure 8 demonstrates that this schemes allows one to reproduce the tracking results up to the dynamic aperture, which is at about $z = 4$ mm and $x = 11$ mm.

Hence we are in position to draw a diagram representing the values of the single-particle tunes as a function of the initial emittances. The latter are varied between 0 and 4π mm mrad along the three lines $\epsilon_x = 0$, $\epsilon_z = 0$ and $\epsilon_x = \epsilon_z$. The resulting tune diagram is depicted in Fig. 9 together with the resonance lines up to order 11. In this figure, 16 crossed resonances of order 7 to 11 can be identified. As an illustration we present the parameters for the six resonances crossed along the diagonal $\epsilon_x = \epsilon_z$ in Table 1.

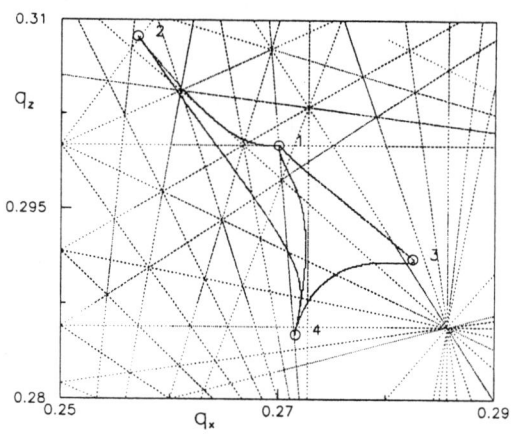

Figure 9: Diagram of the amplitude-dependent particle tune and of all resonance lines up to order 11. The tunes are obtained by the combined 8th-order normal-form analysis / 11th-order Dragt-Finn factorization. The numbered dots are tunes for special values of the start emittances: 1. $\epsilon_x = 0, \epsilon_z = 0$, 2. $\epsilon_x = 0, \epsilon_z = 4\pi$ mm mrad, 3. $\epsilon_x = 4\pi$ mm mrad, $\epsilon_z = 0$, 4. $\epsilon_x = 4\pi$ mm mrad, $\epsilon_z = 4\pi$ mm mrad. The connecting lines correspond to a continuous variation of the initial emittances between these values.

Table 1: Characteristics of high-order resonances in HERA as a function of start amplitudes x and z.

x	z	Resonance	$\partial^2 g/\partial I_x^2$	$\partial^2 g/\partial I_x \partial I_z$	$\partial^2 g/\partial I_z^2$
5.489	1.996	$7Q_x - 3Q_z = 1$	$0.37 \cdot 10^{-2}$	$-0.25 \cdot 10^{-2}$	$-0.34 \cdot 10^{-3}$
6.215	2.260	$-4Q_x + 7Q_z = 1$	$0.35 \cdot 10^{-2}$	$-0.24 \cdot 10^{-2}$	$-0.6 \cdot 10^{-3}$
7.568	2.752	$3Q_x + 4Q_z = 2$	$0.29 \cdot 10^{-2}$	$-0.23 \cdot 10^{-2}$	$-0.17 \cdot 10^{-2}$
9.856	3.584	$2Q_x + 5Q_z = 2$	$-0.85 \cdot 10^{-3}$	$-0.35 \cdot 10^{-2}$	$-0.80 \cdot 10^{-2}$
10.540	3.832	$Q_x + 6Q_z = 2$	$-0.3 \cdot 10^{-2}$	$-0.5 \cdot 10^{-2}$	$-0.12 \cdot 10^{-1}$
10.920	3.972	$7Q_z = 2$	$-0.5 \cdot 10^{-2}$	$-0.6 \cdot 10^{-2}$	$-0.15 \cdot 10^{-1}$

x	z	h	Q_I
5.489	1.996	$1.0 \cdot 10^{-7}$	$1.7 \cdot 10^{-4}$
6.215	2.260	$0.16 \cdot 10^{-6}$	$1.6 \cdot 10^{-4}$
7.568	2.752	$0.56 \cdot 10^{-5}$	$5.6 \cdot 10^{-4}$
9.856	3.584	$0.39 \cdot 10^{-4}$	$3.2 \cdot 10^{-3}$
10.540	3.832	$0.35 \cdot 10^{-4}$	$4 \cdot 10^{-3}$
10.920	3.972	$0.32 \cdot 10^{-5}$	$2 \cdot 10^{-3}$

5 Chaotic Fraction of Phase Space

The total fraction Γ of phase space that becomes chaotic due to a tune modulation of frequency Q_m and amplitude q is given by the product of the total island width and the width of the stochastic layer, summed over all relevant resonances and normalized to the total emittance range:

$$\Gamma(Q_m, q) \equiv \sum_{\text{all resonances } i} 2\Delta I^i_{tot} \cdot w_i(Q_m, q)/\epsilon_{tot}$$

$$= \sum_i \left(\frac{Q_m}{Q_{I,i}}\right)^2 \frac{\sinh\left(\frac{\pi Q_m}{2Q_{I,i}}\right)}{\sinh\left(\frac{\pi Q_m}{Q_{I,i}}\right)} \frac{\sqrt{l_i^2 + k_i^2}(l_i + k_i)8\pi q}{\left|l_i^2 \frac{\partial^2 g_i}{\partial I_z^2} + 2k_i l_i \frac{\partial^2 g_i}{\partial I_x \partial I_z} + k_i^2 \frac{\partial^2 g_i}{\partial I_x^2}\right|} \frac{1}{\epsilon_{tot}}. \quad (19)$$

The subindex i denotes the individual resonances. Each resonance will contribute to the sum mostly near its island tune. Γ can equally be viewed as a sensitivity function, which measures the harmfulness of different modulation frequencies Q_m. It is shown in Fig. 10 as a function of the modulation frequency Q_m for the fixed modulation amplitude $q = 10^{-4}$. The figure suggests that a tune modulation at frequencies above 100 Hz has no harmful effect on the particle dynamics.

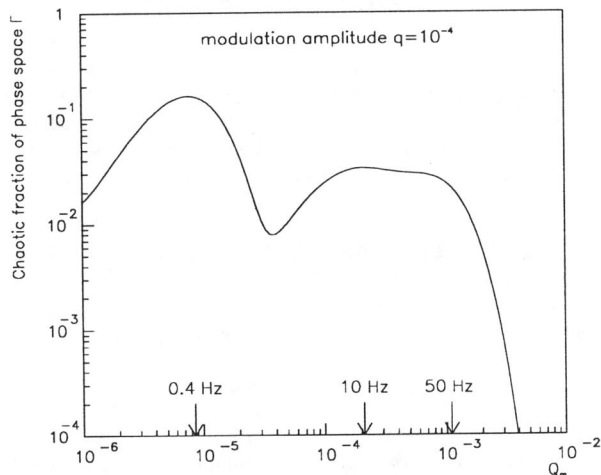

Figure 10: Chaotic fraction of phase space Γ (19) as a function of the modulation frequency.

6 Diffusion Model

If the particle distribution function f in the chaotic regions of phase space depends only on the transverse emittance $\epsilon \equiv \epsilon_x + \epsilon_z$ and obeys a diffusion equation of the form

$$\frac{\partial f}{\partial t} = \frac{\partial}{\partial \epsilon}\left(D(\epsilon)\frac{\partial f}{\partial \epsilon}\right) \quad (20)$$

the diffusion coefficient $D(\epsilon)$ is related to the squared emittance change per time interval by the simple formula

$$D(\epsilon) = \frac{1}{2} \left\langle \frac{\overline{(\Delta\epsilon)^2}}{\Delta t} \right\rangle \qquad (21)$$

and the mean emittance growth rate $< \overline{\Delta\epsilon} > /\Delta t$ satisfies

$$\left\langle \frac{\overline{\Delta\epsilon}}{\Delta t} \right\rangle \approx \frac{\partial}{\partial\epsilon} D(\epsilon). \qquad (22)$$

Local approximations to the diffusion coefficient $D(\epsilon)$ are obtained by averaging over the phase-space region between two adjacent resonances, assuming that regular trajectories show no diffusion and that the emittance growth rate in chaotic areas close to a specific resonance is given by (13). Then our estimate for the emittance-dependent diffusion coefficient reads

$$D(\epsilon_i) = \frac{1}{2} \left\langle \frac{\overline{[\Delta(\epsilon_x + \epsilon_z)]^2}}{\Delta t} \right\rangle_i \approx 2 \cdot \Delta I_{tot}^i \cdot w_i \cdot \left\langle \frac{\overline{[\Delta(\epsilon_x + \epsilon_z)]^2}}{\Delta t} \right\rangle_i \frac{1}{\epsilon_{i+1} - \epsilon_{i-1}}$$

$$= \left(\frac{Q_m}{Q_{I,i}}\right)^2 \frac{\sinh\left(\frac{\pi Q_m}{2Q_{I,i}}\right)}{\sinh\left(\frac{\pi Q_m}{Q_{I,i}}\right)} \frac{256\pi\sqrt{k_i^2 + l_i^2}(k_i + l_i)^2 f_{rev} h_i^2(I_{x,r,i}, I_{z,r,i})}{\left| l_i^2 \frac{\partial^2 g_i}{\partial I_z^2} + 2k_i l_i \frac{\partial^2 g_i}{\partial I_z \partial I_x} + k_i^2 \frac{\partial^2 g_i}{\partial I_x^2} \right|} \frac{1}{\epsilon_{i+1} - \epsilon_{i-1}}. \qquad (23)$$

Note that instead of the total emittance range ϵ_{tot}, the normalization constant is now the distance between two adjacent resonances $(\epsilon_{i+1} - \epsilon_{i-1})/2$. The diffusion coefficient $D(\epsilon)$ in (23) is independent of the modulation amplitude q, which reflects the fact that the fraction of phase space covered with chaotic trajectories increases linearly with the modulation amplitude q, whereas the diffusion rate in this zone decreases as $1/q$ [see (11) and (13)].

7 Emittance Growth Rates in HERA

The emittance-dependent diffusion coefficient $D(\epsilon)$, defined in (23), is shown in Fig. 11 assuming a modulation frequency of 50 Hz. A possible parametrization is

$$D(\epsilon) \begin{cases} \sim 0 & \text{for } \epsilon < \epsilon_0 \\ \sim c \cdot \epsilon^\alpha & \text{for } \epsilon > \epsilon_0. \end{cases} \qquad (24)$$

Here ϵ_0 denotes the threshold for tune-modulation-induced diffusion. It can be considered as an estimate of the dynamic acceptance and has the value (see Fig. 11)

$$\epsilon_0 \approx (1 - 1.5) \pi \text{ mm mrad}. \qquad (25)$$

This is in remarkable agreement with the threshold of chaos found in tracking simulations including tune modulation and with the actually observed acceptance of $\sim 1.2\pi$ mm mrad. The coefficient c is approximately

$$c \approx 5 \cdot 10^{-14} \, (\pi \text{ mm mrad})^{2-\alpha} \, \text{s}^{-1} \qquad (26)$$

and the exponent

$$\alpha \approx 15. \qquad (27)$$

Nearly the same values of ϵ_0 and α are obtained for a modulation frequency of 10 Hz. According to (22) the locally averaged emittance growth is given by

$$\left\langle \overline{\frac{\Delta\epsilon}{\Delta t}} \right\rangle \approx \alpha c \epsilon^{\alpha-1}, \qquad (28)$$

which yield

$$\overline{\frac{\Delta\epsilon}{\Delta t}} \approx \begin{Bmatrix} 10^{-12} \\ 10^{-8} \\ 2\cdot 10^{-4} \end{Bmatrix} \pi\text{ mm mrad s}^{-1} \text{ at } \epsilon = \begin{Bmatrix} 1 \\ 2 \\ 4 \end{Bmatrix} \pi\text{ mm mrad}. \qquad (29)$$

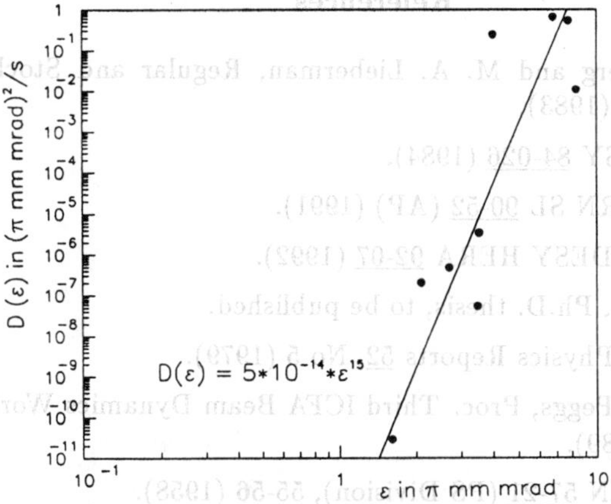

Figure 11: Emittance-dependent diffusion coefficient $D(\epsilon)$ (23) (closed circles) at a modulation frequency $Q_m = 10^{-3}$ for the 18 crossed resonances of order 6 to 11.

8 Summary and Conclusion

The observed dynamic aperture of about 1.2 π mm mrad in the HERA proton ring at 40 GeV is in good agreement with the results of computer simulations and analytical predictions which include a modest tune modulation and the effect of the measured nonlinear field errors of all superconducting magnets.

It was shown that a suitable combination of normal-form and factorization processes allows one to extract parameters of resonances including 11th order inside the whole dynamic aperture for a complex storage ring such as HERA.

A detailed theory describing the effect of an additional tune modulation in a nonlinear Hamiltonian system has been developed, which applies both to the nonlinear magnet errors at injection energy and to the beam-beam interaction. It provides the chaotic fraction of phase space as well as the amplitude-dependent emittance growth rate and the diffusion coefficient.

The calculated amplitude-dependent diffusion rates can be compared with future collimator measurements in HERA. The predicted emittance growth rate may be slightly too pessimistic due to the incorrect 20-pole coefficient of the quadrupole magnets used in the analysis and in the simulation.

Acknowledgements

The author is grateful to Brookhaven National Laboratory and in particular to Dr. A. Ruggiero for the experienced hospitality and for support. Furthermore he wants to thank Dr. F. Willeke for guidance and encouragement and Dr. F. Schmidt for many interesting discussions. Prof. M. Berz and Dr. É. Forest deserve appreciation for providing the differential-algebra package and the Lie-algebra software tools respectively. Also the help of the DESY magnet measuring group, which made available the multipole data of the HERA magnets, is gratefully acknowledged.

References

[1] A. J. Lichtenberg and M. A. Lieberman, Regular and Stochastic Motion, Springer Verlag (1983).

[2] A. Wrulich, DESY 84-026 (1984).

[3] F. Schmidt, CERN SL 90-52 (AP) (1991).

[4] K. Wittenburg, DESY HERA 92-07 (1992).

[5] F. Zimmermann, Ph.D. thesis, to be published.

[6] B. V. Chirikov, Physics Reports 52, No 5 (1979).

[7] T. Chen and S. Peggs, Proc. Third ICFA Beam Dynamics Workshop, Novosibirsk (1989).

[8] A. Schoch, CERN 57-21 (PS Division), 55-56 (1958).

[9] L. R. Evans, J. Gareyte, CERN SPS 82-8 (DI-MST) (1982).

[10] L. R. Evans, CERN 84-15 (1984).

[11] M. Berz, Part. Acc. 24, 109 (1989).

[12] E. Forest, M. Berz and J. Irwin, Part. Acc. 24, p. 91 (1989).

[13] A. J. Dragt and J. M. Finn, J. Math. Phys. 17, 2215-2227 (1976).

SLOW EMITTANCE GROWTH DUE TO MODULATION EFFECTS IN THE PRESENCE OF NONLINEAR FIELDS

Oliver S. Brüning
DESY, 2 Hamburg 52, Germany

Abstract

Looking at the simple model structure of a long FODO cell with one sextupole kick, we study the effect of tune modulation in the presence of sextupole nonlinearities.

1 Introduction

A recent experiment at the SPS [1] concentrated on the nonlinear beam dynamics in the presence of strong nonlinear multipoles and tune modulation. The nonlinearities were generated by strong sextupole magnets, and three different frequencies were used for the tune modulation. An interesting result of the experiment was that the emittance growth due to the tune modulation could be enhanced by using two, rather than only one, modulation frequencies. Because the same net modulation amplitude was used for both cases, the experiment clearly indicates an emittance growth due to the nonlinear character of the beam dynamics.

In this paper we will investigate the combined effect of tune modulation and sextupole nonlinearities. In particular, we are interested in the sideband structure and a possible sideband overlap due to the tune modulation. In our analysis we look at a long FODO cell with only one sextupole kick and assume that all particles lie inside some volume V_{beam} around the origin of the transversal phase space. Our model structure has five dominating resonances in the horizontal phase space, all of which lie outside the volume V_{beam}. In the following, we will assume that the emittance growth is determined entirely by the horizontal resonances, which implies that the vertical emittance is much smaller than the horizontal. In the absence of tune modulation we do not observe any substantial emittance growth during the first $2.0 \cdot 10^6$ passages through the FODO cell. In the presence of tune modulation however, we observe a vertical emittance growth with the same characteristics as those observed in the SPS experiment. The emittance growth for a tune modulation with two frequencies is much larger than in the case of a tune modulation with one frequency only.

The tune modulation leads to resonance sidebands of the five dominant resonances. The emittance growth due to these sidebands depends entirely on the number, strength, and position (overlap) of the sidebands which reach into the volume V_{beam}. For a fast tune modulation $[f = O(100 \text{ Hz})]$, the distance of neighbouring sidebands is large, but the amplitudes decrease very rapidly with increasing order. A fast tune modulation might therefore lead only to a small number of well-separated sidebands which reach into the volume V_{beam}. For a slow tune modulation $[f = O(1 \text{ Hz})]$, one observes the opposite behavior. The distance of neighbouring sidebands is small and the amplitudes decrease only slowly with increasing order. However, the sidebands cannot spread over large distances of the phase space. A slow tune modulation might therefore lead to a rich structure of sidebands, but no sidebands with a significant amplitude which can reach into the volume V_{beam}. Both modulation cases lead therefore only to very slow emittance growth.

For the case of a tune modulation with two frequencies, three effects lead to increased emittance growth. First, in the case of a tune modulation with a fast and a slow frequency, the sidebands of the fast modulation frequency act as resonance seeds for the slow modulation frequency, i.e. the slow tune modulation leads to a rich structure of sidebands around the sidebands of the fast modulation. As a result, we expect a large number of resonance sidebands with non-vanishing amplitudes inside the volume V_{beam}, and hence, increased emittance growth. Second, in the case of a tune modulation with two approximately equal frequencies, the number of resonance sidebands increases and the sidebands of the two modulation frequencies might overlap. As a result, the sidebands that reach into the volume V_{beam} will result in significant emittance growth. Third, a tune modulation with any two frequencies increases the widths of the stochastic layers related to the resonance sidebands.

The paper is organized as follows. First we derive the Hamilton function of a FODO cell with sextupole kick and harmonic tune modulation in action-angle variables [2] and calculate the nonlinear detuning due to the sextupole kick with the help of Deprit perturbation theory [2],[3]. In the second part we construct the corresponding nonlinear map and calculate the mode amplitudes for the resonances that appear in sixth-order perturbation theory or lower. In this calculation, we use the nonlinear map for particle tracking and derive the mode amplitudes from the numerically determined resonance widths. Third, using these mode amplitudes, we analyse the sideband structure in the presence of tune modulation and estimate modulation frequencies that lead to extremely fast emittance growth. We present numerical data that show enhanced vertical emittance growth for a tune modulation with the estimated frequencies.

2 The Hamilton Function

The linear motion of an on-momentum particle in a FODO structure is governed by Hill's equation [4], i.e., linear equations with periodic coefficients:

$$\frac{d^2x}{ds^2} = -K_x(s) \cdot x, \qquad \frac{d^2z}{ds^2} = -K_z(s) \cdot z, \qquad (1)$$

where s is the distance along the equilibrium orbit and x and z are the horizontal and vertical displacements of the particle from the equilibrium orbit, respectively. Introducing action-angle variables via the canonical transformation

$$F_1 = -\sum_{y=x,z} \frac{y^2}{2\beta_y(s)} \cdot \{\tan(\Phi_y + \Phi_{y,0}) + \alpha_y(s)\}, \qquad (2)$$

the corresponding Hamilton function takes the convenient form

$$H = \nu_x \cdot I_x + \nu_z \cdot I_z; \qquad (3)$$

ν_x and ν_z are the horizontal and vertical tunes, and for simplicity we assume constant beta-functions: $\beta_y(s) = 1/\nu_y$, $\rightarrow \alpha_y \equiv 0$ [4] ($y = x, z$).

Introducing the potential for the sextupole kick [5]

$$V(x,z,s) = \delta_L(s) \cdot \frac{\lambda}{6} \cdot (x^3 - 3xz^2), \qquad (4)$$

and a horizontal tune modulation with N frequencies, the Hamilton function (3) takes the form

$$\begin{aligned}
H = & \nu_x \cdot \left[1 + \sum_{p=1}^{N} a_p \cdot \cos\left(\frac{2\pi\Omega_p}{L} \cdot s\right)\right] \cdot I_x + \nu_z \cdot I_z + \\
& \frac{\lambda}{L \cdot 3\sqrt{8}} \cdot \sqrt{\beta_x \cdot I_x}^3 \cdot \sum_{k=-\infty}^{+\infty} \left\{3 \cdot \cos\left(\phi_x + \frac{2\pi}{L} \cdot k \cdot s\right) + \right. \\
& \left. \cos\left(3\phi_x + \frac{2\pi}{L} \cdot k \cdot s\right)\right\} - \\
& \frac{\lambda}{L \cdot 3\sqrt{8}} \cdot \sqrt{\beta_x \cdot I_x} \cdot \beta_z \cdot I_z \cdot \sum_{k=-\infty}^{+\infty} \left\{6 \cdot \cos\left(\phi_x + \frac{2\pi}{L} \cdot k \cdot s\right) + \right. \\
& \left. 3 \cdot \cos\left(\phi_x - 2 \cdot \phi_z + \frac{2\pi}{L} \cdot k \cdot s\right) + 3 \cdot \cos\left(\phi_x + 2 \cdot \phi_z + \frac{2\pi}{L} \cdot k \cdot s\right)\right\}. \quad (5)
\end{aligned}$$

λ is the integrated sextupole strength [$\lambda = (e/P_0) \int_0^L (\partial^2 B_z/\partial x^2)_{x=z=0} ds \rightarrow [\lambda] = (1/\text{m}^2)$], B the magnetic field, e the unit charge, L the length of the

FODO cell, P_0 the particle momentum, and $\delta_L(s)$ is the periodic delta-function $[\delta_L(s) = \sum_m \delta(s - m \cdot L)]$.

Finally, we use the canonical transformation

$$F_2 = \tilde{I}_x \cdot \left[\phi_x - \sum_{p=1}^{N} \frac{a_p \nu_x L}{2\pi \Omega_p} \cdot \sin\left(\frac{2\pi \Omega_p}{L} \cdot s\right)\right], \qquad (6)$$

in order to eliminate the tune modulation from the linear part of (5) and get

$$H = \nu_x \cdot \tilde{I}_x + \nu_z \cdot \tilde{I}_z + \frac{2\varepsilon}{3 \cdot L} \cdot \sqrt{\beta_x \cdot \tilde{I}_x}^{-3} \cdot$$

$$\sum_{k,\vec{n}} \left\{ \left[\prod_{p=1}^{N} J_{n_p}\left(\frac{a_p \nu_x L}{2\pi \Omega_p}\right)\right] \cdot 3 \cdot \cos\left(\tilde{\phi}_x + \frac{2\pi}{L} \cdot (k + \vec{n} \cdot \vec{\Omega}) \cdot s\right) + \right.$$

$$\left. \left[\prod_{p=1}^{N} J_{n_p}\left(\frac{3 a_p \nu_x L}{2\pi \Omega_p}\right)\right] \cdot \cos\left(3\tilde{\phi}_x + \frac{2\pi}{L} \cdot (k + \vec{n} \cdot \vec{\Omega}) \cdot s\right) \right\} -$$

$$\frac{2\varepsilon}{3 \cdot L} \cdot \sqrt{\beta_x \cdot \tilde{I}_x} \cdot \beta_z \cdot \tilde{I}_z \cdot \sum_{k,\vec{n}} \left[\prod_{p=1}^{N} J_{n_p}\left(\frac{a_p \nu_x L}{2\pi \Omega_p}\right)\right] \cdot$$

$$\left\{ 6 \cdot \cos\left(\tilde{\phi}_x + \frac{2\pi}{L} \cdot (k + \vec{n} \cdot \vec{\Omega}) \cdot s\right) + \right.$$

$$3 \cdot \cos\left(\tilde{\phi}_x - 2 \cdot \tilde{\phi}_z + \frac{2\pi}{L} \cdot (k + \vec{n} \cdot \vec{\Omega}) \cdot s\right) +$$

$$\left. 3 \cdot \cos\left(\tilde{\phi}_x + 2 \cdot \tilde{\phi}_z + \frac{2\pi}{L} \cdot (k + \vec{n} \cdot \vec{\Omega}) \cdot s\right) \right\}, \qquad (7)$$

where we have introduced the small parameter $\varepsilon = (\lambda/2\sqrt{8})$, the vectors $\vec{n} = (n_1, ..., n_N)$ and $\vec{\Omega} = (\Omega_1, ..., \Omega_N)$, and where $J_{n_p}(a_p \nu_x L/2\pi\Omega_p)$ are the Bessel functions of the first kind.

Using perturbation theory [2], Hamiltonian (7) can be written in the general form

$$H = \sum_{\alpha,\beta,l,m,k,\vec{n}} \varepsilon^{(\alpha+\beta-2)} \cdot A_{\alpha,\beta,l,m,k,\vec{n}} \cdot \hat{I}_x^{\alpha/2} \cdot \hat{I}_z^{\beta/2} \cdot e^{i[l \cdot \hat{\phi}_x + m \cdot \hat{\phi}_z + \frac{2\pi}{L} \cdot (k + \vec{n} \cdot \vec{\Omega}) s]}. \qquad (8)$$

Neglecting all the vertical Fourier modes but the nonlinear detuning terms and the coupling mode of the horizontal and vertical motion, (8) takes the form

$$H = H_x + H_z + H_{x,z} \qquad (9)$$

with

$$H_x = \nu_x \cdot \hat{I}_x + \frac{1}{2}\varepsilon^2 \cdot \nu_{x,2} \cdot \hat{I}_x^2 + \frac{1}{3}\varepsilon^3 \cdot \nu_{x,4} \cdot \hat{I}_x^3 +$$
$$\sum_{\alpha,k,l,\vec{n}} \varepsilon^{(\alpha-2)} \cdot A_{x,\alpha,l,k,\vec{n}} \cdot \hat{I}_x^{\alpha/2} \cdot e^{i\left[l\cdot\hat{\phi}_x + \frac{2\pi}{L}\cdot(k+\vec{n}\cdot\vec{\Omega})s\right]}, \quad (10)$$

$$H_z = \nu_z \cdot \hat{I}_z + \frac{1}{2}\varepsilon^2 \cdot \nu_{z,2} \cdot \hat{I}_z^2, \quad (11)$$

$$H_{x,z} = \varepsilon^2 \cdot \nu_{xz,2} \cdot \hat{I}_x \cdot \hat{I}_z + \varepsilon^2 \cdot A_{x,z} \cdot \hat{I}_x \cdot \hat{I}_z \cdot \cos\left(2\cdot\hat{\phi}_x - 2\cdot\hat{\phi}_z\right), (12)$$

where we have introduced the notation $\nu_{x,2}$, $\nu_{z,2}$, $\nu_{xz,2}$, and $\nu_{x,4}$ for the nonlinear detuning of the sextupole in second- and fourth-order perturbation theory. It is this Hamilton function that lies at the heart of our paper. In the following analysis we will look at the horizontal motion (10) as the driving term for the growth of the vertical action I_z. This mechanism dominates the emittance growth as long as $I_x \gg I_z$ [6].

For the following analysis we choose

$Q_x = 25.158$	$Q_z = 25.252$	$L = 47.018 \cdot 10^2$ m
$\lambda = 3.0$ m^{-2}		$P_0 = 820$ GeV/c

(13)

with $\nu_x = (2\pi/L) \cdot Q_x$ and $\nu_z = (2\pi/L) \cdot Q_z$. L is 100 times the HERA-p FODO cell length, $P_0 c$ is the HERA-p design energy, and λ is about 8% of the sextupole strength of the sextupole correction coils in these 100 HERA-p FODO cells [7]. Furthermore, we will characterize the horizontal resonances with the corresponding parameter vector $(\alpha, l, k, \vec{n})$.

In order to treat the emittance growth analytically we have to estimate the nonlinear detuning terms, the horizontal resonance amplitudes $A_{x,\alpha,k,l,n}$, and the coupling mode amplitude $A_{x,z}$ in (10) and (11). For the calculation of the detuning terms and the coupling mode amplitude, we used a computer algorithm based on Deprit perturbation theory [2],[3], and get

$\varepsilon^2 \cdot \tilde{\nu}_{x,2} = -9.0$ m^{-2}	$\varepsilon^2 \cdot \tilde{\nu}_{z,2} = -10.0$ m^{-2}
$\varepsilon^2 \cdot \tilde{\nu}_{xz,2} = +11.6$ m^{-2}	$\varepsilon^4 \cdot \tilde{\nu}_{x,4} = -0.12 \cdot 10^7$ m^{-3}
$|A_{x,z}| \approx 35.39$ m^2	

(14)

For the calculation of the horizontal resonance amplitudes, we insert (13) and (14) into $\partial H_x / \partial \hat{I}_x$ and calculate the stable and unstable fixed points $\hat{\phi}_{x_0,k}$ and $\hat{I}_{x_0,k}$ for the resonance of interest. After transforming these coordinates back to the initial variables $I_{x_0,k}$ and $\phi_{x_0,k}$, we use the fixed point coordinates as the initial conditions for an iterative application of the nonlinear map corresponding to (5). From the numerical data we can finally evaluate the resonance amplitudes. In the next section we will construct the nonlinear map necessary for this calculation.

3. Nonlinear Mapping

In this section we will construct the one-turn map corresponding to the Hamilton function (5) and present numerical data obtained by iterative application of this map. The numerical data are used for construction of the surfaces of section, for determination of the resonance amplitudes, and for calculation of the emittance growth under the influence of one and two modulation frequencies.

The linear equation of motion corresponding to the Hamilton function (5) without sextupole kick can be integrated, yielding

$$I_x[(h+1)\cdot L] = I_x(h\cdot L), \qquad I_z[(h+1)\cdot L] = I_z(h\cdot L),$$
$$\phi_x[(h+1)\cdot L] = \phi_x(h\cdot L) + \Delta_x(h\cdot L), \qquad \phi_z[(h+1)\cdot L] = \phi_z(h\cdot L) + \Delta_z,$$
(15)

with

$$\Delta_x(h\cdot L) = \nu_x \cdot L + \sum_{p=1}^{N}\left[\frac{a_p \nu_x L}{\pi \Omega_p} \cdot \sin(\pi\Omega_p) \cdot \cos\left(2\pi\Omega_p \cdot \left(h + \frac{1}{2}\right)\right)\right],$$

$$\Delta_z = \nu_z \cdot L.$$
(16)

Taking into account the sextupole kick (4) and transforming back to the variables x, P_x, z, and P_z, the position after the $(h+1)$th passage through the FODO cell can be expressed as a function of the position after the hth passage through the FODO cell via successive application of the linear transfer map

$$\begin{pmatrix} \tilde{x} \\ \tilde{P}_x \\ \tilde{z} \\ \tilde{P}_z \end{pmatrix}_{h+1} = \begin{pmatrix} a_{1,x} & a_{2,x} & 0 & 0 \\ a_{3,x} & a_{4,x} & 0 & 0 \\ 0 & 0 & a_{1,z} & a_{2,z} \\ 0 & 0 & a_{3,z} & a_{4,z} \end{pmatrix} \cdot \begin{pmatrix} x \\ P_x \\ z \\ P_z \end{pmatrix}_h \qquad (17.a)$$

with

$$a_{1,y} = a_{4,y} = \cos(\Delta_y), \quad a_{2,y} = -a_{3,y}\cdot\beta_y^2 = \beta_y \cdot \sin(\Delta_y), \quad y = x, z,$$

and the sextupole kick

$$\begin{pmatrix} x \\ P_x \\ z \\ P_z \end{pmatrix}_{h+1} = \begin{pmatrix} \tilde{x} \\ \tilde{P}_x \\ \tilde{z} \\ \tilde{P}_z \end{pmatrix}_{h+1} + \begin{pmatrix} 0 \\ -\frac{\lambda}{2}\cdot(\tilde{x}_{h+1}^2 - \tilde{z}_{h+1}^2) \\ 0 \\ \lambda \cdot \tilde{x}_{h+1}\cdot\tilde{z}_{h+1} \end{pmatrix} \qquad (17.b)$$

to the particle coordinates.

We will use this combined map for the construction of the surface of section (SoS) for the horizontal particle motion ($I_z \equiv 0$, $\phi_z \equiv 0$). We choose the surface $s = \kappa \cdot L$ for the surface of section, with $\kappa \in \mathbf{Z}$ imposed by the map (17). For the construction of the SoS, one has to limit the tune modulation to frequencies that satisfy the resonance condition

$$\gamma + \vec{\delta} \cdot \vec{\Omega} = 0; \qquad \gamma, \delta_p \in \mathbf{Z}; \quad p = 1, 2, .., N; \qquad \rightarrow \Omega_p = \frac{q}{r}, \; q, r \in \mathbf{Z}. \tag{18}$$

Rather than taking the particle coordinates after each passage through the FODO structure, as in the case without tune modulation, we take the coordinates now only after each κth passage through the FODO structure, with κ being the smallest common denominator of the Ω_p. Without this restriction, we can construct only a stroboscopic phase-space projection, which does not depict the clear resonance island structure necessary for calculation of the resonance widths.

Figure 1 shows the SoS without tune modulation. The left-hand side of Fig. 1 shows the SoS in the x, P_x variables and the right-hand side shows the SoS in the action angle variables (ϕ_x, I_x). The island structures of the $(5, 7, -175, \vec{0})$, $(6, 8, -201, \vec{0})$, and $(5, 9, -226, \vec{0})$ resonances are clearly recognizable.

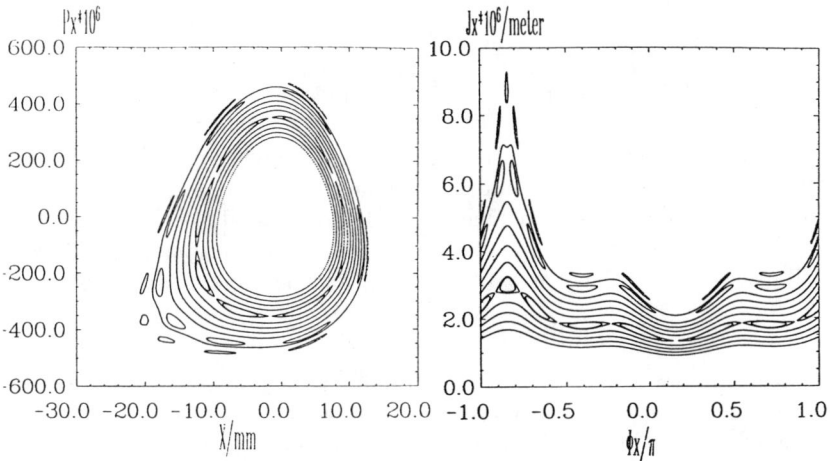

Figure 1: The surfaces of section without tune modulation. Left: The SoS in the P_x and x variables. Right: The SoS in the action angle variables.

With the help of the SoS, we can measure the resonance widths of the resonance sidebands and calculate the resonance amplitudes. For the calculation of the mode amplitudes, we change to resonance variables

$$F_2 = \bar{I}_x \cdot \left(l \cdot \hat{\phi}_x + \frac{2\pi}{L} \cdot (k + \vec{n} \cdot \vec{\Omega}) \cdot s \right) \tag{19}$$

and expand (10) into a Taylor series around a stable resonance fixed point $\bar{I}_{+\text{res}}$ of the map. Recognizing that the absolute value of the resonance half-width $\Delta \bar{I}_{1/2}$ is always much smaller than the fixed point value $\bar{I}_{x,\text{res}}$, we get the following relation between the mode amplitudes and the measured resonance widths:

$$|A_{\alpha,k,l,\vec{n}}| = \left| \frac{\Delta \hat{I}_{1/2}^2 \cdot (\nu_{x,2} + 2 \cdot \hat{I}_x \cdot \nu_{x,4})}{4 \cdot \varepsilon^{(\alpha-2)} \cdot \hat{I}_{x,\text{res}}^{\alpha/2}} \right|. \tag{20}$$

Equations (21) show the resulting resonance amplitudes for resonances up to 6th-order perturbation theory.

Resonances in 1st-order perturbation theory:	
$A_{3,k,1,\vec{0}} = 0.07$ m$^{1/2}$	$A_{3,k,3,\vec{0}} = 0.023$ m$^{1/2}$
Resonances in 2nd-order perturbation theory:	
$A_{4,-151,6,\vec{0}} = 0.05$ m^2	
Resonances in 3rd-order perturbation theory:	
$A_{5,-176,7,\vec{0}} = 60.0$ m$^{7/2}$	$A_{5,-226,9,\vec{0}} = 85.0$ m$^{7/2}$
Resonances in 4th-order perturbation theory:	
$A_{6,-201,8,\vec{0}} = 2.0 \cdot 10^4$ m^5	
Resonances in 6th-order perturbation theory:	
$A_{8,-352,14,\vec{0}} = 1.0 \cdot 10^{10}$ m^8	

(21)

All other resonances are either too far away from the dynamic aperture for their sidebands to reach into the volume V_{beam}, or their amplitudes are so small that we will neglect them in the following analysis.

Before we use the map (17) for calculation of the emittance growth, we will discuss the sideband structure arising from the tune modulation.

4 The Sideband Structure

Now that we have constructed the Hamilton function and its corresponding nonlinear map for our model structure, we are ready to examine the tune modulation and its effect on the emittance growth. First, we will look at the Hamilton function (10) in order to understand the structure of the modulation sidebands. In a second step, we will use the insight gained into the sideband structure and estimate modulation frequencies which result in increased emittance growth, and finally we will illustrate the effects with numerical data from an iterative application of the map (17).

In the following, we will limit our analysis to the case of one and two modulation frequencies, and assume that all particles lie within the volume V_{beam}, given by

$$4.0 \cdot 10^{-7} \text{ m} \leq I_x \leq 10.0 \cdot 10^{-7} \text{ m},$$
$$0.9 \cdot 10^{-7} \text{ m} \leq I_z \leq 1.0 \cdot 10^{-7} \text{ m}. \tag{22}$$

We start our analysis with the Hamiltonian (10) and derive a closed expression for the mode amplitudes. For this, we neglect the $\vec{\Omega}$ term in the denominator

$$\frac{1}{l \cdot \nu_x + (2\pi/L) \cdot \left[k + \vec{n} \cdot \vec{\Omega}\right]},$$

which appears in the Deprit perturbation theory. This approximation is good as long as the linear tune ν_x is not close to a resonance

$$l \cdot L\nu_x + 2\pi \cdot k = 0,$$

and as long as

$$l \cdot L\nu_x + 2\pi \cdot \left[k + \vec{n} \cdot \vec{\Omega}\right] \ll L, \quad \text{for} \quad n_p \lesssim \frac{a_p \cdot \nu_x \cdot L}{2\pi \Omega_p}.$$

In this approximation, we get for the mode amplitudes

$$A_{x,\alpha,k,l,\vec{n}} = \left[\prod_{p=1}^{N} J_{n_p}\left(\frac{l \cdot a_p \cdot \nu_x \cdot L}{2\pi \Omega_p}\right)\right] \cdot A_{x,\alpha,k,l,\vec{0}}. \tag{23}$$

Equation (23) implies that all sidebands with

$$|n_p| > \left|\frac{la_p \nu_x L}{2\pi \Omega_p}\right|$$

have a vanishing amplitude and can hence be neglected in the following analysis. We test (23) for a modulation frequency of 875 Hz ($\rightarrow \Omega = 1/70$). Looking at (23), we expect a vanishing amplitude for the ($l = 7, n = 0$) resonance for $a = 1.9 \cdot 10^{-4}$, because the Bessel function $J_0(z)$ has a root for $z = 2.4$. The numerical data obtained with the nonlinear map (17) yield $a = 1.6 \cdot 10^{-4}$, which is in good agreement with the value predicted by (23).

In order to calculate the resonance values for the horizontal action, we solve the fixed-point equations

$$\frac{\partial H}{\partial \hat{I}_x} = 0, \quad -\frac{\partial H}{\partial \hat{\phi}_x} = 0$$

for $\hat{I}_{x,\text{res}}$ and get

$$\hat{I}_{x,res} = -\left[\nu_x + \frac{2\pi}{l \cdot L} \cdot \left(k + \vec{n} \cdot \vec{\Omega}\right)\right]/\nu_{x,2}. \tag{24}$$

Equation (24) shows that a slow modulation frequency leads to a small, and a fast modulation frequency to a large, spacing of neighbouring sidebands.

First, we discuss the effect of a tune modulation with one fast and one slow modulation frequency. The Hamilton function (10) has five dominating resonances, all of which lie outside the volume V_{beam}:

$$
\begin{array}{|l|l|}
\hline
(\alpha = 4, l = 6, k = -150, \vec{n} = \vec{0}) & \hat{I}_{x,6} \approx 1.0 \cdot 10^{-5} \text{ m} \\
\hline
(\alpha = 5, l = 7, k = -176, \vec{n} = \vec{0}) & \hat{I}_{x,7} \approx 2.0 \cdot 10^{-6} \text{ m} \\
\hline
(\alpha = 5, l = 9, k = -226, \vec{n} = \vec{0}) & \hat{I}_{x,9} \approx 6.3 \cdot 10^{-6} \text{ m} \\
\hline
(\alpha = 6, l = 8, k = -201, \vec{n} = \vec{0}) & \hat{I}_{x,8} \approx 4.4 \cdot 10^{-6} \text{ m} \\
\hline
(\alpha = 8, l = 14, k = -353, \vec{n} = \vec{0}) & \hat{I}_{x,14} \approx 7.5 \cdot 10^{-6} \text{ m} \\
\hline
\end{array}
\quad (25)
$$

The tune modulation leads to resonance sidebands of these five resonances. For a fast tune modulation [$f = O(100 \text{ Hz})$], the distance of neighbouring sidebands is large [see equation (24)], and mode amplitudes decrease very rapidly with increasing order n_p [see equation (23)]. Hence, a fast tune modulation might only lead to a small number of well-separated sidebands which reach into the volume V_{beam}. For a slow tune modulation [$f = O(1 \text{ Hz})$], the behaviour is reversed. The distance of neighbouring sidebands is small, and the amplitudes decrease only slowly with increasing order n_p. A slow tune modulation might lead to a rich structure of sidebands, but no sidebands with a significant amplitude which can reach into the volume V_{beam}. Both modulation cases lead therefore only to very slow emittance growth. For the case of a tune modulation with a fast and a slow frequency, the situation changes drastically. The sidebands of the fast modulation frequency act as resonance seeds for the slow modulation frequency, i.e. the slow tune modulation leads to a rich structure of sidebands around the sidebands of the fast modulation. As a result, we expect a large number of resonance sidebands with non-vanishing amplitudes inside the volume V_{beam}, and hence, increased emittance growth.

In order to illustrate this seeding effect, we add the resonance widths of all modulation sidebands of the five resonances (25) that lie inside the volume V_{beam}. Figure 2 shows the quotient of the total volume covered by the modulation sidebands

$$
V_{\text{res}} = \sum_{I_{x,(\alpha,l,k,\vec{n})} \in V_{\text{beam}}} \Delta_{1/2} I_{x,\alpha,l,k,\vec{n}} \tag{26}
$$

and the volume V_{beam} occupied by the particles as a function of the modulation frequencies. The solid line shows the relative volume for a tune modulation with one frequency only, and the dashed line shows the relative volume for a modulation with two frequencies. For the latter, the slow modulation frequency was kept at 9 Hz (the slowest modulation frequency

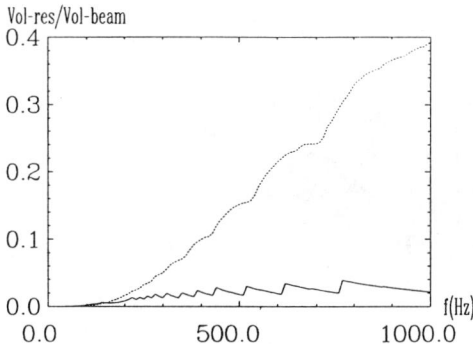

Figure 2: The quotient $V_{\rm res}/V_{\rm beam}$ vs. the modulation frequency.

used in the SPS experiment) and only the second frequency was varied. In both cases, the net modulation depth was $a = 2.0 \cdot 10^{-4}$, which corresponds to $\Delta Q_x \approx 5.0 \cdot 10^{-3}$ and is approximately twice the modulation depth used in the SPS experiment. Figure 2 shows clearly that the relative volume $V_{\rm res}/V_{\rm beam}$ is approximately 14 times larger for a modulation with 9 Hz and 750 Hz than for any single modulation frequency, and suggests a drastic increase of the vertical emittance growth due to an increased number of resonances inside the volume $V_{\rm beam}$ and due to an overlap of resonance sidebands. Qualitatively, one observes the same effect for any modulation depth, but the volume gain becomes smaller with decreasing modulation depth.

In the next step, we will measure the specific effect of the tune modulation on the emittance growth. We use a Gaussian distribution in the action variables and look at 3000 particles with initial conditions inside the volume $V_{\rm beam}$. We track the particles for three different tune modulations through our model structure and assume a horizontal and vertical aperture limitation of ± 40.0 mm for the x and z variables. First, we consider a modulation depth of $a = 2.0 \cdot 10^{-4}$ and a single frequency modulation of 9 Hz ($\rightarrow \Omega_1 = 1/6200$) and 875 Hz ($\rightarrow \Omega = 1/70$). For both frequencies, we have no particle loss due to tune modulation over the first $2.0 \cdot 10^6$ turns, and observe a vertical emittance growth of less than 3%. The left-hand side of Fig. 3 shows the vertical emittance as a function of the number of turns for $f = 875$ Hz. Second, we look at a simultaneous tune modulation with both frequencies, each having a modulation depth $a = 1.0 \cdot 10^{-4}$. Although we have the same net modulation depth as in the previous case, we now lose all 3000 particles

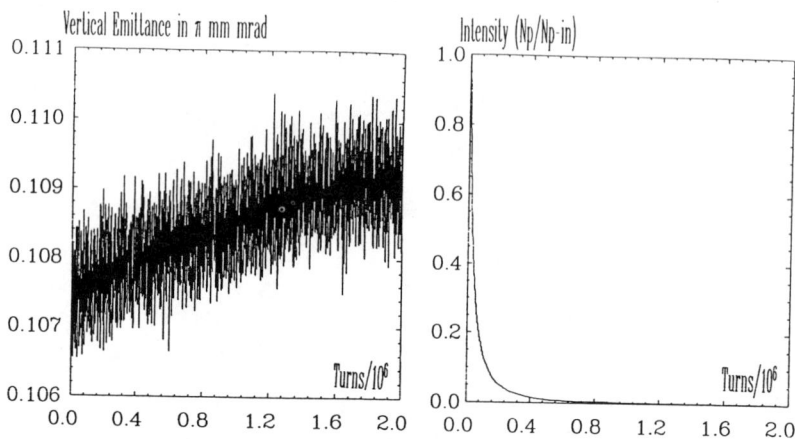

Figure 3: Left: The vertical emittance vs. the number of turns. Right: The normalized number of particles vs. the number of turns.

during the first $2.0 \cdot 10^6$ turns. The right-hand side of Fig. 3 shows the normalized number of particles versus the number of turns. The drastic increase in the particle loss nicely illustrates the seeding effect for a tune modulation with a slow and a fast frequency.

Next, we analyse the tune modulation with two approximately equal frequencies. We choose for example

$$f_1 = 875 \text{ Hz}(\to \Omega_1 = 1/70), \qquad f_2 = 766 \text{ Hz}(\to \Omega_2 = 1/80),$$

and look at the location of the resonance sidebands. For $f = 875$ Hz the smallest non-negative action for the $(4, 6, -151, n)$ resonance according to (24) is given by

$$\hat{I}_{x,(4,6,-151,5)} \approx 2.8 \cdot 10^{-7} \text{ m} \tag{27}$$

and has a sideband spacing of

$$\Delta \hat{I}_{x,\text{res}} \approx 3.0 \cdot 10^{-7} \text{ m}. \tag{28}$$

As a result, only the $(n = 6)$ and $(n = 7)$ sidebands of the $(\alpha = 4, l = 6, k = -151)$ resonance lie inside the volume V_{beam}. For a modulation depth of $a = 4.0 \cdot 10^{-4}$, which corresponds to $\Delta Q_x = O(10^{-2})$ and which is approximately five times the modulation depth of the SPS experiment, the resonance width given by (20) is approximately $3.0 \cdot 10^{-9}$ m. The $(n = 6)$ and $(n = 7)$ sidebands cannot overlap and cover only a small fraction of the volume V_{beam}. Consequently, we expect only a small emittance growth. If we consider on the other hand a simultaneous tune modulation with both frequencies and with the same net modulation depth, the situation is quite different. For the $(4, 6, -151, \vec{n})$ resonance, we get now for the sideband spacing

$$\Delta \hat{I}_{x,\text{res}} \approx 4.0 \cdot 10^{-8} \text{ m}. \tag{29}$$

The small sideband spacing has two main effects. First, the number of sidebands reaching into the volume V_{beam} increases, and second, neighbouring sidebands might now overlap. Because the condition for sideband overlap depends on the resonance widths of the sidebands, we expect the existence of a critical modulation depth above which the resonance overlap occurs. If we neglect the fourth-order detuning term in (20), solve for $\Delta I_{1/2}$, and substitute (23) into (20), we get for the resonance width of the sidebands

$$\Delta I_{1/2} \approx \sqrt{\frac{4\varepsilon^{(\alpha-2)} \cdot I_{x,\text{res}}^{\alpha/2} \cdot J_{n_1}(la_1\nu_x L/2\pi\Omega_1) \cdot J_{n_2}(la_2\nu_x L/2\pi\Omega_2) \cdot A_{\alpha,k,l}}{\nu_{x,2}}}. \tag{30}$$

Two neighbouring sidebands overlap if

$$\Delta I_{1/2}(\alpha, l, k, \vec{n}) + \Delta I_{1/2}(\tilde{\alpha}, \tilde{l}, \tilde{k}, \vec{\tilde{n}}) > \Delta \hat{I}_{res}. \tag{31}$$

For $I_{x,\text{res}} < 1.0 \cdot 10^{-6}$, $a_1 = a_2$, $f_1 = 875$ Hz, and $f_2 = 766$ Hz, we get for example

$$a_{\min} \approx 1.5 \cdot 10^{-4}. \tag{32}$$

For modulation depths smaller than $a_{\min}$, we expect no sideband overlap for any of the five dominating sextupole resonances (21). For modulation depths larger than this critical modulation depth, the modulation sidebands might overlap and we expect a large emittance growth. For the following numerical analysis we choose $a_1 = a_2 = 2.0 \cdot 10^{-4}$. Figure 4 shows the SoS for the horizontal motion in the presence of tune modulation. The left-hand side of Fig. 4 shows the SoS for $\Omega_{\text{mod}} = 1/70$ and $a = 4.0 \cdot 10^{-4}$ and the right-hand side of Fig. 4 shows the SoS for $\Omega_1 = 1/70$, $a_1 = 2.0 \cdot 10^{-4}$ and $\Omega_2 = 1/80$, $a_2 = 2.0 \cdot 10^{-4}$. One clearly sees the appearance of new resonance sidebands and the stochastic layers due to overlap of the sidebands for a tune modulation with two frequencies. Again, we will measure the effect of a tune modulation on the emittance growth for both frequencies. For the tracking, we choose again a Gaussian distribution in the actions with the initial coordinates inside the volume V_{beam}. For the tune modulation with one frequency only and with $a = 4.0 \cdot 10^{-4}$, we do not lose any particles due to tune modulation during the first $2.0 \cdot 10^6$ turns. However, for the simultaneous modulation with both frequencies and with the same net modulation depth, we lose 65% of the particles during the first

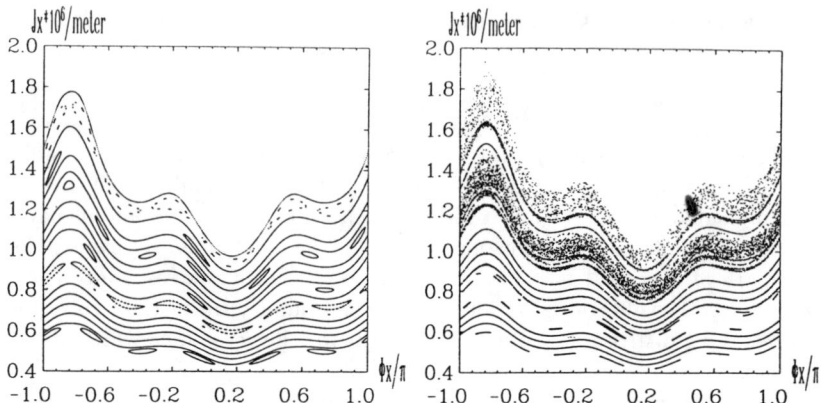

Figure 4: The surfaces of section in the presence of tune modulation. Left: One modulation frequency. Right: Two modulation frequencies.

$2.0 \cdot 10^6$ turns. The right-hand side of Fig. 5 shows the normalized number of particles versus the number of turns for the simultaneous modulation with both frequencies, each with $a_p = 2.0 \cdot 10^{-4}$.

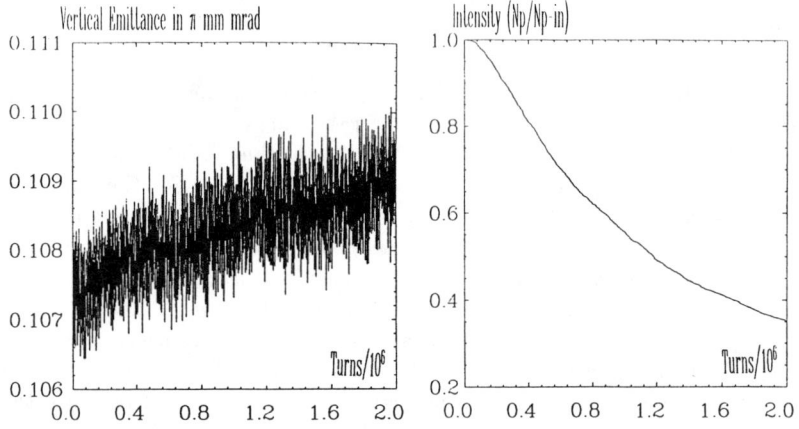

Figure 5: Left: The vertical emittance vs. the number of turns. Right: The normalized number of particles vs. the number of turns.

In order to demonstrate the qualitative difference between the modulation with two approximately equal frequencies and the resonance seeding of a slow and a fast frequency, we look at a simultaneous tune modulation with the same frequencies ($f_1 = 875$ Hz, $f_2 = 766$ Hz), but each with half the modulation depth $a_p = 1.0 \cdot 10^{-4}$, which is just below the critical modulation depth $a_{\min}$. With this smaller modulation depth, the emittance growth is smaller than 3% and in contrast to the modulation with $a_p = 2.0 \cdot 10^{-4}$, we do not lose any particles due to the tune modulation. The left-hand side of Fig. 5 shows the vertical emittance versus the number of turns for $a_p = 1.0 \cdot 10^{-4}$. But the relative volume $V_{\text{res}}/V_{\text{beam}}$ is still of the same order for both cases,

$$V_{\text{res}}/V_{\text{beam}}(a_p = 1.0 \cdot 10^{-4}) \approx 0.1 \ V_{\text{res}}/V_{\text{beam}}(a_p = 2.0 \cdot 10^{-4}) \approx 0.3,$$

which is qualitatively different from the tune modulation with one slow and one fast modulation frequency.

The last effect in our discussion of the tune modulation is the widening of the stochastic layers. Looking at one resonance island of a sextupole resonance sideband, we see that the tune modulation might be in resonance with the island oscillation frequency. The islands of such a resonance are depicted in the left-hand side of Fig. 6. Figure 6 shows the SoS of the horizontal motion in action angle variables and with a modulation frequency $\Omega = 1/84$. Inside the island of the $(5, 7, -151, \vec{0})$ sextupole resonance we see three smaller islands surrounding the stable fixed point of the $(5, 7, -151, \vec{0})$

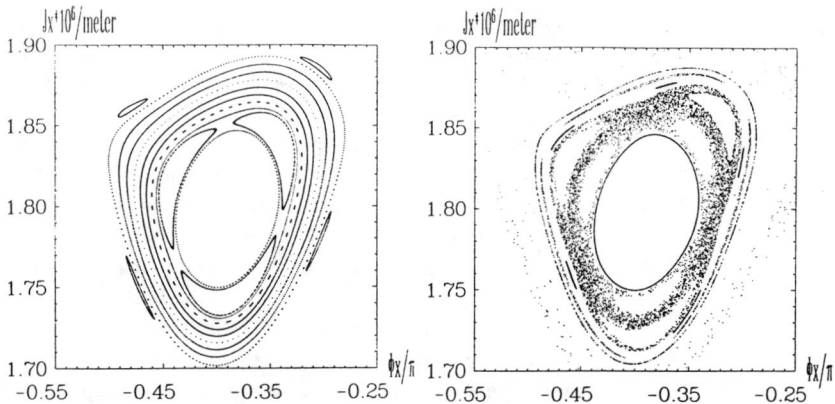

Figure 6: The modulation resonances for $\Omega_1 = 1/84$ and $\Omega_2 = 1/91$. Left: Modulation with $\Omega_1 = 1/84$. Right: Modulation with both frequencies.

resonance. If we take the particle coordinates after each passage through the FODO structure, we obtain a stroboscopic projection of the phase-space trajectory onto the horizontal phase space. In such a stroboscopic projection, the coordinates circulate around the stable fixed point of the $(5, 7, -151, \vec{0})$ resonance. For the fixed points of the three smaller islands, the revolution frequency is related to the modulation frequency by

$$\Omega_o = \Omega_{mod}/3.$$

Therefore, the observed resonance islands in the sextupole resonance correspond to a resonance between the island oscillation frequency of the $(5, 7, -151, \vec{0})$ sextupole resonance and the modulation frequency $\Omega = 1/84$. In the following, we will call a resonance with N islands an Nth-order modulation resonance. Because the period of the island oscillation increases as the trajectory approaches the separatrix of the sextupole resonance ($\Omega_o \to 0$) [2], the number of resonance islands inside the sextupole resonance increases as well. In the vicinity of the separatrix, the resonance islands overlap and lead to a wider stochastic layer as in the case of no tune modulation. For a modulation with more than one frequency, modulation resonances of the different modulation frequencies can overlap even if they are not close to the separatrix of the sextupole resonance. This overlap of modulation resonances of different modulation frequency leads to a widening of the stochastic layers related to the sextupole sideband resonances, and is depicted in the right-hand side of Fig. 6 for $\Omega_1 = 1/84$ and $\Omega_2 = 1/91$.

We want to conclude this section with a final comment. For a tune modulation with a frequency slightly below the free island oscillation frequency, the modulation resonance has only one resonance island around the stable fixed point $I_{0,res}$ of the initial sextupole resonance island (see left-hand side of Fig. 7). In the limit of a tune modulation with the same frequency as the free island oscillation frequency, the stable fixed point $I_{0,res}$ of the sextupole resonance disappears and only the two fixed points of the modulation resonance survive. However, the disappearing of the stable fixed point $I_{0,res}$ does not lead to global chaos within the sextupole resonance island, as suggested in ref. 8 (see right-hand side of Fig. 7). We observed the largest chaotic region for the first-order modulation resonance $\Omega_{island} \lesssim \Omega_{mod}$, when the separatrix of the modulation resonance island intersects with the separatrix of the surrounding sextupole resonance.

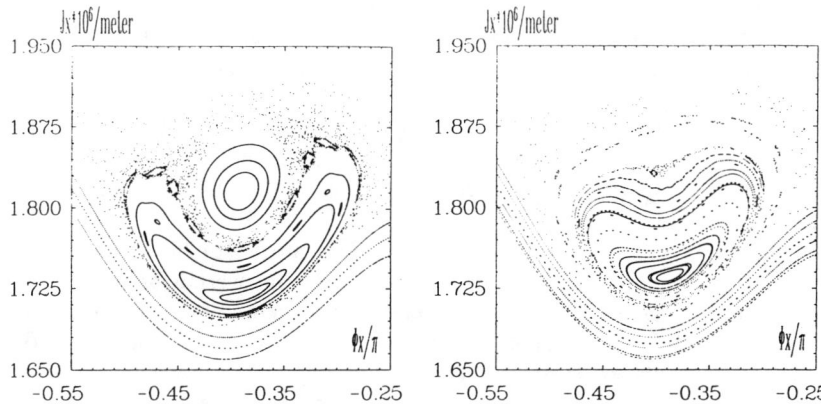

Figure 7: The modulation resonances for a modulation frequency close to the free island oscillation frequency $\Omega_{\text{island}} \approx 1/250$. Left: Modulation with $\Omega_1 = 1/280$. Right: Modulation with $\Omega_1 = 1/250$.

5 Summary

Looking at the simple model structure of a long FODO cell with one sextupole kick, we studied the effect of tune modulation in the presence of sextupole nonlinearities. For modulation frequencies and modulation depths of the same order of magnitude as in the SPS experiment ($f = 9$ Hz $\leftrightarrow$ 800 Hz, $\Delta Q = 1.0 \cdot 10^{-3} \leftrightarrow 5.0 \cdot 10^{-3}$), we were able to reproduce the effects observed in the SPS. For a tune modulation with two frequencies the vertical emittance growth was much larger than for a tune modulation with one frequency only. The simple nature of our model structure could be used to illustrate nicely the different effects which lead to increased vertical emittance growth. Our analysis showed three qualitatively different effects.

In the case of a tune modulation with one fast and one slow frequency, the sidebands of the fast modulation act as resonance seeds for the slow tune modulation. Consequently, the area accessible to the beam is densely covered with resonance sidebands and we expect increased emittance growth. This increase of the number of resonance sidebands reaching into the area accessible to the beam can be easily recognized by looking at the the quotient of the total volume covered by resonance islands (V_{res}) and the volume covered by the beam (V_{beam}). For the case of two extremely different modulation frequencies (for example $f_1 = 9$ Hz and $f_2 = 875$ Hz) the quotient $V_{\text{res}}/V_{\text{beam}}$ was 14 times larger than in the case of two approximately equal frequencies or any single modulation frequency. The expected increase in the vertical emittance was verified by particle tracking.

In the case of a modulation with two approximately equal frequencies, we observed an overlap of neighbouring resonance sidebands which was extremely sensitive to the modulation depth. Below a critical modulation depth, neighbouring resonance sidebands can not overlap and the emittance growth decreases drastically. We illustrated this second effect for $f_1 = 766$ Hz and $f_2 = 875$ Hz with the two net modulation depths $\triangle Q_1 \approx 2.5 \cdot 10^{-4}$ and $\triangle Q_2 \approx 5.0 \cdot 10^{-4}$.

In all cases, we measured a widening of the stochastic layers for a modulation with more than one frequency due to an overlap of modulation resonances. For a modulation resonance, the modulation frequency is an integer multiple of the island oscillation frequency. If one considers more than one modulation frequency, neighbouring modulation resonances might overlap and lead to an increase in the width of the chaotic area.

Even though we are far from explaining the SPS experiment quantitatively with our results, we illustrate three qualitatively different mechanisms that lead to ncreased emittance growth for more than one modulation frequency. A quantitative explanation of the effects observed in the SPS would clearly demand a more elaborate lattice than our simple model structure, but the insight gained from the work presented here could assist in such a quantitative analysis.

References

[1] X. Altuna, C. Arimatea, R. Bailey, T. Bohl, D. Bramdt, K. Cornelis, C. Depas, F. Galluccio, J. Gareyte, R. Giachino, M. Giovannozzi, Z. Guo, W. Herr, A. Hilaire, T. Lundberg, J. Miles, L. Normann, T. Risselada, W. Scandale, F. Schmidt, A. Spinks, and M. Venturini, CERN SL/91-43 (AP), LHC Note 171 (1991).

[2] A.J. Lichtenberg and M.A. Lieberman, *Regular and Stochastic Motion*, Springer.

[3] A. Deprit, *Cel. Mech.* **1**, 12 (1969).

[4] E. Courant and H. Snyder, *Ann. Phys.* **3**, 1-48 (1958).

[5] F. Willeke and G. Ripken, DESY 88-114 (August 1988).

[6] O. Brüning, HEEAC 92, 1992.

[7] B.H. Wiik, DESY Hera 88-05 (April 1988).

[8] T. Chen and S. Peggs, SSC Central Design Group (1989).

SYMPLECTIC INTEGRATORS FOR SPIN MOTION*

S.R. Mane

Brookhaven National Laboratory, Upton, NY 11973

ABSTRACT

The theory of symplectic integrators for the orbital motion in accelerators is fairly well developed. However, the integration of the spin variables, for polarized particle beams, has received much less attention. Two simple examples of second-order symplectic integrators for the spin are presented, together with simple geometrical interpretations of their effect on the spin motion.

INTRODUCTION

The development of symplectic integrators for the orbital motion of particles in accelerators has been extensively studied. Early work was performed by Ruth [1], and a review of the subsequent refinements is given by Forest [2]. The pioneering work of Yoshida [3] showed how to construct symplectic integrators of arbitrarily high order given only a symmetric second-order symplectic integrator. However, all of the above papers only treated the orbital particle motion. Little or nothing has been published on symplectic integrators for the spin motion of a particle in an accelerator. This paper focuses on the spin, and attempts to benefit from the expertise gained in developing orbital symplectic integrators to derive expressions for second-order symplectic spin-orbit integrators.

REVIEW OF FORMALISM

Some basic definitions and terminology will be given first. Suppose that one has a time-independent Hamiltonian H, and the map obtained by integrating the equations of motion for a time t is $M(t)$, and the Poisson bracket of two operators f and g is denoted by the Dragt-Finn notation $:f:g$. Then the equation of motion for M is [4]

$$\frac{dM}{dt} = M :-H:, \qquad (1)$$

and the solution is

$$M(t) = \exp(t :-H:). \qquad (2)$$

This expression is assumed to be too complicated to evaluate explicitly. An nth order *symplectic integrator* is an approximation to $M(t)$, say $T_n(t)$, which equals $M(t)$ up to error terms of order t^{n+1}, i.e.

$$T_n(t) = M(t) + O(t^{n+1}) = \exp[t :-H: + t^{n+1} R_n], \qquad (3)$$

---

*Work performed under the auspices of the U.S. Department of Energy.

where R_n is a remainder term whose detailed form is not important. It is assumed that T_n can be evaluated explicitly. Examples of such integrators are given in the review by Forest [2], for n up to 6.

In a remarkable paper, Yoshida [3] showed how to construct higher-order symplectic integrators given only a second-order symplectic integrator $T_2(t)$. The gist of Yoshida's derivation is that

$$T_{2n+2}(t) = T_{2n}(at)\, T_{2n}(bt)\, T_{2n}(at) \tag{4}$$

with the conditions

$$2a + b = 1, \qquad 2a^{2n+1} + b^{2n+1} = 0, \tag{5}$$

and Yoshida found the explicit solutions

$$a = \frac{1}{2 - 2^{1/(2n+1)}}, \qquad b = -\frac{2^{1/(2n+1)}}{2 - 2^{1/(2n+1)}}. \tag{6}$$

For the above construction to work, the symplectic integrators must be symmetric, i.e. one must have

$$T_{2n}(-t)\, T_{2n}(t) = 1 \tag{7}$$

exactly, not merely to some degree of approximation. As can be seen, this is guaranteed for all n provided T_2 is symmetric. Although the above construction was derived with only the orbital motion in mind, it is quite general, and applies to spin motion as well. This shows that we need only to derive an explicit expression for a symmetric second-order symplectic spin integrator.

It will be adequate to represent the spin of a particle by a three-component classical unit vector $\vec{s}$, and to denote all the orbital variables by z (a six-component vector). The spin obeys the Thomas-Bargmann-Michel-Telegdi (Thomas-BMT) equation [5]

$$\frac{d\vec{s}}{dt} = \vec{\Omega}(z) \times \vec{s} = :-\vec{\Omega}\cdot\vec{s}: \vec{s}, \tag{8}$$

where the spin Hamiltonian is $\vec{\Omega}(z)\cdot\vec{s}$, and the Poisson brackets of the spin variables are

$$:s_i:\, s_j = \epsilon_{ijk}\, s_k, \tag{9}$$

where ϵ_{ijk} is the isotropic totally antisymmetric tensor. Note that one cannot solve for the spin motion independently of the orbit: the spin precession vector $\vec{\Omega}$ depends on the orbit z. Hence we must derive a second-order combined spin-orbit symplectic integrator. Assumimg a symmetric integrator $T_2(t)$ for the orbit alone is known, e.g. from Ref. [2], two possible solutions are

$$T_2^{(1)}(t) = \exp(:-t\vec{\Omega}\cdot\vec{s}/2:)\, T_2(t)\, \exp(:-t\vec{\Omega}\cdot\vec{s}/2:), \tag{10}$$

and
$$T_2^{(2)}(t) = T_2(t/2) \exp(:-t\vec{\Omega}\cdot\vec{s}:) T_2(t/2). \qquad(11)$$

It can be immediately verified that both integrators are symmetric. It is also relatively obvious that the first integrator can be interpreted as performing half the spin rotation at the beginning of an element, and half at the end, i.e. the spin rotation is averaged using the orbital coordinates (and momenta) at the entrance and exit. The second example corresponds to a "drift" for the spin halfway through the element, followed by a "kick" to the spin given by the full spin rotation using the value of the orbit at the midpoint, followed by another "drift" of the spin through the rest of the element. This is analogous to the "drift-kick-drift" type of second-order orbital symplectic integrator based on the Hamiltonian

$$H = A(p) + V(q), \qquad(12)$$

given by

$$T_2(t) = \exp(:-tA/2:) \exp(:-tV:) \exp(:-tA/2:). \qquad(13)$$

Higher-order symplectic spin-order spin integrators can now be built up using the Yoshida technique, although the integrators then lose the simple geometrical interpretations given above.

CONCLUSIONS

The contents of this paper are really a footnote to the pioneering papers on orbital symplectic integrators, especially the work of Yoshida [3], and the developments using Lie group theory reported by Forest [2]. In particular, the use of Lie group theory makes the generalization to include spin motion relatively straightforward, because it should be noted that the spin components (s_1, s_2, s_3) are not canonical variables (e.g. there are three of them), but the Lie algebraic formalism is not affected in any way by this detail. Any derivation explicitly dependent on pairs of canonically conjugate variables would have difficulty accomodating the spin.

ACKNOWLEDGEMENTS

The author thanks J.S. Berg, A. Chao, A.J. Dragt, E. Forest and I. Gjaja for their various pertinent questions and helpful comments during his presentation at the Workshop, and W.T. Weng for supporting this work at the Brookhaven National Laboratory AGS department.

REFERENCES

1. R. Ruth, IEEE Trans. Nucl. Sci. **NS-30** (1983).

2. E. Forest, J. Comp. Phys. **99**, 209 (1992).

3. H. Yoshida, Phys. Lett. A **150**, 262 (1990).

4. A.J. Dragt and E. Forest, J. Math. Phys. **24**, 2734 (1983).

5. L. Thomas, Philos. Mag. **3**, 1 (1927); V. Bargmann, L. Michel and V.L. Telegdi, Phys. Rev. Lett. **2**, 435 (1959).

THE METHOD OF MINIMAL NORMAL FORMS

S.R. Mane and W.T. Weng

Brookhaven National Laboratory, Upton, NY 11973

ABSTRACT

Normal form methods for solving nonlinear differential equations are reviewed and the comparative merits of three methods are evaluated. The concept of the minimal normal form is explained and is shown to be superior to other choices. The method is then extended to apply to the evaluation of discrete maps of an accelerator or storage ring. Such an extension, as suggested in this paper, is more suited for accelerator-based applications than a formulation utilizing continuous differential equations. A computer code has been generated to systematically implement various normal form formulations for maps in two-dimensional phase space. Specific examples of quadratic and cubic nonlinear fields were used and solved by the method developed. The minimal normal form method shown here gives good results using relatively low-order expansions.

1 INTRODUCTION

It is relatively easy and straightforward to find solutions of accelerator lattice design for a linear system [1]. Once the nonlinear elements are introduced, no preferred method has been found. Lately, one-turn maps [2] have been suggested as a useful tool for embodying all perturbations in an accelerator and hence a testing bed for evaluating the comparative merits of various approaches. A good one-turn map representation of an accelerator can provide all necessary information for the lattice description and hopefully also facilitate the evaluation of the long-term stability and dynamic aperture of the accelerator or storage ring.

A powerful techqniue of analyzing nonlinear one-turn maps is the use of normal forms. A recent theoretical development in this area is that of so-called minimal normal forms. Motivated by the successful and apparently superior performance of this method [3] in solving ordinary nonlinear differential equations, this report is a first attempt to apply the same method to the evaluation of maps suitable for accelerator design. A more detailed exposition can be found in Ref. [4]. A rapidly convergent method can provide basic reliable lattice information in a few terms, which is an important contribution in its own right. The larger payoff is in the trustworthiness of the prediction of long-term stability and accuracy in the the calculation of dynamic aperture of a given accelerator. A good general discussion of normal forms is given in Ref. [5], and an example of the application of normal form techniques, but not minimal normal forms, for the LHC (Large Hadron Collider) at CERN, is given in Ref. [6].

---

*Work performed under the auspices of the U.S. Department of Energy.

2 REVIEW OF FORMALISM

The minimal normal form method will be explained below. Many of the technical details can be found in Ref. [4]. First, differential equations will be treated. This will also serve the purpose of defining some notations and preparing the groundwork for an extension of the method to treat discrete maps. A description of the formalism, for differential equations, is also given in Ref. [3], which is not cast in a form directly useful to the development below.

Consider a dynamical system describable by a canonically conjugate (coordinate, momentum) pair (x,p), with $z = x + ip$, which executes a harmonic oscillation with a nonlinear autonomous perturbation

$$\dot{z} = -i\mu z + \sum_{k=1}^{\infty} \epsilon^k Z_k(z, z^*), \qquad (1)$$

where the frequency is μ, ϵ is the small parameter of the perturbation expansion, and the Z_k are homogenous polynomials of degree $k+1$ in z and z^*, given by

$$Z_k = \sum_{p+q=k+1} Z_{pq}\, z^p z^{*q}. \qquad (2)$$

The above equation is solved by the use of a near-identity transformation to a normal form variable u via

$$z = u + \sum_{k=1}^{\infty} \epsilon^k T_k(u, u^*), \qquad (3)$$

with homogenous polynomials

$$T_k = \sum_{p+q=k+1} T_{pq}\, u^p u^{*q}, \qquad (4)$$

which are to be determined. The resulting equation of motion for u is written in the form

$$\dot{u} = -i\mu u + \sum_{k=1}^{\infty} \epsilon^k U_k(u, u^*), \qquad (5)$$

where the U_k are also to be determined, and are obviously related to the T_k. The above equation is made into a normal form by choosing the functions T_k so that the U_k contain resonant monomials only, i.e. monomials of the form $u(uu^*)^p$, $p = 1, 2, \ldots$, which means that $U_k = 0$ for odd k, and

$$\dot{u} = -i\mu u + \sum_k \epsilon^{2k} \tilde{U}_{2k}\, u^{k+1} u^{*k} \equiv -iu\,\Omega(uu^*), \qquad (6)$$

where Ω is an amplitude-dependent frequency. The quantity uu^* is the nonlinear invariant of the motion, and the solution for u is $u = \rho e^{-i\psi}$, where $\rho = |u|$ is constant, and $\psi = \Omega t$. The nonlinear motion has been transformed into a phase-space rotation with an amplitude-dependent frequency.

However, what is *not* obvious from the above discussion is that not all of the coefficients T_{pq} in the normal form transformation, nor the tuneshift coefficients $\tilde{U}_{2k}$, are uniquely determined by demanding that the equation of motion for u contain only resonant monomials. It is shown in Ref. [4] that the coefficients T_{pq} for which $p = q+1$, i.e. T_{21}, T_{32}, etc., are *not* determined by this requirement. Such coefficients will be called "free terms" or "free functions". The above statements do not mean that the normal form is not affected by the free functions, but rather that some additional information must be supplied to fix their values. The criterion chosen for specifying the free functions is where the various methods of calculating normal forms differ from each other. A few possible choices will be described below.

The simplest choice is to put all the free functions to zero. This is perfectly valid, and reduces the computational effort required. Another possibility is to demand that the transformation to the normal form be a canonical transformation. In general, the relation between the Poisson brackets $[z, z^*]$ and $[u, u^*]$ has the form

$$[z, z^*] = [u, u^*] \left\{ 1 + \sum_{k=1}^{\infty} \epsilon^{2k} \tilde{P}_{2k} (uu^*)^k \right\}, \qquad (7)$$

when the equation of motion for u has been brought into the normal form Eq. (6). Choosing the free functions to cancel all the P_{2k} makes the near-identity transformation in Eq. (3) a canonical transformation. This requires nonzero values for the free functions; it also leads to different values for the amplitude-dependent frequency shift terms $\tilde{U}_{2k}$ in Eq. (6). The choice of canonical transformations is often automatic in the accelerator-physics literature, since there is a well-developed theory of such transformations. The present formulation displays the range of some of the other alternatives available for obtaining normal forms.

In this context, the minimal normal form method provides another prescription for choosing the values of the free functions. Instead of cancelling the Poisson bracket terms $\tilde{P}_{2k}$, one chooses the free functions to cancel out the tuneshift corrections, the $\tilde{U}_{2k}$'s. This turns out to be possible for all but the first term $\tilde{U}_2$, i.e. the amplitude-dependent frequency can be reduced to the form

$$\Omega = \mu + i\epsilon^2 \tilde{U}_2 \rho^2. \qquad (8)$$

The reason is as follows. It is shown in Refs. [3] and [4] the free term of order $2k$, viz. $T_{k+1\,k}$, has no effect on the tuneshift term of the same order, $\tilde{U}_{2k}$, but it does affect the value of the next term $\tilde{U}_{2k+2}$. The first term $\tilde{U}_2$ thus cannot be cancelled, but the others can, which therefore leaves us with Eq. (8).

Hence the minimal normal form method also leads to nonzero values for the free functions, but reduces the infinite series of higher-order corrections to the normal form to only one term: this explains the name "minimal" normal form. As a result of this prescription, the *exact* solution for the normal form can be deduced at $O(\epsilon^2)$ already; the subsequent calculation consists only of determining the transformation between the original and new variables (the T_{pq} coefficients) to higher and higher orders of ϵ. The resulting transformation is not canonical: $[u, u^*]$ does not equal $[z, z^*]$. However, it is still symplectic: the value of the Poisson bracket $[u, u^*]$ is constant in time, as is evident from Eq. (7).

The method of minimal normal forms will now be extended to treat discrete maps, as opposed to differential equations in the previously published literature [3]. As explained above, such a reformulation is more suited to accelerator-based applications. Only a two-dimensional phase space will be treated here; the extension to higher dimensions will be given later in this paper. We again define $z = x + ip$, where (x,p) is the coordinate-momentum pair, and suppose that the map equation relating one turn to the next is

$$z_{n+1} = \lambda z_n + \sum_{k=1}^{\infty} \epsilon^k Z_k(z_n, z_n^*), \qquad (9)$$

where n is the turn number, $\lambda = e^{-i\mu}$, and the Z_k are again homogenous polynomials of degree $k+1$ in z and z^*:

$$Z_k = \sum_{p+q=k+1} Z_{pq} z^p z^{*q}. \qquad (10)$$

The near-identity transformation to the normal form variable u is given by

$$z = u + \sum_{k=1}^{\infty} \epsilon^k T_k(u, u^*), \qquad (11)$$

with homogenous polynomials T_k of degree $k+1$ in u and u^*:

$$T_k = \sum_{p+q=k+1} T_{pq} u^p u^{*q}. \qquad (12)$$

The Z_{pq} are known; the T_{pq} are to be determined. This form of the near-identity transformation is essentially the same as that of Scandale et al. [6], who express the homogenous polynomials as Φ_s (summing over s), rather than T_k (summing over k), and is a nonlinear symplectic transformation. Only nonresonant normal forms will be treated in this paper, i.e. it will be supposed that $\mu/(2\pi)$ is irrational, or $\lambda^k \neq 1$ for any $k \geq 1$. As before, the aim is to terminate the tuneshift corrections at a finite order. The equation of motion for u assumes the form

$$u_{n+1} = \lambda u_n + \sum_{k=1}^{\infty} \epsilon^k U_k(u_n, u_n^*), \qquad (13)$$

and the coefficients T_{pq} are chosen so that the r.h.s contains only resonant monomials $u^{k+1} u^{*k}$:

$$u_{n+1} = \lambda u_n \left[1 + \sum_{k=1}^{\infty} \epsilon^{2k} \frac{\tilde{U}_{2k}}{\lambda} (uu^*)^k \right]. \qquad (14)$$

This requirement again leaves undetermined the free functions, which are the T_{pq} for which $p = q+1$. Analogous to the case of differential equations, the map for u has the general form

$$u_{n+1} = e^{-i\Omega(u_n u_n^*)} u_n, \qquad (15)$$

where Ω consists of a sum of amplitude-dependent tuneshifts, and the minimal normal form procedure is to choose the free functions to terminate the series for Ω at a finite order. *Unlike* the case with differential equations, however, the terms in Ω are not simply the $\tilde{U}_{2k}$, hence one *cannot* terminate the series for the tuneshifts in Ω by setting the $\tilde{U}_{2k}$ to zero beyond $O(\epsilon^2)$. Instead the free function $T_{k+1\,k}$ is chosen so that $\tilde{U}_{2k}/\lambda$ equals $(\tilde{U}_2/\lambda)^k/k!$. This then yields the map

$$u_{n+1} = \lambda u_n \left[1 + \sum_k \frac{\epsilon^{2k}}{k!} \left(\frac{\tilde{U}_{2k}}{\lambda}\right)^k (uu^*)^k\right]$$

$$= \lambda u_n \, \exp\left[\epsilon^2 \, (\tilde{U}_2/\lambda) \, uu^*\right], \quad (16)$$

or

$$\Omega = \mu + i\,\epsilon^2 (\tilde{U}_2/\lambda)\, \rho^2, \quad (17)$$

i.e. a single higher-order exponent. The nonlinear invariant of the motion is again $\rho^2 = u_n u_n^*$. This is our generalization of Eq. (8) of the minimal normal form condition as applied to maps. As in the case of differential equations, setting all the free functions to zero would require less effort in calculating the T_{pq} coefficients, but would yield an infinite series of amplitude-dependent tuneshift corrections

$$\Omega = \mu + \epsilon^2\,\mu'\,\rho^2 + \epsilon^4\,\mu''\,\rho^4 + \cdots, \quad (18)$$

while a canonical transformation would yield both nonzero free functions and an infinite series of amplitude-dependent tuneshift corrections.

We note the following caveat: if the $O(\epsilon^2)$ tuneshift vanishes, so that the lowest-order nonzero tuneshift is $\epsilon^4\,\mu''(u_n u_n^*)^2$, then the free functions should be chosen so that the map for u_n has the form

$$u_{n+1} = \lambda u_n \, \exp(-i\epsilon^4\,\mu''\,u_n^2 u_n^{*2}). \quad (19)$$

As a final development of the theoretical formalism in this paper, before turning our attention to numerical examples, the extension of the minimal normal form algorithm to higher phase-space dimensions will be given. We treat four dimensions in this section. Clearly, it is desirable for any such generalization to extend to six dimensions easily, so part of our effort will be to develop a suitable notation for the task.

At the first step, the linear part of the map is diagonalized. The details do not involve nonlinear dynamics, and will be omitted. The four eigenmodes will be denoted by z_1, z_1^*, z_2, and z_2^*. Only z_1 and z_2 need be treated below. In the absence of transverse x-y coupling, $z_1 = x + ip_x$ and $z_2 = y + ip_y$, but such an identification is not required in the general formalism. We define $\lambda_1 = e^{-i\mu_1}$ and $\lambda_2 = e^{-i\mu_2}$ with an obvious notation. The equations of the map are then written in the form, using superscripts i and f for initial and final, respectively,

$$z_1^f = \lambda_1 z_1^i + \sum_{k=1}^{\infty} \epsilon^k\, Z_k^{(1)}(z_1^i, z_1^{i*}, z_2^i, z_2^{i*}),$$

$$z_2^f = \lambda_2 z_2^i + \sum_{k=1}^{\infty} \epsilon^k Z_k^{(2)}(z_1^i, z_1^{i*}, z_2^i, z_2^{i*}). \tag{20}$$

We now introduce a vector index $\mathbf{m} = (m_1, m_2, m_3, m_4)$, and the norm $|\mathbf{m}| = |m_1| + \cdots + |m_4|$. The above map equations can then be reexpressed as

$$z_j^f = \lambda_j z_j^i + \sum_{\mathbf{m}} \epsilon^{|\mathbf{m}|-1} Z_{\mathbf{m}}^{(j)} |\mathbf{m}, z^i\rangle, \tag{21}$$

where $j = 1$ or 2, and

$$|\mathbf{m}, z\rangle = z_1^{m_1} z_1^{*m_2} z_2^{m_3} z_2^{*m_4} \tag{22}$$

is a shorthand notation to avoid a proliferation of symbols in the equations. The near-identity transformation to the normal form variables u_1 and u_2 is written as

$$z_j = u_j + \sum_{\mathbf{m}} \epsilon^{|\mathbf{m}|-1} T_{\mathbf{m}}^{(j)} |\mathbf{m}, u\rangle, \tag{23}$$

where again $j = 1, 2$. Here $|\mathbf{m}, u\rangle$ is defined analogously to $|\mathbf{m}, z\rangle$ with u in place of z. The above expressions are substituted into the map equations, and terms collected in powers of u_1, u_1^*, etc. The solutions for u_1 and u_2 are of the form $u_j = \rho_j e^{-i\psi_j}$, where $\rho_1^2 = u_1 u_1^*$ and $\rho_2^2 = u_2 u_2^*$ are the two nonlinear invariants of the motion.

The map equations for the normal form must now be specified. There is only one tuneshift parameter μ' in two phase-space dimensions, but in four dimensions there are three:

$$\begin{pmatrix} \psi_1^f \\ \psi_2^f \end{pmatrix} = \begin{pmatrix} \psi_1^i \\ \psi_2^i \end{pmatrix} + \begin{pmatrix} \mu_1 \\ \mu_2 \end{pmatrix} + \epsilon^2 \begin{pmatrix} \mu_{11} & \mu_{12} \\ \mu_{21} & \mu_{22} \end{pmatrix} \begin{pmatrix} \rho_1^2 \\ \rho_2^2 \end{pmatrix}, \tag{24}$$

where $\mu_{12} = \mu_{21}$, i.e. the tuneshift parameter μ' generalizes into a symmetric 2×2 matrix. Hence we seek equations of the form

$$\begin{aligned} u_1^f &= \lambda_1 u_1^i \exp\left[-i\epsilon^2 (\mu_{11} u_1^i u_1^{i*} + \mu_{12} u_2^i u_2^{i*})\right], \\ u_2^f &= \lambda_2 u_2^i \exp\left[-i\epsilon^2 (\mu_{21} u_1^i u_1^{i*} + \mu_{22} u_2^i u_2^{i*})\right]. \end{aligned} \tag{25}$$

Finally, the free terms must be identified. They are the coefficients of $u_1 (u_1 u_1^*)^p (u_2 u_2^*)^q$ and $u_2 (u_1 u_1^*)^p (u_2 u_2^*)^q$, respectively, i.e. $T_{p+1,p,q,q}^{(1)}$ and $T_{p,p,q+1,q}^{(2)}$ where $p, q = 0, 1, \ldots$ with $p + q \geq 1$. The free terms are obviously chosen so as to attain the tuneshift formula Eq. (24).

The formalism for six phase-space dimensions is formally exactly the same as in four dimensions. The index $\mathbf{m}$ is extended to six components, and the tuneshift formula to

$$\begin{pmatrix} \psi_1^f \\ \psi_2^f \\ \psi_3^f \end{pmatrix} = \begin{pmatrix} \psi_1^i \\ \psi_2^i \\ \psi_3^i \end{pmatrix} + \begin{pmatrix} \mu_1 \\ \mu_2 \\ \mu_3 \end{pmatrix} + \epsilon^2 \begin{pmatrix} \mu_{11} & \mu_{12} & \mu_{13} \\ \mu_{21} & \mu_{22} & \mu_{23} \\ \mu_{31} & \mu_{32} & \mu_{33} \end{pmatrix} \begin{pmatrix} \rho_1^2 \\ \rho_2^2 \\ \rho_3^2 \end{pmatrix}, \tag{26}$$

where the tuneshift matrix is again symmetric: $\mu_{ij} = \mu_{ji}$, and the nonlinear invariants are $\rho_j^2 = u_j u_j^*$, $j = 1, 2, 3$. More complete mathematical details of the formalism for four and six phase-space dimensions are given in Ref. [4].

3 NUMERICAL RESULTS

The numerical studies below will be restricted to two phase-space dimensions. Results for higher phase-space dimensions, though important, are beyond the scope of this paper. We first study some differential equations, then maps. The first model is that of a Duffing oscillator

$$\dot{z} = -i\mu z + \frac{i\epsilon^2}{8}(z + z^*)^3, \qquad (27)$$

which can be derived from the more familiar equations

$$\begin{aligned} \dot{x} &= \mu p, \\ \dot{p} &= -\mu x + \epsilon^2 x^3. \end{aligned} \qquad (28)$$

We may set $\mu = 1$ without loss of generality. Three methods of solution were used: (1) set all free functions to zero (the "$F = 0$" choice), (2) a canonical transformation (denoted "CT" below), and (3) the minimal normal form (denoted "MNF"). For this perturbation, the $O(\epsilon^2)$ tuneshift parameter can be read off immediately and is

$$\mu' = -\frac{3}{8}, \qquad (29)$$

for all of the above methods. It is only at higher orders that the tuneshift parameters, etc., differ between the methods.

To examine the convergence of the series expansions of the various methods above, D'Alembert's test of convergence [7] suggests that we plot the magnitudes of the kth-order terms $|T_k|$, or $\log|T_k|$, as a function of k. The function actually plotted was $\log(\sum_{i+j=k+1} |T_{ij}|)$, i.e. the sum of the magnitudes of the coefficients of the individual terms in each T_k. The results are shown in Fig. 1, where the curves corresponding to the various choices are labelled $F = 0$, CT, or MNF, respectively. It is striking that the coefficients from the first two methods ($F = 0$ and CT) do not decrease with the order at all, whereas those calculated using the MNF decrease rapidly. This is equivalent to saying that the minimal normal form achieves, at low orders of perturbation theory, a better approximation to the exact solution, thus requiring smaller higher-order corrections.

Another important test of the accuracy of the various perturbative solutions is to compute the value of the energy. Since the equation of motion is autonomous, the Hamiltonian is a first integral of the motion, and the energy should remain constant as a function of time. In general, the value of any first integral of the motion should be computed and checked for constancy as a function of time. For the Duffing oscillator of Eq. (27), the energy is

$$E = \frac{\mu}{2}(p^2 + x^2) - \frac{\epsilon^2 x^4}{4}. \qquad (30)$$

In our analysis, $\mu = \epsilon = 1$. The amplitude was set at $\rho \simeq 0.5$, so as to be well within the separatrix, so that all the methods would converge rapidly. [The unstable fixed

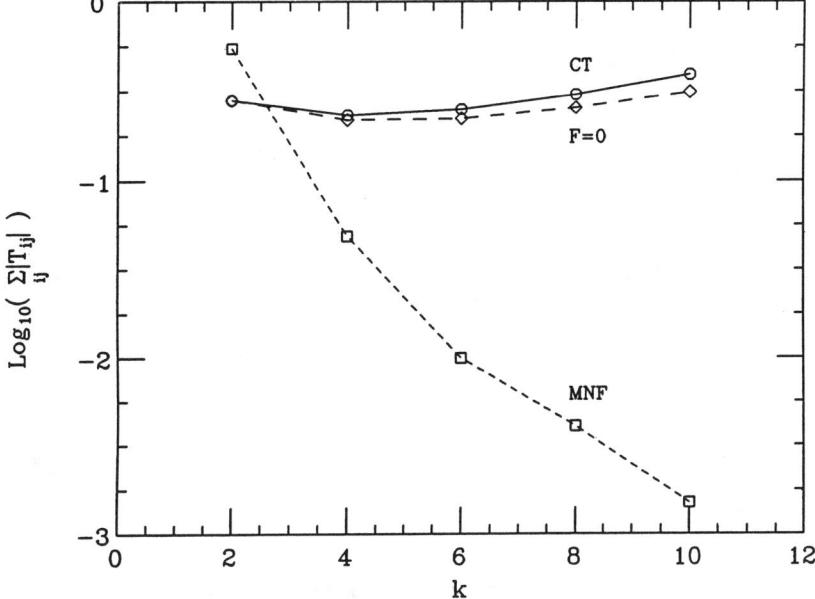

Figure 1: Graph of the magnitudes of the coefficients in the normal form transformation as a function of the order of perturbation theory, for various methods of choosing the free functions for a Duffing oscillator.

points are at $(x,p) = (\pm 1, 0)$, i.e. $z = \pm 1$.] For each method, the calculations were performed to $O(\epsilon^4)$ and $O(\epsilon^6)$, i.e. the near-identity transformation to the normal form was calculated through T_4 or T_6, respectively, and the tuneshifts were also calculated through $O(\epsilon^4)$ or $O(\epsilon^6)$, for the $F = 0$ and CT cases. For the MNF, the tuneshift terminated at $O(\epsilon^2)$, of course. The value of the relative error $(E - E_0)/E_0$, where E_0 is the initial energy, is plotted against the time in Fig. 2. A total of six curves are shown and labelled, viz. dashes for the MNF, solid for the CT, and crosses for the $F = 0$ choice. The results for the $F = 0$ and CT methods are very similar, but that for the MNF is quite clearly better than both. Even at $O(\epsilon^4)$, the MNF yields an error comparable to that from the others at $O(\epsilon^6)$, while when the MNF is used to $O(\epsilon^6)$ the error is almost invisible on the scale of Fig. 2.

On the above evidence, the minimal normal form appears to be a good choice when utilizing the freedom in setting up the near-identity transformation to the normal form, but more evidence is required, using other equations of motion. Use of a quadratic perturbation, viz.

$$\dot{z} = -i\mu z + \frac{i\epsilon^2}{8}(z + z^*)^3, \tag{31}$$

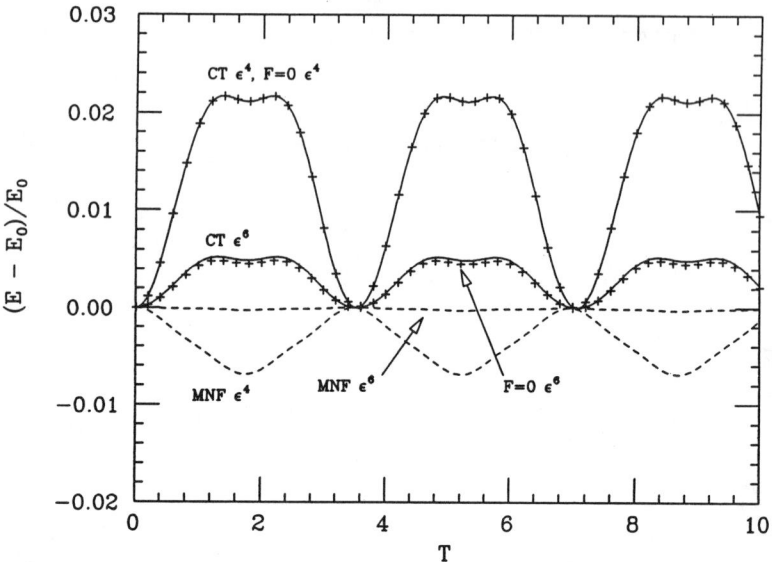

Figure 2: Graph of the relative error $(E - E_0)/E_0$ in the energy as a function of time, for calculations through fourth and sixth order, for various methods of performing the normal form transformation. The equation of motion contained a cubic nonlinear term.

yields results even more impressive than those from the Duffing oscillator. Thus, for two well-known models, the minimal normal form method yields better results than the more traditional techniques. However, there are cases where the use of the minimal normal form as described above is not so effective. Suppose one uses a combination of quadratic and cubic perturbations, e.g. a potential

$$V = -\frac{8}{25}\epsilon x^3 + \frac{\epsilon^2 x^4}{4}, \qquad (32)$$

with $\mu = \epsilon = 1$, for which the $O(\epsilon^2)$ tuneshift parameter is $\mu' \simeq -0.009$, i.e. almost zero. The minimal normal form method works badly in this case. The corresponding graph to Fig. 1 is shown in Fig. 3. The MNF coefficients, this time, are larger than those from the $F = 0$ and CT methods. This behavior is explained in Ref. [4], where it is shown that, to achieve the MNF criterion, the free functions are inversely proportional to μ', and hence become large if μ' is small. The coefficients in the $F = 0$ and CT methods are not directly dependent on the value of μ', hence their behavior is not much different between Figs. 1 and 3.

However, it was shown in Ref. [4] how to cure this problem. As pointed out above, if $\mu' = 0$, then the MNF prescription should be modified to retain the $O(\epsilon^4)$ tuneshift

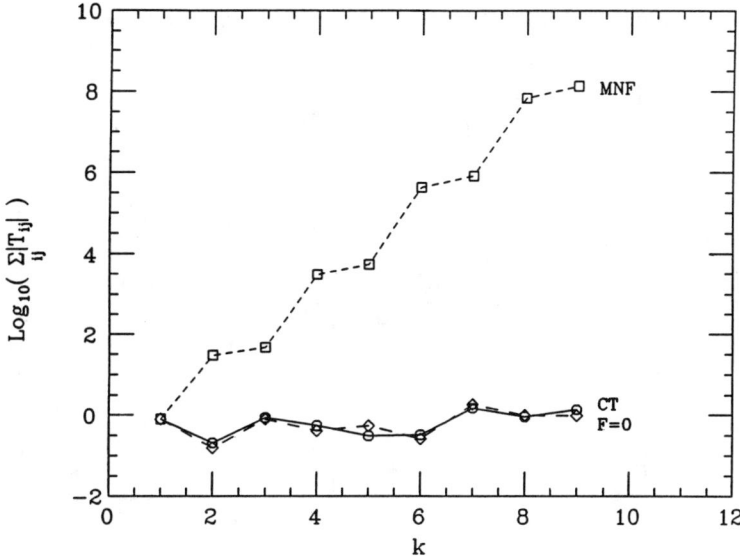

Figure 3: Graph of the magnitudes of the coefficients in the normal form transformation as a function of the order of perturbation theory, with quadratic and cubic nonlinear terms chosen so as to yield a small second-order tuneshift parameter.

parameter μ'', and to use the free functions to achieve a tuneshift of $\epsilon^4 \mu'' \rho^2$ in the normal form. The free functions will then be inversely proportional to μ''. Using the free functions in this way, the magnitudes of the coefficients in the series expansion are shown in Fig. 4, together with the MNF and CT methods, unchanged from Fig. 3. Strictly speaking, the MNF prescription was only applied approximately, i.e. the tuneshift coefficients beyond μ'' were made small but not exactly zero, because the equations to be solved were more complicated and beyond the scope of the present formalism. The MNF coefficients are now an order of magnitude *smaller* than those from the $F = 0$ and CT methods, instead of eight orders of magnitude larger as in Fig. 3. The improvement in the MNF results is dramatic.

Hence we may conclude that the minimal normal form has the potential to be a good method of exploiting the freedom available in the transformation from the original phase-space coordinates to the normal form. It holds out the promise of offering a good approximation to the exact solution using only low orders of perturbation theory, thereby requiring smaller higher-order corrections than other methods such as a canonical transformation, but there are some caveats to the implementation of the method, as noted above, if the magnitude of the lowest-order tuneshift parameter μ' was small, or zero. The merit of the minimal normal form method for discrete maps

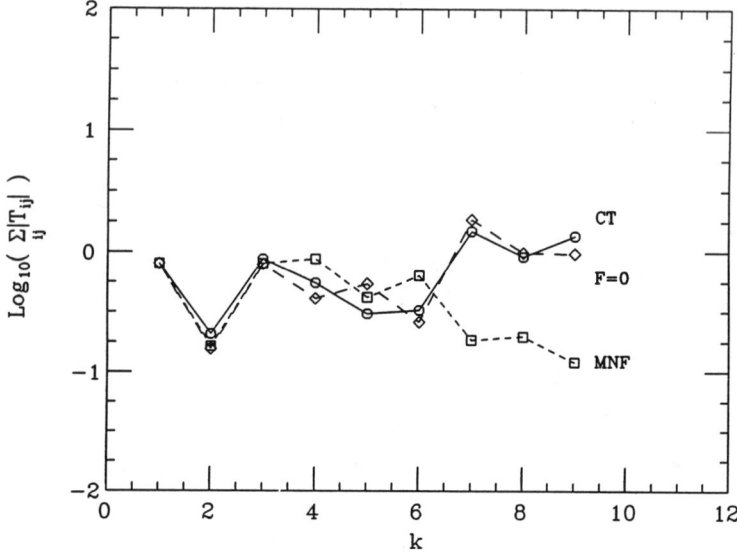

Figure 4: Graph of the magnitudes of the coefficients in the normal form transformation as a function of the order of perturbation theory. Both the second- and fourth-order tuneshift parameters were retained in the minimal normal form method.

will now be investigated.

The calculations of the expressions for the normal form for discrete maps are similar to those for differential equations. The three choices for the normal form will again be denoted $F = 0$, CT and MNF. The maps treated below used a single thin-lens sextupole or octupole, i.e. a quadratic or cubic nonlinearity. A plot of the phase-space trajectories for these maps is shown in Fig. 5. The tunes were set to $\nu = 0.255$ and $\nu = 0.34$ respectively, where $\mu = 2\pi\nu$, and the value of ϵ was chosen so that the separatrices would be at $x^2 + p^2 \simeq 1$ in both cases. This leads to the choices $\epsilon = 0.2$ and 0.1, respectively. For the octupole, the map equation is

$$z_{n+1} = \lambda z_n + \frac{i\epsilon^2 \lambda}{8}(z_n + z_n^*)^3 . \tag{33}$$

The sextupole map would normally be very similar

$$z_{n+1} = \lambda z_n + \frac{i\epsilon \lambda}{4}(z_n + z_n^*)^2 , \tag{34}$$

i.e. the Hénon map [8]. For aesthetic reasons, however, to create phase-space trajectories symmetric around the x axis, the sextupole was located diametrically opposite the observation point, leading to the map equation

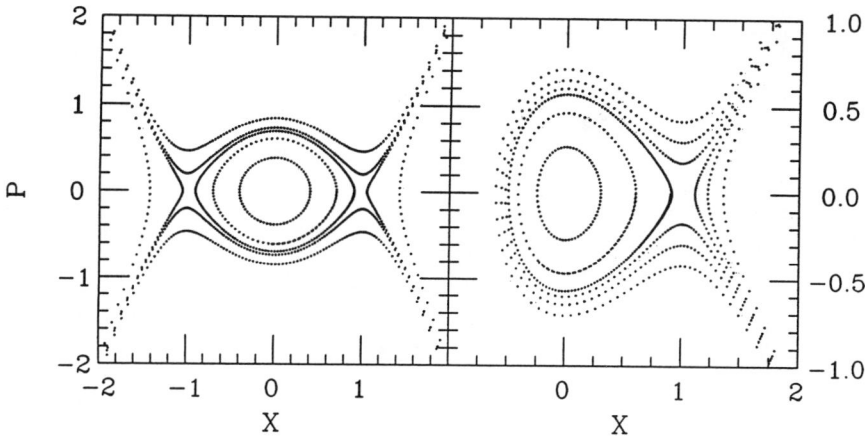

Figure 5: Phase-space trajectories for maps with a thin-lens octupole (left) and sextupole (right). The tunes are labelled in each figure. The nonlinear multipole strength was chosen to place the separatrix at $x^2 + p^2 \simeq 1$ in each case.

$$z_{n+1} = \lambda z_n + \frac{i\epsilon\sqrt{\lambda}}{4}(\sqrt{\lambda}\, z_n + \sqrt{\lambda^*}\, z_n^*)^2, \tag{35}$$

which was used to generate Fig. 5. It has been verified that the location of the sextupole made no difference to the convergence of the series expansion for the normal form. The values of the $O(\epsilon^2)$ tuneshift parameter are

$$\mu' = -\frac{3}{8}, \tag{36}$$

for the thin-lens octupole kick in Eq. (33), and

$$\mu' = \frac{i}{8}\frac{2\lambda^3 + 3\lambda^2 + 3\lambda + 2}{\lambda^3 - 1}, \tag{37}$$

for the thin-lens sextupole kick in Eq. (35).

As explained above, in accordance with D'Alembert's test of convergence [7], the various methods for calculating the normal form were compared by plotting the function $\log(\sum_{i+j=k+1}|T_{ij}|)$ against the order k. The results for the thin-lens octupole map Eq. (33), with a tune of $\nu = 0.255$, are shown in Fig. 6. The results from all three methods are almost the same, and although the MNF result is smaller than the others, the difference is slight. It is not immediately clear why this is so; perhaps it may be due to the presence of a small denominator in the coefficients, caused by the location of large low-order resonance islands close to the origin.

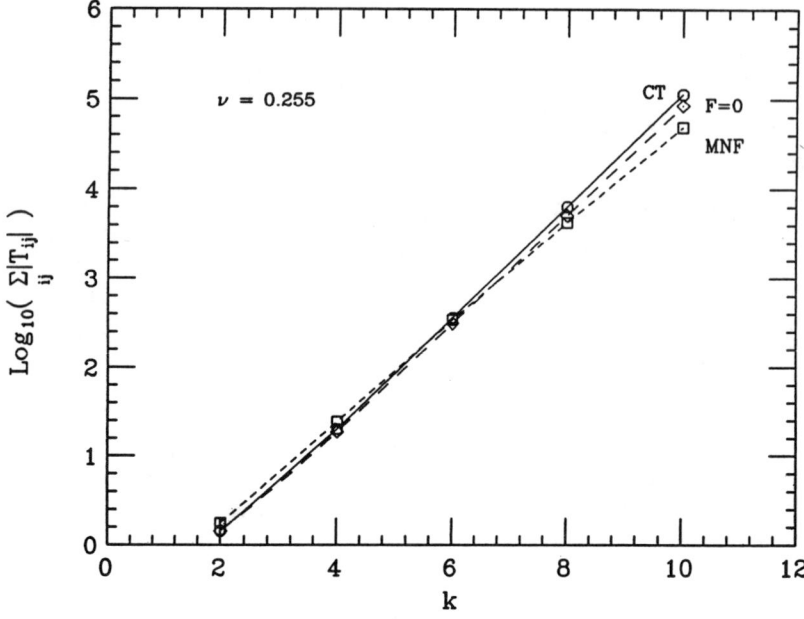

Figure 6: Graph of the magnitudes of the coefficients in the normal form transformation as a function of the order of perturbation theory, for various methods of choosing the free functions, for a map with a single thin-lens octupole, using a tune of $\nu = 0.255$.

In an attempt to approximate the conditions of the differential equation, the calculation was repeated using a tune of $\nu = 0.01$, i.e. a small phase advance per turn. The results are shown in Fig. 7. The MNF result, this time, is distinctly better than the others, the coefficients being smaller by two orders of magnitude at the tenth order.

To examine the behavior for a different map, the thin-lens sextupole map of Eq. (35) was studied next. The results are shown in Fig. 8, for a tune of $\nu = 0.34$. This time, the MNF coefficients for the normal form are clearly smaller than those from the $F = 0$ and CT methods. Use of a combined sextupole and octupole perturbation did not make any significant difference to the above conclusions.

Hence the minimal normal form method, when applied to discrete maps, sometimes yields better results than other methods, but the difference is not as pronounced as when solving differential equations. In all of the maps studied above, the free functions were chosen to cancel the higher-order tuneshift coefficients beyond $O(\epsilon^2)$. However, in those cases where the minimal normal form results were not better than those from other methods, it may be a good idea to retain both the $O(\epsilon^2)$ and $O(\epsilon^4)$ tuneshift terms, or to extend the method to treat resonant normal forms (motivated by the presence of large resonance islands in the phase space of the octupole map). This requires a more sophisticated formalism and computer program, because the equations to be solved are not as straightforward. Future reports will address the issue.

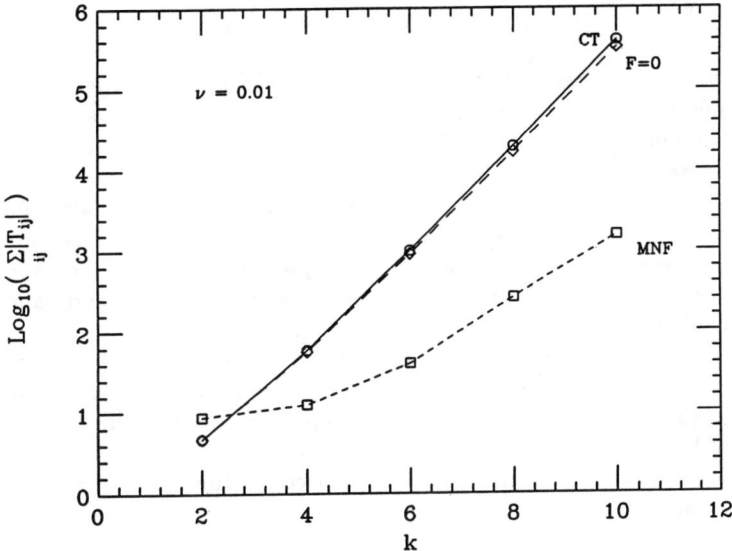

Figure 7: Same as the previous figure, but with a tune of $\nu = 0.01$.

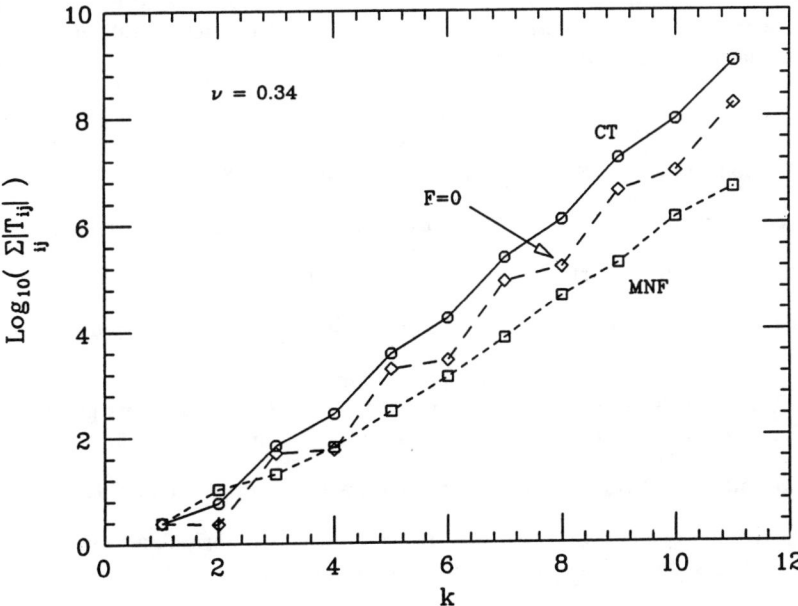

Figure 8: Graph of the magnitudes of the coefficients in the normal form transformation as a function of the order of perturbation theory, for various methods of choosing the free functions, for a map with a single thin-lens sextupole. The small-amplitude tune was $\nu = 0.34$.

4 CONCLUSION

The method of normal forms has been reviewed and the superiority of the minimal normal form method has been demonstrated for ordinary nonlinear autonomous differential equations. The minimal normal form method has also been extended, to treat discrete maps. The application to the evaluation of one-turn maps for accelerators yields mixed results, hence the superiority of the minimal normal form is not as clearly visible. Further studies in investigating the effect of small denominators in the vicinity of low-order resonances will be attempted, to improve the convergence of the perturbation expansion, and numerical model evaluations will be carried out for higher-dimensional phase space.

A major application of the method presented here in the calculation of the one-turn map is to use the map to evaluate the effect of nonlinear perturbations on the lattice functions, such as betatron functions, the dispersion function, chromaticity, and tunes of the accelerator. If such improvements are possible, the minimal normal form method can serve as a standard way of calculating lattice functions of an accelerator or storage ring with small nonlinear imperfections.

ACKNOWLEDGEMENTS

The authors would like to thank Professors P. Kahn and Y. Zarmi for fruitful discussions on the method of minimal normal forms, and Mrs. D. Murray for help on part of the calculations. Discussions with D. Abell, P. Channel, J. Ellison, E. Forest, and A. Sessler are also appreciated.

REFERENCES

1. E.D. Courant and H.S. Snyder, Ann. Phys. **3**, 1 (1958).

2. E. Forest and K. Hirata, KEK Report 92-12 (1992). See also references therein.

3. P.B. Kahn and Y. Zarmi, Physica D **54**, 65 (1991).

4. S.R. Mane and W.T. Weng, Brookhaven National Laboratory Report BNL-48249 (Nov. 1992), and Phys. Rev. E 48(1), 532 (1993).

5. G. Turchetti, Nonlinear Problems in Future Particle Accelerators (World Scientific, Singapore, 1991), eds. W. Scandale and G. Turchetti, p. 16.

6. W.Scandale, F. Schmidt and E. Todesco, Part. Accel. **35**, 53 (1991).

7. I. S. Gradshteyn and I. M. Rhyzik, Table of Integrals, Series, and Products (Academic Press, Inc., New York, 1980), 4th ed., p. 5.

8. M. Hénon, Chaotic Behavior of Deterministic Systems (North-Holland, New York, 1983), eds. G. Iooss et al., p. 53.

PERTURBATION EXPANSION FOR PARTICLE DISTRIBUTION IN HADRON STORAGE RINGS

J. Shi and S. Ohnuma
Department of Physics, University of Houston, Houston, TX 77204-5504

ABSTRACT

Based on the perturbation theory with multiple scales, we have developed a new technique to study the evolution of particle distribution as a function of oscillation amplitudes in hadron storage rings. With a renormalization scheme for the zeroth-order term, a uniformly valid perturbation expansion for the distribution function is obtained. For localized nonlinear perturbations such as the beam-beam interaction at colliding points, this renormalization scheme results in a functional mapping for the particle distribution, and the diffusion processes of particles in the beam can be studied numerically without resorting to the tracking of individual particles. A case involving a single nonlinear kick in the ring is presented to illustrate the method in detail.

1 INTRODUCTION

The understanding of beam dynamics in hadron storage rings has mostly relied on the tracking of a few individual particles. The real beam, however, consists of a large number of particles, typically up to 10^{10} per bunch in large storage rings such as SSC. Consequently it is inconvenient if not impossible to study the real beam behavior by this brute-force tracking which requires many hours of supercomputer CPU time.

Examples of such beam behavior we are interested in are slow particle losses and beam size growth due to either field errors or beam-beam interactions. Observations have shown that the growth of tails of particle distribution is a serious problem as it enhances the background level in detectors. For large hadron storage rings, the slow growth of the beam emittance is closely related to the long-term behavior of the colliding beam. As beams circulate in the storage ring, particles are gradually lost, and the rates of particle loss and beam size growth are more important than the stability of individual particles. Therefore a better understanding of these diffusion processes should be based on the study of the multi-particle system.

One way to describe the multi-particle system is through a use of the single-particle distribution in the transverse phase space. In fact, it has already been done for the study of beam dynamics in high energy electron storage rings but the behavior of the particle distribution in hadron storage rings is not yet understood theoretically. This is due to the different behavior of the particle distribution in the transverse phase space in two cases. For a high energy electron beam, because of the dominant radiation effect, the time scale for a beam to reach the equilibrium distribution is much less than the storage time. Consequently, the study of the beam dynamics can be focused on the behavior of the distribution near its steady states.[1,2] For a hadron beam, however, the damping time scale is substantially larger than the storage time so that the motion of particles can be described by the Hamiltonian formalism. In the presence of nonlinear perturbations due to either field errors or beam-beam interactions, the particle distribution may not reach any steady state within a fraction of the storage time and this makes it important to study the time evolution of the distribution.

In most cases, the strengths of high-order multipole field errors and beam-beam interactions are small enough to be treated as a perturbation. A straightforward approach is the expansion of the distribution function in powers of the perturbation strength. This method has been used to study the evolution of the distribution with Fokker-Planck equation.[3] However, it can be shown that this Poincaré type expansion breaks down in the infinite domain so that the validity of the expansion is limited. In order to obtain a uniformly valid perturbation expansion, we introduce a perturbation expansion with multiple scales[4]. For localized multipole field errors and beam-beam interactions, this treatment results in a renormalization scheme for the zeroth-order term of the expansion. As a result, the evolution of the distribution on amplitudes can be expressed by a functional mapping.

In this paper, we describe the perturbation technique of multiple scales for the study of the evolution of the particle distribution in hadron storage rings. In Section 2, the single-particle distribution function and its Vlasov equation and Fokker-Planck equation are introduced. The failure of the straightforward perturbation expansion is explained in Section 3. In Section 4, we establish a multiple-scale expansion for the Vlasov equation and the Fokker-Planck equation. An illustration for this technique is given in Section 5. In Section 6, we discuss the Gaussian distribution approximation which can simplify the numerical computation. Section 7 contains summary and final remarks. In what follows, "distribution" refers to the single-particle distribution in the transverse phase space.

2 EQUATION FOR THE DISTRIBUTION FUNCTION

In terms of action-angle variables $(\vec{I}, \vec{\phi})$, the Hamiltonian for betatron motions in a ring of circumference C can be written as

$$H(\vec{I}, \vec{\phi}, \theta) = \vec{\nu} \cdot \vec{I} + U(\sqrt{2I_1\beta_1}cos\psi_1, \sqrt{2I_2\beta_2}cos\psi_2, \theta) \ , \qquad (1)$$

where

$$\psi_i = \phi_i - \nu_i\theta + \frac{C}{2\pi}\int_0^\theta \frac{1}{\beta_i}d\theta \ , \quad i = 1, 2 \ . \qquad (2)$$

The independent variable θ is defined as the path length of the reference orbit divided by $(C/2\pi)$. In this paper, vectors are used to denote two-dimensional variables describing the motions in vertical and horizontal planes. Here we assume that there is no linear coupling. $\vec{\beta}(\theta)$ are Courant-Snyder beta functions of the linear lattice. The action-angle variables used here are related to the Cartesian phase-space coordinates (x_i, x_i') by the equations

$$x_i = (2I_i\beta_i)^{1/2} \cos\psi_i \ , \qquad (3)$$

$$x_i' = -(2I_i/\beta_i)^{1/2}\left[\sin\psi_i - \frac{\beta_i'}{2}\cos\psi_i\right] \ , \quad i = 1, 2 \ , \qquad (4)$$

where the prime denotes derivatives with respect to $(\theta C/2\pi)$. When there is a tune modulation, the transverse tunes $\vec{\nu}$ are functions of θ. The nonlinear perturbation U represents either beam-beam interactions or field errors. We assume that the actions of the nonlinear perturbation can be approximated by kicks in the transverse plane, i.e.

$$U = \sum_k U^{(k)}(\sqrt{2I_1\beta_1}cos\psi_1, \sqrt{2I_2\beta_2}cos\psi_2)\delta_c(\theta - \theta_k) \qquad (5)$$

where θ_k are the locations of kicks and

$$\delta_c(\theta - \theta_k) = \sum_n \delta(\theta - \theta_k - 2\pi n) \ . \tag{6}$$

Consider a beam consisting of N particles. If we neglect intra-beam collisions, the phase-space distribution of particles can be described by the single-particle distribution $f(\vec{I}, \vec{\phi}, \theta)$, which satisfies the Vlasov equation

$$\frac{\partial f}{\partial \theta} + \vec{\nu}(\theta) \cdot \frac{\partial f}{\partial \vec{\phi}} = \{U, f\} \ , \tag{7}$$

where $\{\ \}$ is the Poisson bracket.

If the nonlinear perturbation comes from field errors, U is independent of the particle distribution f. In the case of beam-beam interactions with strong-weak model, the strong beam acts as a nonlinear lens located at collision points. As far as the distribution of weak beam f is concerned, U is also independent of f. In such cases we can rewrite the Hamiltonian Eq. (1) in the form

$$H(\vec{I}, \vec{\phi}, \theta) = \vec{\nu} \cdot \vec{I} + U_0(\vec{I}, \theta) + U_1\left(\sqrt{2I_1\beta_1}\cos\psi_1, \sqrt{2I_2\beta_2}\cos\psi_2, \theta\right) \tag{8}$$

where U_0 is that part of U which depends on the amplitude only,

$$U_0(\vec{I}, \theta) = \left\langle U\left(\sqrt{2I_1\beta_1}\cos\psi_1, \sqrt{2I_2\beta_2}\cos\psi_2, \theta\right)\right\rangle_{\vec{\phi}} \ , \tag{9}$$

and

$$U_1 \equiv U - U_0 \ . \tag{10}$$

The Vlasov equation (7) is reduced to

$$\frac{\partial f}{\partial \theta} + \left(\vec{\nu}(\theta) + \frac{\partial U_0}{\partial \vec{I}}\right) \cdot \frac{\partial f}{\partial \vec{\phi}} = \{U_1, f\}$$

$$= \sum_{i=1}^{2}\left\{-\sqrt{\beta_i}u_i\left[(2I_i)^{-\frac{1}{2}}\cos\psi_i\frac{\partial}{\partial \phi_i} + (2I_i)^{\frac{1}{2}}\sin\psi_i\frac{\partial}{\partial I_i}\right] + \frac{\partial U_0}{\partial I_i}\frac{\partial}{\partial \phi_i}\right\}f \tag{11}$$

where $u_i = \partial U(x_1, x_2, \theta)/\partial x_i$. Since $H_0 = \vec{\nu} \cdot \vec{I} + U_0(\vec{I}, \theta)$ is an integrable system, U_1 can be regarded as the perturbation now.

In the case of two strong colliding beams, the distributions $f^{(1)}$ and $f^{(2)}$ of the two beams influence each other according to the Vlasov equation in which the beam-beam perturbation for one beam is a functional of the distribution of the other beam. Two coupled Vlasov equations are given by

$$\frac{\partial f^{(1)}}{\partial \theta} + \vec{\nu}(\theta) \cdot \frac{\partial f^{(1)}}{\partial \vec{\phi}} = \left\{U[f^{(2)}], f^{(1)}\right\}$$

$$= -\sum_{i=1}^{2}\sqrt{\beta_i}\left[(2I_i)^{-\frac{1}{2}}\cos\psi_i\frac{\partial f^{(1)}}{\partial \phi_i} + (2I_i)^{\frac{1}{2}}\sin\psi_i\frac{\partial f^{(1)}}{\partial I_i}\right]u_i[f^{(2)}] \tag{12}$$

together with another equation with indices 1 and 2 exchanged, where $u_i[f] = \partial U(x_1, x_2, \theta)/\partial x_i$ is a linear functional of f. For a head-on collision,

$$U(\vec{x}) = a\int_0^\infty \frac{dq}{q} \int_{-\infty}^\infty d\vec{\xi}d\vec{\xi'} f(\vec{\xi}, \vec{\xi'}, \theta)e^{-|\vec{x}-\vec{\xi}|^2 q^2} \tag{13}$$

where a is a constant, $a = 4e^2N/(\gamma mc^2)$ for p–p colliders.

In order to express Eqs. (11) and (12) in a single form, we define a functional vector $\mathbf{f} = (f^{(1)}, f^{(2)})$ to denote the distribution functions of two colliding beams. Then, the Vlasov equations (11) and (12) can be expressed as

$$\frac{\partial \mathbf{f}}{\partial \theta} + \vec{\omega} \cdot \frac{\partial \mathbf{f}}{\partial \vec{\phi}} = \mathbf{T} \circ \mathbf{f} \ . \tag{14}$$

For field errors or beam-beam interactions in the strong-weak model, $\mathbf{f} = (f, 0)$, $\vec{\omega}(\theta, \vec{I}) = \vec{\nu}(\theta) + \partial U_0/\partial \vec{I}$ and $\mathbf{T} \circ \mathbf{f} = \{U_1, f\}$. For beam-beam interactions in the strong-strong model, $\vec{\omega}(\theta) = \vec{\nu}(\theta)$ and $\mathbf{T}f^{(i)} = \{U[f^j], f^{(i)}\}$ with $i, j = 1, 2, j \neq i$.

If we consider noise and damping, the equation of the distribution is described by the Fokker-Planck equation[5]

$$\frac{\partial \mathbf{f}}{\partial \theta} + \mathbf{L}_{FP} \circ \mathbf{f} = \mathbf{T} \circ \mathbf{f} \tag{15}$$

where $\mathbf{T} \circ \mathbf{f} = \{U, \mathbf{f}\}$ and $\mathbf{L}_{FP}$ is the linear Fokker-Planck operator. In normalized variables,

$$\mathbf{L}_{FP} = \sum_{i=1}^{2} \left\{ \frac{\partial}{\partial \xi_i} \nu_i \eta_i + \frac{\partial}{\partial \eta_i}(-2\alpha_i \eta_i - \nu_i \xi_i) - D_i \frac{\partial^2}{\partial \eta_i^2} \right\} \ , \tag{16}$$

where $\vec{\alpha}$ and $\vec{D}$ are damping and diffusion coefficients respectively. When $\vec{\alpha} = 0$ and $\vec{D} = 0$, Eq. (15) reduces to the Vlasov equation (14) with $\mathbf{L}_{FP} = \vec{\nu}(\theta)$.

If we know the evolution of the distribution by solving the Vlasov equation (14) or the Fokker-Planck equation (15), the rms beam size can be evaluated from

$$\langle \vec{I} \rangle = \int \vec{I} f d\vec{I} d\vec{\phi} = \int \vec{I} \langle f \rangle_{\vec{\phi}} d\vec{I} \ , \tag{17}$$

where $< >_{\vec{\phi}}$ denotes the integral with respect to $\vec{\phi}$. In general, neither Vlasov equation nor Fokker-Planck equation can be solved exactly for the nonlinear system of Eq. (1). By inspecting Eq. (17), however, we see that the particle distribution on amplitudes alone is required for our purpose; the Vlasov equation and the Fokker-Planck equation contain more information than is needed. By removing the unnecessary information, the problem may be simplified and easily handled with perturbation methods.

3 STRAIGHTFORWARD PERTURBATION EXPANSION

In many practical problems, the nonlinear perturbation on the right-hand side of Eqs. (14) and (15) can be treated as small in some sense, i.e. the strength of the perturbation can be used as a small parameter for the perturbation expansion. Thus we assume that the distribution function can be expanded as

$$\mathbf{f}(\vec{I}, \vec{\phi}, \theta) = \sum_{m=0}^{\infty} \epsilon^m \mathbf{f}_m(\vec{I}, \vec{\phi}, \theta) \ , \tag{18}$$

where $\epsilon \sim \| U \|$, and for an initial distribution $\mathbf{f}(\vec{I}, \vec{\phi}, 0) = \mathbf{f}_{in}(\vec{I}, \vec{\phi})$, the initial conditions of $\mathbf{f}_m$ are

$$\mathbf{f}_0(\vec{I}, \vec{\phi}, 0) = \mathbf{f}_{in}(\vec{I}, \vec{\phi}) \ , \tag{19}$$

$$\mathbf{f}_m(\vec{I}, \vec{\phi}, 0) = 0 \quad \text{for } m > 0 \ . \tag{20}$$

In the case of beam-beam interactions in the strong-strong model, because of the linear dependence of U on the distribution [see Eq. (13)], **T** can be also expanded as

$$\mathbf{T} = \sum_{m=0}^{\infty} \epsilon^m \mathbf{T}_m , \qquad (21)$$

where $\mathbf{T}_m = \{U[\mathbf{f}_m], \ \}$. For field errors or beam-beam interactions in the strong-weak model, $\mathbf{T}_0 = \{U, \ \}$ ($\mathbf{T}_0 = \{U_1, \ \}$ for the Vlasov equation) and $\mathbf{T}_m = 0$ for $m > 0$.

Substituting Eqs. (18) and (21) into Eq. (15) and equating coefficients of equal powers of ϵ to zero, we obtain

$$\frac{\partial \mathbf{f}_0}{\partial \theta} + \mathbf{L}_{FP} \circ \mathbf{f}_0 = 0 , \qquad (22)$$

$$\frac{\partial \mathbf{f}_1}{\partial \theta} + \mathbf{L}_{FP} \circ \mathbf{f}_1 = \mathbf{T}_0 \circ \mathbf{f}_0 , \qquad (23)$$

$$\frac{\partial \mathbf{f}_2}{\partial \theta} + \mathbf{L}_{FP} \circ \mathbf{f}_2 = \mathbf{T}_0 \circ \mathbf{f}_1 + \mathbf{T}_1 \circ \mathbf{f}_0 . \qquad (24)$$

$$\cdots$$

Since for any function f

$$\int (\mathbf{T}_q f) d\vec{\phi} d\vec{I} = 0 , \quad \forall q , \qquad (25)$$

the normalization condition of the distribution

$$\int d\vec{I} d\vec{\phi} f(\vec{I}, \vec{\phi}, \theta) = \int d\vec{I} d\vec{\phi} f(\vec{I}, \vec{\phi}, 0) = 1 \qquad (26)$$

is guaranteed in this expansion.

After all $\mathbf{f}_i$'s for $i < (q-1)$ are known, the qth-order equation for $\mathbf{f}_q$ takes the form

$$\frac{\partial \mathbf{f}_q}{\partial \theta} + \mathbf{L}_{FP} \circ \mathbf{f}_q = \mathbf{F}(\vec{I}, \vec{\phi}, \theta) . \qquad (27)$$

For the initial condition $\mathbf{f}_q(\vec{I}, \vec{\phi}, 0)$ in Eqs. (19) and (20), the solution of Eq. (27) is

$$\begin{aligned} f_q(\vec{I}, \vec{\phi}, \theta) &= \int d\vec{I}' d\vec{\phi}' G_2(\vec{I}, \vec{I}', \vec{\phi} - \vec{\phi}', \theta) \mathbf{f}_q(\vec{I}', \vec{\phi}', 0) \\ &+ \int d\theta' \int d\vec{I}' d\vec{\phi}' G_2(\vec{I}, \vec{I}', \vec{\phi} - \vec{\phi}', \theta - \theta') \mathbf{F}(\vec{I}', \vec{\phi}', \theta') \end{aligned} \qquad (28)$$

where G_2 is the Green's function of Eq. (27)[5]

$$G_2(\vec{I}, \vec{I}', \vec{\phi} - \vec{\phi}', \theta - \theta') = G_1(I_1, I_1', \phi_1 - \phi_1', \theta - \theta') G_1(I_2, I_2', \phi_2 - \phi_2', \theta - \theta') \qquad (29)$$

with

$$G_1(I_i, I_i', \phi_i - \phi_i', \theta - \theta')$$
$$= \frac{A_i^{\frac{1}{2}}}{\pi} \exp\left\{ -2A_i \left[I_i + I_i' e^{-\alpha_i(\theta - \theta')} - 2\sqrt{I_i I_i' e^{-\alpha_i(\theta - \theta')}} \cos\tilde{\phi}_i \right] \right\} \qquad (30)$$

where

$$A_i = \frac{\alpha_i}{D_i(1 - e^{-2\alpha_i(\theta-\theta')})} \quad \text{and} \quad \tilde{\phi}_i = \phi_i - \int_{\theta'}^{\theta} \nu_i(\tau)d\tau - \phi'_i \ . \tag{31}$$

When $\vec{\alpha} = 0$ and $\vec{D} = 0$,

$$G_2(\vec{I}, \vec{I'}, \vec{\phi} - \vec{\phi'}, \theta - \theta') = \delta\left(\vec{I} - \vec{I'}\right) \delta\left(\vec{\phi} - \int_{\theta'}^{\theta} \vec{\omega}(\tau)d\tau - \vec{\phi'}\right) \ , \tag{32}$$

and Eq. (28) reduces to

$$\mathbf{f}_q(\vec{I}, \vec{\phi}, \theta) = \mathbf{f}_q\left(\vec{I}, \vec{\phi} - \int_0^{\theta} \vec{\omega}(\tau)d\tau, 0\right) + \int_0^{\theta} \mathbf{F}\left(\vec{I}, \vec{\phi} - \int_{\tau_1}^{\theta} \vec{\omega}(\tau_2)d\tau_2, \tau_1\right) d\tau_1 \ . \tag{33}$$

By solving these expansion equations order by order, we obtain a truncated sequence of **f**. Since

$$\left\langle \mathbf{f}_q(\vec{I}, \vec{\phi}, \theta) \right\rangle_{\vec{\phi}} = \int d\vec{I'} \left\langle G_2(\vec{I}, \vec{I'}, \vec{\phi}, \theta) \right\rangle_{\vec{\phi}} \left\langle \mathbf{f}_q(\vec{I'}, \vec{\phi}, 0) \right\rangle_{\vec{\phi}}$$
$$+ \int d\theta' \int d\vec{I'} \left\langle G_2(\vec{I}, \vec{I'}, \vec{\phi}, \theta - \theta') \right\rangle_{\vec{\phi}} \left\langle \mathbf{F}(\vec{I'}, \vec{\phi}, \theta') \right\rangle_{\vec{\phi}} \ , \tag{34}$$

if there is a secular term $\langle \mathbf{F} \rangle_{\vec{\phi}} \neq 0$, $\mathbf{f}_q$ is proportional to θ for the perturbation in Eq. (5). Consequently, the straightforward expansion of Eq. (18) breaks down when $\theta \sim O(\epsilon^{-1})$ since, beyond that, the condition for a uniform asymptotic sequence $|f_q^{(i)}/f_{q-1}^{(i)}| < \infty$ for $i = 1, 2$ can not be satisfied. With Eqs. (22)–(24) it can be easily shown that the secular terms are generic in these expansion equations. Thus the straightforward expansion is not valid for our problem. As a matter of fact the appearance of the secular terms is a characteristic of nonlinear problems. In order to obtain a proper perturbation expansion, these secular terms must be eliminated systematically.

4 MULTIPLE-SCALE EXPANSION

Eq. (17) shows that for the study of particle loss and beam size growth, only the particle distribution on amplitudes is required. Thus the average with respect to phase $\vec{\phi}$ can be utilized to cure the non-uniform problem due to the secular terms. In order to obtain a truncated expansion valid for all times up to $O(\epsilon^{-M})$, we introduce a set of multiple time scales $t_0, t_1, \ldots, t_M$, where

$$t_m = \epsilon^m \theta \ , \ m = 0, 1, ..., M \ . \tag{35}$$

In general, the time scale t_m is slower than t_{m-1}. Instead of using the expansion (18), we assume that

$$\mathbf{f}(\vec{I}, \vec{\phi}, \theta) = \mathbf{f}(\vec{I}, \vec{\phi}, t_0, t_1, ..., t_M) = \sum_{m=0}^{M-1} \epsilon^m \mathbf{f}_m(\vec{I}, \vec{\phi}, t_0, t_1, ..., t_M) + O(\epsilon^{-M}) \ . \tag{36}$$

Here the truncated expansion is assumed to be valid only for times up to $O(\epsilon^{-M})$. Beyond that, other time scales must be included to keep the expansion uniformly valid. By using the chain rule, the derivative with respect to θ can be transformed into the derivatives with respect to $\{t_m\}$ according to

$$\frac{\partial}{\partial \theta} = \sum_{m=0}^{M} \epsilon^m \frac{\partial}{\partial t_m} . \qquad (37)$$

Substituting Eqs. (36), (37) and (21) into Eq. (15) and equating coefficients of equal powers of ϵ on both sides, we obtain

$$\frac{\partial \mathbf{f}_0}{\partial t_0} + \mathbf{L}_{FP} \circ \mathbf{f}_0 = 0 , \qquad (38)$$

$$\frac{\partial \mathbf{f}_1}{\partial t_0} + \frac{\partial \mathbf{f}_0}{\partial t_1} + \mathbf{L}_{FP} \circ \mathbf{f}_1 = \mathbf{T}_0 \circ \mathbf{f}_0 , \qquad (39)$$

$$\frac{\partial \mathbf{f}_2}{\partial t_0} + \frac{\partial \mathbf{f}_1}{\partial t_1} + \frac{\partial \mathbf{f}_0}{\partial t_2} + \mathbf{L}_{FP} \circ \mathbf{f}_2 = \mathbf{T}_0 \circ \mathbf{f}_1 + \mathbf{T}_1 \circ \mathbf{f}_0 . \qquad (40)$$

$$\cdots$$

Since Eq. (25) is still valid here, the normalization condition Eq. (26) is also guaranteed in the expansion (36). Now, in the qth-order equation, we can choose

$$\sum_{m=1}^{q} \frac{\partial \mathbf{f}_{q-m}}{\partial t_m} = \left\langle \sum_{m=0}^{q-1} \mathbf{T}_m \mathbf{f}_{q-m-1} \right\rangle_{\vec{\phi}} \qquad (41)$$

to eliminate the secular terms. As q is greater than one, the dependence of $\mathbf{f}_m$ on $\{t_m | m = 1, ..., q\}$ is underdetermined so that we can select a particular set of equations

$$\frac{\partial \mathbf{f}_0}{\partial t_q} = \left\langle \sum_{m=0}^{q-1} \mathbf{T}_m \mathbf{f}_{q-m-1} \right\rangle_{\vec{\phi}} , \qquad (42)$$

$$\frac{\partial \mathbf{f}_i}{\partial t_j} = 0 \quad \text{for } i, j > 0 . \qquad (43)$$

Consequently, $\langle \mathbf{f}_m \rangle_{\vec{\phi}} = 0$ for $m > 0$, and the zeroth-order term $\mathbf{f}_0$ is renormalized by including in it all phase-independent parts of the distribution function. If the initial distribution depends on amplitudes only,

$$\langle \mathbf{f} \rangle_{\vec{\phi}} = \sum_{m=0}^{\infty} \epsilon^m \langle \mathbf{f}_m \rangle_{\vec{\phi}} = \mathbf{f}_0 \qquad (44)$$

so that $\mathbf{f}_0$ contains all the information needed for the study of particle losses and beam size growth.

5 AN ILLUSTRATION

To illustrate this theory, we consider a simple case in which $\vec{\alpha} = 0$, $\vec{D} = 0$, and there is only one nonlinear kick located at $\theta = 0$, either due to multipole field errors or beam-beam interaction in the strong-weak model,

$$U = U^{(0)} \left(\sqrt{2I_1 \beta_1} \cos\phi_1, \sqrt{2I_2 \beta_2} \cos\phi_2 \right) \delta_c(\theta) . \qquad (45)$$

The initial distribution is assumed to be phase independent,
$$f(\vec{I}, \vec{\phi}, 0) = f_{in}(\vec{I}) \ . \tag{46}$$
From Eq. (38) we have
$$f_0 = f_0(\vec{I}, t_1, t_2, ...) \ . \tag{47}$$
Inserting this result and Eq. (45) into Eq. (39), we get
$$\frac{\partial f_1}{\partial t_0} + \frac{\partial f_0}{\partial t_1} + \vec{\omega}(\theta) \cdot \frac{\partial f_1}{\partial \vec{\phi}} = -\sum_{i=1}^{2} (2\beta_i I_i)^{\frac{1}{2}} \sin\phi_i \frac{\partial f_0}{\partial I_i} u_i \delta_c(\theta) \ , \tag{48}$$
where $u_i\left(\sqrt{2I_1\beta_1}cos\phi_1, \sqrt{2I_2\beta_2}cos\phi_2\right) = \partial U^0(x_1, x_2)/\partial x_i$, for $i = 1, 2$. Since
$$\langle u_i \sin\phi_i \rangle_{\vec{\phi}} = 0 \ , \tag{49}$$
there is no secular term in Eq. (48), $\partial f_0/\partial t_1 = 0$ and $f_0 = f_0(\vec{I}, t_2, ...)$. Thus, in the first-order perturbation, $f_0(\vec{I}) = f_{in}(\vec{I})$, i.e. there is no beam size growth in the first-order perturbation and we must consider the second-order term. From Eq. (33), f_1 is obtained as
$$f_1 = -\sum_{n=0}^{[\theta/2\pi]} \sum_{i=1}^{2} (2\beta_i I_i)^{\frac{1}{2}} \sin\phi_{ni} \frac{\partial f_0}{\partial I_i} u_i(n) \tag{50}$$
where
$$\vec{\phi}_n = (\phi_{n1}, \phi_{n2}) = \vec{\phi} - \int_{2\pi n}^{\theta} \vec{\omega}(\tau) d\tau \ , \tag{51}$$
and $u_i(n) = u_i\left(\sqrt{2I_1\beta_1}cos\phi_{n1}, \sqrt{2I_2\beta_2}cos\phi_{n2}\right)$. Substituting f_0 and f_1 into Eq. (40) and using Eq. (42) to eliminate the secular terms, we have
$$\frac{\partial f_0}{\partial t_2} = \left\langle \left\{ U^{(0)}, f_1 \right\} \right\rangle_{\vec{\phi}} \delta_c(\theta) \ . \tag{52}$$
Since we did not specify the strength of multipole field errors or the parameter of beam-beam interaction separately from the perturbation $U^{(0)}$, ϵ is used here simply to track the order of the perturbation expansion. In the final expansion, ϵ should be taken as unity. To the second order in the strength of U, we have
$$f_0^{M+1} = f_0^M + \left\langle \left\{ U^{(0)}, f_1^M \right\} \right\rangle_{\vec{\phi}} \tag{53}$$
with
$$\left\langle \left\{ U^{(0)}, f_1^M \right\} \right\rangle_{\vec{\phi}} = \sum_{n=0}^{M} \sum_{j=1}^{2} \left\{ \beta_j \frac{\partial f_0^M}{\partial I_j} \cos\mu_{nj} \langle u_j(n)u_j \rangle_{\vec{\phi}} \right.$$
$$+ \sum_{i=1}^{2} \frac{\partial f_0^M}{\partial I_i} \left[\beta_j \sqrt{2\beta_i I_i} \sin\mu_{nj} \langle \sin\phi_{ni} u_{ij}(n) u_j \rangle_{\vec{\phi}} \right.$$
$$+ \sqrt{2\beta_i I_i}\sqrt{2\beta_j I_j} D_{ij} \langle \sin\phi_j \cos\phi_{nj} u_i(n) u_j \rangle_{\vec{\phi}}$$
$$- \sum_{k=1}^{2} \sqrt{2\beta_i I_i}\sqrt{2\beta_j I_j}\sqrt{2\beta_k I_k} D_{jk} \langle \sin\phi_j \sin\phi_{ni} \sin\phi_{nk} u_{ik}(n) u_j \rangle_{\vec{\phi}} \right]$$
$$\left. + \sum_{i=1}^{2} \sqrt{2\beta_i I_i}\sqrt{2\beta_j I_j} \frac{\partial^2 f_0^M}{\partial I_i \partial I_j} \langle \sin\phi_j \sin\phi_{ni} u_i(n) u_j \rangle_{\vec{\phi}} \right\} \tag{54}$$

where $\vec{\mu}_n = \int_{2\pi n}^{\theta} \vec{\omega}(\tau)d\tau$, $D_{ij} = -(\partial U_0/\partial I_i \partial I_j)(\theta - 2\pi n)$, and $u_{ij} = \partial u_i/\partial x_j$. $M = [\theta/2\pi]$ is the number of turns and the superscript M denotes the distribution function after M turns. Eq. (53) is a functional mapping for the evolution of the particle distribution on amplitudes. For a given initial distribution, the evolution of the distribution on amplitudes as well as the evolution of beam size can be studied by iterating this map with computer.

From Eq. (53), the physical meaning of the technique of multiple scales is quite clear. Each time the beam passes through a weak nonlinear kick, its distribution is perturbed slightly. However, only the phase-independent part of this change in the distribution is relevant to the problem of particle loss and beam size growth. Therefore, after each kick we use this phase-independent part of the change in the distribution to renormalize the zeroth-order term of its expansion, f_0. The phase-dependent part is eliminated by taking average over $\vec{\phi}$. A uniformly valid perturbation expansion is, therefore, obtained by re-expanding the distribution with this update zeroth-order term after each kick. This renormalization scheme results in the functional mapping (53) in which only action variables $\vec{I}$ are involved. In comparison with the direct simulation of particle distribution in phase space[6], this scheme greatly simplifies the numerical computation.

It should be pointed out that the result to the second order in the strength of U is only valid for the "time" up to $O(\|U\|^{-2})$. For the beam-beam interaction, the beam-beam parameter is typically of the order of 10^{-3} in larger colliders such as SSC. This corresponds to 10^5—10^6 turns and, beyond that, the next-order terms must be taken into account.

In order to check the formalism, we consider a beam near the difference resonance of a single sextupole kick, i.e.

$$\nu_1 - 2\nu_2 = l \qquad (55)$$

where $l \geq 0$ is an integer. The nonlinear perturbation of the Hamiltonian can be approximated by the resonance Hamiltonian

$$U^{(0)} = \epsilon I_1^{\frac{1}{2}} I_2 \cos(\phi_1 - 2\phi_2)\delta_c(\theta) \ . \qquad (56)$$

Since $2I_1 + I_2$ is a constant of the motion for $U^{(0)}$ of Eq. (56), the rms beam sizes in two directions also satisfy

$$2\langle I_1 \rangle + \langle I_2 \rangle = \text{constant} \ . \qquad (57)$$

Assume that the beam is a round Gaussian beam initially. After substituting Eq. (56) into Eq. (53), the evolution of rms beam sizes, as shown in Fig. 1, and the change of the particle distribution on amplitudes, as shown in Fig. 2, were easily obtained. It can be seen that Eq. (57) is indeed held here. Fig. 1 also shows a good agreement between the result of the mapping (53) and the result of multi-particle tracking.[7]

6 GAUSSIAN DISTRIBUTION APPROXIMATION

Experimental observations show that the particle distribution in hadron colliders remains approximately Gaussian if it is initially Gaussian.[8] As the beam circulates in the ring, this distribution is gradually distorted with a growth of the distribution tail. In this case we can further simplify the functional mapping (53)

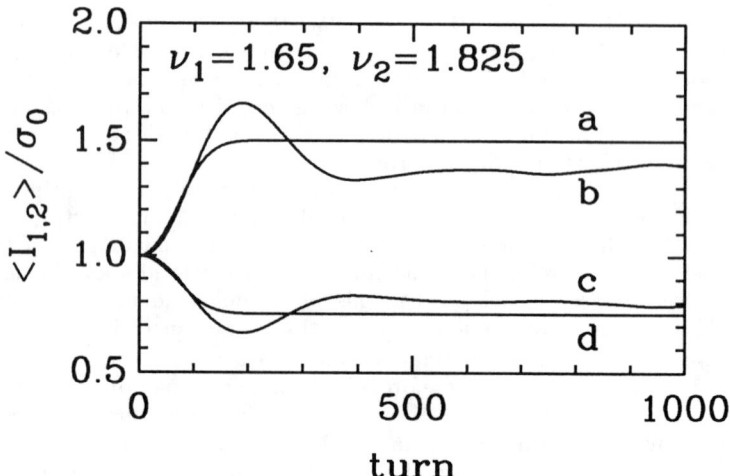

Fig. 1. The evolution of rms beam sizes when the beam is on a difference resonance. The beam is a round Gaussian beam with a rms beam size σ_0 initially. The upper two curves are $\langle I_2 \rangle$ and the lower two $\langle I_1 \rangle$. Curve a and curve d are the result of the map (53) with the nonlinear perturbation of Eq. (56). Curve b and curve c are from the tracking of 5000 particles with a single sextupole kick. $\epsilon \sigma_0^{1/2} = -0.001$.

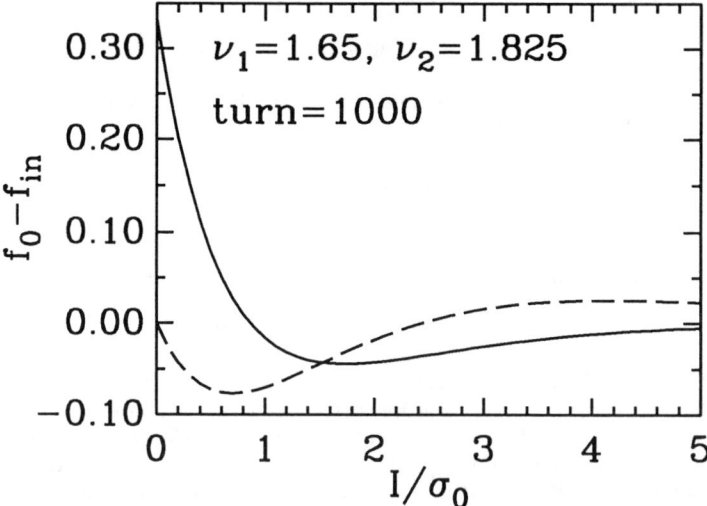

Fig. 2. The change in the particle distribution as a function of the amplitudes for the same case as in fig. 1. The solid (dash) curve is the difference between the distribution on I_1 (I_2) at 1000 turns and its initial distribution.

for numerical computation by approximating f_0^M in Eq. (54) as a Gaussian distribution

$$f_0^M \simeq \frac{1}{(2\pi)^2 \sigma_1^M \sigma_2^M} \exp\left(-\frac{I_1}{\sigma_1^M} - \frac{I_2}{\sigma_2^M}\right) \tag{58}$$

where

$$\vec{\sigma}^M = \int \vec{I} f_0^M d\vec{I} \ . \tag{59}$$

The mapping for the evolution of the distribution on amplitudes can be greatly simplified for the numerical computation, since we now have

$$\frac{\partial f_0^M}{\partial I_i} = -\frac{1}{\sigma_i^M} f_0^M , \tag{60}$$

$$\frac{\partial^2 f_0^M}{\partial I_i \partial I_j} = \frac{1}{\sigma_i^M \sigma_j^M} f_0^M \ . \tag{61}$$

7 SUMMARY

Using perturbation expansion with multiple scales, we have solved the Vlasov equation and the Fokker-Planck equation in the time domain for the particle distribution in hadron storage rings. In order to eliminate secular terms in the expansion, we renormalize the zeroth-order term of the expansion by including the phase-independent part of the perturbed distribution. The phase-dependent part of the distribution function is eliminated by taking average with respect to angle variables. As a result, the zeroth-order term represents the particle distribution of amplitudes. For localized nonlinear perturbations, this renormalization scheme is reduced to a functional mapping for the evolution of the distribution of amplitudes. In this mapping, only action variables are involved so that the evolution of the particle distribution and the beam size can be easily studied by numerically iterating the mapping. One advantage of this method is that we can treat beam-beam interactions in a self-consistent manner.

When the system is close to major resonances, the perturbation expansion of the distribution function may not converge. In this case, the beam is strongly perturbed and the particle distribution in phase space is determined predominantly by the resonance. The perturbed distribution may deviate too far from its unperturbed form to be considered within the frame-work of perturbation theory. Large storage rings, however, are generally operated far from all major resonances. The important problem of our concern here is the slow particle loss and the beam size growth due to weak nonlinear perturbation, which comes primarily from high-order resonances, and the description based on the particle distribution is more meaningful.

ACKNOWLEDGMENTS

J. Shi would like to thank Drs. Leo Michelotti and Andrei Gerasimov of Fermilab, and Drs. Alex Chao and Ken Kauffmann of SSCL for stimulating discussions and valuable suggestions. Work is supported by the SSC National Fellowship and the U.S. Department of Energy under Grant DE-FG05-87ER40374.

REFERENCES

1. A. Chao and R. Ruth, Particle Accelerators **16**, 201(1985).

2. K. Hirata, Phys. Rev. Lett. **58**, 25(1987).
3. S. Kheifets, Particle Accelerators **15**, 67(1984); **15**, 83(1984); **15**, 137 (1984); and **15**, 153(1984).
4. Ali Nayfeh, Perturbation Methods (John Wiley & Sons, New York, 1973).
5. H. Risken, The Fokker-Planck Equation (Springer-Verlag, New York, 1989).
6. S. K. Dutt, J. A. Ellison, T. Garavaglia, S. K. Kauffmann, T. Sen, M. J. Syphers, and Y. T. Yan, Proc. of Fifth Advanced ICFA Beam Dynamics Workshop, Corpus Christi, 1991.
7. J. Shi, Y. Huang and S. Ohnuma, Proc. of the IEEE 1991 Particle Accelerator Conf., p. 407 (1991).
8. G. Goderre, private communication.

A STUDY OF EFFECTS OF TUNE MODULATION IN NONLINEAR MAPS

Armando Bazzani
Dip. di Fisica, Universitá di Bologna – INFN Sez. di Bologna, Italy
Modesto Pusterla, Marco Venturini
Dip. di Fisica, Universitá di Padova – INFN Sez. di Padova, Italy

Abstract

A simple model of betatronic dynamics is studied in which a modulation in tune is present, in order to describe the noise in a magnet power supply. Analytical and numerical results are presented.

1 Introduction

Ripple in a magnet power supply has been recognized as one of the possible causes of beam instability in accelerators when highly nonlinear effects are present. The Accelerator Physics Group of CERN has started a research program on this subject which involves experiments on real machines and numerical simulation on realistic models [1]. As a part of this effort, it has been proposed that a useful insight could be given by the study of simple models where effects could be more easily investigated.

In this papers we consider the simplest map describing transverse dynamics in accelerators in which both nonlinearity (in kick approximation) and slow and small modulation in tune are present. The modulation of the tune is a direct consequence of fluctuation in the quadrupolar magnet currents whose values enter the expression of focusing function.

This map is a direct generalization of the famous quadratic map first studied by Henon [2] and has the following explicit form:

$$\begin{pmatrix} x_{n+1} \\ p_{n+1} \end{pmatrix} = R\left(2\pi Q_n^{(\epsilon)}\right) \begin{pmatrix} x_n \\ p_n + f(x_n) \end{pmatrix} \quad (1)$$

where R is a rotation 2×2 matrix of angle $2\pi Q_n^{(\epsilon)}$, $Q_n^{(\epsilon)} = Q_0 \left[1 = \epsilon \sin(\Omega n)\right]$, and Q_0 is the linear tune. Tune modulation has a slow frequency, $\Omega \ll 2\pi Q_0$, and a small amplitude ϵ. In physical situations it is expected that $\Delta Q = Q_0 \epsilon \sim 10^{-4}$ to 10^{-3}.

The function $f(x_n)$ is analytical; a polynomial expression of the form $f(x_n) = \sum_m a_{m-1} x_n^{m-1}$ corresponds to a dynamics in which magnetic multipoles of order $2m$ are present. Canonical variables x, p are intended to be the normalized Courant-Snyder coordinates. For the connection between maps such as (1) and the dynamics in FODO cells in accelerators, see Ref. [3].

Results presented in this paper are both analytical and numerical.

2 Analytical Results

First we apply a step of perturbation theory in order to estimate the norm of the remainder terms with respect to the normal form, and we establish conditions of existence for invariant surfaces (KAM tori) in the enlarged phase space of the map.

The stability domain is delimited by the last invariant surface that is preserved under perturbation. This is strictly related to the dimensionality of the phase space since, in general, for a number N of dimensions, $N \geq 4$, surfaces that are homomorphic to N-tori do not separate the phase space. This is the main reason why the behavior of particle dynamics in a wider space is expected to be qualitatively different and perhaps worse. The instability related to the topological structure of phase space is known as "Arnold diffusion" [4].

The following results hold for a class of maps that are slightly larger than the one represented by (1) and are better expressed in terms of polar variables:

$$\begin{aligned}
\phi_{n+1} &= \phi_n + \Omega, \\
\phi_{n+1} &= \phi_n + 2\pi Q_0 + \epsilon \cos(\phi_n) + h(r_n, \theta_n), \\
r_{n+1} &= r_n + g(r_n, \theta_n),
\end{aligned} \quad (2)$$

where $h(0, \theta) = g(0, \theta) = 0$ (ϕ works as a time variable).

We now state our results without dwelling on the technical aspects, which can be found in Ref. [5] together with a detailed demonstration.

We recall only that essential hypotheses are analyticity of the functions h and g, and a nonresonance condition on frequencies $2\pi Q_0$ and Ω. Nevertheless, when resonances are present, interesting phenomena happen for which there is no full theory applicable to the study of our model: indeed this will be the subject of the numerical investigations presented in Section 3.

2.1 Estimate of Remainder Terms

A domain $0 < r < \bar{r}$ does exist in which a symplectic transformation reduces the map (2) to the form

$$\begin{aligned}
\phi'_{n+1} &= \phi'_n + \Omega, \\
\theta'_{n+1} &= \theta'_n + 2\pi Q_0 + \alpha(r'_n) + \Phi(r'_n, \theta'_n), \\
r'_{n+1} &= r'_n + \Psi(r'_n, \theta'_n),
\end{aligned} \quad (3)$$

where Φ and Ψ are the remainder terms. The quantity $\alpha(r)$ is known as a "detuning function"; note that explicit dependence on ϕ has been removed from the equation in θ. The following relation holds:

$$|\Psi| \leq A \left(\frac{\bar{r}}{R}\right)^2 \exp\left[-\frac{\Omega}{C + \epsilon (\bar{r}/R)^{1/(1+\mu)}} \frac{B}{\left(1 - \frac{\log 4}{|\log(\bar{r}/R)|}\right)}\right]. \quad (4)$$

A, B, C, μ are numerical constants that depend on the particular choice of the functions h and g [see Eq. (2)], and R fixes the analytical domain in r of the map (3). The main feature of this relation is the exponential dependence on the inverse of the perturbation amplitude ϵ, which recalls the Nechoroshev estimate [4]. A similar dependence, apart from a small logarithmic term, is exhibited by the second perturbative parameter r/R that weights the nonlinearity of the system.

The right side of Eq. (4) is related to the minimum time (i.e. number of iterations of the map) necessary for the particle to cover a finite distance along the radial direction.

Suppose that a particle is initially in position r'_1; then, in the less favorable hypothesis, the number of iterations n_1 necessary to go from r_1 to $\bar{r}$ is not lower than

$$n_1 = 1 + \frac{|r' - r'_1|}{|\Psi|}. \tag{5}$$

To test this relation numerically, we can fill a region around $r', r' < r_1$, with N points, then iterate the map and compute the number n_f of iterations necessary for a fraction of N (say half of initial points), to reach $\bar{r}$. Because of relations (4) and (5) we expect that (D is also a constant)

$$\log n_f \geq -\frac{\Omega}{C + \epsilon} \frac{B}{(\bar{r}/R)^{1/(1+\mu)}} \left(1 - \frac{\log 4}{|\log(\bar{r}/R)|}\right) + D. \tag{6}$$

Numerical results are shown in Fig. 1, where 8 different distributions are considered, corresponding to different values of the initial radial variable. Points are disposed over invariant orbits of the unperturbed (Henon) map ($\epsilon = 0$). The abscissa is $1/\epsilon$ and the ordinate, n_f. We have chosen a function f in Eq. (1) of the form $f(x) = x^2$ (sextupolar magnets).

For the values of ϵ considered in Fig. 1, the orbits of the Henon map are destroyed, and iteration of the map rapidly brings particles forward to a great distance from the origin of the phase space. We can see that for large enough values of ϵ relation (6) seems to be fulfilled at a minimum value and the points are disposed roughly on straight lines. For smaller ϵ, or higher values of the initial radial variable, sudden jumps take place; this means that the rate of gain in stability is above the lower bound given by Eq. (6). (A similar plot, but without a complete interpretation, was given in Ref. [6].)

2.2 Invariant Surfaces

The second analytical result is the existence of invariant surfaces. As expected from general KAM theory, it can be proved that there is a radial domain in which such surfaces indeed exist. The projection of the surface into the coordinate plane of the original coordinate x, p is a thick crown: If a particle has an initial position inside, its trajectory will be forever bounded. This analytical result is not of great utility since the technique of demonstration requires too small a value for r_s, whereas numerical inspection shows that invariant surfaces seem to exist even for higher values.

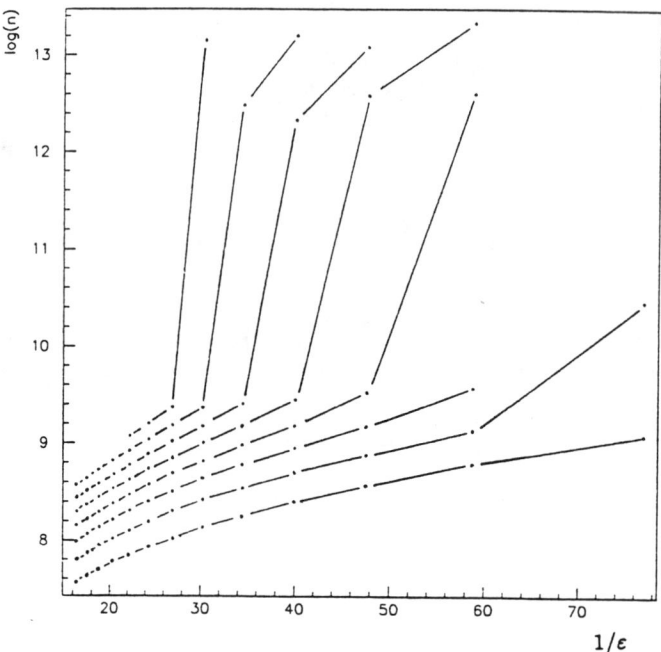

Fig. 1: Stability time vs. amplitude of tune modulation ($Q_0 = 0.618, \Omega = 0.0001$).

3 Numerical Study

When resonances are present in phase space the previous theory is no longer applicable. We refer to the situation in which the tune (with nonlinear correction) has values like m/n (m, n are integers). According to the Chiricov principle [4,7], if the modulation frequency is small enough, we expect the creation of a stochastic layer where motion is chaotic. The width of the layer is of order $\epsilon \left(d^2\alpha/dr^2 \right)$, where α is the detuning function in Eq. (3).

In particular we have investigated the role of the "trapping" phenomenon, and the possibility of defining a diffusion coefficient.

When Ω is very low, the position of stable and unstable fixed points which localize resonances in phase space varies so smoothly that the whole characteristic island structure of resonances is approximately conserved during the action of modulation of the tune. The only evident effect seems to be periodic movement of isles from inner regions of phase space to outer ones (and vice versa) and periodic beat of separatrices. If a particle is initially placed inside an island, it is possible that it remains locked inside the fluctuating resonance. This means that the real tune felt by the particle (that is, the average angular velocity around the origin of the phase space) is "locked" on the resonant value m/n whereas the tune of the map undergoes a full excursion between $Q_0^{(\epsilon)} - \Delta Q$ and $Q_0^{(\epsilon)} + \Delta Q$.

A critical frequency Ω_l can be numerically defined, and locking takes place if $\Omega \ll \Omega_l$; but for $\Omega \sim \Omega_l$, the locking can work in only one direction during the oscillation: this is the worst situation for stability of the motion since this partial trapping can act as a very fast transport mechanism toward the external region of space phase.

The critical quantity Ω_l can be easily computed. We have chosen $Q_0 = 0.618$ and calculated numerically the dependence of Ω_l on the amplitude ϵ by considering particles inside the resonance of order 8. Results fit quite well with a law of the type

$$\Omega_l \sim 1/\epsilon$$

(in Fig. 2 the product $\Omega_{l\epsilon}$ is plotted for different ϵ). Such a dependence can be predicted if we describe the resonance of the Henon map approximately as an invariant resonant Hamiltonian (pendulum-like system: see Ref. [8]).

Moreover we have noted that there is quite strong dependence of Ω_l on the initial phase space of the particle inside the resonance.

For $\Omega \ll \Omega_l$ the average mobility of particles is much lower than in the case $\Omega \sim \Omega_l$, since the radial variable after a period of perturbation assumes more or less the same value as at the beginning. An exact maximization of the difference $|\Delta r|$ between the two positions has not yet been achieved for our model and is one of our aims for the future.

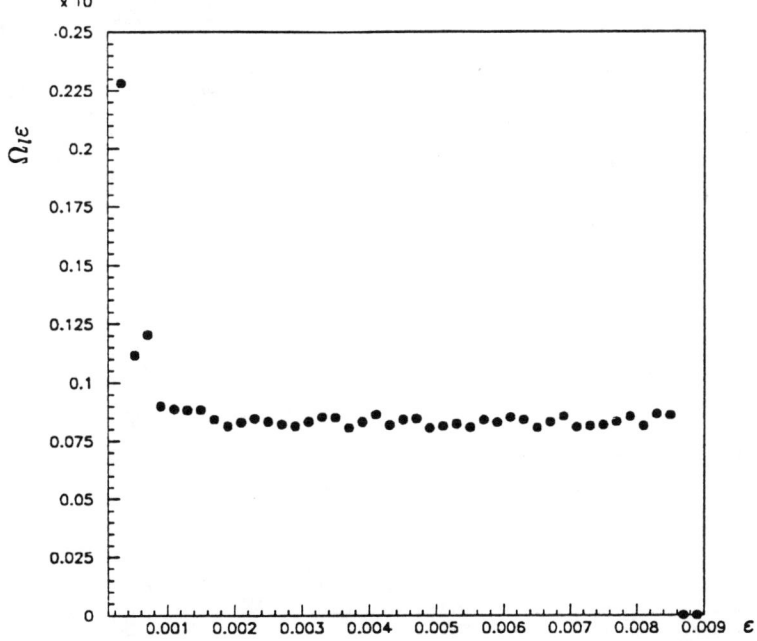

Fig. 2: Trapping critical frequency vs. amplitude.

Nevertheless, we expect it to be possible to extend certain results that have been stated, not rigorously, for general resonant dynamic systems in the adiabatic approximation [9] (see also Ref. [10] for direct application to a physical situation). More recently some results have been found rigorously in Refs.[11-13] but only for some special cases of integrable Hamiltonian systems.

In particular in Ref. [9] a dependence of $|\Delta r|$ was proposed like

$$|\Delta r| \sim 1/\epsilon^{1/2}. \tag{7}$$

If we suppose that there is no correlation between the phases of particles after a whole period of modulation, we expect that the algebraic sign of Δr is random. In this case the walk of the particle along the radius direction would be described as a Wiener process. In this hypothesis a diffusion coefficient can be defined (see Ref. [7]):

$$D_n < \Delta r_n^2 > /n \tag{8}$$

where <> represents an average over initial conditions. In the case of the random walk the quantity $\lim_{n\to\infty} D_n$ exists, and it differs from 0 if no boundary is present. This is not however our case since the stochastic layer, in general, will be surrounded by regular orbits which act as barriers.

In order to check the validity of the definition of D_n we have computed the evolution of the variance $< \Delta r_n^2 >$ and plotted it versus the number of iterations of the maps. In Fig. 3 the variance has been plotted for three different values of the amplitude ϵ; we see that we can define a number of iterations $n = n_{\max}$ up to which the curves in the plot grow linearly with a certain approximation.

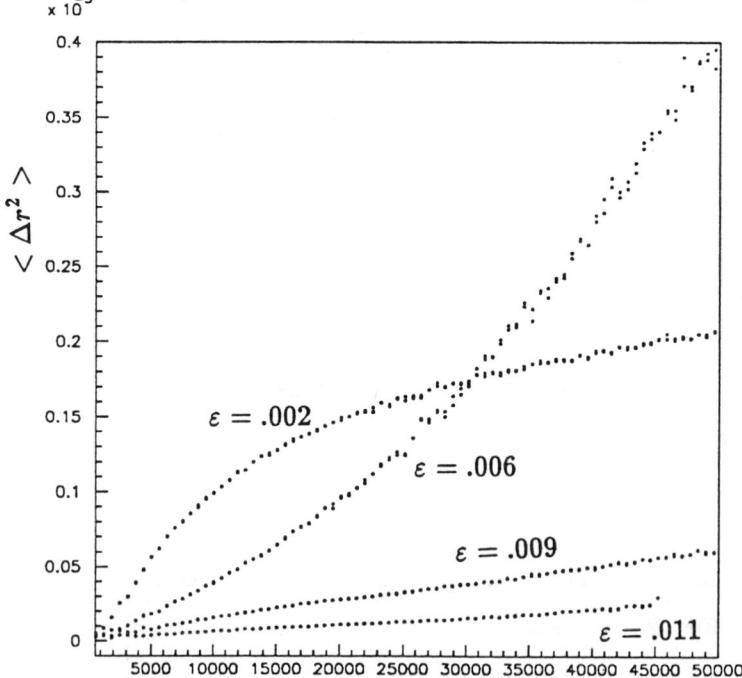

Fig. 3: Variance $< \Delta r^2 >$ vs. number of map iterations.

The behavior of the curves relative to $\epsilon = 0.002$ for higher values of n has to be interpreted as a progressive filling of the stochastic region accessible to the particles. This observation is confirmed by the other curves ($\epsilon > 0.002$), which do not tend to stop growing since the stochastic region is proportionally larger ($\alpha\epsilon$). The curve $\epsilon = 0.011$ is uncompleted since some particles of the initial distribution have been lost before 5×10^4 iterations: this is because the depth of the stochastic layer is so large that it "touches" the external border of the stability domain of the map.

Calculations have been performed by considering a population of 1000 particles placed on an orbit of the unperturbed map very close to (but not exactly inside) the resonance. The quantity Δr_n^2 has been determined as the distance between the previous orbit and position after n iterations. We have chosen $Q = 0.6187$ and a resonance of order 8.

In Fig. 4 we have plotted the quantity D_n versus the amplitude ϵ. As far as we can see from the figure, the "diffusion coefficient" has roughly a decreasing behavior with ϵ, which is particularly evident for higher values of the modulation amplitude.

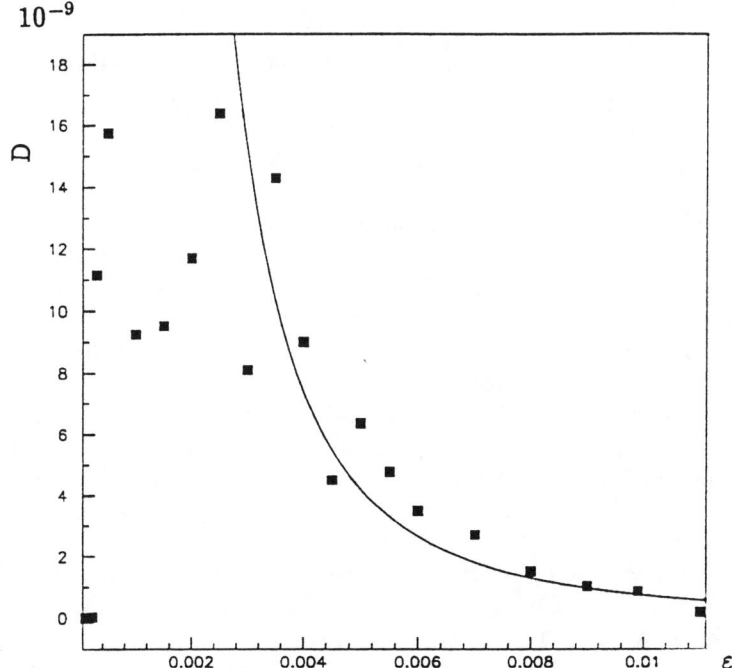

Fig. 4: Diffusion coefficient vs. tune modulation amplitude.

The points at the lower left corner of the plot have $D_n = 0$ since ϵ is so small that the distribution of points we have considered is not touched by resonance during its fluctuation in phase space. The continuous curve is the best fit done with trial functions of the form

$$f = k_1 / \left(\epsilon^2 - k_2^2\right)^q$$

with $q = 1.2$ (in the fitting we have not considered the points with the lowest ϵ). This function should be compared to $D_n \sim 1/\epsilon$ [see Eq. (7)], and the accord is only qualitative.

Acknowledgments

We thank Drs. M. Giovannozzi and W. Scandale for very fruitful stimulating discussions. M. Venturini thanks Dr. J. Gayrete for his interest in this subject and his hospitality in the CERN Accelerator Physics Group.

References

1. D. Brandt et al., Influence of Power Supply Ripple on the Dynamic Aperture of the SPS in the Presence of Strong Nonlinear Fields, CERN SL/90-67 (AP); LHC Note 126 (1990).
2. M. Henon, Numerical study of quadratic area preserving mapping, Quart. J. Appl. Math. 27(3), 291-312 (1969).
3. A. Bazzani, P. Mazzanti, G. Servizi, G. Turchetti, Normal forms for Hamiltonian maps and nonlinear effects in a particle accelerator, Il Nuovo Cimento 102, 51-80 (1988).
4. B.V. Chiricov, A universal instability of many-dimensional oscillator systems, Physics Reports 52, 263-379 (1979).
5. M. Venturini, Modelli per lo studio della diffusione in acceleratori adronici: Tesi di laurea, Universitá di Padova, Anno. Acc. 1991/92.
6. M. Malavasi, Chaos transition and diffusion in symplectic map: some numerical results, in Nonlinear Problems in Future Particle Accelerators, Eds. W. Scandale, G. Turchetti, World Scientific, 1991.
7. A.J. Lichtenberg, M.A. Lieberman, Regular and Stochastic Motion, Springer, New York, 1983.
8. B.V. Chiricov, M.A. Lieberman, D.L. Shepelyansky, F.M. Vivaldi, A theory of modulational diffusion, Physica 14D, 289-304 (1985).
9. A. Schoch, Theory of Linear and Nonlinear Perturbations of Betatron Oscillations in Alternating Gradient Synchrotrons, CERN 57-21 (1957).
10. L.R. Evans, The Beam-Beam Interaction, CERN SPS/83-38 (1983).
11. A.I. Neishtadt, Change in adiabatic invariant at separatrix, Sov. J. Plasma Phys. 12(8), 568-573 (1986).
12. A.I. Neishtadt, Averaging, capture into resonances, and chaos in nonlinear systems, Soviet-American Perspectives on Nonlinear Science, AIP, 261-273 (1990).
13. A.I. Neishtadt, On the change in the adiabatic invariant on crossing a separatrix in systems with two degrees of freedom, PMM U.S.S.R., 51(5), 586-592 (1987).

Exact Physical Model for Magnets in Storage Rings*

D.Maletić, A.G.Ruggiero

Brookhaven National Laboratory, Upton, NY 11793

1 Motivation

It is common practice to perform computer simulation to determine the stability of motion of charged particles circulating in storage rings over long periods of time. For this purpose several computer codes have been written.[1] In the past, great effort was spent to insure that the representation of the motion used during simulation was symplectic;[2] that is, for conservative systems of particles, like protons, antiprotons and heavy ions, the equations of motion not only are to be derived from a Hamiltonian, but they also have to satisfy a variety of momentum and energy conservation laws. If these should not become fully satisfied, the results of such computer trackings, especially long ones, could be invalid. Simulated motion could show instability where it should not occur, and vice versa.

A particle encounters several types of elements during its motion in a storage ring. Some of them have linear properties, like drifts and pure quadrupoles. In these cases, the approximate status of motion of the particle at the end of the particular element can be easily given as an integrable function of the status of motion at the beginning of that element. Nevertheless, even in these linear or quasi-linear elements the exact equations of motion have kinematic terms that are difficult to integrate.

Other elements encountered by the particle during its motion in a storage ring are nonlinear, like sextupoles, intentionally used to correct chromatic effects, and the nonlinear imperfections of magnets, especially superconducting magnets. The motion in these nonlinear elements cannot be generally integrated. Therefore one has to use some numerical technique to perform integration by computer.

One method is to lump each nonlinear element into a thin lens of zero length. Motion through such an element is then represented by a kick. The

---
*Work performed under the auspices of the U.S. Department of Energy

position coordinates of the particle remain unchanged, while velocity receives a sudden change that depends on the particle position and on the properties and strength of the nonlinear lens. The numerical codes that make use of this method are called *kick codes* or *exact codes*. Since the motion they represent with a kick is truly symplectic, their approximation of reality is physically correct. TEAPOT[3] is one of these codes which also represents integrable elements, like dipoles and quadrupoles, as a series of thin lenses and drifts. Since the motion in a drift can be calculated exactly, it seems that the representation of any accelerator element as a sequence of drifts and thin lenses, linear and nonlinear, is indeed very powerful.

It should be understood that in order to predict the stability of the motion over very long periods of time it is mandatory that all kinematic terms be properly included in the model. Neglecting some of them may invalidate the results of very time–consuming exercises on the computer.

Adopting the thin lens representation requires discussion of one more issue. Forces acting on a particle are to be derived exactly and consistently from Maxwell's equations. Generally, ordinary kick codes do not enforce a complete description of the fields, and this is to be corrected. Not only do magnets exhibit edge effects due to their discontinuity in space, but the thin lens model of either linear or nonlinear elements is a three-dimensional field representation that requires careful estimation of the field components. Usually only the longitudinal component of the vector potential is retained for the field estimate. Since this has discontinuous behaviour, transverse components are also present, especially at the edges of the elements; these are usually neglected. If the transverse components of the field are neglected, the divergence of the vector potential does not vanish and the divergence and the curl of the magnetic field differ from zero, though the corresponding equations of motion remain symplectic. This, again, may invalidate results of long computer simulations. Thus, we consider it very important that the field representation be exact and complete and that it satisfy Maxwell's equations.

In this report we try to make estimates of both kinematic and field effects on the stability of a particle motion, by employing a truly Maxwellian representation of the magnetic field in exact equations of motion. For this purpose we adopt a simple FODO cell model, which repeats periodically to infinity. This model includes only quadrupoles and drifts, leaving out the bending magnets to avoid the problem of the trajectory curvature. We think this model is a physically consistent approximation of a storage ring. We derive several models with different levels of approximation and compare them by evaluating the importance of these effects. The relevance to long-term stability is being investigated in the meantime by comparing the different models with extensive computer simulations. The results will be shown in a subsequent report.

2 The Test Model

To determine the importance and the magnitude of kinematic effects, and of the effects of the field components required for complete Maxwellian representation of the field, we adopt the following test model. It consists of an infinitely long sequence of FODO cells, all identical, as shown in Figure 1.

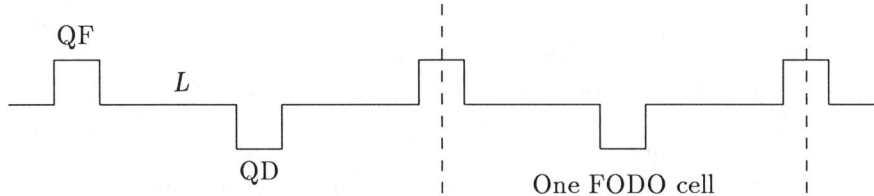

Figure 1: The test model as an infinite sequence of identical FODO cells

The FODO cell has the following structure:

$$\text{QF}/2 \quad L \quad \text{QD}/2 \quad \text{QD}/2 \quad L \quad \text{QF}/2 \tag{1}$$

which begins from the middle of the horizontally focusing quadrupole (QF) and ends in the middle of the following focusing quadrupole. The vertically focusing quadrupole (QD) is located halfway. The quadrupoles have the same length l and are separated by a drift of length L. Quadrupole magnets are described by the field gradient. The gradient is $+G$ in QF and $-G$ in QD. The only other parameter required for calculating the motion of a charged particle of charge q and momentum p is the particle magnetic rigidity

$$B\rho = pc/q \tag{2}$$

where c is the speed of light.

Our test model is made only of drifts and quadrupoles. For simplicity we did not include dipole magnets to avoid the issue of the curvature they introduce.

The reference line is the straight line that coincides with the symmetry axis of the quadrupole. We assume that quadrupoles are perfectly aligned so that their axes coincide. The quadrupole fields vanish on this axis.

This model is basically an approximation of a storage ring where the reference orbit is a straight line. We introduce a longitudinal coordinate s along the reference line and a pair of rectangular transverse coordinates, x

for horizontal and y for vertical, which measure the distance of a particle trajectory from the reference line; that is, on the reference line $x = y = 0$.

3 Equations of Motion

To describe the motion of a particle it is convenient to consider the plane perpendicular to the main axis of motion (the reference orbit) which includes the particle position at time t. It is on this plane that the distances x and y from the reference axis are evaluated at time t. To the same instant t one can associate the longitudinal coordinate s, which is given by the interception of the $x-y$ plane with the reference orbit. Since there is a unique correspondence between s and t, it is possible and actually more convenient to take the longitudinal coordinate s as the independent variable, rather than the time t. The displacements are then taken as functions of s, that is, $x = x(s)$ and $y = y(s)$.

In the following we use a prime to denote the derivative with respect to s, and we denote the components of the magnetic field as B_x, B_y and B_s.

It can be proven[4] that the exact equations of motion for the FODO test model are

$$x'' = -\frac{q}{cp}\sqrt{1 + x'^2 + y'^2}[B_y(1 + x'^2) - B_x x'y' - B_s y'], \qquad (3)$$

$$y'' = \frac{q}{cp}\sqrt{1 + x'^2 + y'^2}[B_x(1 + y'^2) - B_y x'y' - B_s x']. \qquad (4)$$

It can also be proven[4] that the following relation holds:

$$x'x'' + y'y'' = \frac{q}{p}(1 + x'^2 + y'^2)^{3/2}(y'B_x - x'B_y), \qquad (5)$$

which may be found useful in a variety of situations.

4 A Symplectic Integration

The general integration of the system consisting of Eqs. (3) and (4) is very difficult. It may not be possible to prove their integrability in general. Two features that contribute to the difficulty of the problem are:

1. There are kinematic terms represented by the x' and y' factors which are usually neglected.

2. There are field contributions, represented by the components B_x, B_y and B_s. Usually the longitudinal component is neglected as well as the edge effects.

We need to find a symplectic integration method that allows us to solve the equations of motion, satisfying all energy and momentum conservation laws. Conventional methods, like Newton, Lagrange, Runge–Kutta, etc., may not be valid for this purpose. We find the "TEAPOT–like" method of replacing each element with a sequence of thin lenses and drifts, as shown in Figure 2, a valid one.

Figure 2: Replacing a quadrupole with a sequence of thin lenses and drifts

The length l of the quadrupole magnet is divided into N steps, each consisting of a drift of length $\lambda = l/N$ with a thin lens quadrupole in the middle. Each of these N thin lenses is a quadrupole magnet with zero length. In the limit $\lambda \to 0$ or $N \to \infty$, one can recover the original model of the full–length quadrupole.

In a drift $B_x = B_y = B_s = 0$, so that the equations of motion are simply

$$x'' = y'' = 0 \qquad (6)$$

with solutions represented by straight–line trajectories.

The integration through the thin lens quadrupoles is shown in Figure 3. The trajectory of the particle consists of straight lines before and after the thin lens. The trajectory is deflected by the thin lens, leaving the position of the particle unchanged and altering the velocity by an amount that depends on x, y, x' and y' as well as on the field gradient in the magnet.

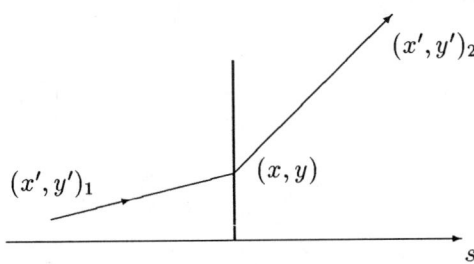

Figure 3: The trajectory of a particle going through a thin lens

5 Exact Field Expression

We wish to calculate the exact expression for the magnetic field distribution of a finite–length n–pole magnet, including the distribution at the edges. For this purpose we introduce two functions: the dimensionless function $f(s)$ which defines the longitudinal shape of the magnet and

$$g_n(x,y) = (a_n + ib_n)(x+iy)^n \tag{7}$$

which describes the longitudinal field potential inside the n–pole magnet. It is easily verified that

$$\nabla^2 g_n(x,y) = 0 \;. \tag{8}$$

Let $\mathbf{B}$ and $\mathbf{A}$ be respectively the magnetic field and the magnetic vector potential. They satisfy

$$\nabla \mathbf{A} = 0 \;, \tag{9}$$

$$\nabla^2 \mathbf{A} = 0 \;, \tag{10}$$

and

$$\mathbf{B} = \nabla \times \mathbf{A} \;. \tag{11}$$

Since we are using a rectangular coordinate system (x,y,s) Eq. (10) is equivalent to

$$\nabla^2 A_x = \nabla^2 A_y = \nabla^2 A_s = 0 \;. \tag{12}$$

It can be proven by direct substitution that the following expressions solve Eqs. (9) and (10) for any chosen functions $f(s)$ and $g_n(x,y)$:

$$A_s(x,y,s) = g_n \sum_{m=0}^{\infty} C_m (x^2+y^2)^m f^{(2m)} \;, \tag{13}$$

$$A_x(x,y,s) = -g_n(x+iy) \sum_{m=0}^{\infty} \frac{C_m}{2(m+n+1)} (x^2+y^2)^m f^{(2m+1)} \;, \tag{14}$$

$$A_y(x,y,s) = -g_n(y-ix) \sum_{m=0}^{\infty} \frac{C_m}{2(m+n+1)} (x^2+y^2)^m f^{(2m+1)} \;, \tag{15}$$

where the coefficients C_m are related to each other according to

$$C_{m-1} = -4m(m+n)C_m , \qquad (16)$$

$$C_0 = 1 , \qquad (17)$$

and

$$f^{(m)} = \frac{d^m f(s)}{ds^m} . \qquad (18)$$

Inserting Eqs. (13) to (15) into Eq. (11) yields

$$B_x = ig_n \sum_{m=0}^{\infty} C_m[(2m+n)x - iny](x^2+y^2)^{m-1} f^{(2m)} , \qquad (19)$$

$$B_y = ig_n \sum_{m=0}^{\infty} C_m[(2m+n)y + inx](x^2+y^2)^{m-1} f^{(2m)} , \qquad (20)$$

$$B_s = ig_n \sum_{m=0}^{\infty} C_m(x^2+y^2)^m f^{(2m+1)} . \qquad (21)$$

As a special case let us consider a quadrupole magnet in normal orientation, that is, $n=2$, $b_2=0$ and $a_2=-G/2$. After taking the imaginary part to obtain the real field representation, we have

$$B_x = Gyf^{(0)} + G \sum_{m=1}^{\infty} C_m(x^2+y^2)^{m-1}[(2m+1)x^2y + y^3]f^{(2m)} , \qquad (22)$$

$$B_y = Gxf^{(0)} + G \sum_{m=1}^{\infty} C_m(x^2+y^2)^{m-1}[(2m+1)xy^2 + x^3]f^{(2m)} , \qquad (23)$$

$$B_s = G \sum_{m=0}^{\infty} C_m xy(x^2+y^2)^m f^{(2m+1)} . \qquad (24)$$

If we retain only the linear terms in x and y, we get

$$B_x = Gyf(s) , \qquad (25)$$

$$B_y = Gxf(s) , \qquad (26)$$

$$B_s = 0 . \qquad (27)$$

6 Linear Model

Let us neglect all the kinematic terms in Eqs. (3) and (4) and retain only the linear terms in the field expansion, that is, Eqs. (25) to (27), where $f(s)$ is a step function equal to 1 in the quadrupole and zero everywhere else.

The resulting equations of motion in a quadrupole are

$$x'' = -Kx , \qquad (28)$$

$$y'' = Ky , \qquad (29)$$

where, for convenience,

$$K = \frac{q}{cp}G = \frac{G}{B\rho} . \qquad (30)$$

In the drifts, between quadrupoles, the equations of motion are

$$x'' = y'' = 0 . \qquad (31)$$

These equations are linear. They can be easily solved and their solution is usually represented by the matrix notation which involves the lattice functions[5] β_H and β_V as well as the phase advance functions[5] Ψ_H and Ψ_V.

As an example we take the following values, which approximate regular cells in the arcs of RHIC[6]:

$$B\rho = 840\ Tm , \qquad (32)$$
$$G = 81\ T/m , \qquad (33)$$
$$L = 14\ m , \qquad (34)$$
$$l = 1\ m . \qquad (35)$$

The lattice functions for this system, obtained with the SYNCH code,[7] are plotted in Figure 4. The quadrupole gradient has been chosen to make the phase advances across one FODO cell equal to 90° in both planes.

7 The TEAPOT Model

Let us now examine the approximation of a quadrupole magnet by a sequence of thin lens magnets and drifts, as shown in Figure 2. We still adopt the linear approximation, as in the TEAPOT code.[3] The only difference from the equations of motion of the linear model is that the function $f(s)$ is a delta function at each thin lens, positioned at $s = s_i$,

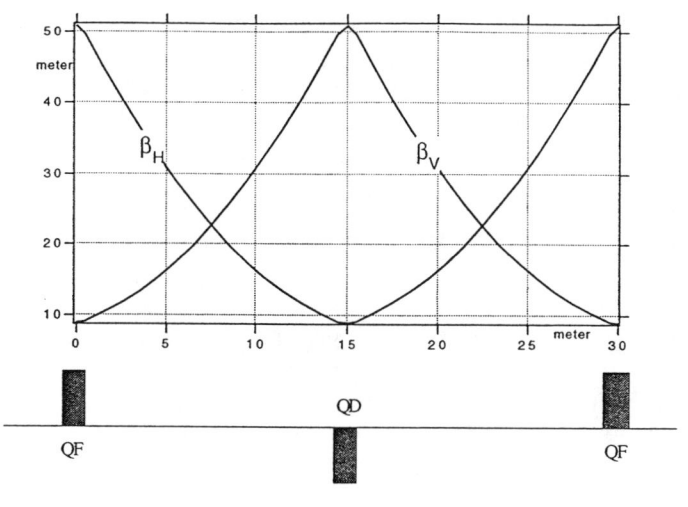

Figure 4: The amplitude lattice functions for the FODO cell

$$f(s) = \lambda \delta(s - s_i) \tag{36}$$

where $\lambda = l/N$ is the length of a step approximated by one thin lens and a drift.

The position coordinates x and y of a particle crossing the thin lens are unchanged, that is

$$\Delta x = \Delta y = 0 . \tag{37}$$

At the same time the angles x' and y' receive a change

$$\Delta x' = -K\lambda x , \tag{38}$$

$$\Delta y' = K\lambda y . \tag{39}$$

This model can be easily integrated and investigated. The results are given for comparison in Table 1. The maximum and minimum values of the amplitude functions are reported as well as the phase advance across the FODO cell, $\Delta\Psi/2\pi$. The results for the linear model in Table 1 were derived with the SYNCH code.[7] The results of the approximation of the whole quadrupole by one thin lens were computed manually, since for this case the following relations hold:

$$\Lambda = \frac{L+l}{2} , \tag{40}$$

$$\sin(\frac{\Delta\Psi}{2}) = \frac{\Lambda l |K|}{2} , \tag{41}$$

Table 1: Comparison of different linear models

MODEL	β_{max}	β_{min}	$\Delta\Psi/2\pi$
Linear Model	50.80254	8.86074	0.24992
Thin Lens	51.75140	8.31240	0.25734
TEAPOT (2)	50.71639	8.77329	0.25176
TEAPOT (4)	50.78029	8.83885	0.25038
TEAPOT (10)	50.79895	8.85724	0.24999
TEAPOT (20)	50.80164	8.85987	0.24994
TEAPOT (100)	50.80231	8.86005	0.24992
TEAPOT (200)	50.80250	8.86071	0.24992
Simpson-Bode (2)	51.15743	8.73205	0.25115
Simpson-Bode (3)	50.78102	8.83882	0.25038
Simpson-Bode (4)	50.75739	8.84859	0.25027

$$\beta_{max} = \frac{2}{l|K|}\sqrt{\frac{2 + \Lambda l|K|}{2 - \Lambda l|K|}}, \qquad (42)$$

$$\beta_{min} = \frac{2}{l|K|}\sqrt{\frac{2 - \Lambda l|K|}{2 + \Lambda l|K|}}. \qquad (43)$$

The other entries in Table 1 were obtained with a TEAPOT type of code, for several different numbers of sequences into which each quadrupole is subdivided. The last three entries correspond to the Simpson–Bode[8] method of integration, where magnets are subdivided into thin lenses each carrying a different weighted strength. Several different numbers of subdividing steps (shown in parentheses) were used.

It can be seen that the single thin lens model provides results only a few percent from those of the exact linear model. Subdividing the quadrupole into more thin lenses gives a closer approximation. Nevertheless, even if we divide the magnet into 200 thin lenses the results are off by an error of 10^{-5} for the amplitude function.

It has been assumed that a more effective way of integrating the motion through a magnet is the Simpson–Bode method. This method is commonly used in tracking particle motion through nonlinear elements. Its effectiveness cannot be easily proven except in a linear case, where the results can be compared to the exact solution. The results in Table 1 show that the Simpson–Bode method has better convergence than the method using TEAPOT procedure.

Let (x, x') and (y, y') be the particle position and angle in the middle of any of the two quadrupoles, QF and QD. Then the following quantities are invariant[5]

$$\epsilon_H = \frac{x^2}{\beta_H} + x'^2 \beta_H , \qquad (44)$$

$$\epsilon_V = \frac{y^2}{\beta_V} + y'^2 \beta_V . \qquad (45)$$

The values of the amplitude functions β_H and β_V vary somewhat from model to model as shown in Table 1. In the following we take the linear model as the reference one.

8 The Kinematic Model

The kinematic model is defined by the set of exact equations (3) and (4), where kinematic terms are retained to any order, but the field expansion is truncated beyond the linear terms, Eqs. (25) to (27). This model is truly symplectic.

Investigation of this model is relevant to determining the magnitude and the importance of the effects of the kinematic terms. This can be done by comparison with the linear model.

We continue to approximate a quadrupole with a sequence of thin lenses and drifts. Indeed this is the only method we know for numerical symplectic integration.

The equations of the motion through a thin quadrupole now become

$$x'' = -K\lambda\sqrt{1 + x'^2 + y'^2}\,(x + xx'^2 - yx'y')\delta(s) , \qquad (46)$$

$$y'' = K\lambda\sqrt{1 + x'^2 + y'^2}\,(y + yy'^2 - xx'y')\delta(s) \qquad (47)$$

and can be integrated exactly, yielding kicks

$$\Delta x' = -K\lambda\sqrt{1 + x'^2 + y'^2}\,(x + xx'^2 - yx'y') , \qquad (48)$$

$$\Delta y' = K\lambda\sqrt{1 + x'^2 + y'^2}\,(y + yy'^2 - xx'y') \qquad (49)$$

where x, x', y and y' on the right correspond to the instant just before entering the lens, as proven in the Appendix. At the same time x and y remain unchanged.

We have not been able to derive invariants similar to those given for the linear model. Nevertheless, it can be seen that the kinematic terms introduce coupling between horizontal and vertical motion. The coupling disappears only if the motion is purely horizontal ($y = y' = 0$) or purely vertical ($x = x' = 0$). In these cases the equations of the motion reduce to either

$$x'' = -K\lambda x(1 + x'^2)^{3/2}\delta(s) , \qquad (50)$$

$$y'' = 0 , \qquad (51)$$

or

$$x'' = 0 , \qquad (52)$$

$$y'' = K\lambda y(1 + y'^2)^{3/2}\delta(s) . \qquad (53)$$

Let us take, for instance, the purely horizontal motion. In this case we can decouple our second–order differential equation of motion, Eq. (50), into two first–order differential equations:

$$x' = \frac{p_x}{\sqrt{1 - p_x^2}} , \qquad (54)$$

$$p_x' = -K\lambda x\delta(s) . \qquad (55)$$

which can be derived from the Hamiltonian

$$H = 1 - \frac{K\lambda}{2}x^2\delta(s) - \sqrt{1 - p_x^2} . \qquad (56)$$

9 The Maxwellian Field Model

The Maxwellian field model is obtained by neglecting all the kinematic terms in Eqs. (3) and (4) but retaining the exact field series representation, Eqs. (22) to (24). Only B_x and B_y field components enter the equations:

$$x'' = -\frac{q}{cp}B_y , \qquad (57)$$

$$y'' = \frac{q}{cp}B_x . \qquad (58)$$

Inserting Eqs. (22) to (24) we obtain

$$x'' = -K\lambda\{x\delta(s) + \sum_{m=1}^{\infty} C_m(x^2+y^2)^{m-1}[(2m+1)xy^2 + x^3]\delta^{(2m)}(s)\}, \quad (59)$$

$$y'' = K\lambda\{y\delta(s) + \sum_{m=1}^{\infty} C_m(x^2+y^2)^{m-1}[(2m+1)x^2y + y^3]\delta^{(2m)}(s)\}, \quad (60)$$

where again a thin lens representation has been assumed.

The integration of Eqs. (59) and (60) is cumbersome, but it can be exactly derived. If we retain only the linear, $m = 1$ and $m = 2$, terms we obtain

$$\Delta x' = -K\lambda(x - \frac{1}{2}xy'^2 - yy'x' - \frac{1}{2}xx'^2 +$$

$$\frac{5}{16}xy'^4 + \frac{5}{4}yy'^3x' + \frac{5}{8}xy'^2x'^2 - \frac{1}{4}yy'x'^3 - \frac{3}{16}xx'^4), \quad (61)$$

$$\Delta y' = K\lambda(y - \frac{1}{2}yx'^2 - xx'y' - \frac{1}{2}yy'^2 +$$

$$\frac{5}{16}yx'^4 + \frac{5}{4}xx'^3y' + \frac{5}{8}yx'^2y'^2 - \frac{1}{4}xx'y'^3 - \frac{3}{16}yy'^4). \quad (62)$$

The particle coordinates x, x', y and y' on the right again have values describing a particle just before it enters the thin lens.

The factors in x' and y' which appear in these equations do not originate from the kinematic terms. They are introduced by the integration across the derivatives of the delta function.

Note that higher-order terms in the field expansion cause higher-order terms in the equations of motion. Therefore, raising the truncation in the expansion of the field beyond $m = 2$ will not change the terms in the equations of motion up to the fifth order but will only generate terms higher than fifth order.

We like to point out that this model is also symplectic, whatever the order of truncation is in the expression of the magnetic field.

10 Comparison and Conclusion

Let us introduce the general model, where the equations of motion retain the exact representation of the kinematic terms, and include the Maxwellian field series expansion to a sufficiently high order. In particular we retain only the field expansion up to and including the $m = 2$ terms of Eqs. (22) to (24).

Integration through a thin lens then leads to the following changes of the particle velocity:

$$\Delta x' = -K\lambda\sqrt{1 + x'^2 + y'^2}\,(x + \frac{1}{2}xx'^2 - 3yy'x' - \frac{3}{2}xy'^2 +$$

$$\frac{13}{4}yy'^3 x' + \frac{25}{8}xy'^2 x'^2 + \frac{3}{4}yy'x'^3 + \frac{13}{16}xy'^4 - \frac{3}{16}xx'^4)\,, \qquad (63)$$

$$\Delta y' = K\lambda\sqrt{1 + x'^2 + y'^2}\,(y + \frac{1}{2}yy'^2 - 3xx'y' - \frac{3}{2}yx'^2 +$$

$$\frac{13}{4}xx'^3 y' + \frac{25}{8}yx'^2 y'^2 + \frac{3}{4}xx'y'^3 + \frac{13}{16}yx'^4 - \frac{3}{16}yy'^4)\,. \qquad (64)$$

One can see that the square–root term is not expanded but retained exactly to preserve the nature of the kinematic terms. This model also preserves the symplectic properties. The terms within parentheses are proper to the field series representation and truncation.

We have thus developed four models given by the sets of equations (38)–(39), (48)–(49), (61)–(62) and (63)–(64). Each of these models represents different approximations, and they can be easily tested against each other to determine the importance of either neglecting or including the kinematic as well as the field expansion terms for the stability of particle motion over long periods of time.

It is easy to estimate the correction to the usual linear approximation for each of the models. Retaining only the lowest–order part without cross terms, we have

$$\Delta x'_1 = -K\lambda x(1 + \frac{3}{2}x'^2)\,, \qquad (65)$$

$$\Delta x'_2 = -K\lambda x(1 - \frac{1}{2}x'^2)\,, \qquad (66)$$

$$\Delta x'_3 = -K\lambda x(1 + x'^2)\,, \qquad (67)$$

corresponding respectively to the kinematic, the Maxwellian and the general model. These equations are not to be used in real simulation but only to estimate the magnitude of the correction introduced, as they are not symplectic.

According to the linear model

$$\Delta x' = -K\lambda x\,. \qquad (68)$$

Using parameters typical of the arc FODO cell in RHIC[6] each model gives a correction of magnitude

$$\frac{\Delta x'_i - \Delta x'}{\Delta x'} \approx 10^{-8} - 10^{-9} \ . \tag{69}$$

These correction factors are not important in determining the motion of only a few revolutions and for ring design. They may be very important and cannot be ignored in determining the stability of motion of a particle for a long period of time.

Appendix

We are interested in the integration of the following equation:

$$x'' = F[x(s), x'(s)]\delta(s) \tag{A.1}$$

where F is an analytical function in a sufficiently large interval around $s = 0$.

Let us replace the delta function with a step function $H(s, \epsilon)$, which is zero everywhere except in the interval from $s = -\epsilon$ to $s = +\epsilon$, where it is equal to one,

$$x'' = F(x, x')\frac{H(s, \epsilon)}{2\epsilon} \ . \tag{A.2}$$

The exact solution of the original equation is obtained by taking the limit $\epsilon \to 0$.

In particular

$$\Delta x' = \lim_{\epsilon \to 0} \frac{1}{2\epsilon} \int_{-\epsilon}^{\epsilon} F(s)ds \ . \tag{A.3}$$

This integral can be estimated by subdividing the 2ϵ interval into N substeps of length $h = 2\epsilon/N$, so that

$$\Delta x' = \lim_{\epsilon \to 0} \{\frac{1}{2\epsilon} \lim_{N \to \infty} h[\sum_{n=1}^{N} F(-\epsilon + nh)]\} \ . \tag{A.4}$$

Using the analytical characteristics of the $F(s)$ in the interval of integration, we can expand:

$$F(-\epsilon + nh) = F(-\epsilon) + F'(-\epsilon)nh + \frac{F''(-\epsilon)}{2}(nh)^2 + \ldots \ . \tag{A.5}$$

Substituting this expansion into Eq. (A4) yields

$$\Delta x' = \lim_{\epsilon \to 0} \lim_{N \to \infty} \frac{1}{N}[NF(-\epsilon) + hF'(-\epsilon)\sum_{n=1}^{N} n + ... + \frac{F^{(m)}}{m!}h^m \sum_{n=1}^{N} n^m + ...] . \quad \text{(A.6)}$$

As $\sum_{n=1}^{N} n^m$ is a polynomial in N of order $m+1$, in the limit $N \to \infty$ we derive

$$\Delta x' = \lim_{\epsilon \to 0}[F(-\epsilon) + 2\epsilon F'(-\epsilon) + ... + \frac{(2\epsilon)^m}{m+1}\frac{F^{(m)}(-\epsilon)}{m!} + ...]. \quad \text{(A.7)}$$

In the limit $\epsilon \to 0$ finally

$$\Delta x' = F(x, x') \quad \text{(A.8)}$$

where F is evaluated just at the entrance of the integration interval.

The generalization to a system of two second–order differential equations is straightforward.

References

1. F.C. Iselin, Computer Programs for Accelerators, CAS Proceedings, CERN, Geneva (1987).

2. H. Wiedemann, PEP Note 220 (1976).

3. R. Talman, L. Schachinger, SSC-52 (1985).

4. K.L. Brown, SLAC–75, Stanford (1972).

5. E.D. Courant, H.S. Snyder, Theory of the Alternating Gradient Synchotron, Ann. Phys. 3, 1-48 (1958).

6. Conceptual Design of RHIC, BNL 52195 (1989).

7. E.D. Courant, A.A. Garren, A.S. Kenney, M.J. Syphens, A Users Guide to SYNCH, BNL (1985).

8. Handbook of Mathematical Functions, Eds. M. Abramowitz and A. Stegun, NBS, Washington (1970).

A SIMULATION OF MODULATIONAL DIFFUSION FOR THE FERMILAB TEVATRON

T. Satogata
Northwestern University, Evanston, IL 60205
and
Fermi National Accelerator Laboratory, Batavia, IL 60510

S. Peggs
Brookhaven National Laboratory, Upton, NY 11973

ABSTRACT

A summary of the requirements for modulational (thick-layer) diffusion to exist in a particle synchrotron is presented and applied to a simple tune-modulated collider model of the Fermilab Tevatron where the only nonlinearities present are two beam-beam kicks. For certain realistic tune modulation parameters and single-particle base tunes, amplitude growth is observed over timescales appropriate to diffusive models. The character of this growth has qualitative features that are similar to those predicted by modulational diffusion models, but is significantly different in that the amplitude growth is exponential in time, not root-time as in classical diffusion. Some possible explanations for this effect are briefly noted, and impact of the possible existence of such a mechanism on future Fermilab collider upgrades is mentioned.

1 CHARACTERISTICS OF MODULATIONAL DIFFUSION

Modulational diffusion has been the subject of many investigations in the past ten years, since it provides a particle loss mechanism in many-dimensional dynamical systems such as particle accelerators over timescales that are longer than those from pure resonant loss (typically hundreds of turns), but shorter than the timescales of Arnol'd, or thin-layer, diffusion (typically hundreds of millions of turns). Most of the salient features and quantitative analysis can be found in assorted publications[1-3]; here we highlight the requirements for modulational diffusion to exist in a synchrotron and what characteristics such diffusion might have.

First, consider motion of resonant particles in only the horizontal transverse dimension. It has been shown that with a suitable variation of particle tune with amplitude, and within a one-dimensional resonance influenced by tune modulation, particle trajectories can be characterized as either regular or stochastic.[4-6]

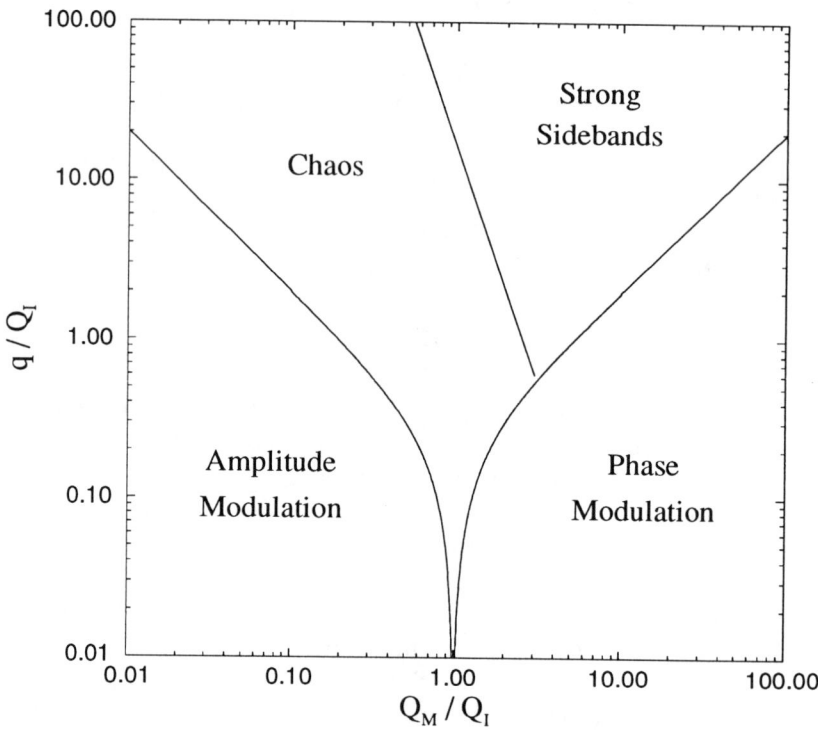

Figure 1: The tune modulation (q, Q_M) parameter plane.

This is most clearly seen within a single-resonance tune-modulation model that produces four phases within the dynamical tune modulation parameter space (q, Q_M), where the tune modulation is given by

$$Q_x = Q_{x0} + q\sin(2\pi Q_M t) , \qquad (1)$$

t is time measured in machine turns, and Q_{x0} is the unperturbed horizontal base tune. (See Figure 1.) Note that in this figure both q and Q_M scale with the parameter Q_I, which is the frequency of small oscillations around locally stable fixed points inside the unmodulated resonance islands, or the so-called "island tune".

The relevant portion of this parameter space for modulational diffusion is the region marked "Chaos" — in this region sidebands around the primary resonance which are created by the tune modulation overlap (the so-called Chirikov overlap), and a thick layer of bounded chaotic motion is formed for appropriate tune modulation strength q and frequency Q_M. (See Figure 2 — this is a stark

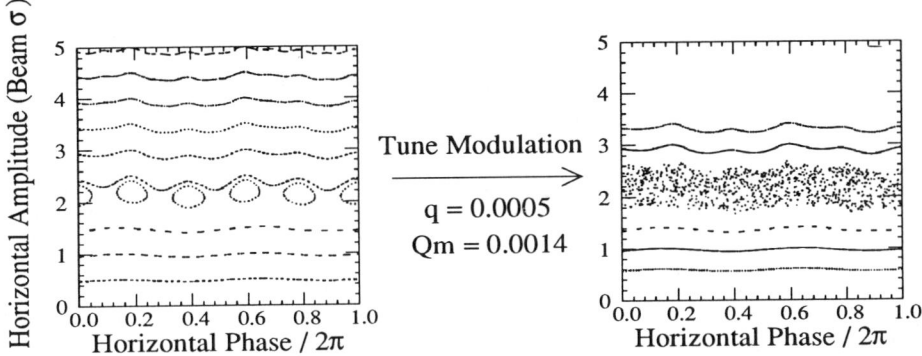

Figure 2: 1-dimensional phase space on the $Q_x = 3/5$ resonance, for moderately realistic parameters in the Fermilab Tevatron. Particle tracking is done with the beam-beam B0D0 lattice, with a linear beam-beam tuneshift of 0.0005. Resonance island structures are apperent without tune modulation, and tune modulation creates a thick stochastic band.

contrast to the thin-layer stochasticity generated by webs of interacting higher-order resonances seen in Arnol'd diffusion.[3]) In modulational diffusion models this layer of chaos serves as a noise source for vertical motion that is coupled to the horizontal phase through a weak nonlinear coupling resonance.

Examine what would be nominally regular motion in the vertical dimension, influenced by one or more of these "weak" coupling resonances. (For these purposes such resonances are considered to be "weak" if their amplitudes are much smaller than that of the primary resonance in the horizontal plane that drives the horizontal stochasticity.) The motion in this plane is now that of a very weakly driven oscillator, where the driving force is chaotic due to the weak coupling to the horizontal stochastic motion. Such motion is similar to that of a random-walk problem; the stochastically driven vertical motion can "diffuse" out to large amplitudes in finite time and therefore be lost from the machine.

The Hamiltonian for the standard modulational diffusion model[1] can be written as

$$H_1(\theta_x, I_x, \theta_y, I_y) = \frac{1}{2}I_x^2 - \epsilon \cos[(k+1)\theta_x + \lambda \sin \Omega t] + \frac{1}{2}I_y^2 - \mu \cos[k\theta_x + \theta_y], \quad (2)$$

where the (θ, I) variables are action-angle variables in each plane. Equivalences can easily be drawn between this model and that of the resonant beam-beam model used here to investigate modulational diffusion, which also gives the phase diagram in Figure 1. For example, $\epsilon \rightleftharpoons (2\pi Q_I)^2$ (scaling of island tune), $\lambda \rightleftharpoons$

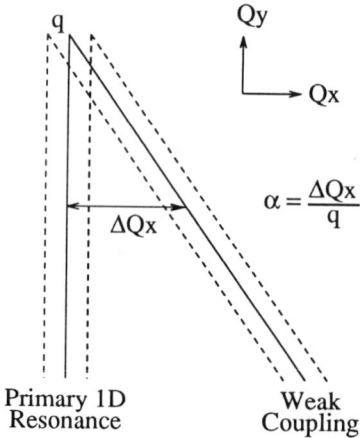

Figure 3: Resonance structure for modulational diffusion. α is the horizontal tune distance between the primary driving resonance and the secondary weak coupling resonance, scaled by the tune modulation depth q.

(q/Q_M) (modulation strength), and $\Omega \rightleftharpoons (2\pi Q_M)$ (modulation frequency).

One prediction of modulational diffusion theory is that this diffusive growth in the vertical dimension will scale as the square root of turn number. A diffusion coefficient can also be defined,

$$D \equiv \frac{\langle[\Delta I_y(t)]^2\rangle}{2T},\qquad(3)$$

where ΔI_y is the vertical action excursion from the initial action and T is the time width of the averaging in turns. The averaging should be performed over a time T short compared to the vertical diffusion time [so $\Delta I_y/I_y(t=0)$ is small] but long compared to the timescales of horizontal motion across the thick chaotic band. As the vertical tune is varied along the horizontal one-dimensional resonance, the proximity of the weak coupling resonance changes, as given by the dimensionless quantity

$$\alpha \equiv \frac{|Q_x(\text{weak coupling resonance}) - Q_x(\text{primary resonance})|}{q}.\qquad(4)$$

A plot of the logarithm of the diffusion coefficient D versus this scaled proximity α shows a series of descending plateaus and sudden drops[1]. This strange structure has sharp drops in D at even integer values of α for the case where both the driving resonance and the coupling resonance are modulated, and it

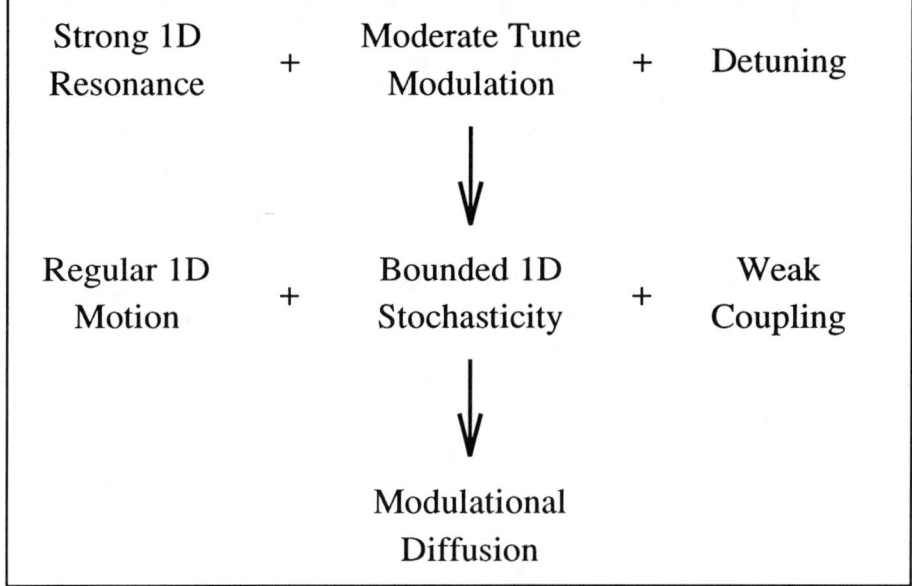

Figure 4: A diagrammed summary of the requirements for modulational diffusion.

is this sort of structure we attempt to qualitatively reproduce here within the operational framework of the Fermilab Tevatron.

2 THE TEVATRON SITUATION AND AN OPERATIONAL MODEL

In the Fermilab Tevatron during the 1992 collider run with separators, there were two strong beam-beam interactions every turn — one at the CDF experimental site at ring location B0, and one at the D0 experimental site. The operating estimate of the linear beam-beam tune shift ξ is approximately $\xi \approx 0.005$ per interaction, and with planned upgrades including the Fermilab Main Injector, this value may very well rise even further.[7] With the exceptions of these beam-beam kicks and chromaticity-correction sextupoles (which are neglected for the sake of simplicity of the tracking model), the Tevatron is quite a linear machine, and so its transverse dynamics in this situation can be modeled simply using only linear phase advances and beam-beam kicks. Typical operating parameters for the 1992 collider run are listed in Table 1.

The base tunes of the Tevatron in typical collider run circumstances are $Q_{x0} \approx 20.586$ and $Q_{y0} \approx 20.575$, running at a horizontal tune between the 12th- and 5th-order resonances. For the purposes of this study, however, a worst-case scenario is investigated, where the driving resonance for the horizontal stochasticity necessary for modulational diffusion is the 5th-order resonance

Table 1. Typical Fermilab Tevatron 1992 Operational Parameters at 900 GeV

Horizontal and vertical chromaticities	(ξ_x, ξ_y)	3.0
Typical momentum offset	$\Delta p/p$	0.0003
Synchrotron (modulation) frequency	Q_M	0.00078
Beam-beam linear tune shift per crossing	ξ	0.005
Revolution frequency	f_{rev}	47.7 kHz

and single particles are launched at a variety of vertical tunes along this resonance. If ξ ever exceeds .009 with two collisions in the Tevatron, the available space between the 12th and the 5th becomes too small for the entire beam, and a significant portion of the beam could be strongly affected by one of these resonances. The relevant portion of the tune plane diagram and the strange shape of the beam-beam footprint are shown in Figure 5.

The beam-beam force used here (within the tracking program Evol) uses the weak-strong approximation and assumes both beams have round Gaussian distributions of equal transverse size σ. For the horizontal beam-beam kick,

$$\frac{\Delta x'}{\sigma} = \frac{-4\pi\xi}{\beta_x^\star R^2}\left[1 - e^{-R^2/2}\right]\frac{x}{\sigma}, \tag{5}$$

where x is the transverse position relative to the opposing beam center, $x' \equiv dx/ds$, $\beta_x^\star$ is the beta function at the interaction point, and R is the distance from the center of the opposing beam scaled to the beam size σ:

$$R \equiv \sqrt{\left(\frac{x}{\sigma}\right)^2 + \left(\frac{y}{\sigma}\right)^2}; \tag{6}$$

a kick similar to (5) is seen in the vertical plane. Salient features relevant to this study can be noted:

- The variation of tune with amplitude (detuning) given by the beam-beam force is nonlinear and strongly coupled.

- From the form of R, even-order resonances are driven to first order in ξ. Odd-order resonances of order N are driven as even-order resonances of order $2N$.

- Resonance strengths vary with particle amplitude, or action.

- There is no beam-beam tuneshift or resonance driving at infinite amplitudes, so global motion of the unperturbed beam-beam system is stable.

The detuning is drastically different from the model of Equation 2, where there is explicitly no coupling other than the weak resonance, and the explicit variation of resonance strength with particle amplitude is also a difference between

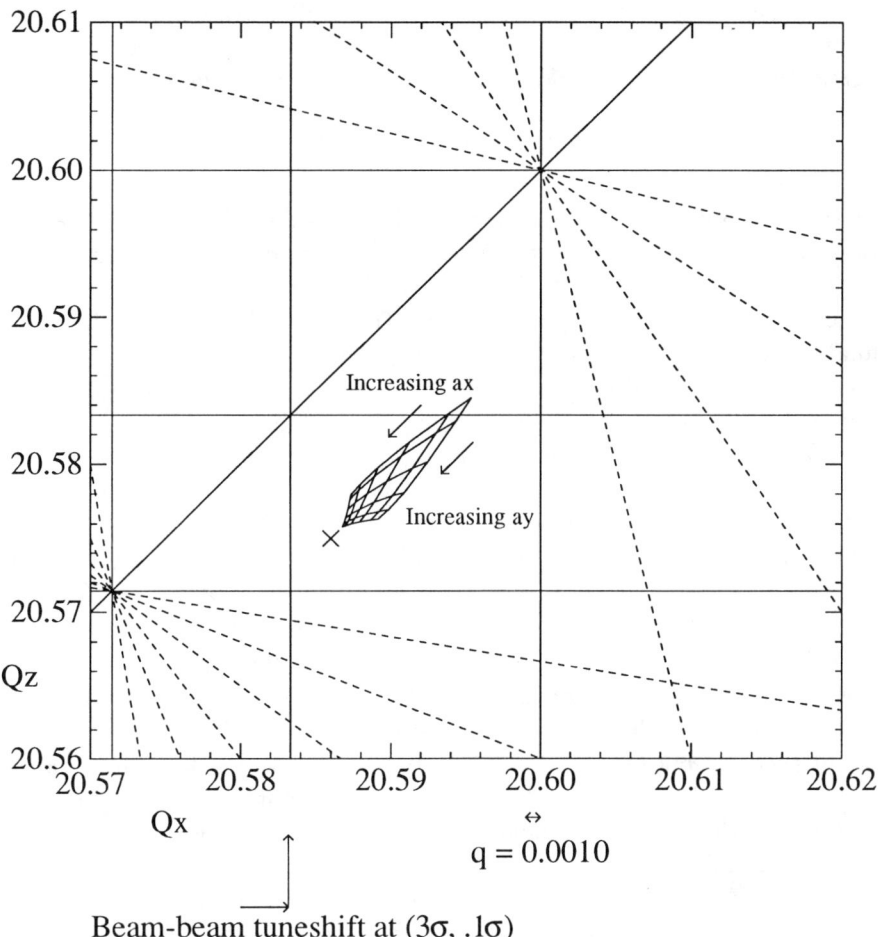

Figure 5: The tune plane for typical Fermilab Tevatron 1992 collider operations, showing the 5th, 7th and 12th-order resonances. The nominal operating tunes are indicated by the cross, and the beam-beam footprint is shown for $\xi = 0.005$ with two collisions. Footprint contours of constant amplitude range from 0.1σ to 5.1σ in 1σ increments.

these two models. As vertical amplitude grows in the beam-beam situation, one of two mechanisms will halt modulational diffusion: the vertical amplitude growth will either pull the horizontal tune off the primary driving resonance or it will suppress the coupling resonance strength. It remains to be conclusively shown whether such vertical amplitude growth will cause significant particle loss; however, even collective vertical amplitude growth without loss will raise the vertical beam emittance and result in luminosity degradation.

3 SIMULATION RESULTS

The tracking program Evol was used for all simulations.[8] In order to drive the 5th-order resonance strongly for the worst-case scenario, a small horizontal 0.1σ beam-beam offset was included; closed-orbit alignment errors of this magnitude at the collision points are quite possible. Tracking this lattice with no tune modulation with the beam-beam tuneshift given in Table 1 on the $Q_x = 20.6$ resonance finds an island tune of $Q_I = 1.51 \times 10^{-3}$. Since the synchrotron frequency of the Tevatron at this energy is approximately $Q_M = 7.8 \times 10^{-4}$ (with a period $T_M = 1/Q_M = 1280$ turns), the chaotic region of the tune modulation parameter space in Figure 1 is quite accessible for moderate tune modulation depths q. In this study a tune modulation depth $q = 0.0010$ was used, present only in the horizontal plane for comparison to the results originating in the similarly modulated Hamiltonian of Equation 1. This tune modulation amplitude corresponds to a horizontal chromaticity of about 3 units with a momentum offset $\Delta p/p$ of 3×10^{-4}, realistic values for the Tevatron.

To establish the timescales of the relevant amplitude growth mechanisms, the maximum vertical amplitude was recorded for single particles launched at the above initial conditions over a mesh on the tune plane, for tracking times ranging from 10 to 10^4 synchrotron periods. The tune mesh limits used were the same as those shown in the tune plane diagram, Figure 5, and the results of this tracking are shown in Figures 6a-6d. A quite definite vertical amplitude growth is seen near the intersections of the $Q_x - Q_y$ and $5Q_x$ resonances that evolves over timescales of thousands of synchrotron periods, consistent with the timescales of modulational diffusion. Such growth is completely absent with modulation turned off ($q = 0$), where only amplitude growth on the $Q_x - Q_y$ resonance is seen due to energy exchange between the unbalanced horizontal and vertical amplitudes; this is a conclusive indication that the modulation drives this vertical amplitude growth. The growth also seems to have a structure along the horizontal resonance, consistent with the modulational diffusion model expectations of the dependence of the amplitude growth rate on distance from the nearest coupling resonance. Some growth also appears on the $3Q_x + 2Q_y$ resonance; however, the structure along this resonance is quite minimal in comparison to the vertical amplitude growth near the previously mentioned intersection of $Q_x - Q_y$ and $5Q_x$ resonances.

Once the timescales of amplitude growth have been established, there remains the question of how the vertical amplitude evolves with time. It has already been mentioned that classical diffusion predicts that the vertical amplitude will grow

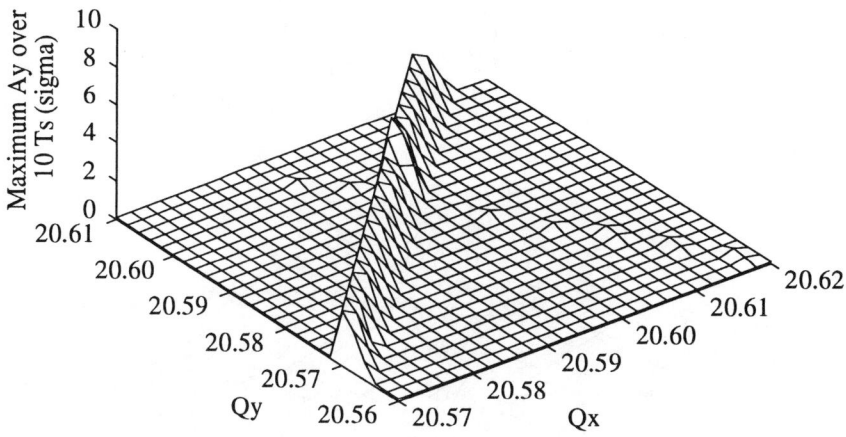

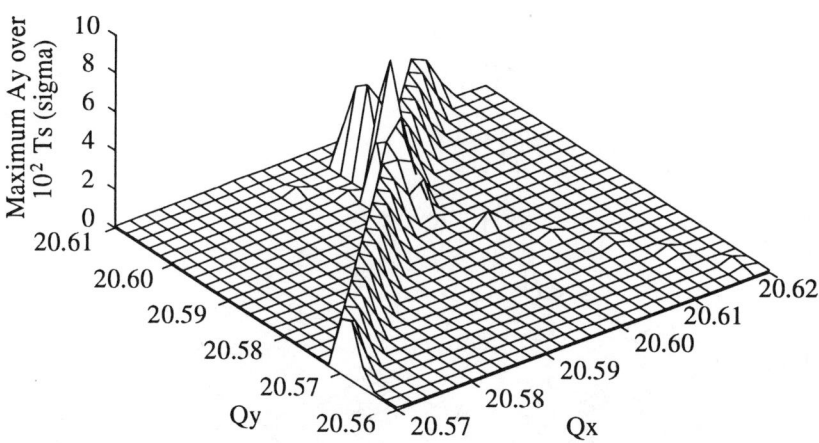

Figures 6a and 6b: Maximum vertical amplitudes of single particles with initial vertical amplitudes 0.1σ, tracked over 10 and 100 synchrotron periods. Horizontal particle amplitude is 3σ, inside a chaotic band.

A Simulation of Modulational Diffusion

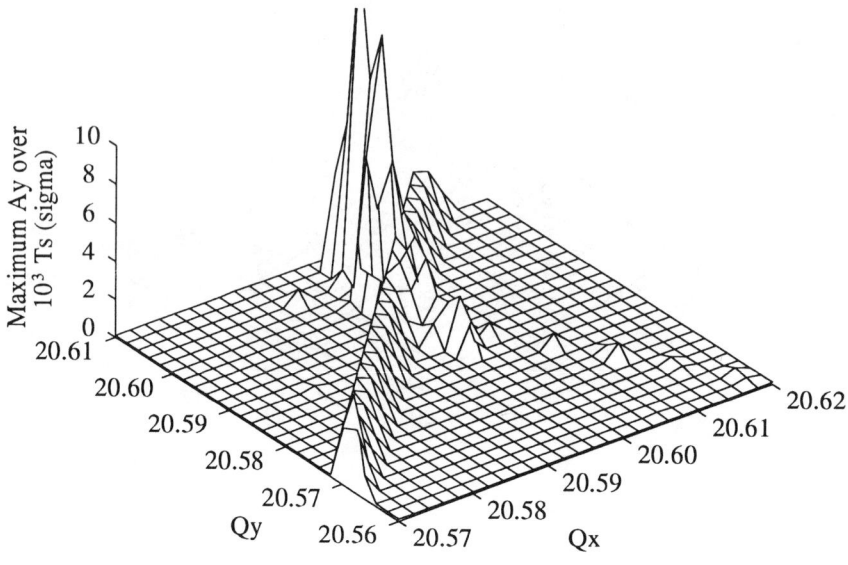

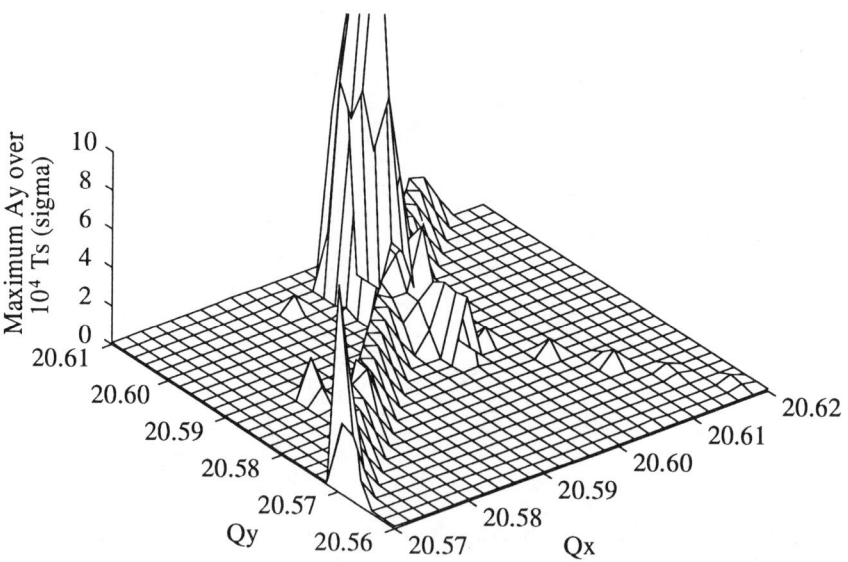

Figures 6c and 6d: Maximum vertical amplitudes of single particles with initial vertical amplitudes 0.1σ, tracked over 10^3 and 10^4 synchrotron periods.

proportionally to $t^{1/2}$ — if this is the case, a plot of $\log a_y$ versus $\log t$ should be a straight line with a slope of one half. However, if the vertical amplitude grows exponentially with time,

$$a_y(t) = a_{y0}\, e^{\gamma t}, \tag{7}$$

the plot of $\log a_y$ versus t, not $\log t$, should grow linearly, and the slope of this line is the exponential growth rate γ. This rate has units of inverse synchrotron periods, because the natural time unit for problems involving direct modulation is the modulation period.

To investigate the vertical amplitude evolution, particles were launched with horizontal amplitude of 3 σ, with base tune $Q_{x0} = 20.597$ (directly on the $5Q_x$ resonance with the beam-beam tune shift) and vertical amplitude of 0.1 σ with various base tunes. Tracking was stopped when either 10^4 synchrotron periods had been tracked (corresponding to nearly 5 minutes of real particle evolution), or the vertical amplitude had reached 1 σ. The one-sigma vertical cutoff was introduced because the influence of the vertical motion on the horizontal motion within the stochastic band was expected to become non-negligible at moderate vertical amplitudes; both the beam-beam detuning and resonance strengths decrease as vertical amplitudes increase.

Figure 7 shows three examples of vertical amplitude evolution over relatively long timescales, each plotted on log-linear and log-log scales. It is clear from examining these evolutions (as well as those of many other particles at different distances α from the nearby weak coupling resonance $4Q_x+Q_y$) that the vertical amplitude is growing as an exponential of time, not a power law as one would expect from standard diffusion phenomenology. It has been suggested[9] that this behavior may be explained by the dependence of resonance strengths on particle amplitude — the change in amplitude creates a changing resonance strength which feeds back upon the amplitude growth, creating exponential growth.

The exponential growth coefficient γ can now be plotted versus the scaled distance to the weak coupling resonance as one varies the vertical tune along the $5Q_x$ resonance to investigate whether there is any structure present. Since γ is expected to vary over many orders of magnitude, we instead plot $\log \gamma$ versus α; α can be directly determined from the vertical base tune Q_{y0} via

$$\alpha = \frac{20.60 - Q_{y0}}{5q} \tag{8}$$

when considering the $4Q_x + Q_y$ resonance to be the source of weak coupling. Other resonances such as $Q_x - Q_y$ and $3Q_x + 2Q_y$ are also nearby, but are farther away in horizontal tune distance α than this resonance, as can be seen in Figure 5. The exponential growth rate γ is measured from a standard linear fit of tracked $\log a_y$ versus time data. Figure 8 shows this data; note the two distinct "plateaus" and the sudden drops in the growth rate at $\alpha = 2$ and $\alpha = 3$.

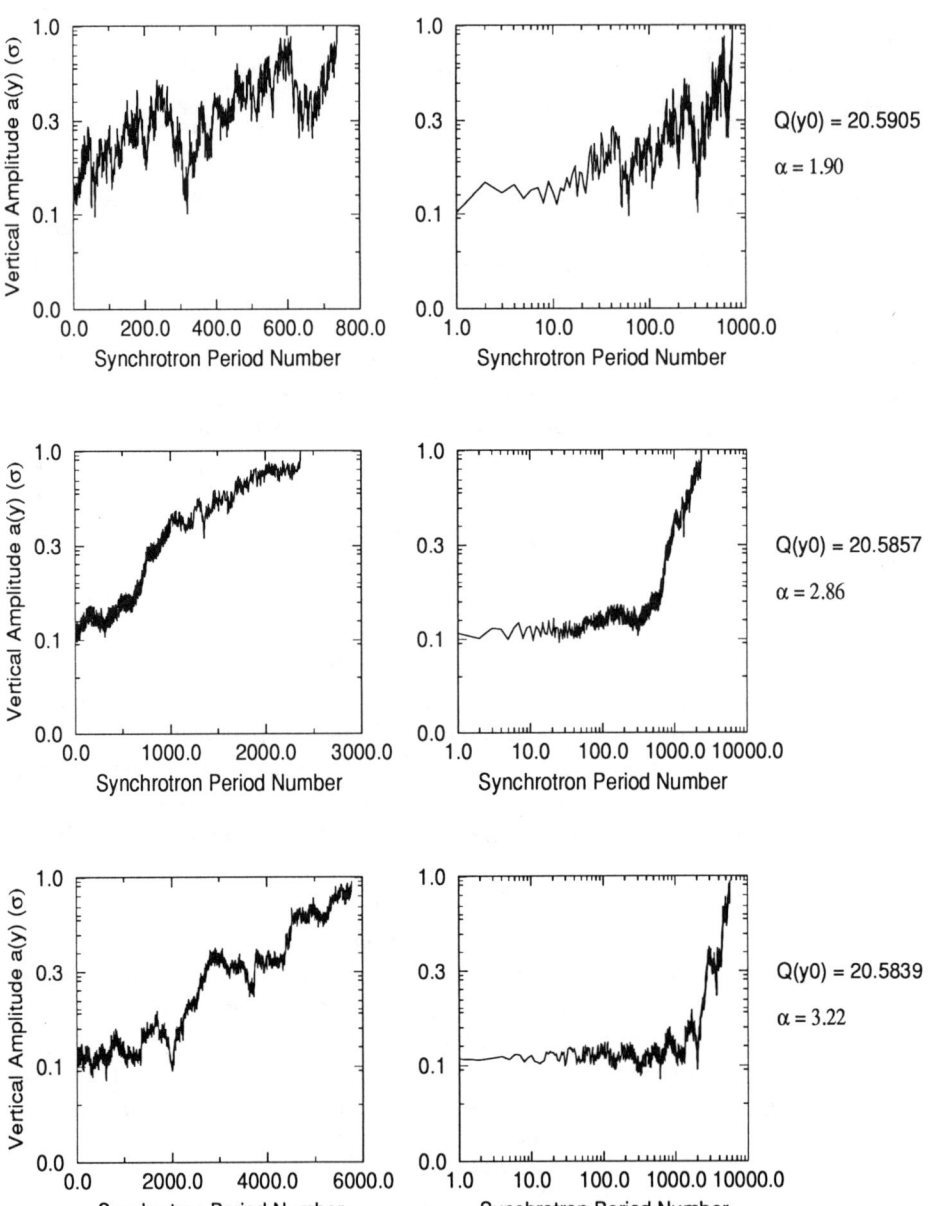

Figure 7: Character of vertical amplitude growth for particles launched within a horizontal stochastic band, and at initial vertical amplitude 0.1σ. Tracking was stopped when the vertical amplitude reached 1σ; note the differing growth timescales.

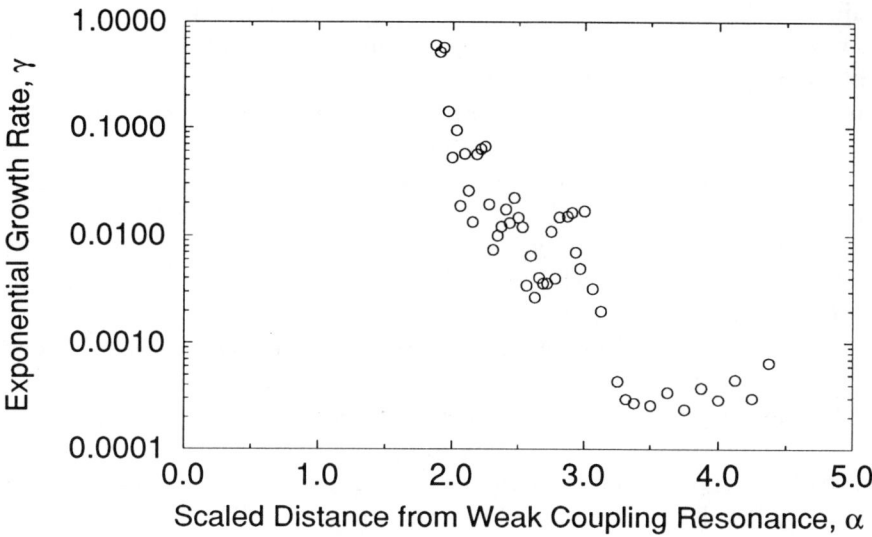

Figure 8: Exponential vertical amplitude growth rate γ plotted versus the scaled distance α from the $4Q_x + Q_y$ resonance.

4 CONCLUSIONS AND FUTURE DIRECTIONS

Modulational diffusion has been investigated within a simple model of the beam-beam interaction in the Fermilab Tevatron collider. Realistic operational parameters indicate that particles subject to horizontal stochasticity, or naively those that are within the tune modulation depth distance in horizontal tune of the $5Q_x$ resonance, experience modulational diffusion that causes their vertical amplitudes to grow exponentially over timescales of thousands of synchrotron periods, or millions of turns, leading to possible long-term particle loss. The rate of this amplitude growth is also dependent on proximity of nearby coupling resonances, and shows a structural dependence similar to those of previous modulational diffusion studies. However, the vertical amplitude growth is not root-time as naively predicted in these models where resonance strengths are not action-dependent.

Under current operating conditions in the Fermilab Tevatron collider, no particles are expected to be affected by the $5Q_x$ resonance this severely unless the horizontal tune drifts upwards, dragging particles into the fifth, or the linear beam-beam tune shift ξ increases. However, with future luminosity upgrades, this tune shift per crossing will almost certainly rise, and the operational space used in past runs may not be large enough to accomodate the entire tune spread of the beam, with a significant portion of the beam becoming influenced by the horizontal 7th and 5th and vertical beam blowup as a result. This could lead to luminosity degradation and intensity loss over a collider store as the beam size grows. Also, these effects, were they present in an actual collider, could be difficult to diagnose due to their slow growth nature.

Future studies should be twofold. First, a concrete theoretical structure of modulational diffusion should be investigated to conclusively show that in the case of amplitude-dependent coupling resonance strengths, vertical amplitude growth is exponential instead of root-time as in classical models. Secondly, the statistical nature of this amplitude growth should be investigated to see what observable effects such a mechanism could have on the beam size (and thus luminosity) evolution over time.

Another tracking model might be used to investigate the amplitude growth mechanism once a theoretical framework is in place, to avoid the rather complex detuning coupling of the beam-beam force. A lattice with sets of octupoles (which control detuning independently to first order in their strength in each transverse plane) and a decapole (which drives the horizontal $5Q_x$ resonance to first order) has already been developed for other tune modulation investigations.[5] Tracking with this lattice has several distinct advantages — motion at large particle amplitudes is no longer stable, so no ad hoc aperture needs to be introduced, and parameters for a Hamiltonian description as in Equation 1 can easily be found to first order in the individual magnet strengths.

REFERENCES

1. B.V. Chirikov, M.A. Lieberman, F. Vivaldi and D.L. Shepelyanski, "A Theory of Modulational Diffusion", Physica 14D, 298 (1985).
2. F. Vivaldi, "Weak Instabilities in Many Dimensional Hamiltonian Systems", Rev. Mod. Phys. 56, 737 (1984).
3. A.J. Lichtenberg and M.A. Lieberman. Regular and Stochastic Motion (Springer, New York, 1983).
4. S. Peggs, "Hamiltonian Theory of the E778 Nonlinear Dynamics Experiment", SSC-175 (April 1988).
5. T. Satogata, Ph.D. Thesis, Northwestern University (1993).
6. T. Satogata et. al. , "Driven Response of a Trapped Particle Beam", Phys. Rev. Lett. 68, 1838 (1992).
7. Steven D. Holmes, "Achieving High Luminosity in the Fermilab Tevatron", Proceedings of the 1991 IEEE PAC, 2986 (1991).
8. S. Peggs, "Hadron Collider Behavior in the Nonlinear Numerical Model Evol", Particle Accelerators 17, 11 (1985).
9. O. Brüning, private communication at this workshop.

THE APPLICABILITY OF DIFFUSION PHENOMENOLOGY TO PARTICLE LOSSES IN HADRON COLLIDERS*

A. Gerasimov
Fermi National Accelerator Laboratory
P.O. Box 500, Batavia, IL 60510

ABSTRACT

An analytic approach is developed for solving the inhomogeneous diffusion equation with the diffusion intensity being a fast-growing function of the betatron energy. When applied to the survival data of particle tracking for LHC and SSC, the method shows that these data are inconsistent with (any) diffusion phenomenology. A similar inconsistency is observed in the data from CERN diffusion experiments.

1 INTRODUCTION

Understanding the limits of the stability of particles at large betatron amplitudes presents one of the major accelerator physics challenges in the design of new supercolliders such as SSC and LHC [1,2]. The reliability of the tracking codes that are used for this purpose needs to be checked in experiments with the existing colliders. In this conjunction it is also important to understand the nature of the slow transport of particles below the dynamical aperture due to lattice nonlinearities and/or power supply ripples. The natural description of such transport is the diffusion process with amplitude-dependent diffusion coefficient, as used in the Fermilab [3] and CERN [4] diffusion experiments.

In the present paper, we analyze the properties of this diffusion model as applied to three different studies:
I) density profile measurements in Fermilab experiments [3],
II) beam intensity time dependence in CERN scraper retraction experiments [4],
III) escape time spreads versus initial amplitude as produced in tracking for SSC [5] and LHC.

The basis of the analysis is the strong inhomogeneity, i.e. fast growth with the amplitude, of the diffusion observed in I and III.

2 ASYMPTOTIC ANALYSIS OF INHOMOGENEOUS DIFFUSION

The diffusion model that was used for the data analysis of the Fermilab diffusion experiment and which we will use to model both the tracking data and CERN scraper retraction data, has the following form [3]:

$$\frac{\partial \rho}{\partial t} = \frac{\partial}{\partial I}\left(D(I)\frac{\partial \rho}{\partial I}\right) \quad (1)$$

---

* FERMILAB-PUB-92/185; CERN SL/92-30 (AP).

where I is the betatron action (energy) $I = (x^2/2) + (x'^2/2)$, $D(I)$ is the action-dependent diffusion coefficient, and ρ is the density distribution function. We will develop now an asymptotic approach to solving the diffusion equation (1) when the diffusion is strongly inhomogeneous, i.e. grows very fast with increasing I : $D'D \to \infty$.* Apart from assuming that (D'/D) is large, we will consider only the case when the logarithm of $D'(I)$ does not change much over the range of I, corresponding to one decade of variation of D. This restriction, which can be explicitly written as $ff' \ll (f')^2$ [where $f(I) = \ln D(I)$], is quite mild, and, as we will see in a number of examples later, is usually satisfied quite well. Within the class of functions $D(I)$ thus defined we can always use the "local exponential approximation"

$$D(I_0 + \Delta I) \approx D(I_0) \exp\left(\frac{D'(I_0)}{D(I_0)} \Delta I\right) \qquad (2)$$

which corresponds to the linear term of the Taylor expansion of the function $\ln D(I)$, implying $\Delta I \lesssim (D/D')$. In the reference example $D = AI^n$ with $n \gg 1$, that proved a good model for the Fermilab diffusion data [3], one gets $D \approx AI_0^n \exp\left(\frac{n}{I_0} \Delta I\right)$.

The basic idea in the solution of the diffusion equation (1) for the class of functions $D(I)$ (2), is that, for the fast growing functions $D(I)$ and at each given moment of time, the particles with initial conditions above a certain value $I_e(t)$ are all lost at the absorbing boundary $I_{ab} > I_e(t)$, while those below the value $I_e(t)$ are not affected by diffusion at all. The distribution function in the transition area should have then some universal properties due to the narrowness of that region. Thus, we will seek the general time-dependent solution in the form

$$\rho(I,t) = \rho_0(I) g\left(\frac{I - I_e(t)}{\lambda}\right) \qquad (3)$$

where $\rho_0(I)$ is an (arbitrary) initial distribution $\rho(I,0) = \rho_0(I)$ and g is a fixed function of one variable $S = (I - I_e(t))/\lambda(t)$ qualitatively shown in Fig. 1. We expect λ to be asymptotically small when D'/D tend to infinity.

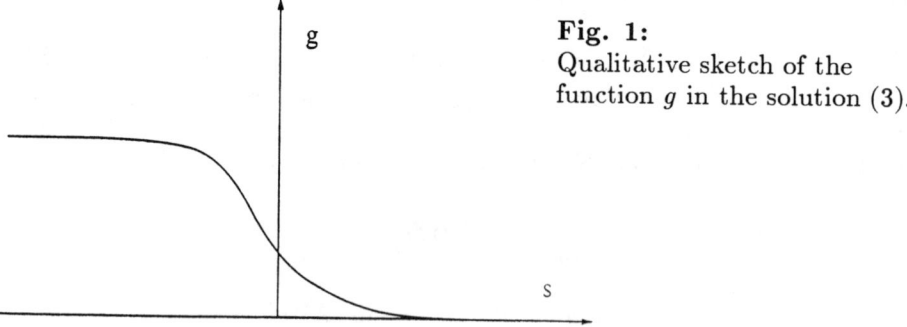

Fig. 1:
Qualitative sketch of the function g in the solution (3).

---

* In the following, we reserve the prime for differentiation by action I, and the dot • for differentiation by time t.

The distribution (3) therefore is a product of initial distribution ρ_0 and the "kink" function g, with moving $I_e(t)$ and varying width $\lambda(t)$.

To find the three unknown functions $g(s), I_e(t), \lambda(t)$ we plug the ansatz (3) in the equation (1), using the local approximation (2) with $I_0 = I_e(t)$ and do not differentiate the slow-varying function $g(I)$. The resulting equation is

$$-g'(s)\left[s\dot{\lambda}(t) + \dot{I}_e(t) + D'(I_e(t))\exp[k(I_e(t))\lambda(t)s]\right]$$
$$= \frac{D(I_e(t))}{\lambda(t)}\exp[k(I_e(t))\lambda(t)s]\,g''(s) \qquad (4)$$

where we introduced the notation $k(I_0) = D'(I_0)/D(I_0)$.

Looking at equation (4) more closely, it is not difficult to realize that the only way of "untangling" the variables s and t and obtaining separate equations for the functions of different arguments is to assume that $\lambda(t) = 1/k(I_e(t))$. Note that an arbitrary constant that can be put in the denominator can be shown to simply rescale g and thus disappears from the final result. Having the expression for $\lambda(t)$, one immediately realizes that the first term in the **large** brackets in equation (4) is much smaller than the second one. This follows from the inequality $k'/k^2 \ll 1$ that is equivalent to the condition of applicability of the "local exponential approximation" (2) $ff' \ll (f')^2$. Dropping thus that term, we arrive at

$$-\frac{g''(s)}{g'(s)} = 1 + \frac{\dot{I}_e(t)}{D'(I_e(t))}e^{-s}. \qquad (5)$$

The final untangling of variables s and t can be achieved now only through the choice of the function $I_e(t)$ to satisfy the equation $\dot{I}_e(t) = -D'(I_e(t))$. An arbitrary constant that could be multiplying the right-hand side can be shown to disappear from the final result. The initial condition for that equation is $I_e(0) = I_{ab}$.

The function g is found then to be
$$g(s) = 1 - \exp\left[-e^{-s}\right]. \qquad (6)$$

Thus, by determining the functions g, I_e and λ we completely defined the solution (3). Due to the smallness of λ, the qualitative image of that solution is that of moving boundary $I_e(t)$, with the distribution ρ_0 vanishing above this value.

3 MEAN ESCAPE TIME ANALYSIS

One of the common ways of representing the tracking data for long-term stability of particles is through so-called "survival plots," where escape times to a certain boundary for particles started at different amplitudes are shown. One example of "survival plots" for the tracking study of SSC [5] is shown in Fig. 2. We will address the issues of whether such distributions of escape times can appear in the diffusion model (1) and how to extract the diffusion intensity $D(I)$ from the survival data. The idea of using the magnitudes of the spreads of escape times in the survival plots as a compatibility test with the diffusion model was originally introduced by J. Cary [7]. Our approach to the realization of this idea differs though from that of J. Cary [8]: we take advantage of the approximately 'locally exponential' character of the escape time dependence on the action $\pi(I)$ and employ the analytic method of Section 2 to calculate the escape time spreads.

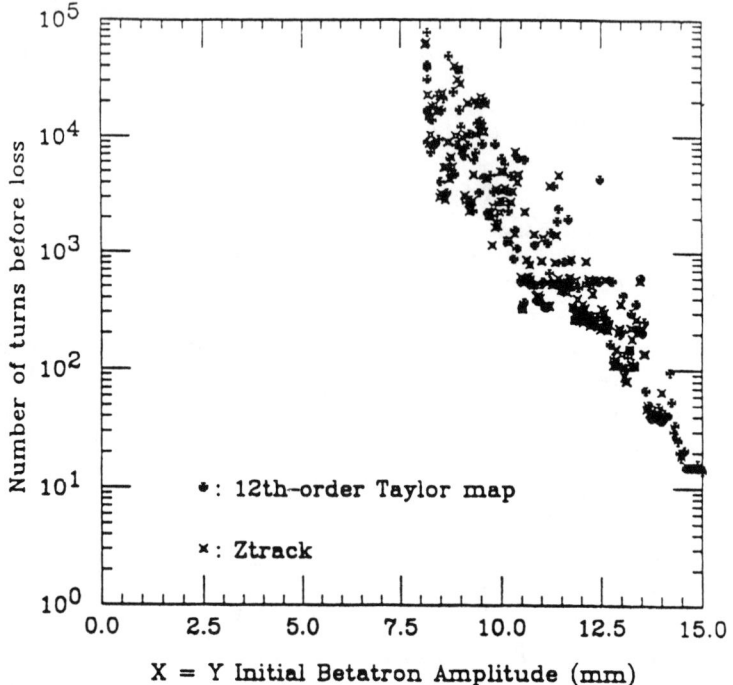

Fig. 2: 100.000-turn survival plots for 2-TeV SSC injection lattice. [5]

According to the more general theory of escape in diffusion processes with drifts [6], the probability $G(I, t)$ of surviving, or not escaping to an absorbing boundary, within time t starting from the initial condition I for the diffusion process (1) satisfies the diffusion equation

$$\frac{\partial G}{\partial t} = \frac{\partial}{\partial I}\left(D(I)\frac{\partial G}{\partial I}\right) \qquad (7)$$

with the initial condition $G(I, 0)$ and boundary condition $G(I_{ab}, t) = 0$ at the absorbing wall. The mean escape time defined as $T(I) = -\langle t(\partial G/\partial t)\rangle = \langle G \rangle$ (average is over t), can then be shown [6] to be the solution of the equation

$$\frac{d}{dI}\left(D(I)\frac{dT}{dI}\right) = -1 \qquad (8)$$

with the boundary condition $T(I_{ab}) = 0$.

The second boundary condition at $I = 0$ in our case of positive-definite coordinate I is that of a reflective wall yielding [6] $D(I)(dT/dI)_{|I=0} = 0$.

The escape time is found then to be

$$T(I) = \int_{I}^{I_{ab}} dI_1 \frac{I_1}{D(I_1)}. \qquad (9)$$

The escape time data in the survival plots provide the dependence of the escape time on the action $T(I)$, which defines thus the diffusion intensity $D(I) = -(I/(dT/dI))$. An important question then is to understand whether the observed spreads of escape times, which are quite large (Fig. 2), are compatible with the diffusion model as deduced from the mean escape time. The natural measure of these spreads is the r.m.s. width of the escape time distribution $\Delta T = \sqrt{\langle t^2 \rangle - \langle t \rangle^2}$. While the first moment $\langle t \rangle$ is just T, the second moment $T_2 = \langle t^2 \rangle = -\langle t(\partial G/\partial t)\rangle$ is known from the general theory [6] to satisfy the equation

$$-T(I) = \frac{d}{dI}\left(D(I)\frac{dT_2}{dI}\right) \tag{10}$$

with the same boundary conditions as for $T(I)$. The second moment thus is explicitly found to be

$$T_2(I) = \int_I^{I_{ab}} \frac{dI_1}{D(I_1)} \int_\delta^{I_1} T(I_2)\, dI_2 \tag{11}$$

where the reflecting wall was artificially placed at $I = \delta$ rather than at $I = 0$.

Let us try now to extract the quantity $T_2(I)$ from the survival plots of Fig. 2. The mean escape time $T(I)$ on this plot decreases (with an increasing I) exponential-like, so that the dominant contributions to both internal and external integrals in (11) come from the lower limits of integration. It appears also that the function $T(I)$ and the corresponding diffusion intensity $D(I) = -(I/(dT/dI))$ satisfy the "local exponential approximation" (2), since the slope of the curve $f(I) = \ln T(I)$ doesn't change much when f changes by about unity. From this approximation, one can explicitly find the quantity T_2 (11) to be

$$T_2(I) = -\frac{1}{D'(I)}\frac{T^2(\delta)}{T'(\delta)} = -\frac{T(I)}{I}\cdot\frac{T^2(\delta)}{T'(\delta)} \tag{12}$$

where the condition of the distances $I - I_{ab}$ and $I_1 - \delta$ being large relative to the characteristic scale $T(I)/T'(I)$ was used. This formula indicates that the second moment T_2 at the point I is very sensitive to the behavior of diffusion intensity at small actions. In particular, the quantity $T_2(I)$ diverges whenever the escape time $T_2(I_1)$ behaves at small $I_1 \to 0$ as $T(I_1) \sim 1/I_1^k$ with any $k > 1$.

The quantity T_2 is thus inconvenient for comparison with the "survival plots" since large T_2 in many cases can (and in fact does) account only for the long "tail" of the distribution of escape times $f(I,t) = -(\partial G(I,t)/\partial t)$. The preferable quantity of choice then is the width of this distribution Δ defined as the half-max width. To find it, one can use the same asymptotic approach of Section 2 for the solution of the diffusion equation (7) for the function $G(I,t)$. The estimates of the half-max width then can be obtained by requiring that the argument of the function g in the solution (3) change by unity, yielding

$$\Delta(I) \sim \frac{\lambda(I)}{D'^2(I)} = \frac{D(I)}{D'^2(I)} = \frac{T^2(I)}{I(dT/dI(I))}. \tag{13}$$

The dimensionless quantity $\Delta/T = \lambda(I)/I$ [where I is the characteristic scale of growth of the function $T(I): \lambda = T/T'$] for the curve $T(I)$ in Fig. 2

can be easily seen then to be much smaller than unity. This clearly contradicts the wide spreads (over one decade at least) of the escape times observed in the survival plot in Fig. 2.

Thus, we arrive at the negative and conceptually important conclusion: the statistics of the escape times as observed in tracking simulation are not compatible with any diffusion process. Therefore more refined models are needed to describe the dynamical processes involved.

4 SCRAPER MANIPULATION ANALYSIS

The measurements of the transport of the particles in CERN SPS were performed [4] by kicking the beam to rather large amplitudes, moving the scraper in to the edge of the core of the beam, and retracting it in a short time by just a millimeter or two. The subsequent evolution of the beam intensity was observed over a period of about 10 to 20 minutes.

The characteristic result of such measurement appears as shown in Fig. 3. The basic features of this curve are:
1) the presence of a flat section with zero loss,
2) rather sharp shoulder of transition to non-zero loss,
3) quite linear behavior after the shoulder for the time at least just as long as the length of the flat section.

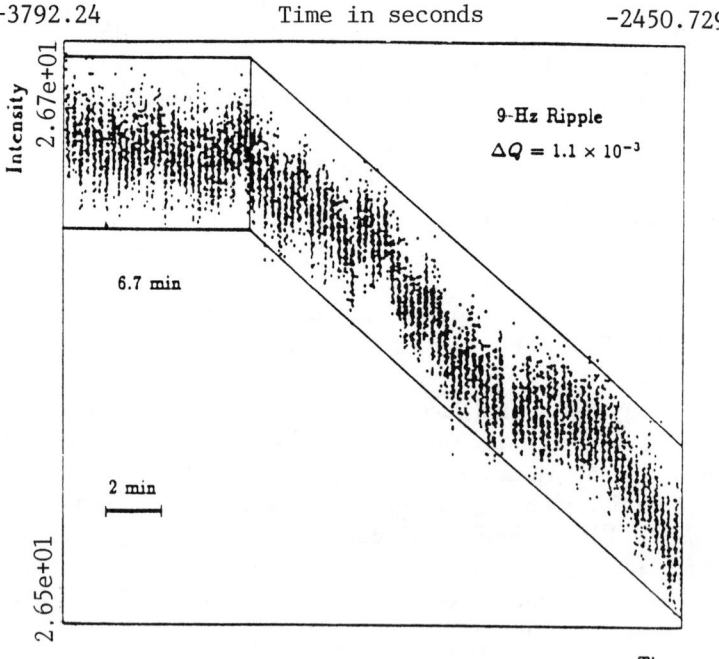

Fig. 3: After the retraction of both scrapers in the horizontal and vertical plane the particles need 6.7 minutes to reach the first (vertical) scraper. Thereafter a constant loss in intensities sets in.

The question then is whether these features can be accounted for in a diffusion model. The setup of the initial distribution in the scraper retraction experiment is shown in Fig. 4. I_s is the position of the scraper as it was moved in, and I_r is where it was retracted to.

We will consider the loss of particles at the wall $I = I_r$ in two extreme situations: first, when the diffusion intensity $D(I)$ changes very little at the distances of the order $I_r - I_s$ around the point $I = I_s$ and second, when it changes by several decades, so that $\lambda = D(I_s)/D'(I_s) \ll I_r - I_s$.

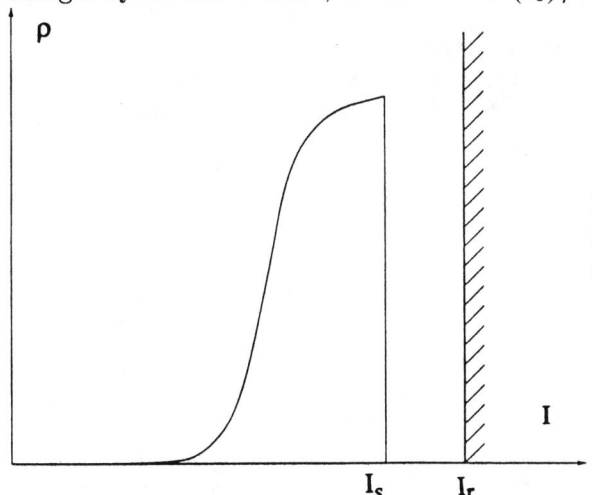

Fig. 4:
Initial distribution after the scraper retraction.

In the first case, one can obtain an explicit solution of the diffusion equation with a constant diffusion coefficient if one assumes a simple initial distribution that equals unity for $I < I_r$ and zero for $I > I_r$,

$$\rho(I,t) = \frac{2}{\sqrt{\pi}} \int_{\frac{I-I_r}{\sqrt{Dt}}}^{0} e^{-s^2/2} ds. \tag{14}$$

The total "intensity of the beam" to be compared with the experimental curve of Fig. 3 can be calculated (apart from the irrelevant constant contribution) as

$$L(t) = \int_{0}^{I_r} (\rho(I,t) - 1) \, dI \tag{15}$$

and is explicitly found to be proportional to $-\sqrt{Dt'}$.

The intensity $L(t)$ for the realistic initial distribution, which equals unity for $I < I_s$ and zero for $I > I_s$, will naturally have a transient with characteristic time scale $\Delta t \sim (I_r - I_s)^2 / D(I_s)$ and then approach the asymptotic behavior $\Delta L \sim -\sqrt{Dt}$ as shown in Fig. 5.

This type of behavior obviously does not possess the most important feature, a sharp shoulder, of the experimental graph of Fig. 3.

In the second case $\lambda = D(I_s)/D'(I_s) \gg I_s$ we will assume again that the diffusion intensity $D(I)$ satisfies the "local exponential approximation" (2) in the range $|I - I_s| \sim I_r - I_s$. The motion of the center of the "kink" g in the solution (3) then is $I_e(t) \approx I_s - \lambda \ln(tD_0/\lambda^2)$ [where $D_0 = D(I_s)$]. Since the initial condition ρ_0 of Fig. 4 is not smooth as it was assumed in the derivation of Section 2, the beam intensity dependence on time will have some transient over time of the order $t_{tr} = \lambda^2/D_0$ after which the loss rate $r(t)$ will approach its asymptotic time dependence $r(t) \approx \dot{I}_e(t) \sim (1/t)$.

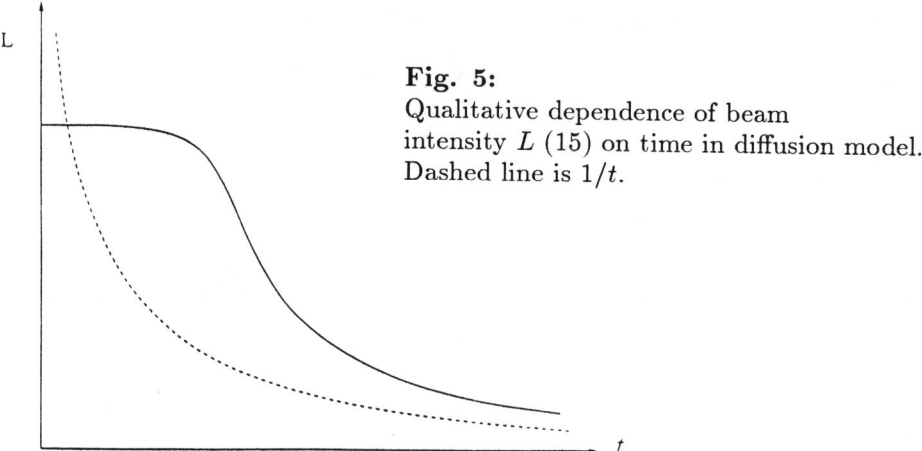

Fig. 5:
Qualitative dependence of beam intensity L (15) on time in diffusion model. Dashed line is $1/t$.

Since the only time scale involved is t_{tr}, the presence of a "shoulder" as in the experiment graph of Fig. 4 is not possible, and qualitatively the beam intensity dependence on time will be similar to the one shown in Fig. 5. Thus, in the second case as well as in the first one, the loss of particles in the diffusion process is distinctly different from what is observed in the experimental graph of Fig. 3.

5 DISCUSSIONS AND CONCLUSIONS

The results of the Fermilab diffusion [3] experiment indicate that the diffusion intensity is a fast growing function of betatron energy that can be locally approximated by an exponential function. The same property is observed in the survival plots in particle tracking for LHC and SSC.

We presented an analytic approach to describing the evolution of the distributions of particles in the presence of absorbing walls in the case of such fast growing diffusion intensities. This evolution demonstrates a major degree of universality: the initial distribution remains unaffected for betatron energies I less than a certain time-dependent value $I_e(t)$ while the density is completely depleted from the energies I larger than $I_e(t)$. The transient region of energies around I_e is narrow and the distribution function there is also universal (does not depend on initial distribution except for the normalization).

Asymptotic analysis of density evolution was applied to the "survival data" from particle tracking for SSC and LHC and it was demonstrated that the observed spreads of survival times are much larger than what they should be from the diffusion model. That proves that these survival data are not compatible with any diffusion model, and more refined statistical models of dynamics are required. One candidate for such a model can be the more general Markov process with jumps, where the evolution of the density distribution is defined by [6]

$$\frac{\partial \rho}{\partial \tau} = \frac{\partial}{\partial I}\left(D(I)\frac{\partial \rho}{\partial I}\right) + \int dI' \left[\rho(I')W(I',I) - \rho(I)W(I,I')\right]. \quad (16)$$

The last term in the right-hand side describes the effect of jumps with the probability $W(I,I')$. The idea of including jumps in the statistical description seems appealing also because of the commonly observed "intermittency" of slow and fast motion in tracking [2]. It would be interesting to try to fit the statistics of trajectories in tracking into some jump and diffusion model (16).

In Section 4, we discussed the character of particle loss in the scraper retraction diffusion experiment at CERN. Simple qualitative-type considerations also indicate the incompatibility of the observed decay of the beam intensity with that of diffusion models in the same setup.

In view of the incompatibilities of the diffusion models with the CERN diffusion experiment data and tracking "survival" data one could naturally ask why the Fermilab diffusion experiment data [3] were basically quite successfully fitted with diffusion models. The answer to that question can be conjectured to be the somewhat different quantities analyzed in these different approaches. Indeed, in the Fermilab experiment, the measured quantities were the substantial changes of the density distribution (corresponding to the intensity loss of 20 to 80 percent) as obtained by "flying wire" monitors. In the CERN diffusion experiment, the beam intensity changes very little, so the measured loss rates are defined by the small probability of escape to the retracted scraper. Similarly it is the escape process characteristics that are represented in the "survival plots" of tracking. It may very well be that the major changes in the density distributions are accounted for by the diffusive part of the random process, while the escape processes with sufficiently far removed boundaries are dominated by other properties like jumps in the model (16).

6 ACKNOWLEDGEMENTS

I gratefully acknowledge many useful discussions with F. Schmidt on the subject of the present paper.

References

[1] Y. Yan, G. Bourianoff and L. Schachinger, A typical 'Ztrack' long-term tracking result, in Proc. Workshop on Nonlinear Problems in Future Particle Accelerators, Capri, Italy (April 1990).

[2] F. Galluccio, F. Schmidt, Towards a Better Understanding of Slow Particle Losses in Large Hadron Colliders, CERN SL/91-44 (AP), LHC Note 172 (1991).

[3] T. Chen et al., Measurements of diffusion in Hamiltonian system, *Phys. Rev. Lett.* **68**, 33 (1992).
[4] X. Altuna et al., CERN SL/91-43 (AP), LHC Note 171 (1991).
[5] Y. Yan, Applications of differential algebra to single particle dynamics in storage rings, SSC preprint SSCL-500 (1991).
[6] K. Gardiner, Handbook of Stochastic Methods, Spring Series in Synergetics (Springer, Berlin, 1985).
[7] J. Cary, Report of the working group on nonlinear analysis of beams, in Nonlinear Dynamics and Particle Accelerations, AIP Conf. Proc. No. 230, p. 282 (1991).
[8] S. Hendrikson, T. Antousen, J. Cary, Diffusion model for transverse particle loss in transport structure, *Bull. Am. Phys. Soc.* **36** (5), 1573 (1991).

INVARIANT MANIFOLDS AND STABILITY: SOME RESULTS FOR 1-D MAPS

M. Giovannozzi

CERN-SL Division CH-1211 Geneva 23

and

University of Bologna, v. Irnerio 46 Bologna, Italy

ABSTRACT

To give an analytic estimate of the stability domain of non-linear systems, like large accelerators, has been a subject of intense theoretical studies for many years.

Up to now, even for the simple case of the one-dimensional Hénon map, one could not determine the border of the stability domain using analytic tools. On the other hand this estimate can easily be found by tracking through the map.

A promising new attempt is presented here, to estimate the dynamic aperture for the Hénon map via following the invariant manifolds of its hyperbolic fixed point.

The same technique is then applied to a different map and an attempt to generalize the method presented in this paper to generic 1-D polynomial maps is briefly discussed.

1 INTRODUCTION

The standard approach to find the dynamic aperture of an accelerator with non-linearities, is to use element by element tracking. The dynamic aperture specifies the maximum amplitude up to which the motion stays bounded. In the one-dimensional case the concept of the dynamic aperture is well defined as the regular orbits of a Hamiltonian system form closed curves so that the two-dimensional phase space is divided into parts that are completely separated. The tracking has the main drawback of giving little theoretical insight into the causes for the particle losses. Analytical tools are therefore investigated to understand these losses. However, even the most simple case of a linear lattice plus a single sextupole in the one-dimensional case (i.e. the Hénon map [1,2]) is not fully understood up to now.

In this paper an analytical method is presented (see also [3,4,5]) which allows to determine the stability border of the Hénon map following the invariant manifolds that emanate from the hyperbolic fixed point of the map. The same method is then applied to a cubic polynomial map, showing the generality of this approach.

In Section 2 we define the concept of fixed point and its classification, introducing the concept of invariant manifolds and explaining how to construct them numerically. In Section 3 we introduce the Hénon map, giving a brief review of its properties; Section 4 deals with a different model: the cubic map. The results are presented in Section 5, where we briefly discuss the possibility of generalizations to arbitrary polynomial maps in one dimension.

2 FIXED POINTS AND INVARIANT MANIFOLDS

The fixed points of a generic map F are the solutions of the equation

$$F(\vec{x}_0) = \vec{x}_0. \tag{1}$$

We can also define cycles of order m: they are simply fixed points of the mth iterate of the map F:

$$F^m(\vec{x}_0) = \vec{x}_0. \tag{2}$$

In the case of one–dimensional systems these points are classified according to the eigenvalues (λ_1, λ_2) of the linearized map. Owing to the area–preserving property they must fulfill the condition $\lambda_1 \lambda_2 = 1$ so that only three different situations are allowed:

$$\begin{array}{lll} \lambda_1, \lambda_2 \in \mathbb{C} & \lambda_1 = \lambda_2^* & \text{elliptic case} \\ \lambda_1, \lambda_2 \in \mathbb{R} & \lambda_1 \neq \lambda_2 & \text{hyperbolic case} \\ \lambda_1, \lambda_2 \in \mathbb{R} & \lambda_1 = \lambda_2 = 1 & \text{parabolic case} \end{array}$$

Around a fixed point the dynamics of the map F is determined by its linear part:

$$F(\vec{x}_0 + \vec{\delta}) = \vec{x}_0 + L_F(\vec{x}_0)\vec{\delta} + \mathcal{O}(||\vec{\delta}||^2). \tag{3}$$

In the hyperbolic case the eigenvectors of $L_F(\vec{x}_h)$ define two lines $\mathcal{W}_L^u(\vec{x}_h)$ and $\mathcal{W}_L^s(\vec{x}_h)$, along which the motion induced by the linearized map has an expanding (superscript u for *unstable*) or a contracting (superscript s for *stable*) behaviour. Also for the global non–linear map we can define two subspaces emanating from the unstable fixed point such that these sets, called $\mathcal{W}^u(\vec{x}_h)$ and $\mathcal{W}^s(\vec{x}_h)$, show the same expanding or contracting behaviour. These manifolds are uniquely defined and have the important feature of being invariant sets for the map:

$$F\big(\mathcal{W}^{u,s}(\vec{x}_h)\big) = \mathcal{W}^{u,s}(\vec{x}_h). \tag{4}$$

Obviously, we can also define invariant manifolds for hyperbolic cycles of order m. In this case, all the previous definitions and properties will hold, replacing the map F with its mth iterate F^m.

From the definition it follows that the invariant manifolds have at least one intersection which is the hyperbolic fixed point. An additional intersection, $\vec{x}_{hom}$ is called *homoclinic* or *heteroclinic* depending on whether the two intersecting manifolds emanate from the same hyperbolic fixed point or not.

The existence of one such point implies that there exist others, in fact an infinite number of them, as this point belongs to several manifolds for each of which the invariant property (4) has to be fulfilled. Unless the two manifolds coincide completely (integrable case), they will oscillate around each other (Fig 1). When approaching the hyperbolic fixed point the distance between successive intersections decreases as the period of the motion tends to zero. Due to the area–preserving character of the map, the area enclosed between two successive intersections has to stay constant, leading to larger and larger oscillations near the unstable fixed point.

To construct the whole sets $\mathcal{W}^{u,s}(\vec{x}_h)$, it is sufficient to iterate many times an ensemble of initial conditions belonging to a small part of these manifolds around the hyperbolic fixed point. Moreover it turned out that these initial conditions can be chosen on the eigenvalues of the linearized map as long as one stays within sufficiently small distance to the hyperbolic fixed point.

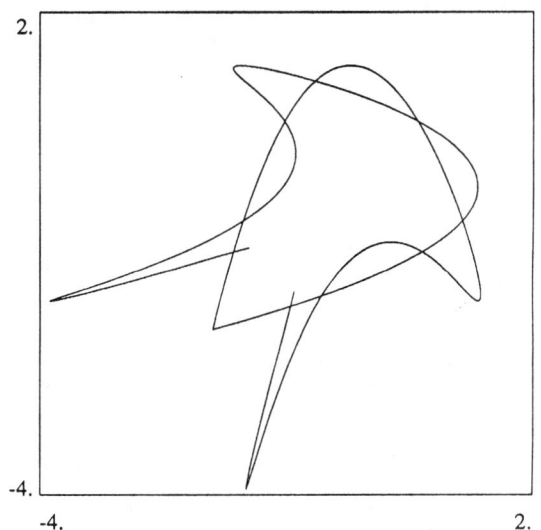

Figure 1:

Example of homoclinic intersections for the invariant manifolds of the Hénon map

The invariant manifolds depicted refer to the case $\omega/2\pi = 0.2450$. Nine homoclinic points are shown which are the result of the oscillations of the manifolds due to the area–preserving property of the Hénon map. It is also apparent that near the hyperbolic fixed point the two manifolds resemble the separatrix of an integrable system. It must be pointed out that here and in the following plots the coordinates used are the normalized Courant–Snyder coordinates.

3 GENERAL PROPERTIES OF THE HENON MAP

The Hénon map,[1] first introduced in 1969, is a quadratic area–preserving polynomial map. It can be written in the following form:

$$H_2 : \begin{pmatrix} x_1 \\ x_1' \end{pmatrix} = R(\omega) \begin{pmatrix} x_0 \\ x_0' + x_0^2 \end{pmatrix}, \qquad (x_0, x_0') \in \mathbb{R}^2 \; ; \; \omega \in [0, \pi], \qquad (5)$$

where $R(\omega)$ is a rotation matrix. Owing to the area–preserving property it is possible to define a Hamiltonian whose Poincaré section coincides with the Hénon map. This Hamiltonian has the well-known expression [2]

$$\mathcal{H}(x, x'; s) = \omega^2 \frac{(x^2 + x'^2)}{2} - \omega \frac{x^3}{3} \sum_{n=-\infty}^{+\infty} \delta(s-n). \tag{6}$$

It is easy to see that (5) coincides with the expression of the transfer map of a simple FODO cell with one sextupolar nonlinearity considered in the kick approximation [2]: the rotation represents the effect of the linear part of the cell (in the Courant-Snyder coordinates), whereas the quadratic part comes from the sextupolar nonlinearity.

One peculiar property of the map is the fact that it is symmetric with respect to a line with a slope that depends on the frequency of the map. In fact H_2 can be rewritten in the following form:

$$H_2(\vec{x}) = \mathcal{I}_1 \circ \mathcal{I}_2(\vec{x}) \tag{7}$$

where $\mathcal{I}_1$ is defined as

$$\mathcal{I}_1 : \begin{pmatrix} x_1 \\ x_1' \end{pmatrix} = \begin{pmatrix} \cos\omega & \sin\omega \\ \sin\omega & -\cos\omega \end{pmatrix} \begin{pmatrix} x_0 \\ x_0' \end{pmatrix} \tag{8}$$

and $\mathcal{I}_2$ is

$$\mathcal{I}_2 : \begin{pmatrix} x_1 \\ x_1' \end{pmatrix} = \begin{pmatrix} 1 & 0 \\ 0 & -1 \end{pmatrix} \begin{pmatrix} x_0 \\ x_0' \end{pmatrix} - \begin{pmatrix} 0 \\ x_0^2 \end{pmatrix}. \tag{9}$$

The two transformations $\mathcal{I}_1, \mathcal{I}_2$ are called *involutions*, i.e. their square is the identity transformation: $\mathcal{I}_2$ is simply a vertical shift, whereas $\mathcal{I}_1$ represents a reflection around the line of equation

$$x' = x \tan \frac{\omega}{2}. \tag{10}$$

This explains the reflection symmetry of the orbits of H_2. The factorization (7) is fundamental for the properties of the fixed points of (5). In fact it can be shown that the solutions of

$$\begin{aligned} \mathcal{I}_1(\vec{x}) &= \vec{x} \\ \mathcal{I}_2(H_2^n(\vec{x})) &= H_2^n(\vec{x}) \end{aligned} \tag{11}$$

satisfy $H_2^{2n+1}(\vec{x}) = \vec{x}$, while the roots of the systems

$$\begin{aligned} \mathcal{I}_1(\vec{x}) &= \vec{x} \\ \mathcal{I}_1(H_2^n(\vec{x})) &= H_2^n(\vec{x}) \end{aligned} \quad \text{or} \quad \begin{aligned} \mathcal{I}_2(\vec{x}) &= \vec{x} \\ \mathcal{I}_2(H_2^n(\vec{x})) &= H_2^n(\vec{x}) \end{aligned} \tag{12}$$

are cycles of even period, i.e. $H_2^{2n}(\vec{x}) = \vec{x}$. These special fixed points are called *symmetric fixed points*, as they are based on the symmetry transformations $\mathcal{I}_1, \mathcal{I}_2$.

For instance the fixed points of the Hénon map are the solutions of (11) with $n = 0$. They have coordinates

$$\vec{x}_e = \begin{pmatrix} 0 \\ 0 \end{pmatrix}, \quad \vec{x}_h = \begin{pmatrix} -2\tan\frac{\omega}{2} \\ -2\tan^2\frac{\omega}{2} \end{pmatrix}. \tag{13}$$

the origin is obviously elliptic, while the second one is hyperbolic. Moreover their stability type does not change over the whole range of frequencies. The linearized map around $\vec{x}_h$ has the following expression:

$$H_L : \begin{pmatrix} x_1 \\ x'_1 \end{pmatrix} = \begin{pmatrix} \cos\omega - 2x_h \sin\omega & -\sin\omega \\ \sin\omega + 2x_h \cos\omega & \cos\omega \end{pmatrix} \begin{pmatrix} x_0 \\ x'_0 \end{pmatrix}, \tag{14}$$

with eigenvalues

$$\lambda_1 = 1 + 2\sin^2\frac{\omega}{2} + 2\sin\frac{\omega}{2}\sqrt{1 + \sin^2\frac{\omega}{2}}, \quad \lambda_2 = \lambda_1^{-1} \tag{15}$$

and its eigenvectors are given by

$$\vec{e}_1 = \begin{pmatrix} \sin\omega \\ \cos\omega + 4\tan\frac{\omega}{2}\sin\omega - \lambda_1 \end{pmatrix}, \tag{16}$$

$$\vec{e}_2 = \begin{pmatrix} \sin\omega \\ \cos\omega + 4\tan\frac{\omega}{2}\sin\omega - \lambda_2 \end{pmatrix}. \tag{17}$$

4 A SECOND MODEL: THE CUBIC MAP

We can generalize the previous model introducing a new map with a cubic non-linearity:

$$H_3 : \begin{pmatrix} x_1 \\ x'_1 \end{pmatrix} = R(\omega) \begin{pmatrix} x_0 \\ x'_0 + \alpha_1 x_0^2 + \alpha_2 x_0^3 \end{pmatrix}, \quad \alpha_1, \alpha_2 \in \mathbb{R}\,; \omega \in [0, \pi], \tag{18}$$

using the same symbols as in the previous section. In this case H_3 depends on two parameters α_2, ω, as α_1 can be set to one by simply rescaling the coordinates. Also in this case the map (18) can be derived from a Hamiltonian,

$$\mathcal{H}(x, x'; s) = \omega^2 \frac{(x^2 + x'^2)}{2} - \omega(\frac{x^3}{3} + \frac{\alpha_2 x^4}{4}) \sum_{n=-\infty}^{+\infty} \delta(s - n), \tag{19}$$

and represents the transfer map of a FODO cell with a sextupole and an octupole. The decomposition (7) holds with a modified $\mathcal{I}_2$ given by

$$\mathcal{I}_2 : \begin{pmatrix} x_1 \\ x'_1 \end{pmatrix} = \begin{pmatrix} 1 & 0 \\ 0 & -1 \end{pmatrix} \begin{pmatrix} x_0 \\ x'_0 \end{pmatrix} - \begin{pmatrix} 0 \\ x_0^2 + \alpha_2 x_0^3 \end{pmatrix}. \tag{20}$$

In the previous discussion it was pointed out that in the quadratic map the stability of the fixed points is independent of the values of the parameter ω. In the new map the fixed points are strongly influenced by changes in ω and in α_2. The equation $H_3(\vec{x}) = \vec{x}$ has three solutions, $\vec{x}_1$ is the origin, while $\vec{x}_2$ and $\vec{x}_3$ are given by

$$\vec{x}_2 = \begin{pmatrix} \frac{-1+\sqrt{1-8\alpha_2 \tan \frac{\omega}{2}}}{2\alpha_2} \\ \tan \frac{\omega}{2} \left(\frac{-1+\sqrt{1-8\alpha_2 \tan \frac{\omega}{2}}}{2\alpha_2} \right) \end{pmatrix}, \quad \vec{x}_3 = \begin{pmatrix} \frac{-1-\sqrt{1-8\alpha_2 \tan \frac{\omega}{2}}}{2\alpha_2} \\ \tan \frac{\omega}{2} \left(\frac{-1-\sqrt{1-8\alpha_2 \tan \frac{\omega}{2}}}{2\alpha_2} \right) \end{pmatrix} \quad (21)$$

where $\vec{x}_2, \vec{x}_3$ are real if $1 - 8\alpha_2 \tan \frac{\omega}{2} \geq 0$.

The results of the stability analysis can be summarized as follows: if $\alpha_2 \leq 0$ then $\vec{x}_2, \vec{x}_3$ are both hyperbolic $\forall \omega \in [0, \pi]$. If $\alpha_2 > 0$ then $\vec{x}_2$ is hyperbolic and $\vec{x}_3$ is elliptic $\forall \omega \in [0, \hat{\omega}]$, where $\hat{\omega}(\alpha_2) = 2 \arctan \frac{1}{8\alpha_2}$.

For the analysis of this map it is useful to consider the problem of the existence of periodic solutions for $H_3^2(\vec{x})$. A possibility is to exploit the equations (12) for $n = 1$. The first one provides only fixed points of $H_3(\vec{x})$ (as can be easily checked directly) that are trivial cycles of period two. The second one has at least one real, non-trivial solution for each value of ω and α_2. This fact can be understood from the geometrical meaning of equation (12). If $\vec{x}$ is a fixed point of $\mathcal{I}_2$ this means that

$$H_3(\vec{x}) = \mathcal{I}_1(\vec{x}), \quad (22)$$

i.e. the map applied to $\vec{x}$ will simply reflect it around the symmetry line. In order for $\vec{x}$ to be a fixed point of H_3^2, $\mathcal{I}_1(\vec{x})$ must be a fixed point of $\mathcal{I}_2$. To summarize: $\vec{y}$ is a solution of (12) if it belongs to $\mathcal{I}_2(\vec{x}) = \vec{x}$ together with its symmetric companion. The curve

$$y = -\frac{x^2 + \alpha_2 x^3}{2} \quad (23)$$

is the locus of fixed points of $\mathcal{I}_2$. The presence of the cubic term gives the right symmetry to (23) in order to fulfill the conditions imposed by (12). This, incidentally, explains why the quadratic map does not have real fixed point of period two.

The actual position of these points, together with their stability, has been computed numerically for some values of ω and α_2. In all the cases the fixed points turned out to be hyperbolic.

5 RESULTS

The new method presented here allows to determine the dynamic aperture in the whole range of frequencies, simply following the invariant manifolds of the hyperbolic fixed point of the map and computing the minimal distance from the origin of these sets. It exploits the fact that the manifolds are connected via heteroclinic intersections to the unstable cycles that actually determine the border of the stability domain. Two examples are shown in Fig 2,3. The invariant manifolds of

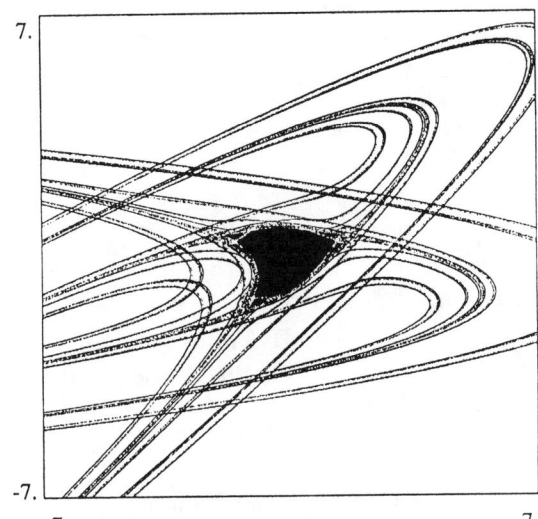

Figure 2:
Invariant manifolds of the Hénon map near the one–third resonance

The invariant manifolds for $\omega/2\pi = 0.404$ are shown together with the tracking data (black area). In this case 1.5×10^5 initial conditions are used, iterated for 5×10^2 turns. The border of the stability domain is reproduced to a very high precision by following the invariant manifolds.

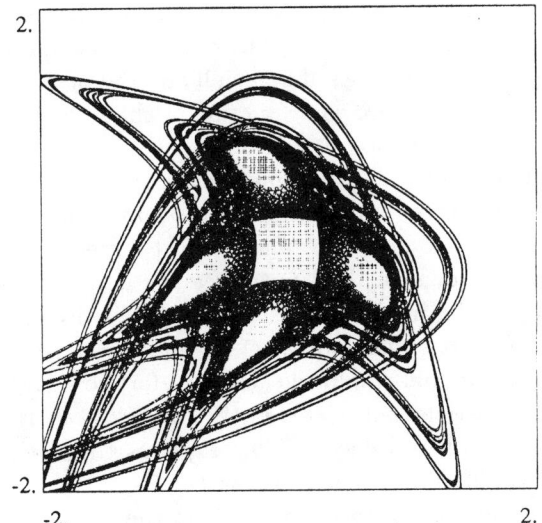

Figure 3:
Invariant manifolds of the Hénon map near the one–fourth resonance

The invariant manifolds for $\omega/2\pi = 0.255$ are shown together with the tracking data (grey area). In this case 1.5×10^5 initial conditions are used, iterated for 1×10^3 turns. Also in this case following the invariant manifolds allows to recover the complicated structure of phase space close to the border of the stability.

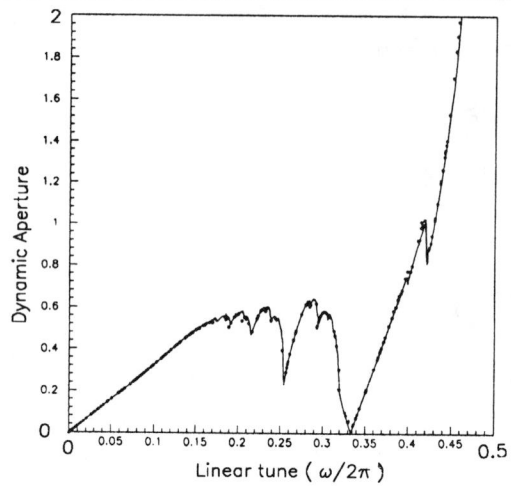

Figure 4:
Dynamic aperture for the Hénon map

The dynamic aperture is depicted for 2000 values of ω (solid line) tracked for 1×10^4 iterations. The circles superimposed represent the minimum distance from the origin of the invariant manifolds for 250 values of ω. The agreement is excellent over the whole range of frequencies.

the hyperbolic fixed point are compared with the stability region as computed by direct tracking through the map for two different values of the frequency. In both cases the agreement is excellent down to the very details, including the chains of stable islands due to the subresonances. Moreover it is apparent that this result is totally independent of the topology of the border itself, especially of the presence of islands. A comparison of the dynamic aperture as a function of the frequency between this new method and tracking is presented in Fig 4.

The drawback is clearly the need in cpu–time: a tracking run for one frequency requires roughly 2.8 s on a Cray X/MP, while this method takes approximately between 8 s and 24 s. The time needed also depends very much on the frequency: close to the regime of the first–order resonance the time necessary is naturally very short ($\omega/2\pi = 0.50$ in Fig 5A); while in the case close to the fourth-order resonance the passing to the manifolds of the fourth-order cycle leads to an intermediate regime of a slowly decreasing minimal amplitude and therefore to a large time ($\omega/2\pi = 0.255$ in Fig5B). The particles move extremely slow close to the hyperbolic fixed point especially for small frequencies. This explains the initial plateaus and the difference of their lengths in the two parts of Fig 5.

The same technique has been applied to the cubic map. From the discussion of the properties of the fixed points it should be clear that there are two different situations.

In the case $\alpha_2 < 0$ the two hyperbolic fixed points of $H_3(\vec{x})$ can be used to determine the border of the stability domain. The invariant manifolds that emanate from them are connected through heteroclinic intersections, and via these

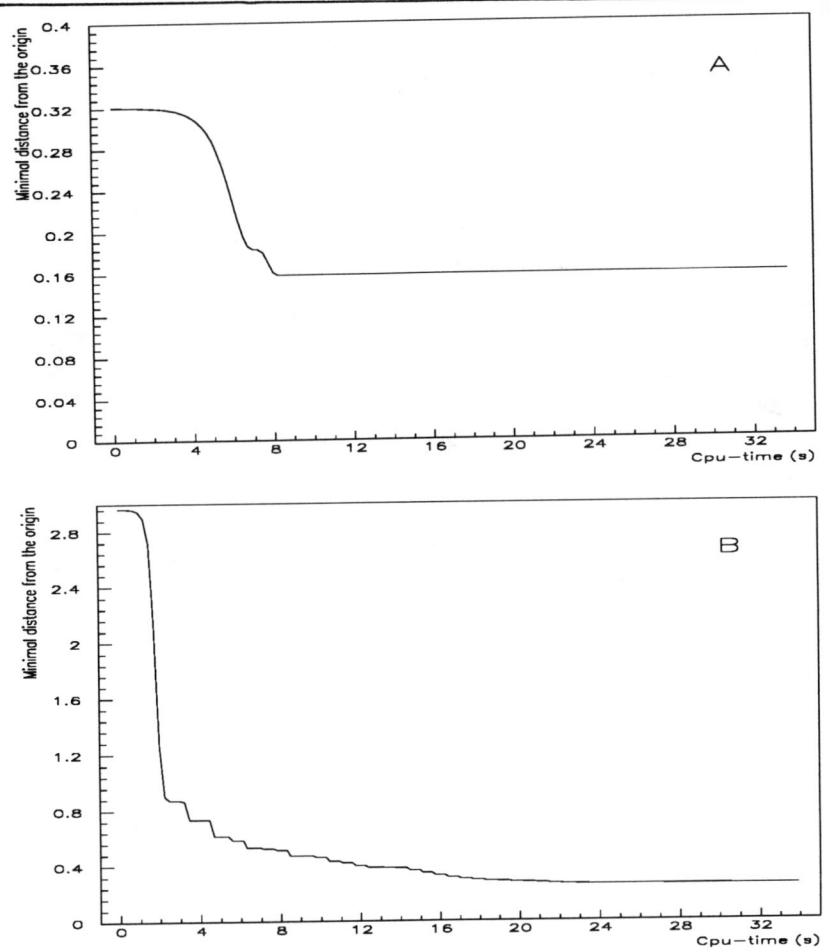

Figure 5: Plot of the minimal distance of $\mathcal{W}^u(\vec{x}_h)$ from the origin

The minimal distance of the invariant manifolds for $\omega/2\pi = 0.050$ (part A) and $\omega/2\pi = 0.255$ (part B) is shown as a function of the cpu-time of the Cray X/MP needed for the calculations.

---

intersections they reach the dynamic aperture. Once again the invariant manifolds allow to compute the stability domain. In Fig 6 are shown the results obtained for two different values of the coefficient α_2. Using the invariant manifolds it is possible to reconstruct the dynamic aperture for all values of ω.

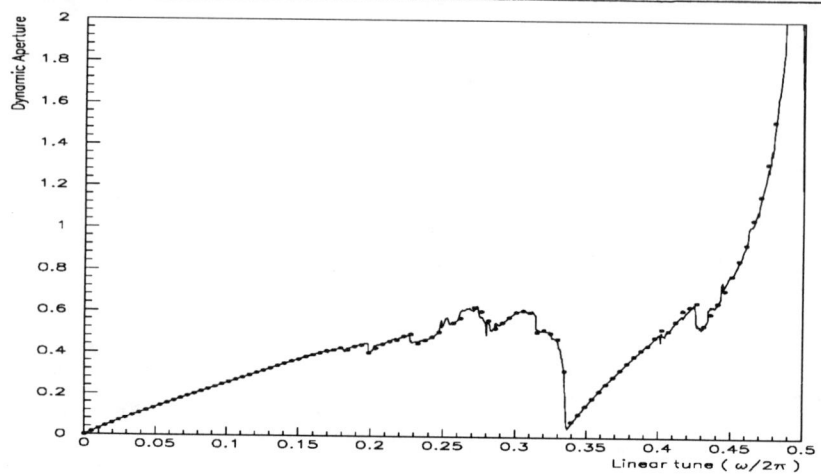

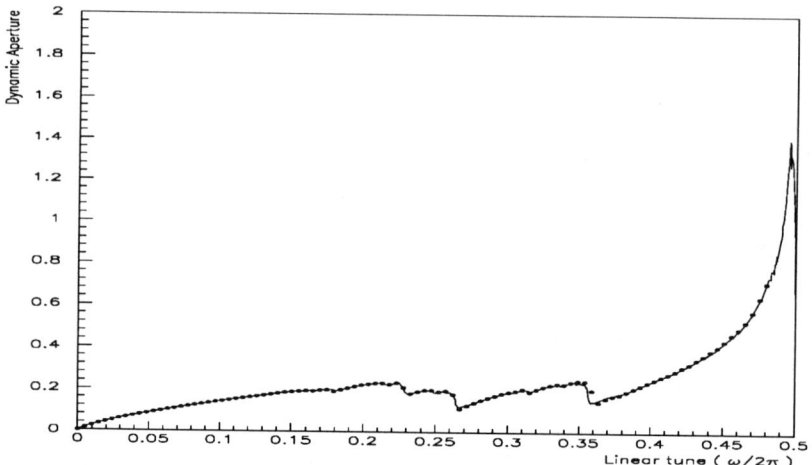

Figure 6: Dynamic aperture for the cubic map

The dynamic aperture is depicted for 2000 values of ω (solid line) tracked for 1×10^4 iterations. The circles superimposed represent the minimum distance from the origin of the invariant manifolds. The upper part refers to $\alpha_2 = -1$, while the lower one to $\alpha_2 = -10$. The agreement is excellent in both cases over the whole range of frequencies.

In the other case, when $\alpha_2 \geq 0$, the method needs some modifications. In fact the unique hyperbolic fixed point $\vec{x}_2$ disappears, together with $\vec{x}_3$, for $\omega \geq \hat{\omega}(\alpha_2)$. Moreover, even in the case $\omega < \hat{\omega}(\alpha_2)$ the phase space has a topology of a *figure eight*, due to the presence of the two elliptic fixed points, with the hyperbolic one in between. This specific situation does not exclude the existence of closed curves outside the invariant manifolds. The unstable fixed point and the related manifolds can not provide information on the stability domain of the map. One must exploit the cycles of period two. As already discussed, there is at least one such point for every value of the parameters and the numerical evidence suggests that it is always unstable. It is possible, then, to construct the invariant manifolds that emanate from this cycle trying to reconstruct the dynamic aperture. As an example the results of this method are shown in Fig 7. It refers to the case $\omega/2\pi = .21$ and $\alpha_2 = 10$. The agreement with the tracking is impressive. The same agreement has been found for different values of the parameters.

Using the results obtained for the two models studied, it is possible to conjecture an extension of the method presented in this paper to explain the dynamic aperture for a class of polynomial maps of the form

$$H_n : \begin{pmatrix} x_1 \\ x'_1 \end{pmatrix} = R(\omega) \begin{pmatrix} x_0 \\ x'_0 + P_n(x_0) \end{pmatrix}, \qquad \omega \in [0, \pi], \tag{24}$$

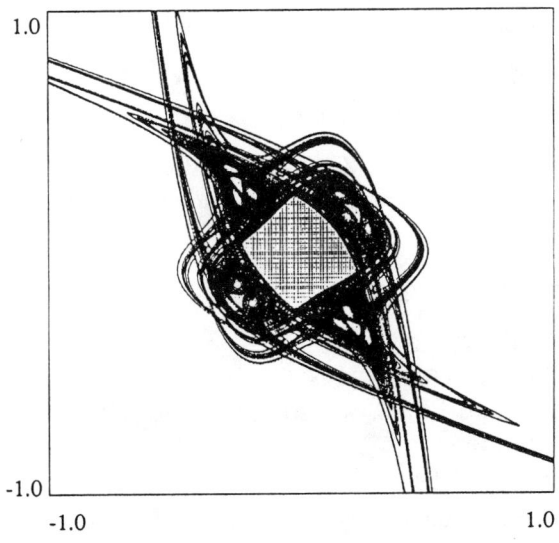

Figure 7:
Invariant manifolds of the cubic map near the one–fourth resonance

The invariant manifolds for $\omega/2\pi = 0.255$ emanating from the unstable two cycle of the cubic map are shown together with the tracking data (grey area). In this case 2×10^5 initial conditions are used, iterated for 2×10^3 turns. The value of α_2 is -10. Also in this case following the invariant manifolds allows to recover the complicated structure of phase space close to the border of the stability.

where $P_n(x_0) = \sum_{j=2}^{n} p_n x_0^n$ is a polynomial function. According to the parity of the degree n of the map, the dynamic aperture will be computed using the unstable fixed points of H_n or the hyperbolic cycles of period two. In case $n = 2\ell$ the outermost unstable fixed point has to be considered. On the other hand if $n = 2\ell + 1$ one has to distinguish two cases according to the sign of $p_{2\ell+1}$. If it is negative one has to take into account the outermost unstable fixed point of the map, otherwise the cycle of period two has to be considered.

6 CONCLUSIONS

The method outlined here allows to connect the stability region with the invariant manifolds of the hyperbolic fixed point for two different type of 1–D maps. To the knowledge of the author, it is the first time that an analytic method is capable to determine the dynamic aperture to this high precision. It must be noted, however, that due to the large amount of cpu–time needed, the method is not yet ready for practical applications.

It is well known that in two–dimensional maps the concept of dynamic aperture is not well defined, at least from a theoretical point of view, as in principle particles close to the origin can escape to infinity due to Arnold diffusion.[7] Nevertheless, in practical cases like accelerators with a time scale of 10^9 iterations, such a diffusive phenomenon is too slow to be noticeable. Experience from tracking[8] has shown that one can specify a dynamic aperture for that time scale. The method outlined here is expected to sample the same barrier of stability as the tracking.

7 ACKNOWLEDGEMENTS

I want to thank G. Turchetti, W. Scandale and F. Zimmermann for many stimulating and helpful discussions. I want also to express my gratitude to F. Schmidt for a patient reading of the manuscript.

REFERENCES

1. M. Hénon, Quart. of Appl. Math. **27**, 291 (1969).
2. A. Bazzani et al., Il Nuovo Cimento **102**, 51 (1988).
3. A. Bazzani et al., Proceedings Third EPAC Conference, Berlin (1992).
4. A. Bazzani et al., Physica D (in press).
5. M. Giovannozzi, CERN SL/ 92-23 (AP) (1992).
6. R. DeVogelaere, in Contributions to the Theory of Nonlinear Oscillations **4**, 53 (1958).
7. V.I. Arnold, Russian Math. Surveys **18**, 85 (1964).
8. F. Galluccio et al., Proceedings Third EPAC Conference, Berlin (1992).

APERTURE DETERMINATION OF RHIC92 FROM RANDOMLY GENERATED INITIAL COORDINATES[*]

G.F. Dell

Brookhaven National Laboratory, Upton, NY 11973, USA

ABSTRACT

Results obtained by tracking 100 particles for 1000 turns when initial coordinates are selected randomly, with the requirement that the total emittance be constant, are compared to results from 1000-turn and 10^6-turn runs when initial coordinates satisfy $\epsilon_x(i) = \epsilon_y(i)$ and $X'_i = Y'_i = 0$. For studies of ten distributions of magnetic field errors, the 100-particle results give apertures equivalent to those from 10^6-turn runs, have an aperture distribution of considerably less width, and yet require only one tenth the computer time.

1 INTRODUCTION

Aperture determinations are intended to probe phase space in a direction defined by the initial coordinates. In general, most investigators have used initial coordinates defined by $\epsilon_x(i) = \epsilon_y(i)$ and $X'_i = Y'_i = 0$. In the following text this set of coordinates is denoted by the expression "single-particle launch."

Tracking on lattices, such as the AGS Booster and RHIC at BNL and various SSC lattices, indicates there is repetitive transfer of emittance between the horizontal and vertical planes throughout a tracking run. This transfer depends on sextupoles used for chromaticity correction and random field errors and can be complete to either or both planes.

The relation between ϵ_x and ϵ_y changes throughout the run; Figure 1 shows normalized phase plots from a 300-turn run at $\beta^* = 2$ m when seed 9 is used to generate random field errors, and the initial coordinates are those defined for the single-particle launch. The variation of $\sqrt{\epsilon_x}$ and $\sqrt{\epsilon_y}$ is shown by the smear plot of Figure 2(a), where the solid line indicates the locus of $\sqrt{\epsilon_x + \epsilon_y} = \sqrt{\epsilon_t(0)}$. Had there been no coupling from magnetic field errors and sextupoles used for chromaticity correction, all turns would have been represented by a single point at (1.0,1.0). The time dependences of ϵ_x and ϵ_y are shown in Fig. 2(b); ϵ_x and ϵ_y are unequal most of the time, and hence the particle is not probing in the direction defined by its initial coordinates. Figures 1 and 2 show results from the first 300 turns of the worst case run for which the particle survived 10^6 turns. It is an extreme case in that the emittance transfer is nearly complete in both directions.

A more representative example is shown in Figure 3(a). There is complete emittance transfer from the horizontal to the vertical plane. Figure 3(b) gives the time dependence of ϵ_x and ϵ_y. The emittances become unequal within the

---

[*] Work performed under the auspices of the U.S. Department of Energy.

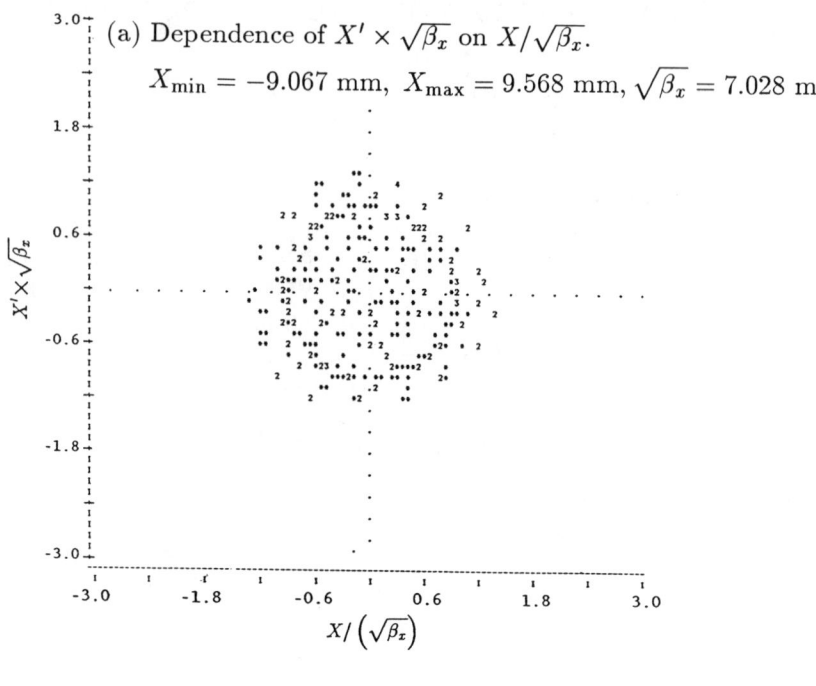

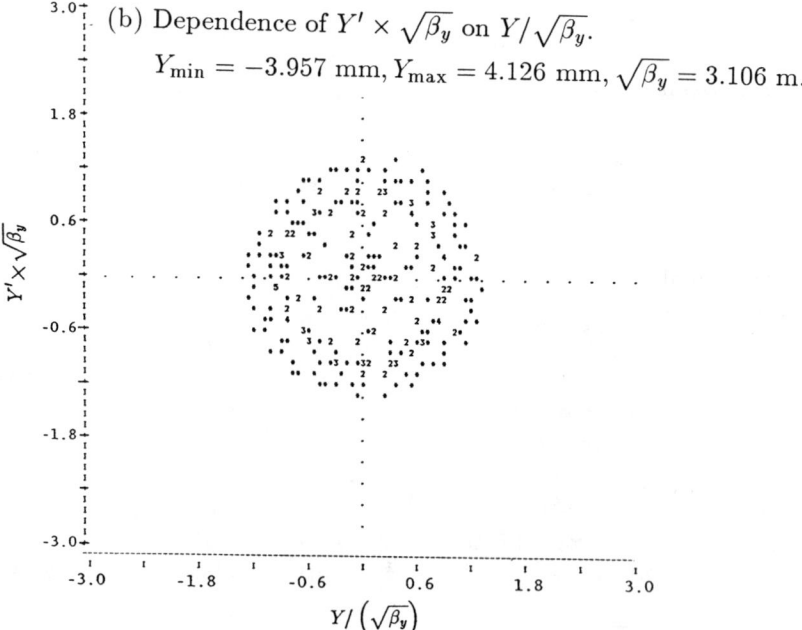

Figure 1: RHIC92, $\beta^* = 2$ m, seed 9: worst case of Figure 4(b). Phase plots from 300-turn run with $\epsilon_{x(0)} = \epsilon_{y(0)} = 0.871\ \pi$ mm mmradians. Emittance transfer is nearly 100% between the horizontal and vertical planes.

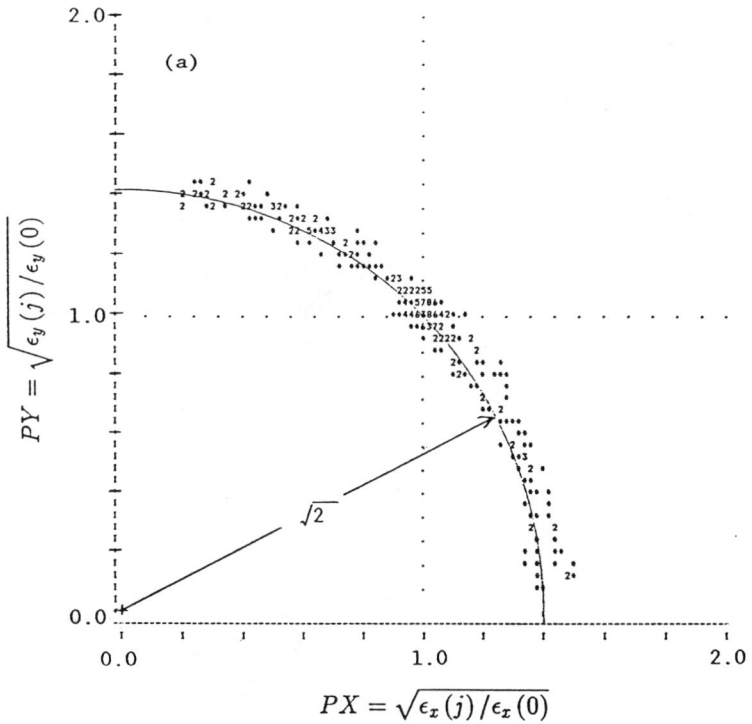

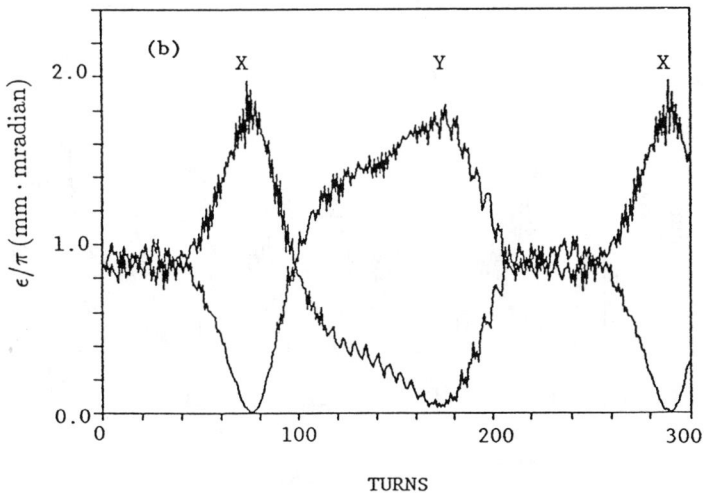

Figure 2: RHIC92, $\beta^* = 2$ m, seed 9. First 300 turns of tracking run: (a) relation between $\sqrt{\epsilon_x}$ and $\sqrt{\epsilon_y}$; (b) time dependence of ϵ_x and ϵ_y.

400 Aperture Determination of RHIC92

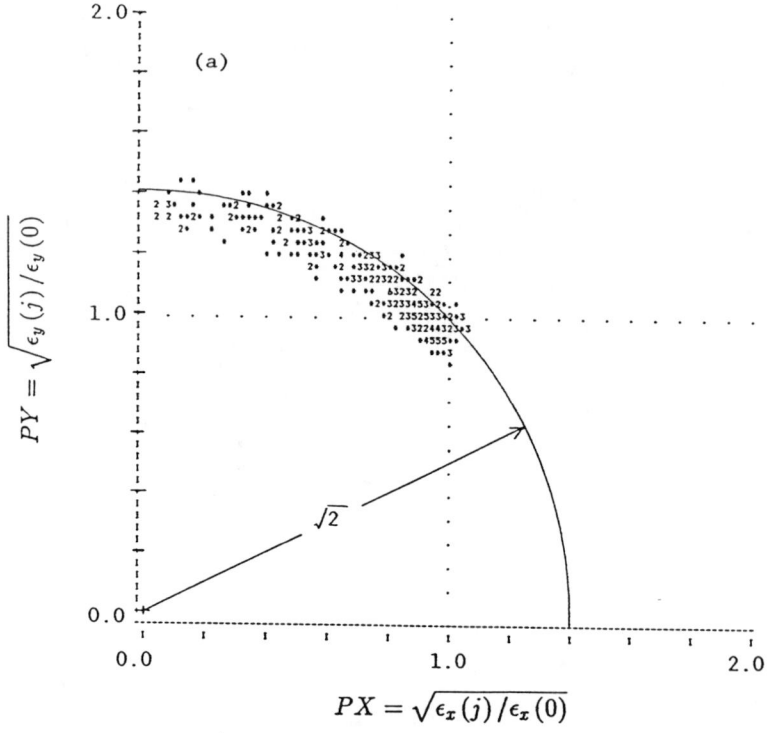

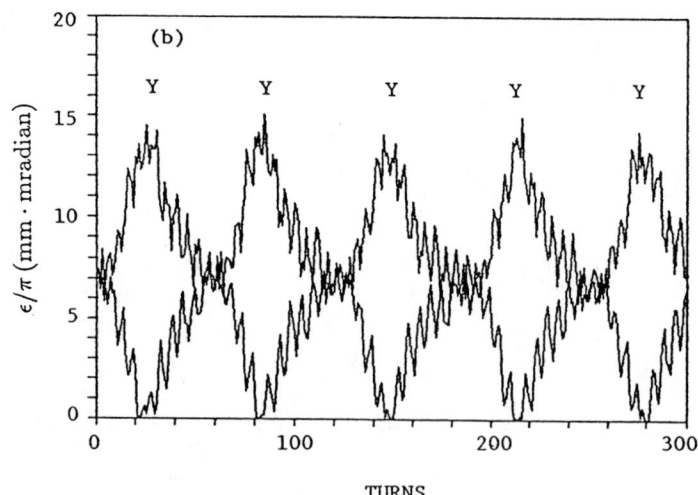

Figure 3: RHIC92, $\beta^* = 6$ m, seed 4. First 300 turns of tracking run: (a) relation between $\sqrt{\epsilon_x}$ and $\sqrt{\epsilon_y}$; (b) time dependence of ϵ_x and ϵ_y.

first 20 turns and are only approximately equal about 25% of the time. Again, the intent of probing in a prescribed direction is not realized.

In this paper a method that is patterned after the dependence shown in Figure 2(a) is used for selecting the initial coordinates of the test particles. The total emittance $\epsilon_t(0)$ can be distributed in any way that satisfies the relation $\epsilon_t(0) = \epsilon_x(i) + \epsilon_y(i)$, and also X_i' and Y_i' can be nonzero. One hundred particles having randomly selected $\epsilon_x(i)$, $\epsilon_y(i)$, X_i, X_i', Y_i' and Y_i' are tracked for 1000 turns. The term "multiparticle launch" is used to denote this method of selecting initial coordinates.

2 PROCEDURE

A. Setup

The study was done using the RHIC92(0,0) lattice when $\beta^* = 2$ m and $\beta^* = 6$ m. The tunes were $\nu_x = 28.827$ and $\nu_y = 28.823$. The chromaticity was corrected to zero, and $\Delta P/P = 0$. All dipole, quadrupole, and sextupole elements of RHIC92 were included in the lattice used for tracking. The effects of nonlinear fields were represented by thin-lens kicks located at the center of all quadrupoles and at both ends and the center of all dipoles, where they were given weights of 1/6 and 2/3, respectively. Multipole expansions of random field errors were generated from the rms errors $(\sigma a_n, \sigma b_n)$ of Herrera et al.[1] according to a Gaussian distribution that was truncated at $\pm 3\sigma$. The expansion was made with $2 \leq n \leq 16$ for dipoles and $2 \leq n \leq 10$ for quadrupoles. Aperture determinations were made for ten sets of random field errors generated by using different seeds to initialize the random number generator. No systematic errors were included. Tracking was performed on the NERSC CRAY.C at LLNL using a special version of PATRICIA.[2] The test particles were always launched at the center of Q9 quadrupole at the beginning of an inner arc. The amplitude of the test particle was checked at every element to assure it remained within the vacuum chamber.

B. Generation of Initial Coordinates

The steps used in generating the coordinates for multiparticle launching are as follows.

1. Define the initial total emittance $\epsilon_t(0)$ of the test particle.
2. Use the randomly generated number RANF to select the initial horizontal emittance:

$$\epsilon_x(i) = \epsilon_t(0) \times \text{RANF}.$$

3. Determine $\epsilon_y(i)$ from $\epsilon_t(0)$ and $\epsilon_x(i)$:

$$\epsilon_y(i) = \epsilon_t(0) - \epsilon_x(i).$$

4. Determine the initial coordinate X_i randomly:

   (a) $X_{\max} = \sqrt{(\epsilon_x(i) \times \beta_x)}$;   (b) $X_i = X_{\max}(1. - 2. \times \text{RANF})$.

5. Solve the Courant-Snyder relation for emittance to determine the initial X'_i associated with X_i:

(a) $X'_i = \left(-\alpha_x \times X_i \pm \sqrt{(\epsilon_x(i) \times \beta_x - X_i^2)}\right)/\beta_x$;

(b) select the sign of $\sqrt{(\epsilon_x(i) \times \beta_x - X_i^2)}$ randomly.

6. Repeat steps 4 and 5 when X is replaced by Y.

The random number generator was not reset prior to generating initial coordinates, and hence the one hundred sets of initial coordinates are different for each seed.

C. Tracking

One hundred particles were launched and tracked in sequence. If any particle failed, the motion was considered unstable, and the run was terminated. If no particles failed, the motion was considered stable. The total emittance $\epsilon_t(0)$ was varied until all particles survived the specified number of turns. The results are expressed in terms of an equivalent X defined as $X = \sqrt{(\epsilon_t(0) \times \beta_x/2)}$ and are thus consistent with the convention used for the single-particle launch for which $\epsilon_x(0) = \epsilon_y(0)$.

3 RESULTS

The results obtained by multiparticle tracking are compared with results obtained using the single-particle launch with tracking times of 1000 and 10^6 turns.

1. <u>Single-particle launch with 1000 turns</u>: With $\epsilon_x(0) = \epsilon_y(0)$, X_i was incremented by 0.2 mm (measured where $\beta_x = 50$ m). The aperture quoted is the smallest X_i of the ten measurements for which the test particle remained within the vacuum chamber at amplitude X_i and hit the vacuum chamber at amplitude $X_i + \Delta X$. The distributions of X_i's are shown as histograms in Figure 4(a) for $\beta^* = 2$ m and Figure 5(a) for $\beta^* = 6$ m. The number in each cell identifies which seed was used to generate the random field errors; the width of the cell indicates the amplitude at which the particle survived (left side) and failed (right side). The worst case scenario gives an aperture of $X_i = 8.3$ mm at $\beta^* = 2$ m and $X_i = 16.3$ mm at $\beta^* = 6$ m.

2. <u>Single-particle launch with 10^6 turns</u>: With $\epsilon_x(0) = \epsilon_y(0)$, X_i was decreased in increments of $\Delta X = 0.2$ mm (measured where $\beta_x = 50$ m) from a large value until the test particle first stayed within the vacuum chamber for 1 million turns. The results are shown in Figure 4(b) for $\beta^* = 2$ m and Figure 5(b) for $\beta^* = 6$ m.

3. <u>Multiparticle launch</u>: One hundred particles having initial coordinates X_i, X'_i, Y_i, Y'_i satisfying $\epsilon_t(0) = \epsilon_x(i) + \epsilon_y(i)$ were tracked for 1000 turns.

(a) One particle (10^3 turns)

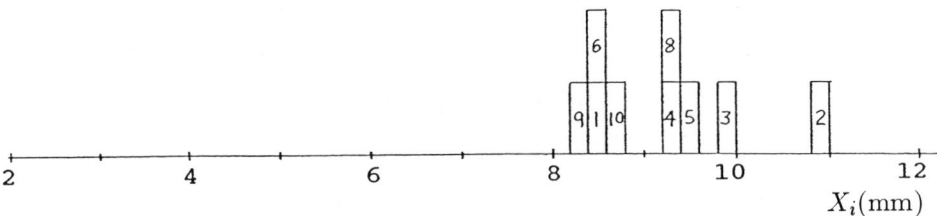

(b) One particle (10^6 turns)

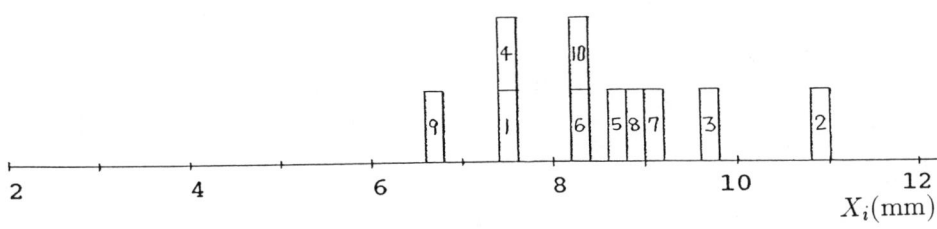

(c) Multipart (100×10^3 turns)

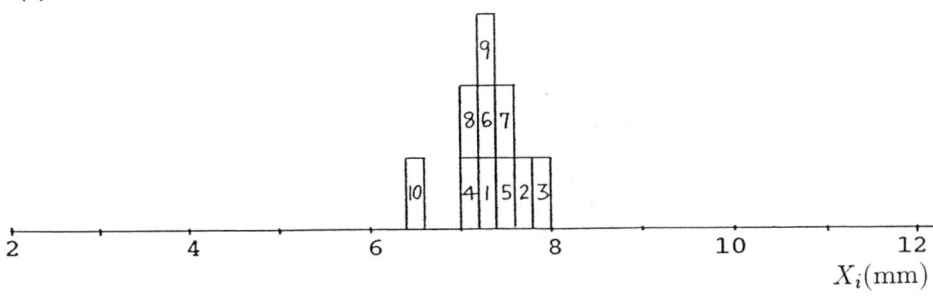

Figure 4: RHIC92, $\beta^* = 2$ m. Aperture determinations for: (a) one particle/seed tracked 1000 turns, (b) one particle/seed tracked 10^6 turns; (c) 100 particles/seed tracked 1000 turns.

404 Aperture Determination of RHIC92

(a) One particle (10^3 turns)

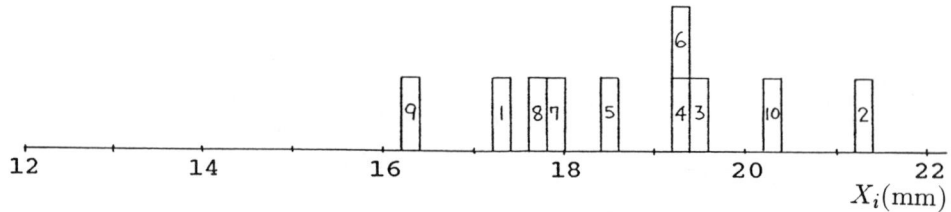

(b) One particle (10^6 turns)

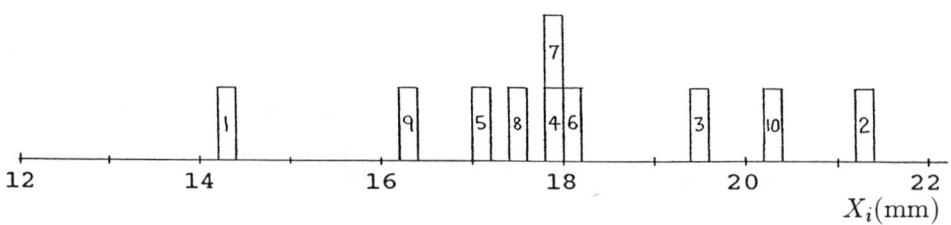

(c) Multipart (100×10^3 turns)

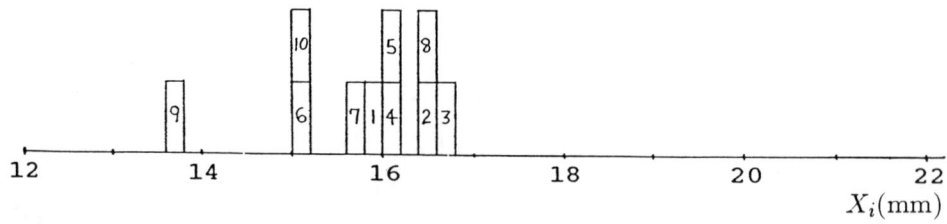

Figure 5: RHIC92, $\beta^* = 6$ m. Aperture determinations for: (a) one particle/seed tracked 1000 turns, (b) one particle/seed tracked 10^6 turns; (c) 100 particles/seed tracked 1000 turns.

The total emittance was varied in steps such that $X = \sqrt{(\epsilon_t(0) \times \beta_x/2)}$ was incremented by 0.2 mm (measured where $\beta_x = 50$ m). The total emittance was reduced until all 100 particles remained within the vacuum chamber throughout the run. Histograms of the aperture determinations are shown in Figure 4(c) for $\beta^* = 2$ m and Figure 5(c) for $\beta^* = 6$ m.

4 DISCUSSION

A. Aperture

The apertures, defined by the worst case scenario from Figures 4 and 5, are listed in Table 1. It is seen that:
1. The apertures defined by the worst case values from 10^6-turn runs and multiparticle runs are essentially equal.
2. The spread in the distribution of results from multiparticle tracking is smaller than that from the single-particle launch.

Table 1: Aperture Determinations for RHIC92
X_i (mm)$=\sqrt{(\epsilon_t(0) \times \beta_x/2)}$ where $\beta_x = 50$ m

	Standard launch TURNS		Multiparticle launch TURNS
β^*	1000	10^6	1000
2	8.3 ± 0.1	6.7 ± 0.1	6.5 ± 0.1
6	16.3 ± 0.1	14.3 ± 0.1	13.7 ± 0.1

B. Effectiveness of the Multiparticle Launch

1. Use of 100 particles tracked for 1000 turns requires approximately one tenth the computer time needed for 1-million-turn studies and thus enables more varied studies for a given computer budget.
2. Results for seed 2 in Figure 4(b) and (c) show marked differences determined with the single- and multiparticle launch. The multiparticle launch searches more trajectories and is frequently able to locate the aperture limits in fewer turns than required when using a single trajectory.
3. Not only have recent studies on the single-particle launch shown a correlation between the degree of emittance transfer and the worst and best case aperture, but also the plots of the relation between $\sqrt{\epsilon_x}$ and $\sqrt{\epsilon_y}$ throughout these runs show not much difference between plots for the first 300 turns of a one-million-turn run and those for runs made for 300 turns beyond a million turns. Examples of this are shown in Figure 6 for seed 9, the set of random field errors that gave the smallest aperture value, and in Figure 7 for seed 2, the set of random field errors that gave the largest aperture value. For these seeds it is also found, with initial amplitudes for which runs are stable for 1000 turns or more, that the degree of emittance transfer has only a weak dependence on the initial emittance.

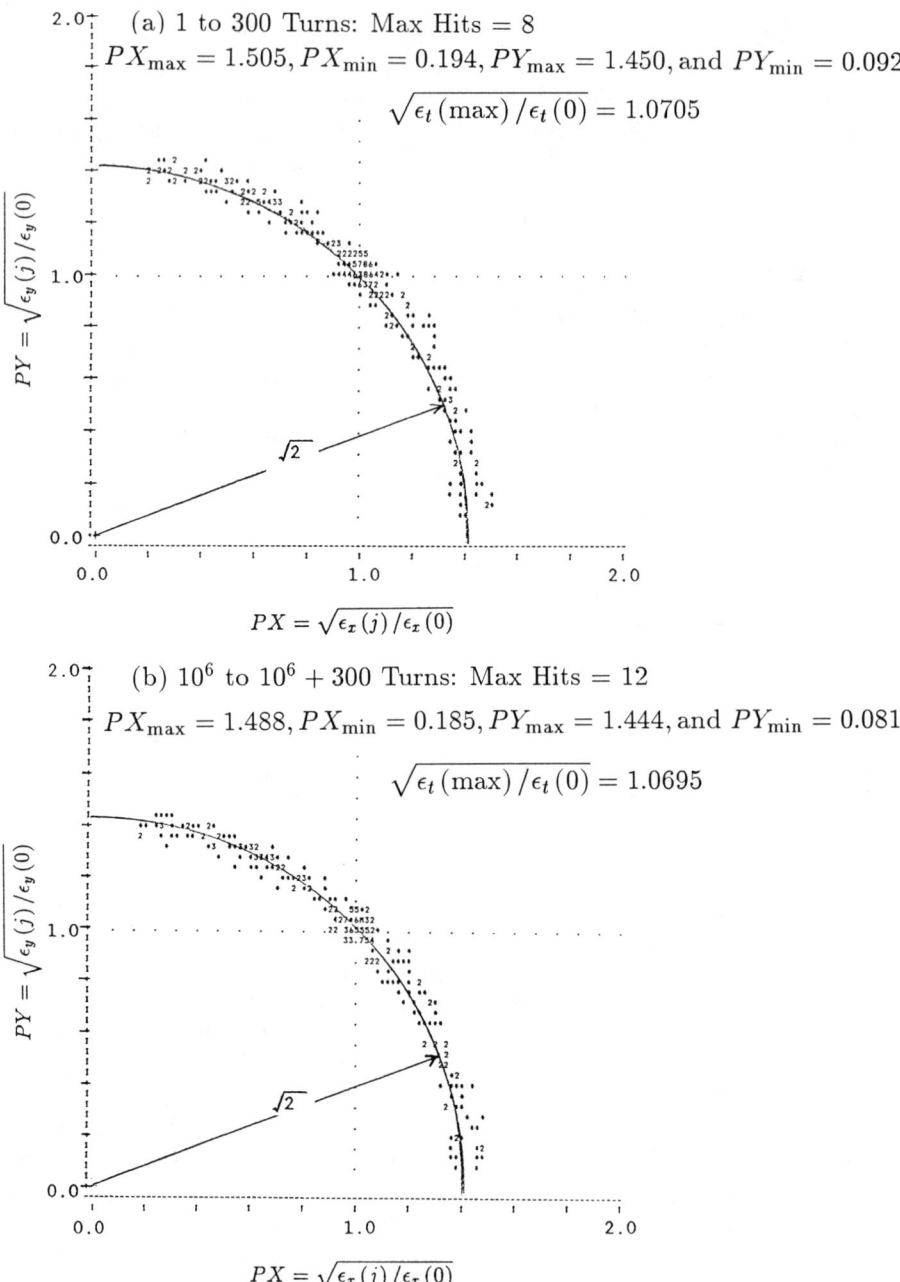

Figure 6: RHIC92, $\beta^* = 2$ m, seed 9: worst case in Fig. 3(b). Plots of ϵ_y vs ϵ_x: (a) first 300 turns of 10^6-turn run; (b) continuation from 10^6 to $10^6 + 300$ turns. Plots show little change during 10^6-turn run.

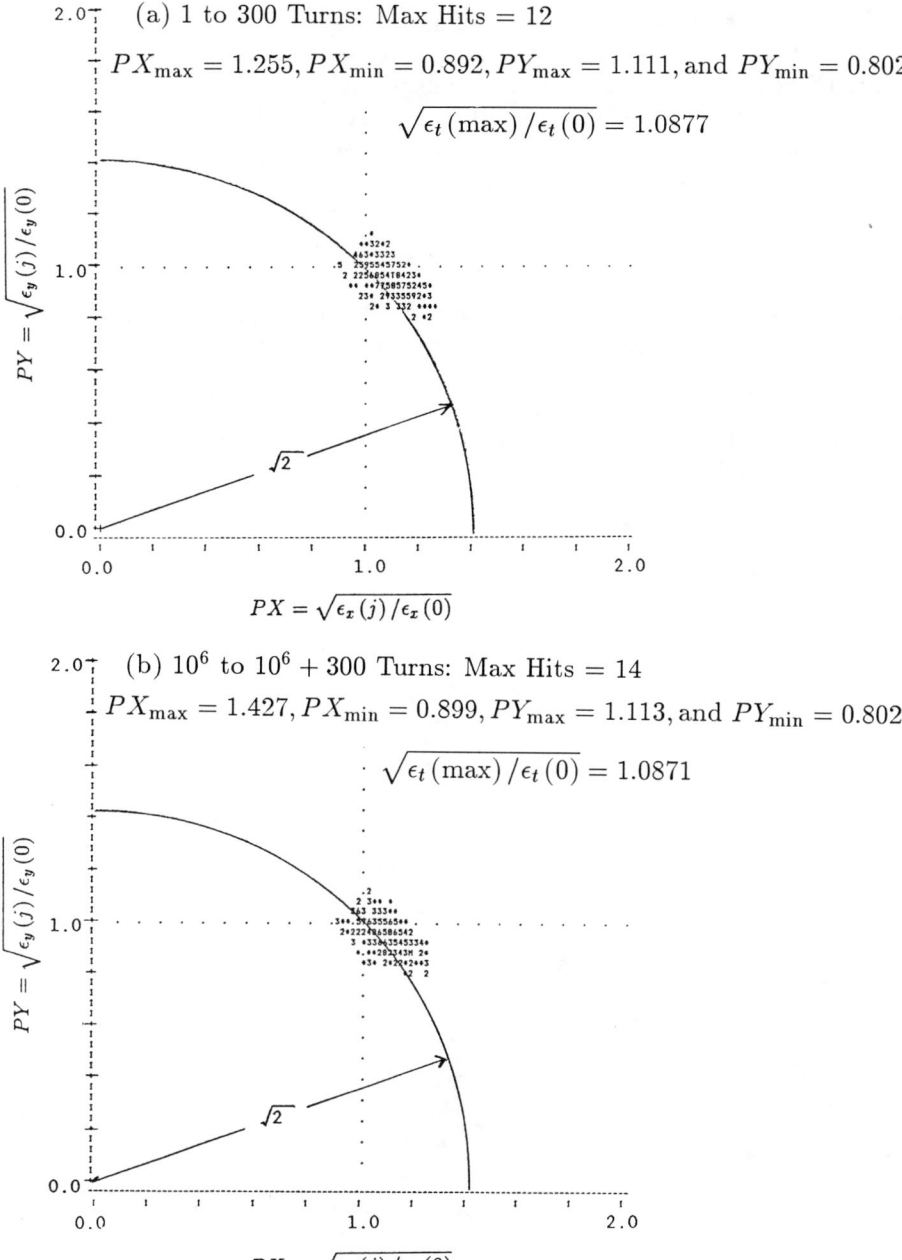

Figure 7: RHIC92, $\beta^* = 2$ m, seed 2: best case of Figure 3(b). Plots of ϵ_y vs ϵ_x: (a) first 300 turns of 10^6-turn run; (b) continuation from 10^6 to $10^6 + 300$ turns. Plots show little change during 10^6-turn run.

These results suggest that the use of many sets of randomly generated field errors may just be a method of finding a case for which large emittance transfers occur. In contrast, the multiparticle launch forces exploration of many combinations of (ϵ_x, ϵ_y) that satisfy the relation $\epsilon_x + \epsilon_y = \epsilon_t(0)$. This is thought to be the explanation for the small spread of aperture determination in Figures 4 and 5 obtained with multiparticle tracking.

5 REFERENCES

1. J. Herrera, R. Hogue, A. Prodell, P. Thompson, P. Wanderer and E. Willen, IEEE PAC, Washington, DC, March 1987, pp. 1477-1479.
2. PATRICIA, 1980 version by H. Wiedemann, SLAC, modified for multipoles by S. Kheifets, SLAC; further modifications at BNL by G.F. Dell.

THE EFFECT OF SYNCHROBETATRON COUPLING ON THE DYNAMIC APERTURE*

G. Parzen
Brookhaven National Laboratory, Upton, NY 11973, USA

1 INTRODUCTION

The effect of synchrobetatron coupling on the dynamic aperture was studied by comparing the dynamic aperture for a particle with a large fixed $\Delta p/p$, no synchrotron oscillations present, with the dynamic aperture for a particle, with a synchrotron oscillation amplitude, of the same $\Delta p/p$. The particle with the synchrotron oscillation present was found to have a smaller dynamic aperture than the particle with the fixed $\Delta p/p$. It is suggested that this reduction in dynamic aperture may be due to non-linear coupling between the longitudinal and transverse motions.

For RHIC, whose lattice and rf were used in this study, the longitudinal phase space is much larger than the transverse phase space, by a factor of several thousand, and a small amount of coupling can cause considerable growth in the transverse motion.

The effect is most pronounced at lower energies in RHIC, where larger momentum spread and transverse amplitudes are required. At $\gamma = 30$ in RHIC, with a synchrotron oscillation amplitude of $\Delta p/p = 0.005$, the dynamic aperture is reduced by about 6 mm by the presence of the synchrotron oscillations. The effect may be more important for RHIC than for other superconducting proton colliders, because of its relatively low energy and because of the importance of intrabeam scattering for heavy ions. This results in larger dynamic aperture requirements for RHIC both in transverse space and in momentum spread.

2 TRACKING RESULTS

The largest effect is seen at $\gamma = 30$ for RHIC, where the beam momentum spread can grow to $\Delta p/p = \pm 0.005$ after 10 hours. Results are shown in Fig. 1 for the dynamic aperture as a function of the size of the synchrotron oscillation amplitude, $\Delta p/p$. The dynamic aperture drops from $A_{SL} = 14.5$ mm at $\Delta p/p = 0$ to $A_{SL} = 8.5$ mm at $\Delta p/p = 0.005$. Also shown in Fig. 1 is the dynamic aperture found when the particle momentum is held fixed at some level, and no synchrotron oscillations are present. This curve is relatively flat and at, $\Delta p/p = 0.005$, there is a difference of 6 mm between the two curves.

---

* Work performed under the auspices of the U.S. Department of Energy.

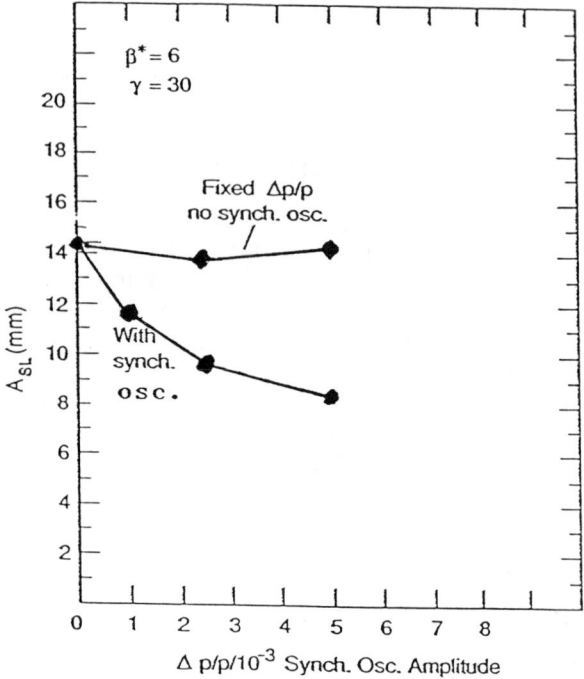

Figure 1: Comparison of results with and without synchrotron oscillations.

These tracking runs were done for 800,000 turns at $\gamma = 30$ using the RHIC91 lattice with 6 insertions having $\beta^* = 6$ m. One rf cavity is present when the synchrotron oscillations are included. All the results shown in Fig. 1 were found for the same distribution of field errors, which gave the smallest dynamic aperture of 10 distributions of errors studied at $\Delta p/p = 0$.

The effect of the synchrotron oscillations on the dynamic aperture is studied further in Figs. 2 and 3. In Fig. 2, the survival time in turns is plotted against the initial betatron amplitude x_0 for four different synchrotron oscillation amplitudes, $\Delta p/p = 0$, 0.001, 0.0025, and 0.005, for a particular distribution of field errors which gave the smallest dynamic aperture of 10 distributions studied at $\Delta p/p = 0$. The particle is started out with $p_{x0} = p_{y0} = 0$ and $\epsilon_{x0} = \epsilon_{y0}$. The lattice is the same as used for the results in Fig. 1. One sees that the survival curve moves towards lower x_0 as the synchrotron oscillation amplitude increases. Note that x_0 is the initial betatron oscillation, defined as the initial x minus the closed-orbit position for the initial $\Delta p/p$, which is computed for a fixed $\Delta p/p$.

Fig. 3 compares survival plots, for this same worst distribution of field errors, for two cases. In one case, the rf is on and the particle has a certain synchrotron oscillation amplitude $\Delta p/p$. In the second case the rf is off, and the particle momentum is fixed. At $\Delta p/p = 0$, the curves for the two cases do not differ greatly. For $\Delta p/p = 0.005$, the two cases differ considerably; the x_0 to survive 800,000 turns is $x_0 = 8.5$ mm for the case with the synchrotron oscillation amplitude $\Delta p = 0.005$, compared to $x_0 = 14.5$ mm for the case with no synchrotron oscillations and $\Delta p/p$ fixed at $\Delta p/p = -0.005$.

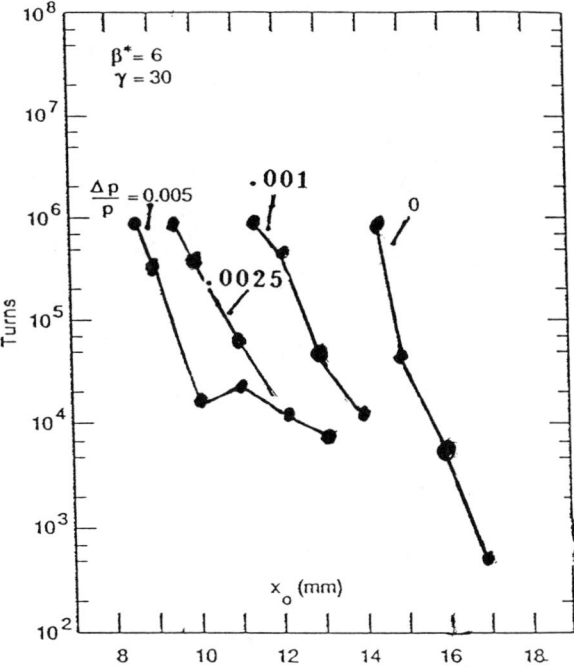

Figure 2: Comparison of results for different $\Delta p/p$.

2.1 Comments on the Tracking

It is important that the tracking be symplectic. To achieve this, the OR-BIT program was changed to allow the use of point magnets. The methods used are similar to those used in the TEAPOT[1] program, with some modifications. The RHIC91 lattice was used in the above studies. One 160-MHz rf cavity with an rf voltage of 4.5 MV was placed between Q8 and Q9 in one insertion region. The bucket height is $\Delta p/p = 6 \times 10^{-3}$ at $\gamma = 30$ and $\Delta p/p = 2 \times 10^{-3}$ at $\gamma = 100$. Random and systematic field errors were present in each magnet at the level expected for RHIC. Field error multipoles up to order 10 were included. The nominal operating tune is $\nu_x = 28.826$, $\nu_y = 28.821$. To establish the dynamic aperture, the stability limit for 800,000 turns was examined for ten different distributions of the random field errors.

An interesting question is, at what synchrotron oscillation amplitude, $\Delta p/p$, does the synchrobetatron coupling start to have an appreciable effect on the dynamic aperture, A_{SL}? The tracking studies indicate that it is the magnitude of $\Delta p/p$, rather than how close $\Delta p/p$ is to the edge of the rf bucket,

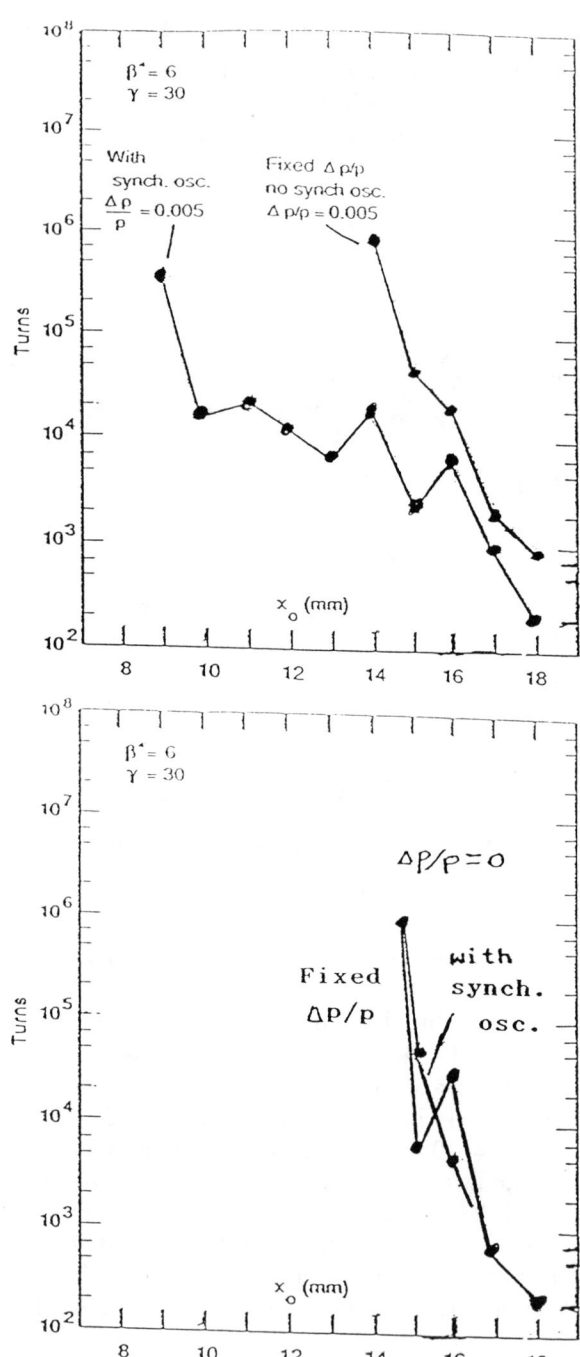

Figure 3: Comparison of results with and without synchrotron oscillations.

that is important. The tracking results suggest the following criterion for the $\Delta p/p$ where the synchrobetatron coupling becomes significant,

$$X_p \frac{\Delta p}{p} \simeq A_{SL} \tag{1}$$

where X_p is the horizontal dispersion at the location in a normal cell where A_{SL} is measured. In RHIC, $X_p \simeq 1.5$ m. Equation (1) says that the synchrobetatron coupling becomes important when the particle transverse displacement due to $\Delta p/p$ is about equal to the betatron oscillation amplitude. The loss in dynamic aperture due to the presence of synchrotron oscillations appears to be roughly given by $X_p \, \Delta p/p$.

3 REFERENCE

1. L. Schachinger and R. Talman, TEAPOT, A Thin Element Tracking Program, SSC-52 (1985).

V. CONTRIBUTIONS WORKING GROUP C

METHODS

The Modern Approach to Single-Particle Dynamics for Circular Rings

Etienne Forest
Center for Beam Physics, Lawrence Berkeley Laboratory, Berkeley, CA 94720

Leo Michelotti
Fermi National Accelerator Laboratory, Batavia, IL 60510

Alex J. Dragt
Department of Physics, University of Maryland, College Park, MD 20742

J. Scott Berg
Stanford Linear Accelerator, Stanford University, Stanford, CA 94309

With Foreword by J. Bengtsson

Contents

I Foreword
II Introduction: The Hybrid Approach Versus the Traditional Global Approach
  II.1 Motivation and Definitions
    II.1.1 Our Motivation: Benefits
    II.1.2 Definitions
  II.2 Contrasting the Hybrid and Traditional Approaches
    II.2.1 Objective of Single-Particle Dynamics in Rings
    II.2.2 Traditional Strategy from the Hybrid Point of View
    II.2.3 The Hybrid Point of View
  II.3 Conclusion
III Description of the Flow in Terms of Functional Maps
  III.1 From the Flow to the Map
    III.1.1 Differential Equations for $\mathcal{M}_{\vec{\xi}_{0 \to s}}$
    III.1.2 The Group Properties of $\mathcal{M}_{\vec{\xi}_{0 \to s}}$: The BCH Theorem
    III.1.3 Transforming $\vec{\xi}_{0 \to s}$ with $\vec{\omega}_s$: What Is the New Map Equation?
  III.2 Lie Algebras and Charts
    III.2.1 Why Are We Stuck with the Lie Operators?
    III.2.2 Charts: the Traveler Analogy
  III.3 Conclusion
IV The Hybrid Approach: Analytic Calculations
  IV.1 Computation of the Perturbed Map
    IV.1.1 A Localized Perturbation: Effect on the One-Turn Map
    IV.1.2 A Localized Perturbation: Effect on the One-Turn Floquet Map
    IV.1.3 Rules for Computation of the Total Map
  IV.2 Normalization of the Perturbed Map
    IV.2.1 The Canonical Transformation $\mathcal{B}$

IV.2.2 The Manipulation Rules of Normalization
IV.2.3 The Eigenbasis of the Linear Motion
IV.2.4 The Sextupole Example Revisited
IV.2.5 Phase Advance, Matching and Hamiltonian-Free Theory
IV.3 Conclusion
V New Results
　V.1 Symplectic Integration
　　V.1.1 Ruth's Original Idea
　　V.1.2 Beyond Ruth
　V.2 Fitted Maps: Computation and Normalization
　　V.2.1 The Concept of a Fitted Map
　　V.2.2 Fitting a Generating Function Map
　　V.2.3 Conclusion on Fitted Maps
　　V.2.4 Normalizing by Fitting Techniques
　　V.2.5 Conclusion on Fitting Techniques
　V.3 Symplectified Taylor Series Maps for Tracking
　　V.3.1 Generating Functions
　　V.3.2 Jolt Tracking
　　V.3.3 Monomial Tracking
　　V.3.4 Conclusion on Taylor Series Symplectification
　V.4 Connecting Ray and Wave Optics
　　V.4.1 The Equation of Motion for Wave Optics
　　V.4.2 Relation between Hamilton and Schrödinger
　V.5 Conclusion
Appendix　The Green's Function
References

Acknowledgments

We would like to acknowledge the contributions of John Irwin to the modern approach. His map calculus has contributed to the understanding of high-order effects in final focusing systems. Robert Warnock and Ronald Ruth have been instrumental in the development of fitted map techniques. Ruth's contribution to the rebirth of symplectic integration is well known beyond our field.

In addition, we would like to thank Dr. Swapan Chattopadhyay for his continuous intellectual support and Dr. David Sutter for his encouragement. E. Forest is grateful to Dr. Shinichi Kurokawa for financing his stay at KEK, during which writing on the global/local approach started.

Abstract

In this paper, we describe the hybrid local/global approach applied to single-particle ring dynamics. We contrast this modern approach to the traditional methods used in accelerator theory. We describe how this approach unites numerical and analytical aspects. Due to lack of space, we emphasize only the analytical (global) aspect of the modern approach. Work supported in part by DOE.

I Foreword

In the preparation of this document I was asked to give my personal views on the subject. I will limit my remarks to the modelling aspects of circular accelerators. References to earlier work will be given but there is no intention to be exhaustive. The quoted references should be viewed as illustrations of previous work and, at times, confusions. Let me make clear that in the following I will take a highly critical point of view. The purpose is not to critisize pioneering work and ideas, but rather to illuminate and clarify implicit approximations as well as limitations of traditional methods. Towards the end I will briefly try to point out why and how these methods can be improved, which is precisely the purpose of this paper.

My first reflection concerns the explicit lack of appreciation of a coherent view among many practitioners of the field, in spite of significant advances made recently in this field and existing well-established techniques already used in other branches of physics and engineering, e.g. celestial mechanics, modern nonlinear theory, wave optics, etc. It is also crucial that numerical simulations and complex analytical calculations (to high order) be done preferably by the same person having appreciation and understanding of both.

The wide range of phenomena observed in accelerators covers not only classical but also quantum electrodynamics. Examples are single-particle dynamics, synchrotron radiation (e.g. classical radiation and quantum fluctuations), coherent phenomena (e.g. wakefields), intrabeam scattering, beam-beam scattering, residual gas scattering, etc.

However, in the case of particle accelerators, it is certainly time to appreciate rare but precise statements like the following by Dragt [36]: "There are two sanguine (perhaps almost to the point of naïveté) and yet remarkably tacit assumptions made in the design and construction of accelerators and storage rings. The first, made by machine builders, is that if two machines are nearly the same, their performance (including long-term orbit stability) should be nearly the same. Without this assumption, it would be impossible to proceed since construction errors are unavoidable at some level. The second, made by accelerator theorists, is that analytical/numerical models of machine behavior, despite their approximate nature, still have relevance for real machines. Without this assumption it would be impossible to design machines. From a mathematical perspective, what is being assumed in either case is that if two symplectic maps are close (in some not yet precisely defined sense) then the behavior (including long-term behavior) of systems described by these maps should be nearly the same."

In the case of analytical calculations for single-particle dynamics, many of the studies are done with highly simplified models, often limited by the use of outdated tools (e.g. low-order perturbation theory using von Zeipel's procedure). We point out, in contrast, that there exist modern tools like Deprit's algorithm, used heavily in celestial mechanics, which could be more enlightening.

The engineering (construction and design) of accelerators includes aspects like electrical and mechanical engineering, microwave technology, control theory and stability of nonlinear systems. Examples are magnet design, mechanical alignment, feedback systems, lattice design, rf systems, phase-space cooling (e.g. stochastic cooling and electron cooling), etc. In the case of lattice design (generally based on single-particle dynamics), one should never use tracking codes as black boxes, with the checking done by comparing numerical results from different codes without a detailed knowledge of either the equations of motion (or equivalently, the Hamiltonian) or the integration method. Also, I have the impression that, in general, there is a tendency to rely on "brute force" numerical methods, overlooking the qualitative understanding that only an analytical approach can give. An analytical approach allows one to study a simplified system which includes only the dominating terms of the dynamics. The validity of this simplified system is verified by numerical simulation. I strongly believe that only by such a combined approach is it possible to reach a deeper understanding and thereby find designs with improved performance that are easier to operate and commission.

This paper is going to present what the authors and I see as the current status of the tools used for single-particle beam dynamics in the case of circular accelerators. In particular, it claims the existence of a self-consistent model that allows numerical simulations as well as analytical calculations to be performed in a coherent and effective manner. I see this as a major achievement, allowing validation from first principles rather than by application of "brute force" methods (i.e., deriving the relevant equations of motion from a locally defined $\vec{F} = d\vec{p}/dt$ and *not* Hill's equation). Examples of where this latter situation exists include incorrect or misused fringe-field models, spurious emittance growth from non-symplectic integrators, artificial distinction created between "tracking" and "mapping," and that only recently has a correct model for the "small ring" case with strong and extended bending fields been fully developed. (More precisely, how to apply explicit symplectic integration to the correct ideal Hamiltonian of a ring [47].) Let me begin by reviewing what I see as traditional methods that have been developed and applied in this field. The paper will explain how, by taking a different approach, it is possible to simplify and improve these models and calculations dramatically, with the possibility of extending them far beyond what has traditionally been feasible.

Strong focusing was originally studied in the classic paper by Courant and Snyder [5], by parametrizing the motion around a linear ring in terms of action-angle variables. Application of linear control theory (see for example [6,7]) allowed them to study the stability of this system. Since nonlinear terms are also important (in particular if sextupoles are added for linear chromaticity correction) these effects must also be included. However, for the nonlinear case there is no complete theory for stability. The elegance of linear control theory is that the question of stability, controllability, etc., amounts to computation of the eigenvalues and the rank of the map (the linear one-turn matrix for circular accelerators). In other words, the question of stability can be answered by analyzing the functional form of the map. For the general nonlinear case this is not possible. It is therefore also necessary to evaluate the effect of the map acting on particular initial conditions (tracking).

In the case of circular accelerators we are dealing with a highly complex system with an s-dependent restoring force. This system is generally described by a nonlinear, nonintegrable s-dependent Hamiltonian. The linearized Hamiltonian has solutions in the form of the pseudo-harmonic oscillator ansatz studied by Courant and Snyder. For the general nonlinear case* only the following two types of systems can be solved:
1. An integrable Hamiltonian which is a nonlinear extension of the system studied by Courant and Snyder. In this case we find N invariants and quasi-periodic motion of the system which is topologically deformable into an N-dimensional torus, e.g. perturbation theory away from resonances.
2. A Hamiltonian which includes one chain of islands (single resonance theory, pendulum $\times$ $N-1$ torus) which is exactly solvable, e.g. the longitudinal motion in the adiabatic limit and resonant extraction.
All simple solvable models belong to one or the other of these two cases.

If the effects of the nonlinearities are small, they can be viewed as distortions of the invariant tori. Analytical calculations therefore attempt to transform the initial system into one of these two solvable cases (through perturbation theory and normal form methods). This approach is clearly incomplete, since chaos is excluded. The analytical approach is motivated by the KAM theorem [1] which states that, for sufficiently small and smooth perturbations of an integrable Hamiltonian system, most of the phase space is occupied by regular orbits. In other words, if the perturbation is small enough, the first case is a reasonable approximation; however, chaos does exist, so this model is clearly incomplete. The utility of the analytical approach is based on the fact that chaos is precisely what careful accelerator design seeks to avoid. The analytical approach can therefore be a valid tool to describe the regular motion.

There are, quite generally, two different formulations of perturbation theory: time-independent and time-dependent. [In ring dynamics, the independent variable is a spatial coordinate along the ring rather then time.] In the first case one tries to force the Hamiltonian (or the one-turn map) into either of the mentioned solvable forms using canonical transformations. This leads to new (pseudo-) invariants for the perturbed system. Using the second approach one tries to find an expression for the map and study it in the limit of large times (i.e. the variation of the old invariants due to the perturbation). Necessary assumptions on the convergence of certain terms force the system into one or the other of the solvable forms. The two formalisms are mathematically equivalent when carried through analytically.

---

* It is assumed that, near the origin, the linear motion is stable with distinct eigenvalues on the unit circle. We neglect the case of the coasting beam ring where the longitudinal phase space can be brought into a Jordan normal form.

A complement to the analytical approach is to study (in parallel) an exactly symplectic numerical model:
- Find approximate numerical solutions which are expanded in the independent variable of the non-integrable Hamiltonian by using a symplectic integrator (i.e. tracking with a "kick" code). Even though this approach allows for chaos, it is still approximate in the independent variable, and the integrator might introduce additional chaos. This approach is normally also incomplete due to limited resources (man- and computer-power).

Hamiltonian perturbation theory was initially developed to allow for the correct application of perturbation theory in celestial mechanics. It has been applied, essentially without any modifications, to accelerators since the construction of the first strongly focusing synchrotrons [2,3]. There is a long list of papers about applications of time-independent perturbation theory to accelerators and I list a few for illustration [4,11-13]. More modern and powerful techniques have been developed in other fields since then, which are highly relevant to dynamics in a ring and which cannot be ignored.

The Hamiltonian can be Fourier-expanded since it is periodic for a circular accelerator. In extreme cases all but a single term are neglected (one-resonance theory) leading to the simplest possibility of case 2, i.e. neglecting the $N-1$ tori [14]. The one-resonance theory has been applied for simple correction schemes (where a single resonance dominates) and for resonant extraction. This approach is incomplete in either case since we have no (analytical) model for the dynamics with the dominating resonance compensated and, in the case of resonant extraction, coupling, islands, etc., are ignored. However, it is possible to include the $N-1$ tori [15,16] but this is normally not done because, until now, the necessary formalism was thought to be too cumbersome or simply not understood. If all terms are kept, one still Fourier-expands, due to the search of a periodic generating function. It is possible to evaluate the sum over harmonics analytically if the magnets are approximated with thin kicks. Solutions are normally found by applying the Poincaré-Von Zeipel procedure [17] (i.e. find a generating function F_2 in action-angle variables defining a canonical transformation that transforms the Hamiltonian into case 1):

$$(\phi_{1x}, J_{1x}, \phi_{1y}, J_{1y}) \stackrel{\vec{\omega}}{\mapsto} (\phi_{2x}, J_{2x}, \phi_{2y}, J_{2y}) \ . \tag{1}$$

Due to the implicit nature of the canonical transformations

$$\phi_{2x} = \frac{\partial F_2(\phi_{1x}, J_{2x}, \phi_{1y}, J_{2y})}{\partial J_{2x}}, \qquad J_{1x} = \frac{\partial F_2(\phi_{1x}, J_{2x}, \phi_{1y}, J_{2y})}{\partial \phi_{1x}},$$

$$\phi_{2y} = \frac{\partial F_2(\phi_{1x}, J_{2x}, \phi_{1y}, J_{2y})}{\partial J_{2y}}, \qquad J_{1y} = \frac{\partial F_2(\phi_{1x}, J_{2x}, \phi_{1y}, J_{2y})}{\partial \phi_{1y}},$$

$$K(J_{2x}, J_{2y}) = H(\phi_{2x}, J_{2x}, \phi_{2y}, J_{2y}) + \left. \frac{\partial F_2(\phi_{1x}, J_{2x}, \phi_{1y}, J_{2y})}{\partial s} \right|_{(\phi_{2x}, J_{2x}, \phi_{2y}, J_{2y})}$$

$$\tag{2}$$

where (J_{2x}, J_{2y}) are the new invariants, this approach is rarely applied beyond first order. Solving for (ϕ_{1x}, ϕ_{1y}) by, e.g., successive approximations gives the distortions of the invariant tori. The amplitude-dependent tune shifts are given by the new Hamiltonian

$$\nu_x = \frac{\partial K}{\partial J_{2x}}, \quad \nu_y = \frac{\partial K}{\partial J_{2y}}. \tag{3}$$

It is also rare to find explicit expressions for the s-dependent solutions

$$x(s) = \sqrt{2 J_{1x}(s) \beta_x(s)} \cos\left[\mu_x(s) + \phi_{1x}(s)\right] \tag{4}$$

for the same reason. For realistic problems it is almost always necessary to use computer algebra systems (e.g. Reduce, Macsyma, Mathematica, etc.) for the analytical calculations. However, the complexity of the calculations has also led to conceptual mistakes. For example, in Ref. [18], although this was pioneering work, by neglecting the perturbation of the phase $\vec{\phi}$, the problem of inverting the implicit transformation did not appear. However, the treatment is inconsistent since the missing terms also contribute to leading order, explaining the disagreement with tracking results [19].

The use of action-angle variables also implies that the canonical transformation of Eq. (2) is not analytic at the origin, leading to surprising (but correct) results for sextupoles, such as six-fold islands in second order [20,21]. A general solution to these problems can be obtained by the application of Deprit's algorithm, initially developed for celestial mechanics [22,23] and first applied to accelerators by Michelotti [80,24,25]. However, the non-analytical implicit canonical transformation seems to be the most practical choice for fitted map calculations [26-28].

Some readers might be familiar with Collins' distortion functions. These were initially derived in a completely intuitive way, as a generalization of closed-orbit distortions [29]. Even though a useful concept, the lack of mathematical rigor makes the treatment somewhat confusing. It is simply a study of the perturbations of the linear invariants (i.e. the distortions of the invariant tori).

In the framework of time-dependent perturbation theory (variation of constants) one attempts to compute the perturbation of the (linear) invariants. In short, the Courant and Snyder ansatz is generalized into the form of Eq. (4), i.e. the action-angle variables are nonlinear functions of the initial conditions. Straightforward application gives the same results as the time-independent perturbation theory in the case where all harmonics are kept and one sums over the harmonics [30, 19, 31].

Attempts have been made to parametrize long-term stability in terms of smear (the rms variation of the linear invariant) and amplitude-dependent tune shifts [8]. This, of course, implicitly assumes that these quantities are correlated with long-term stability. Even though this may be a reasonable assumption, such a correlation would clearly be highly lattice dependent and, even for a given lattice, sensitive to the detailed structure of the topology of phase space (resonances, islands, chaos, etc). Questions such as the right definition of smear, tolerable amounts of smear and tune shifts, etc., are therefore almost meaningless. Comparison of lattices and correction schemes based solely on these quantities is clearly inadequate. However, when studying the performance, correction schemes, etc., for a given lattice, these concepts can be useful as a quick guide for selecting cases for further studies by tracking.

Often one would like to track with a compact analytic form of the one-turn map. For this purpose the matrix formalism has been generalized to include terms of 2nd order and above in Taylor series maps [32-34]. However, when this approach is applied to study long-term stability for circular accelerators, spurious changes of emittance generally appear due to violation of the symplectic condition. If care is taken to symplectify the matrix evaluation, this "phenomenon" disappears. Pioneering work using a generating function was done by Douglas, Dragt, Forest and Ryne.* Recently, Irwin has proposed a jolt-factorization of the Taylor series using Lie methods [35, 36, 38].

Implementation of truncated power-series algebra on a computer allows for easy calculation of Taylor series to arbitrary order. This is more generally known as automatic differentiation [43, 9, 10] in computer science. This was first applied to accelerators by Berz [39]. However, it has been incorrectly advertised by Berz as differential algebra techniques [40]. This was first pointed out by Garczynski [42]. The underlying algebra in reference [40] is simply the algebra of truncated power-series with standard partial derivatives as stated in an earlier paper by Berz [39]. On the other hand, automatic differentiation is also useful for implementing normal-form methods on a computer [44]. In this case the set of relevant Lie operators (no linear terms) form a differential algebra.

There are numerous examples of incorrect treatments of fringe fields, particularly in attempts to compute the chromaticity for small rings, for example, uncritical use of the thin quadrupole model for a bend (i.e. hard-edge fringe field and edge focusing), leading to incorrect chromatic dependence. Correct pioneering work can be found [51, 52]. A correct treatment for the general ideal bend is given by Healy's modular approach [46]; recently this has been reproduced by a more general technique based on the exact solution for an ideal sector bend [47, 48]. Another example is the failure to recognize the contribution from the longitudinal field component and to neglect the impulsive change of the position in the hard-edge fringe field [49, 50]. Correct treatments can be found in refs. [64, 65].

I would now like to make a few comments about numerical integration (tracking). By partitioning the lattice into segments, each described by a map

---

* The implementation of truncated power-series algebra has permitted the calculation of generating functions to arbitrary order. To my best knowledge this was first done by Berz (see [41]) at the SSC conceptual design group in 1987 with the help of a partial inversion routine (by Lagrange inversion). Later Forest and Neri checked the complex formulae linking the Lie operators and the generating functions initially used by Dragt's group (unpublished).

using local coordinates, it is possible to allow for general multipoles with correct fringe fields and general misalignments (the 3-dimensional Euclidian group) [47, 53]. The advantage of using (arbitrary) local coordinates connected by Euclidian transformations (translations and rotations) instead of a fixed Cartesian system [54] should be clear. One should also be careful of inappropriate use of expanded equations of motion (or Hamiltonians). On one hand, approximate equations of motion are inappropriate in the case of "small rings" and, on the other hand, "exact" equations of motion are an overkill for "large rings" where approximations are justified from a dynamical point of view and permit more efficient modeling.

One should realize that Hamiltonian perturbation theory, attempting to find s-dependent solutions, is generally an overkill, making calculations for realistic cases, as we will see, unnecessarily complicated. However with careful identification, the relevant quantities (i.e. global properties of the lattice) can be computed with high precision by modern methods of perturbation theory.

In the following chapters, an outline will be given on how traditional perturbative calculations can be formulated as transformation of the map to the normal form [44, 55, 56]

$$\mathcal{M} = \mathcal{A}^{-1} \mathcal{R} \, \mathcal{A}$$
$$\mathcal{R} = \exp\left(: K :\right) \Rightarrow \begin{cases} \text{case 1)} K \text{ is a function of } \vec{J} \text{ only} \\ \text{case 2)} K \text{ has a single resonance,} \end{cases} \quad (5)$$

covering both case 1 and 2. The distortions of the invariant tori of $\mathcal{R}$ as well as all linear and nonlinear global quantities can be computed from the canonical transformation $\mathcal{A}$ and $\mathcal{R}$. [$\mathcal{R}$ moves particles on circles in case 1 (J_i=constant). The tori of $\mathcal{M}$ are given by $\mathcal{A}^{-1} J_i$.] Amplitude-dependent tune shifts and isolated resonances are described by a pseudo- Hamiltonian K (incidentally the same operator "discovered" in some of the references [58, 31]). The use of Lie operators allows for a coordinate-independent formulation (an obvious advantage known from, e.g., differential geometry, general relativity, etc.) and enforces the symplectic condition. Parallel transport is introduced, by analogy with tensor calculus, through use of similarity transformations. This allows the effect of perturbations and correctors to be computed at the same point in the lattice (making, e.g., higher-order achromats trivial [59]). Invariants are defined as quantities that commute with the pseudo-Hamiltonian. It should become clear to the reader that these calculations can be computerized by using an integrator and automatic differentiation, resulting in routine calculations of global properties (such as amplitude-dependent tune shift, nonlinear chromaticity, nonlinear dispersion, nonlinear momentum compaction, single resonance, etc.) to arbitrary order [60-63]. Some other possible approaches that can be used when the mentioned normal form breaks down (e.g. close to the dynamical aperture) will also be shown. It is the central theme of this paper to show how analytical calculations and numerical simulations can be done within the same integrated framework.

Johan Bengtsson, Chairman of Section C

II Introduction:
The Hybrid Approach Versus Traditional Global Approach

There are two complementary pictures in which dynamics can be studied. The first, "flows," treats the dynamics as a continuous process and is characterized by vector fields and differential equations; the second, "maps," does not follow an orbit continuously but samples it discretely. We think that a complete picture of circular rings cannot be achieved without merging the two pictures. On one hand, accelerators are characterized by very localized fluctuations in the flow, namely the magnets. Such a localized Hamiltonian is best described by local variables which are defined by a magnet and its immediate surrounding. On the other hand, the physical questions arising in circular rings are global in nature and are best described by global variables. Therefore a suitable description must strike a proper balance between the local (integration) and global (map analysis) aspects of circular ring theory—hence the term "hybrid."

We will discuss in this paper the theoretical foundation of the hybrid approach. We will not go into the local aspect of the hybrid approach, but we will take a glance at the "map-calculus" which allows us to bypass the flow picture entirely in the computation of global quantities. Global quantities are tunes, beta functions, equilibrium sizes, etc. Local quantities are particle positions, moments of a distribution, etc. See Ref. [53] for explanations of the local/global dichotomy. [In future work, we may replace the word "hybrid" by the word "dichotomous" which reflects the apparent paradox of a local/global approach.]

II.1 Motivation and Definitions

II.1.1 Our Motivation: Benefits

The message carried by this paper is not an abstraction held only by theorists with no contact with reality. The authors of this article span the full spectrum: from the control room to the academic universities. What we would like to communicate to the readers are certain convictions on matters pertaining to the computation and analysis of single-particle dynamics in accelerators. Why should we bother? Why should we waste time introducing a theory based on maps which is totally equivalent to standard perturbation theory? Why should we introduce concepts which generalize "symplectic integration" when most of the time it boils down to the usual "kick code"? These are reasonable questions, especially for a student confronted with our iconoclastic advice: relegate the venerable references such as "Courant-Snyder" to the history of science, where they belong, and start right away with a modern point of view.

So, what do we hope to accomplish? What are the fruits to be harvested by investing time in what we call the study of the hybrid approach? We believe that there are three benefits in adopting the modern map approach which we list from the very tangible to the more intangible:

1. A theory that is simpler, free of unnecessary complications and more direct, will allow the computations of quantities that are thought to be too hard to extract in a conventional setting. Some of the authors of this article are confronted

daily with computations that would be extremely tedious in a Hamiltonian setting but are routinely done in a map-related theory with a very low probability of mistakes. This advantage of a map-based theory is very tangible. For example, anyone who tries to compute a third-order momentum compaction or an equilibrium beam size for a realistic model immediately appreciates its power.
2. We believe there are numerous complex problems in accelerators that are not yet understood: some involving single-particle dynamics, some involving beams, and some involving the full electro-magnetic interaction of a beam with its surroundings. To attack these problems *effectively* and *realistically*, we believe it is necessary to master completely the known single-particle dynamics. One example is the beam-beam interaction: it might be necessary in B-factories to introduce better models for the ring between each beam-beam kick. It is a sad state of affairs if our lack of computational abilities prevents the production of more realistic models. The fitted techniques of Warnock are another example: they are held back if the users of such methods have a poor understanding of map techniques in a more traditional Taylor series setting.
3. Finally, there are advantages to the map methods which are not immediately seen. As pointed out earlier, our methods are computationally powerful because they are free of extraneous complications. With a simpler and deeper theory, it might be possible to uncover some connections between complex concepts that could lead to the solutions of problems for which no solutions are yet known. To illustrate this, we can use "symplectic integration" as an example: Lie methods and their application on maps have extended the original ideas of Ruth in a way essentially impossible to see from the original paper. [70-74] This is an example of how new results can emerge from a more powerful approach (see Section V).

II.1.2 Definitions

Words carry meanings and nuances that may or may not be appropriate in a scientific discussion. We feel obligated to provide a lexicon of the vocabulary used in this paper.
- *Approach*: This refers to a certain conceptualisation of the mathematical definition of a circular ring. Such conceptualisations can have a serious impact on the problem-solving ability of the scientist using them. Hence it is appropriate to question the merits and demerits of various approaches even when it is proclaimed that they are totally equivalent mathematically. [Feynman's formulation of quantum mechanics versus Tomonaga/Schwinger's is a good example of two mathematically equivalent approaches. No one contests that the conceptual and computational power of Feynman's method made it an irreplacable tool.]
- *Map*: This word has been very badly used and the authors share the blame for this. We will try to have a consistent set of notations throughout this paper.
    - *Maps*: If nothing else is said, a map is a function f from a set V to a set W.

$$V \to W, \quad x \in V \mapsto f(x) \in W. \tag{6}$$

So, the reader will have to guess by the context what V and W are. Otherwise, here are the main "maps" you will find in this paper.

— *Coordinate Transformations*: A coordinate transformation $\vec{\xi}$ (also known as a transfer map if it represents some physical object) is a map from state space [in this paper, we use this expression unconventionally to denote the total parametric dependence of the system] to phase space; typically

$$R^{2n} \times R^p \to R^{2n}, \quad \left.\begin{array}{c}\vec{x} \in R^{2n} \\ \vec{\Delta} \in R^p\end{array}\right\} \mapsto \vec{\xi}\left(\vec{x}, \vec{\Delta}\right) \in R^{2n}. \quad (7)$$

Here, R^p is a set or subset of parameters governing the system under study. For example, we may want to include magnet strengths or other control parameters in a calculation. [Of course, when we talk of inverting the map, we think of it as a map in R^{2n} with parametric dependence in R^{2p}.]

— *Functional Map*: To a coordinate transformation $\vec{\xi}$, one can associate a functional map $\mathcal{M}_{\vec{\xi}}$ which acts on functions of phase space:

$$\mathcal{M}_{\vec{\xi}}: C_\infty \to C_\infty, \ f \in C_\infty \mapsto \mathcal{M}_{\vec{\xi}} f = f \circ \vec{\xi} \in C_\infty, \ \left(f \circ \vec{\xi}\right)(\vec{x}) = f\left(\vec{\xi}(\vec{x})\right). \quad (8)$$

[The notation C_∞ refers to the set of infinitely differentiable functions. Some applications, notably fitted maps, involve functions with a finite number of derivatives.] From now on, functional maps are denoted by a curly capital letter following a notation of Dragt. A trivial corollary on the inverse map follows from Eq. (8):

$$\mathcal{M}_{\vec{\xi}}^{-1} = \mathcal{M}_{\vec{\xi}^{-1}}. \quad (9)$$

— *Functional Operators or Simply Operators*: These also transform functions into functions. The first important example is the derivative of a function ∂_k which is defined as follows:

$$\begin{array}{ccc} C_\infty & \longrightarrow & C_\infty \\ f & \to & \partial_k f \\ f(\vec{x}) & \mapsto & (\partial_k f)(\vec{x}) = \lim_{\varepsilon \to 0} \left[f(\cdots, x_k + \varepsilon, \cdots) - f(\cdots, x_k, \cdots)\right]/\varepsilon \end{array} \quad (10)$$

For example, we will see that a functional map $\mathcal{M}_{\vec{\xi}}$ associated to a symplectic transfer map $\vec{\xi}$ can often be rewritten as

$$\mathcal{M}_{\vec{\xi}} = \prod_k \exp\left(: g_k :\right). \quad (11)$$

The functional operator $: g_k :$ associated to the function g_k will be defined in Section III in terms of the derivative*. For those familiar with Dragt's "maps,"

---

* For those who can't wait— $: g_k : f = [g, f]$ where $[,]$ denotes the Poisson bracket which is itself defined in terms of the derivative i.e. $[g, f] = \sum_{k,m} \partial_k f \, J_{km} \partial_m f$.

it is clear that they transform functions into functions and therefore should not be confused with the transfer map.

– *Super Operators*: It is possible to imagine operators that transform functional operators into functional operators. The most important one is the *commutator*. For example,

$$\mathcal{C}_\infty \times \mathcal{C}_\infty \to \mathcal{C}_\infty \text{ where } \mathcal{C}_\infty \equiv \text{ set of functional maps,}$$

$$\left(\widehat{f}, \widehat{g}\right) \in \mathcal{C}_\infty \times \mathcal{C}_\infty \mapsto \left\{\widehat{f}, \widehat{g}\right\} = \widehat{f}\widehat{g} - \widehat{g}\widehat{f} \; ; \tag{12}$$

here $\widehat{f}$ or $\widehat{g}$ can be $\mathcal{M}_{\vec{\xi},\cdot} : g_k :$ or $\vec{F} \cdot \vec{\nabla}$. The symbol $\widehat{}$ is used to denote functional operators. In particular, in the case of Hamiltonian systems, there exists a very useful homomorphism between the commutator $\{,\}$ and the Poisson bracket $[,]$:

$$\begin{array}{c} \{: f :, : g :\} \\ \downarrow \\ : [f,g] : \end{array} \tag{13}$$

– *Adjoint Operators:* Of course, it is possible to construct a super operator transforming a single functional map into another one using the commutator:

$$\#\widehat{f}\#\widehat{g} = \left\{\widehat{f}, \widehat{g}\right\} . \tag{14}$$

– $\mathcal{I}$: $\mathcal{I}$ is the functional identity map.
– $\vec{I}$: $\vec{I}$ is the identity transfer map:

$$R^{2n} \to R^{2n}, \qquad \forall \vec{x} \in R^{2n} \mapsto \vec{I}(\vec{x}) = \vec{x} \; . \tag{15}$$

• *Hybrid Approach*: This is a view of circular ring dynamics where one conceptualizes the ring as a finite collection of functional maps. The maps are the results of integration through locally defined coordinates, and the analysis is based on a globally defined one-turn map. The word *"hybrid"* refers to this local/global dichotomy which is emphasized in Ref. [53].

• *Traditional Approach*: Defined within the language of the *hybrid* approach, this is a view of circular ring dynamics where one conceptualizes the ring as an infinite collection of infinitesimal functional maps. The set of these infinitesimal maps is called the flow. In the symplectic case, these infinitesimal maps are associated to the Hamiltonian operator which is in turn associated to the Hamiltonian. For this conceptualization to be fruitful, the Hamiltonian must be expressed in terms of global variables (see below).

$$\begin{array}{c} H \in \mathcal{C}_\infty \\ \downarrow \\ : -H : \\ \downarrow \\ \mathcal{M}_{s,s+ds} = \mathcal{M}_{\vec{\xi}_{s,s+ds}} = \mathcal{I} + ds : -H : \end{array} \tag{16}$$

In the *traditional* language, we would say that we are dealing with a differential equation for the coordinate transformation $\vec{\xi}_{0\to s}$:

$$\vec{x}_0 \mapsto \vec{\xi}_{0\to s}(\vec{x}_0), \qquad \frac{d\vec{\xi}_{0\to s}}{ds} = \left[\vec{\xi}_{0\to s}, H\right] \qquad (17)$$

with the initial condition $\vec{\xi}_{0\to 0} = \vec{I}$.

The reader should note the notation $\vec{\xi}_{0\to s}(\vec{x}_0)$ instead of $\vec{\xi}(\vec{x}_0; 0, s)$. The notation $\vec{\xi}_{0\to s}$ is used to emphasize that $\vec{\xi}_{0\to s}$ and $\vec{\xi}_{0\to s'}$ are two different coordinate transformations. This mathematical distinction is very important.

- *Charts*: The particular set of phase-space variables used to describe phase space at a given time (or s in our case).
- *Surface of Section*: A surface of section (or a Poincaré surface of section), is simply a single chart. One-turn maps (or Poincaré return maps) are maps from a chart back to itself.
- *Atlas*: The collection of all the charts necessary to describe the motion of a particle through the system.
- *Local Hamiltonian*: The atlas of the Hamiltonian of the ring is a collection of charts that contain discontinuties as we go along some closed curve around the ring. In this approach, these discontinuities are unavoidable if we try to connect the local Hamiltonian with the "real" 3-dimensional world (i.e. x, y, z). It is forced upon us by a description based on the details of individual magnets. The coordinate frames are attached to the magnets and their definition is motivated by purely local considerations such as the form of the magnetic field. (See Section II.2.2.)
- *Global Hamiltonian*: The atlas of the Hamiltonian of the ring is a collection of charts that correspound usually to the Frenet-Serrat system of coordinates defined by some closed reference curve around the ring. It is a description based on some global reference curve. The three basis vectors on each chart (tangent, normal and bi-normal) are continuous and differentiable in the path length of the reference curve. In this case, we can relate the Hamiltonian to the "real" world by some continuous transformation within a tube around the reference curve. (Again see Section II.2.2.)

We are now ready to give a pictorial and schematic view of the difference between the "traditional" and "hybrid" approaches.

II.2 Contrasting the Hybrid and Traditional Approaches

II.2.1 Objective of Single-Particle Dynamics in Rings

The central objective of single-particle dynamics in circular rings is to study the behavior of a particle as it goes around a ring. If we denote the coordinate transformation for one turn around the ring $\vec{\zeta}$ [the s in $\vec{\zeta}_s$ is dropped whenever convenient], then the central goal of single-particle dynamics is to investigate what happens to initial conditions within a set A called the aperture upon iteration by $\vec{\zeta}$ for a large number of turns. Mathematically, we say that we are interested in the image of A under $\vec{\zeta}$:

$$A \subset R^{2n}, \qquad A^k = \left\{ \vec{x} \in R^{2n} \mid \vec{x} = \vec{\zeta}^k(\vec{y}) \ \forall \vec{y} \in A \right\} \tag{18}$$

where $\vec{\zeta}^k = \underbrace{\vec{\zeta} \circ \vec{\zeta} \circ \ldots \circ \vec{\zeta}}_{k \text{ times}}$.

The number of turns k may be a few turns or an extremely large number depending on the purpose of the study. [Incidentally, we may be interested in a partition of phase space, i.e. to study many sets A. For example, if we are interested in the evolution of each individual particle.] In general, the transfer map $\vec{\zeta}$ must be evaluated by a combination of numerical and analytic techniques. Having established our goal, let us see how the "traditional" approach sets up its strategy for the numerical and analytical study of $\vec{\zeta}^k$.

II.2.2 Traditional Strategy from the Hybrid Point of View

We start with the Hamiltonian of the ring H. It is instructive to imagine how complex the Hamiltonian H can be:

$$H = \sum_{i=1}^{Elements} H_i, \qquad H_i = \sum_{j=1}^{Patches} h_i^j \ . \tag{19}$$

Each element has a Hamiltonian H_i which itself might be representable by a collection of patches h_i^k. These patches can be made of various regions of an element, say a magnet, such as fringe fields or constant body fields. However, a patch can also be a coordinate transformation connecting a magnet with the frame of reference of the next magnet (see [47] for details). We call this description of the ring *local* since the Hamiltonian (patches and all) depends only on the magnet and its two adjacent companions in the beam line as explained in Refs. [53, 47].

At this point, one is ready for a brute force tracking study on the ring using Eqs. (17), (18), and (19). We denote by $\vec{\xi}_{0 \to s}$ the coordinate transformation from location 0 to location s:*

$$\frac{d\vec{\xi}_{0\to s}}{ds} = \left[\vec{\xi}_{0\to s}, H\right] = \left[\vec{\xi}_{0\to s}, \sum_{i=1}^{Elements} \left\{\sum_{j=1}^{Patches} h_i^j\right\}\right]$$

$$= \sum_{i=1}^{Elements} \left\{\sum_{j=1}^{Patches} \left[\vec{\xi}_{0\to s}, h_i^j\right]\right\}$$

(20)

with $\vec{\xi}_{0\to 0} = \vec{I}$.

Integration consists in pushing particles through each patch one by one using

$$\text{if } s \text{ is in the } (i,j)\text{th patch} \quad \frac{d\vec{\xi}_{0\to s}}{ds} = \left[\vec{\xi}_{0\to s}, h_i^j\right] \quad (21)$$

— it is clear that the variable s need be only a local variable. Now that we know how complex H can be, let us go back to the general discussion of the traditional approach.

Following Figure 1, we describe the various processes. First one starts with the *known model*. This may include all sorts of effects: fringe fields, correct geometry, and other complications. This model is usually simplified into the *chosen model*. From now on, we assume that the *chosen model* is "reality." A local Hamiltonian can be written which describes correctly the chosen model. This is the Hamiltonian H of Eq. (19). This process is schematically shown as arrow 2. Tracking, that is the computation of $\vec{\xi}$, can be performed for the *chosen model*. This is true in the *traditional* as well as in the *hybrid* approach by simply using Eqs. (20) and (21).

Brute force tracking is not the only manipulation done by accelerator physicists. Often, we want to simplify the model under study by using a canonical transformation. Therefore, instead of studying the motion described by $\vec{\xi}_{0\to s}$ and its generator H_s, we define a new variable through a coordinate transformation $\vec{\omega}_s$. [Again s refers to a patch, i.e. is defined "modulo the circumference" of the ring. Therefore implicitly, all the maps are periodic in s.] The map $\vec{\Xi}_{0\to s}$ which describes the motion of the new variable can be expressed as follows:

$$\text{if the old variable } \vec{x} = \vec{\omega}_s(\vec{y}) \text{ at position } s$$
$$\Downarrow$$
$$\vec{\Xi}_{0\to s} = \vec{\omega}_s^{-1} \circ \vec{\xi}_{0\to s} \circ \vec{\omega}_0.$$

(22)

---

* The variable s is not a smooth measure of distance along some reference curve. Actually, it is easiest to define it locally. Locally, if we are inside the (i,j)th patch, then there is a smooth variable representing a length or an angle which we call $s_{(i,j)}$. So, in this paper, when we say from "0 to s" for the total Hamiltonian H, the variable s should be interpreted as an actual location within some patch.

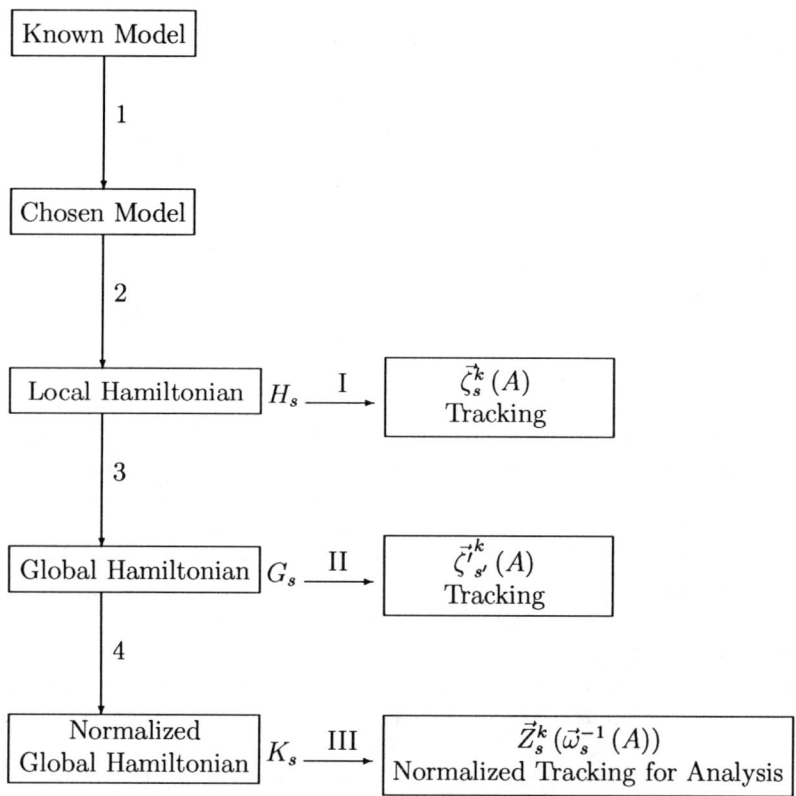

Figure 1: Schematic view of the traditional approach.

The expression $\vec{\Xi}_{0\to s} = \vec{\omega}_s^{-1} \circ \vec{\xi}_{0\to s} \circ \vec{\omega}_0$ has a very intuitive meaning: the new transfer map $\vec{\Xi}_{0\to s}$ is obtained by first transforming into the old variable at $s=0$ using $\vec{\omega}_0$, then we propagate to location s with the old map $\vec{\xi}_{0\to s}$, and finally we extract the coordinate in the new variable at location s using $\vec{\omega}_s^{-1}$.

In the traditional approach, we also ask: what is the Hamiltonian K_s which generates the coordinate transformation $\vec{\Xi}_{0\to s}$?

$$\begin{array}{ccc} & \text{3 and 4} & \\ H_s & \longrightarrow & K_s \\ & & \\ \text{I} \Downarrow & \overset{\vec{\omega}_s}{\longrightarrow} & \text{III} \Downarrow \\ & & \\ \vec{\xi}_{0\to s} & \longrightarrow & \vec{\Xi}_{0\to s} = \vec{\omega}_s^{-1} \circ \vec{\xi}_{0\to s} \circ \vec{\omega}_0 \end{array} \quad (23)$$

The variable s can still be a local variable defined in each patch. In practice, the derivation of K is practically impossible unless we have a smooth Hamiltonian with at least one continuous derivative in some variable s. This Hamiltonian is called a *global* Hamiltonian, denoted by G. Hence, in reality, in the traditional approach, Eq. (23) looks like this:

$$
\begin{array}{ccccc}
H_s & \xrightarrow{3} & G_{s'} & \xrightarrow{4} & K_{s'} \\
\text{I} \Downarrow & \longrightarrow & \text{II} \Downarrow & \xrightarrow{\vec{\omega}'_{s'}} & \text{III} \Downarrow \\
\vec{\xi}_{0\to s} & \longrightarrow & \vec{\xi}'_{0\to s'} & \longrightarrow & \vec{\Xi}'_{0\to s'} = \vec{\omega}'^{-1}_{s'} \circ \vec{\xi}'_{0\to s'} \circ \vec{\omega}'_0
\end{array}
\qquad (24)
$$

The discontinuities of the original Hamiltonian of Eq. (17) have been hidden in transformation number 3. In general, this is a transformation of the coordinates with a redefinition of the variable s into a new variable s'. Typically, s' would measure a path length around the design orbit of some ideal Hamiltonian, and the local reference system on the patch is made of vectors perpendicular to this design orbit (Frenet-Serrat). For systems of any complexity (for example misalignments...), the reference "orbit" is not an orbit of the system. The definition of the local phase space at each s', i.e. the local chart, would typically smoothly transform under translation in s'. All standard textbooks and lectures in accelerator physics advocate the Frenet-Serrat curvilinear system for the Hamiltonian $G_{s'}$. In addition it is clear that s' is a continuous variable defined all around the ring — it does make sense to talk about maps from $s = s_1$ to $s = s_2$. We will assume that all the coordinate transformations done on maps generated by $G_{s'}$ are periodic in s' with a period equal to L'—the circumference of the ring in the variable s'.

In reality, the present discussion and the textbooks are extremely misleading. Either they start with a situation where H_s is already made of simple curvilinear patches (no realistic fringe field and simplified bending magnets valid only for large rings) or they never really carry transformation 3 to its fullest. (The approximate treatment of L. Smith for wigglers is a typical example [79, 66]). Finally, it should be said that process 4, which is a normalization process on $G_{s'}$, is extremely difficult to perform on a complex Hamiltonian. If feasible at all, the Hamiltonian must be a smooth and simple function of s'. The generator of the normalization map $\vec{\omega}'_{s'}$ must be periodic in s', hence it is convenient to Fourier transform the Hamiltonian G and the said generator to facilitate the search for a periodic solution. This is done even in the presence of extremely localized perturbations. For example, magnets are accurately represented by periodic Dirac delta function $\delta_p(s' - a)$ in the variable s' which one must then Fourier transform to handle the discontinuities. This is a problem which the hybrid approach totally avoids. In conclusion, it is fair to say that normalization procedures (i.e. arrow 4) are not done self-consistently on a complex system when the s-dependent theory is used.

What motivates our new approach is the realization that in complex systems the processes 1 and 4 impose contradictory demands on the mathematical description of the Hamiltonian. Process 1 can be carried out exactly using a reference frame which is tailored to the geometry of each individual magnet. Process 4 requires a smooth definition of the reference frames under changes in s'. Hence, performing both calculations self-consistently is not possible without compromising in process 1 or/and 4. The hybrid approach resolves this problem and the price one pays, as we will see, is not significant in accelerator physics.

II.2.3 The Hybrid Point of View

In the hybrid approach we do not try to perform a transformation from local H to smooth G. We seek a method that allows us to bypass G and K. We want to avoid completely the shaded area on Figure 2. This grey area represents the unnecessary steps usually taken by the traditional approach, as previously explained. Following Figure 2, we compute an approximate* one-turn map $\underline{\vec{\zeta}}_s$. For example, using the tools of automatic differentiation (see for example [9, 43]), it is straightforward to compute the map $\underline{\vec{\zeta}}_s$ as a Taylor series in the state-space variables $(\vec{x}, \vec{\Delta})$ as defined by Eq. (7):

$$\underline{\vec{\zeta}}_s = \underline{\vec{\zeta}}_s^{[1]} + \underline{\vec{\zeta}}_s^{[2]} + \cdots + \underline{\vec{\zeta}}_s^{[No]} + \cdots O\left(\left|(\vec{x}, \vec{\Delta})\right|^{No+1}\right) \tag{25}$$

In Eq. (25), $\underline{\vec{\zeta}}_s^{[k]}$ is a kth-degree homogeneous polynomial in the state variable $(\vec{x}, \vec{\Delta})$. For our purpose we assume that it is expressed around the fixed point of the real map $\vec{\zeta}_s$.

In Figure 2, we indicate that, from the tracking loop of a code equipped with automatic differentiation, we can extract the approximate map $\underline{\vec{\zeta}}_s$ which we normalize [a process that depends on the purpose of the study] into the approximate map $\underline{\vec{Z}}_s$ using $\vec{\omega}_s$.

At this point the following remarks are important:

Features of the Hybrid Approach

1. The approximate one-turn map $\underline{\vec{\zeta}}_s$ is easily computed with automatic differentiation to any degree of accuracy.

---

* The approximate maps extracted using automatic differentiation are underlined. The reader will notice that we do not underline the canonical transformation $\vec{\omega}_s$, because, if applied to the tracking data as in Figure 2, the resulting treatment is still as exact as the underlying tracking through the patches. On the other hand, the approximate transfer maps can only be part of an approximate treatment.

436 Single-Particle Dynamics for Circular Rings

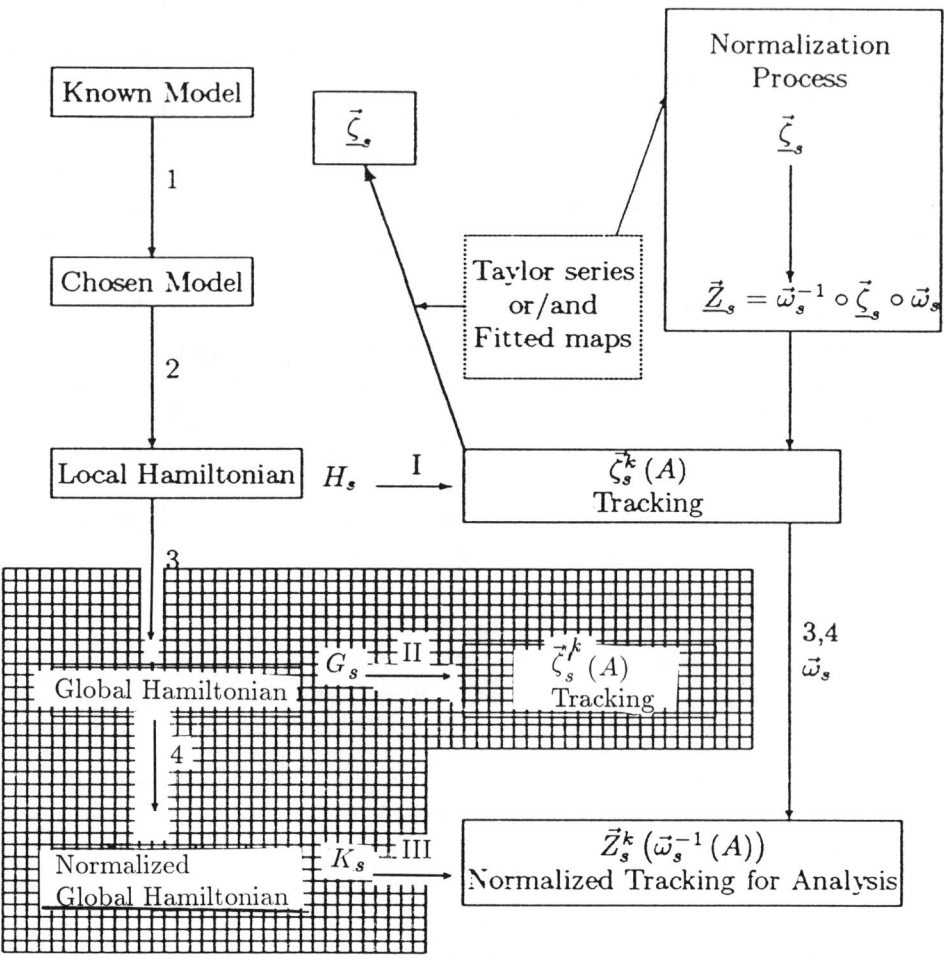

Figure 2: Schematic view of the new hybrid approach with a shaded traditional approach.

2. Using automatic differentiation, we can easily compute the map from s to s': $\vec{\xi}_{s \to s'}$.
3. Hence we can compute the approximate one-turn map at s' by similarity transformation

$$\vec{\zeta}_{s'} = \vec{\xi}_{s \to s'} \circ \vec{\zeta}_s \circ \vec{\xi}_{s \to s'}^{-1} \qquad (26)$$

and therefore we can compute its normalizing transformation $\vec{\omega}_{s'}$.
4. If we imagine perturbing the ring by some parameter $\vec{\Delta}$, then everything we said still applies because we can include the parameters in the definition of the map as indicated by Eq. (7).

5. If we imagine perturbing the ring by a potential depending on some parameter $\vec{\Delta}$ and normalizing the ring to order $\left|\vec{\Delta}\right|^k$ analytically (i.e. without the help of a parameter-dependent Taylor series map) then the following is true:
(a) One needs to compute $\vec{\omega}_s$ with a parameter-free Taylor series.
(b) One needs to compute the effect of $\vec{\omega}_s$ on the maps $\vec{\xi}_{s \to s'}$:

$$\vec{\underline{Z}}_{s \to s'} = \vec{\omega}_{s'}^{-1} \circ \vec{\underline{\xi}}_{s \to s'} \circ \vec{\omega}_s. \tag{27}$$

(c) The actual analytical normalization of the perturbed ring will be done by a calculus containing up to kth-order diagrams similar to the Feynman diagrams of quantum mechanics. These diagrams will be functions of k maps of the type $\vec{\underline{Z}}_{s \to s'}$ and k transformations of the type $\vec{\omega}_{s'}$.
6. If we normalize the approximate one-turn map at a given location s, we do not need to know the functional form of $\vec{\omega}_s$ at another value s' nor do we know $\vec{\zeta}_s$ at some other s'.

Of all the features we listed only one aspect is a potential drawback from the traditional approach, namely item 6. The fact that we do not need to know the functional form of the map in the variable s is certainly a plus. But what about the converse? Is it possible to do an s-dependent calculation with maps? The answer is that we can still do an analytical calculation if we first compute the map keeping the s-dependence. There are some technical reasons, in highly smooth systems, where it is wiser to go directly to the Hamiltonian if one needs to do a high-order calculation. [For example, it is customary to do high-order calculations in celestial mechanics. However if an explicitly Hamiltonian map has to be produced first, it is highly inefficient due to the complexity of the BCH formula. See Section III.] But, for all practical accelerator calculations, we must live with our patchy Hamiltonian. Therefore it is fair and accurate to say that we do not lose anything by using maps. In summary, the reasons are:
• In practice, it is not possible to normalize the full local Hamiltonian H correctly.
• In practice (both theory and experiment), we are interested only in a finite number of surfaces of section. At most, the number of maps equals the number of integration steps in our tracking code. From an experimental point of view, we are also limited to a finite number of observation points (e.g. limited number of beam position monitors, limited number of knobs, etc.).
• Even when we derive an analytical result for a simple system, it is easy to turn sums over the perturbative diagrams into integrals and regain the Hamiltonian results. The reverse is very tricky.

II.3 Conclusion

We conclude this chapter by re-iterating what was said in Sections II.1 and II.2. The goals of the new conceptualization which we call the hybrid approach are:

- Simplify the theory.
- Do not compromise on account of the complexity of the Hamiltonian.

The reasons why such goals are attainable are:
- In a practical setting, the number of surfaces of section which are looked at is always finite.
- Approximate maps which enter in the usual perturbation theory can be extracted automatically using automatic differentiation.
- Analytical calculations can be performed using the "Hamiltonian-free" theory of Refs. [55, 56].

III Description of the Flow in Terms of Functional Maps, or How Do Lie Algebras Sneak into Physics

III.1 From the Flow to the Map

In Sections II.2.2 and II.2.3, we have have seen that the traditional approach and the new hybrid approach have a common starting point, namely the local Hamiltonian H. This was clearly displayed in Figure 2. In this section, we exploit the common trunk of the two approaches. Our goal is to display the various properties of the functional map. We will also derive an explicitly Hamiltonian equation for the transformed map $\vec{\Xi}_{0\to s}$ obtained from $\vec{\xi}_{0\to s}$ using the canonical transformation $\vec{\omega}_s$. Essentially it is a discussion of Hamiltonian perturbation theory in the language of maps.

Some readers will say: "Fine, but why don't you give examples to give this rephrased *traditional approach* a more palatable taste?" There is an easy answer to this question: if it is true that the *hybrid approach* is more powerful, then it is hopeless to try to connect the two methods in practice. They can be connected only in the abstract. (Remember Schwinger versus Feynman?)

In Section IV, we will get to the powerful computational tools; in the meantime, bear with us through this section.

III.1.1 Differential Equation for $\mathcal{M}_{\vec{\xi}_{0\to s}}$

The reader will recall Eq. (16), which is the hybrid approach equivalent of Eq. (17). It is clear that we need to define the operator $:-H:$ which is associated to our Hamiltonian H. For more generality, let us start with a set of first-order differential equations which need not be Hamiltonian:

$$\vec{x}_0 \mapsto \vec{\xi}_{0\to s}(\vec{x}_0), \qquad \frac{d\vec{\xi}_{0\to s}}{ds} = \vec{F}_s\left(\vec{\xi}_{0\to s}\right) \qquad (28)$$

with the initial condition $\vec{\xi}_{0\to 0} = \vec{I}$. Our goal is to derive a first-order differential equation for the functional map $\mathcal{M}_{\vec{\xi}_{0\to s}}$ associated to the transfer map $\vec{\xi}_{0\to s}$ as defined by Eq. (8). We start by applying $\mathcal{M}_{\vec{\xi}_{0\to s}}$ to an arbitrary function f

using Eq. (8) (repeated indices are summed over):

$$\frac{d}{ds}\left(\mathcal{M}_{\vec{\xi}_{0\to s}}f\right) \quad \overset{\text{According to Eq. (8)}}{=} \quad \frac{d}{ds}\left(f\circ \vec{\xi}_{0\to s}\right)$$

$$\overset{\text{Chain rule}}{=} \quad \left(\partial_k f \circ \vec{\xi}_{0\to s}\right)\frac{d\vec{\xi}_{k;0\to s}}{ds}$$

$$\overset{\text{According to Eq. (28)}}{=} \quad \left(\partial_k f \circ \vec{\xi}_{0\to s}\right) F_k\circ\vec{\xi}_{0\to s}$$

$$\overset{\text{Definition of }\circ}{=} \quad (F_k\partial_k f)\circ \vec{\xi}_{0\to s}$$

$$\overset{\text{According to Eq. (8)}}{=} \quad \mathcal{M}_{\vec{\xi}_{0\to s}} F_k\partial_k f$$

$$\forall f \Downarrow$$

$$\frac{d}{ds}\mathcal{M}_{\vec{\xi}_{0\to s}} = \mathcal{M}_{\vec{\xi}_{0\to s}} \vec{F}\cdot\vec{\nabla} \tag{29}$$

The differential equation for the map $\mathcal{M}_{\vec{\xi}_{0\to s}}$ is similar to Schrödinger's equation for the unitary map in quantum mechanics. In the case of a Hamiltonian system, we can exploit the special nature of the vector function $\vec{F}$. It is given by the formula $\vec{F} = J\vec{\nabla}H$ where J is the symplectic form of the Poisson bracket. We can substitute this in Eq. (29):

$$\frac{d}{ds}\mathcal{M}_{\vec{\xi}_{0\to s}} \quad = \quad \mathcal{M}_{\vec{\xi}_{0\to s}} J\vec{\nabla}H\cdot\vec{\nabla}$$
$$\text{According to footnote after Eq. (11)} \tag{30}$$
$$= \quad \mathcal{M}_{\vec{\xi}_{0\to s}} : -H : .$$

The operator $: -H :$ is a Poisson bracket waiting to act on something and it has several notations. For example, as pointed out after Eq. (11), Dragt uses $: -H :$ while others use L_H or even $[-H,\ ]$. It is clearly a functional operator because it is made with the derivative.

III.1.2 The Group Properties of $\mathcal{M}_{\vec{\xi}_{0\to s}}$: The Baker-Campbell-Hausdorff (BCH) Theorem

Now, we would like to investigate the properties of $\mathcal{M}_{\vec{\xi}_{0\to s}}$ under the assumption that it is generated by Eq. (29). If the vector field $\vec{F}\cdot\vec{\nabla}$ is s-independent, then a formal solution is immediate:

$$\mathcal{M}_{\vec{\xi}_{0\to s}} = \exp\left(s\vec{F}\cdot\vec{\nabla}\right) \tag{31}$$

but Eq. (31) is not true if the operator $\vec{F}_s \cdot \vec{\nabla}$ depends on the parameter s. In that case, we want to prove that the functional map $\mathcal{M}_{\vec{\xi}_0 \to s}$ is the exponential of a functional map $\widehat{H}$ if it is near the identity and, second, that $\widehat{H}$ belongs to the Lie algebra under commutation generated by the set $\left\{ \vec{F}_s \cdot \vec{\nabla} \mid s \in \text{Ring} \right\}$ and, as a corollary, that it is a vector field $\vec{H} \cdot \vec{\nabla}$. This is done in three steps:
1. Get a formula for $\widehat{H}$ to justify its "asymptotic" existence for maps near the identity $\mathcal{I}$.
2. Prove that the commutator of two vector fields is a vector field.
3. Express $\widehat{H}$ only in terms of *adjoint operators* from which it is explicit that $\widehat{H}$ is made of commutators (Lie algebra).

Proof of Property 1

We need to compute $\mathcal{M}_{\vec{\xi}_0 \to s}$ as a single exponent operator:

$$\mathcal{M}_{\vec{\xi}_0 \to s} = \exp\left(\widehat{H}\right),$$
$$\widehat{H} = \log\left(\mathcal{I} + \left(\mathcal{M}_{\vec{\xi}_0 \to s} - \mathcal{I}\right)\right) = \sum_{n=1}^{\infty} (-1)^{n+1} \left(\mathcal{M}_{\vec{\xi}_0 \to s} - \mathcal{I}\right)^n. \quad (32)$$

The convergence of Eq. (32) is a tricky issue. In general, the series is asymptotic and, if a few terms are kept, it represents very well a map near the identity. This is due to the infinite dimension of the Lie algebra generated by $\left\{ \vec{F}_s \cdot \vec{\nabla} \mid s \in \text{Ring} \right\}$. In the linear case, the Lie algebra is finite, and therefore we have convergence within a certain domain.

Proof of Property 2

Before computing $\widehat{H}$, we now show that the commutator of two vector fields is a vector field. This is the essential motivation for computing the single exponent representation of a functional map:

$$\forall f \in C_\infty \to \left\{ \vec{F} \cdot \vec{\nabla}, \vec{G} \cdot \vec{\nabla} \right\} f = \vec{F} \cdot \vec{\nabla} \left(\vec{G} \cdot \vec{\nabla} f \right) - \vec{G} \cdot \vec{\nabla} \left(\vec{F} \cdot \vec{\nabla} \right) f$$

but $\vec{F} \cdot \vec{\nabla} \left(\vec{G} \cdot \vec{\nabla} f \right) = F_a \partial_a (G_b \partial_b f) = F_a (\partial_a G_b) \partial_b f + F_a (G_b \partial_a \partial_b f)$

Exchanging F and G
$$\left\{ \vec{F} \cdot \vec{\nabla}, \vec{G} \cdot \vec{\nabla} \right\} = (F_a (\partial_a G_b) - G_a (\partial_a F_b)) \partial_b = \vec{H} \cdot \vec{\nabla}$$
(33)

where $\vec{H} = \left\langle \vec{F}, \vec{G} \right\rangle \equiv F_a \left(\partial_a \vec{G} \right) - G_a \left(\partial_a \vec{F} \right)$.

From (33), we see that the commutator of two vector fields is a vector field. Furthermore, we can define a Lie bracket on the vector functions $\vec{F}$ and $\vec{G}$, which we denote by $\langle, \rangle$. In the case of a Hamiltonian vector field, there exist two functions f and g, such that

$$\vec{F} \cdot \vec{\nabla} = J \vec{\nabla} f \cdot \vec{\nabla} =: -f: \quad \text{and} \quad \vec{G} \cdot \vec{\nabla} = J \vec{\nabla} g \cdot \vec{\nabla} =: -g: \quad (34)$$

which leads us to the following homomorphisms as in Eq. (13):

$$\begin{array}{c} \{\vec{F}\cdot\vec{\nabla}, \vec{G}\cdot\vec{\nabla}\} \\ \downarrow \\ \langle \vec{F}, \vec{G} \rangle \\ \downarrow \\ [f, g]. \end{array} \quad \Leftarrow \quad \left\{ \begin{array}{c} \text{because } \exists (f, g) \\ \text{such that} \\ \vec{F} = J\vec{\nabla} f \text{ and } \vec{G} = J\vec{\nabla} g \end{array} \right. \quad (35)$$

Proof of Property 3

We are now ready for the main result of this section, the computation of $\widehat{H}$ in terms of commutators. We start with the equation of motion for the map $\mathcal{M}_{\vec{\xi}_0 \to s}$:

$$\frac{d}{ds}\mathcal{M}_{\vec{\xi}_0 \to s} = \mathcal{M}_{\vec{\xi}_0 \to s}\vec{F}_s \cdot \vec{\nabla}$$

$$\mathcal{M}^{-1}_{\vec{\xi}_0 \to s}\left\{\frac{d}{ds}\mathcal{M}_{\vec{\xi}_0 \to s}\right\} = \vec{F}_s \cdot \vec{\nabla}$$

$$\text{assuming } \mathcal{M}_{\vec{\xi}_0 \to s} = \exp\left(\alpha \widehat{H}\right)$$
$$\Downarrow$$

$$\frac{d}{d\alpha}\underbrace{\left\{\mathcal{M}^{-1}_{\vec{\xi}_0 \to s}\left\{\frac{d}{ds}\mathcal{M}_{\vec{\xi}_0 \to s}\right\}\right\}}_{\mathcal{K}} = \frac{d}{d\alpha}\left\{e^{(-\alpha \widehat{H})}\left\{\frac{d}{ds}e^{(\alpha \widehat{H})}\right\}\right\}; \quad (36)$$

$$\text{solving for } \mathcal{K} \text{ using } \mathcal{K}_{\alpha=0} = 0 \Downarrow$$

$$\mathcal{K} = \underbrace{\text{iexp}\left(-\alpha \# \widehat{H} \#\right)\frac{d}{ds}\alpha \widehat{H}}_{\text{See Eq.(14)}}$$

where $\text{iexp}(x) \equiv (e^x - 1)/x$.

After equating the last line of Eq. (36) with its second line and setting α to one, we get an equation for $\widehat{H}$:

$$\text{iexp}\left(-\#\widehat{H}\#\right)\frac{d}{ds}\widehat{H} = \vec{F}_s \cdot \vec{\nabla}$$
$$\Downarrow \qquad\qquad\qquad (37)$$
$$\widehat{H} = \int_0^s ds' \text{ inviexp}\left(-\#\widehat{H}\#\right)\vec{F}_{s'}\cdot \vec{\nabla}$$
$$\text{where inviexp}(x) \equiv 1/\text{iexp}(x).$$

Essentially, the last line proves the BCH theorem. The operator $\widehat{H}$ can be solved by iteration using the integral in Eq. (37). Because this integral involves only the *adjoint operator* $\#\widehat{H}\#$ and the functional operator $\vec{F}_s \cdot \vec{\nabla}$, the final expression for $\widehat{H}$ is made of multi-fold commutators of $\vec{F}_s \cdot \vec{\nabla}$ with $\vec{F}_{s'} \cdot \vec{\nabla}$. And, by property 2, it is a vector field.

Corollaries

Corollary 1: The familiar BCH theorem with two operators can be derived using adjoint operators. In fact, it follows from Eq. (37):

$$\underbrace{\left.\begin{array}{c} \text{if for } s \in [0,1] \quad \vec{F}_s \cdot \vec{\nabla} = \widehat{A} \\ \text{and} \\ \text{if for } s \in [1,2] \quad \vec{F}_s \cdot \vec{\nabla} = \widehat{B} \end{array}\right\} \Rightarrow \mathcal{M}_{0 \to 2} = \exp\left(\widehat{A}\right) \exp\left(\widehat{B}\right)}_{\Downarrow} \quad (38)$$

$$\mathcal{M}_{0 \to 2} = \exp\left(\widehat{A}\right) \exp\left(\widehat{B}\right) = \exp\left(\widehat{A} + \widehat{B} + \tfrac{1}{2}\{\widehat{A}, \widehat{B}\} + ...\right).$$

Corollary 2: There is an interesting corollary to the usual BCH theorem of Eq. (38), namely that a similarity transformation on a vector field is a vector field if the functional map $\mathcal{A}$ of the similarity transformation is produced by a vector field. It goes as follows:

$$\begin{aligned} \mathcal{A} \exp\left(\vec{F}_s \cdot \vec{\nabla}\right) \mathcal{A}^{-1} &= \exp\left(\mathcal{A}\vec{F}_s \cdot \vec{\nabla}\mathcal{A}^{-1}\right) \\ \text{but, using BCH, } \exists \vec{F}^\dagger \text{ such that} & \\ = \exp\left(\vec{F}^\dagger \cdot \vec{\nabla}\right) &\equiv \exp\left(\overrightarrow{\mathcal{A}\vec{F}_s} \cdot \vec{\nabla}\right). \end{aligned} \quad (39)$$

If we construct the transfer map $\vec{\alpha}$ associated with $\mathcal{A}$ by acting on the identity map then we can express $\overrightarrow{\mathcal{A}\vec{F}_s}$ in terms of $\vec{\alpha}$:

$$\text{General case}\begin{cases} \mathcal{A}\vec{F}_s \cdot \vec{\nabla}\mathcal{A}^{-1} = \overrightarrow{\mathcal{A}\vec{F}_s} \cdot \vec{\nabla} \\ \left.\overrightarrow{\mathcal{A}\vec{F}_s}\right|_k = \left(\vec{F} \cdot \vec{\nabla}\alpha_k^{-1}\right) \circ \vec{\alpha} \end{cases} \quad (40)$$

$$\text{Hamiltonian case}\begin{cases} \mathcal{A} : -H_s : \mathcal{A}^{-1} =: -\mathcal{A}H_s : \\ \mathcal{A}H_s = H_s \circ \vec{\alpha} \end{cases}$$

End of Proofs

In general, it is not true that the map $\mathcal{M}_{\vec{\xi}_{0 \to s}}$ is near the identity, hence it cannot be written as the exponential of a vector field. [Eq. (37), if solved by iteration, leads, in general, to an asymptotic series.] However, it can often be written as a product of exponentials of vector fields:

$$\mathcal{M}_{\vec{\xi}_{0 \to s}} = \prod_{\vec{H}^i \in \{\text{Lie algebra of } \vec{F}_s\}} \exp\left(\vec{H}^i \cdot \vec{\nabla}\right). \tag{41}$$

For example, it is often possible to rewrite $\mathcal{M}_{\vec{\xi}_{0 \to s}}$ as follows:

Dragt-Finn $\to$ $\mathcal{M}_{\vec{\xi}_{0 \to s}} = e^{(\vec{H}^0 \cdot \vec{\nabla})} e^{(\vec{H}^1 \cdot \vec{\nabla})} e^{(\vec{H}^2 \cdot \vec{\nabla})} \dots e^{(\vec{H}^k \cdot \vec{\nabla})} \dots$

or modified Dragt-Finn (42)

$$\mathcal{M}_{\vec{\xi}_{0 \to s}} = e^{(\vec{H}^0 \cdot \vec{\nabla})} e^{(\vec{H}^1 \cdot \vec{\nabla})} e^{(\vec{H}^{k \geq 2} \cdot \vec{\nabla})}$$

where $\vec{H}^k$ is a vector of homogeneous polynomials of degree k in the state-space variables and $\vec{H}^{k \geq 2}$ is of degree 2 and higher.

III.1.3 Transforming $\vec{\xi}_{0 \to s}$ with $\vec{\omega}_s$: What Is the New Map Equation?

We start with the computation of the effect of $\vec{\omega}_s$ on the Hamiltonian H of Figure 2. Of course, we use the functional map point of view. So, let us assume that $\mathcal{A}_{\vec{\omega}_s}^{-1}$ [the "famous A-script" of Dragt and Forest, which is nothing more than the functional map for a normalizing canonical transformation] is generated by a functional operator $\vec{\Omega}_{s,\sigma}^{-1} \cdot \vec{\nabla}$,

$$\frac{d}{d\sigma} \mathcal{A}_{\vec{\omega}_{s,\sigma}}^{-1} = \mathcal{A}_{\vec{\omega}_{s,\sigma}}^{-1} \vec{\Omega}_{s,\sigma}^{-1} \cdot \vec{\nabla}$$

with periodic boundary conditions $\begin{cases} 1 & \vec{\omega}_{s;\sigma=0} = \vec{I} \text{ and } \vec{\omega}_{s;\sigma=1} = \vec{\omega}_s \\ 2 & \vec{\omega}_{s+L} = \vec{\omega}_s \end{cases}$ (43)

The purpose of the map $\mathcal{A}_{\vec{\omega}_s}^{-1}$ is to simplify the Hamiltonian H into the Hamiltonian K. We would like to sketch here the derivation of K. Here, we assume that the coordinate transformation $\vec{\omega}_{s,\sigma}$ is generated by a σ-dependent vector field $\vec{\Omega}_{s,\sigma}$. Of course, the notation $\vec{\Omega}_{s,\sigma}^{-1}$ refers to the vector field of the transfer map $\vec{\omega}_{s,\sigma}^{-1}$. To obtain the functional map $\mathcal{A}_{\vec{\omega}_s}^{-1}$, we must integrate Eq. (43) from $\sigma = 0$ to $\sigma = 1$ using the first set of boundary conditions in Eq. (43). Because we leave the σ dependence arbitrary, in principle the canonical transformation can have several forms. [Dragt-Finn, modified Dragt-Finn, superconvergent, expanded time-ordered exponential, etc.; see Eq. (69).] In addition, as said in Eq. (43), there are periodic boundary conditions in the variable s: if we go around the ring $s \to s + L$ the map $\vec{\omega}_s$ must go back to its original value. These are the second set of conditions in Eq. (43). They are very difficult to satisfy in the traditional Hamiltonian theory and often require the use of a periodic set of basis functions Fourier expanded in the variable s. This can be avoided with the introduction of a Green's function which ties the Hamiltonian and the map treatment (see Appendix).

Returning to our problem, we start with the functional map generated by K which, according to Eq. (22), is given by

$$\mathcal{N}_{\Xi_{0\to s}} = \mathcal{N}_{\vec{\omega}_s^{-1}\circ\xi_{0\to s}\circ\vec{\omega}_0} = \mathcal{A}_{\vec{\omega}_0}\mathcal{M}_{\xi_{0\to s}}\mathcal{A}_{\vec{\omega}_s^{-1}} = \mathcal{A}_{\vec{\omega}_0}\mathcal{M}_{\xi_{0\to s}}\mathcal{A}_{\vec{\omega}_s}^{-1}. \qquad (44)$$

We proceed by taking the s derivative of Eq. (44):

$$\begin{aligned}
\frac{d}{ds}\mathcal{N}_{\Xi_{0\to s}} &= \frac{d}{ds}\left(\mathcal{A}_{\vec{\omega}_0}\mathcal{M}_{\xi_{0\to s}}\mathcal{A}_{\vec{\omega}_s}^{-1}\right) \\
&= \mathcal{A}_{\vec{\omega}_0}\left(\frac{d}{ds}\mathcal{M}_{\xi_{0\to s}}\right)\mathcal{A}_{\vec{\omega}_s}^{-1} + \mathcal{A}_{\vec{\omega}_0}\mathcal{M}_{\xi_{0\to s}}\left(\frac{d}{ds}\mathcal{A}_{\vec{\omega}_s}^{-1}\right) \\
&= \mathcal{A}_{\vec{\omega}_0}\underbrace{\mathcal{M}_{\xi_{0\to s}}\vec{F}_s\cdot\vec{\nabla}}_{\text{Using Eq. (29)}}\mathcal{A}_{\vec{\omega}_s}^{-1} + \mathcal{A}_{\vec{\omega}_0}\mathcal{M}_{\xi_{0\to s}}\left(\frac{d}{ds}\mathcal{A}_{\vec{\omega}_s}^{-1}\right) \\
&= \mathcal{N}_{\Xi_{0\to s}}\mathcal{A}_{\vec{\omega}_s}\vec{F}_s\cdot\vec{\nabla}\mathcal{A}_{\vec{\omega}_s}^{-1} + \mathcal{N}_{\Xi_{0\to s}}\mathcal{A}_{\vec{\omega}_s}\left(\frac{d}{ds}\mathcal{A}_{\vec{\omega}_s}^{-1}\right) \\
&= \mathcal{N}_{\Xi_{0\to s}}\left\{\underbrace{\left(\mathcal{A}_{\vec{\omega}_s}\vec{F}_s\right)\cdot\vec{\nabla}}_{\text{Eq. (40)}} + \underbrace{\mathcal{A}_{\vec{\omega}_s}\left(\frac{d}{ds}\mathcal{A}_{\vec{\omega}_s}^{-1}\right)}_{\text{This better be a vector field!}}\right\}. \qquad (45)
\end{aligned}$$

As pointed out in Eq. (45), the second term must be a vector field according to the group properties discussed in Section III.1.2. The expression for this vector field can be found by using *adjoint operators* as we did in Eqs. (36) and (37). We quote the result which, for example, is derived in Appendix A of Ref. [55] in the Hamiltonian case (See also Cary's review paper [77]):

$$\mathcal{A}_{\vec{\omega}_s}\left(\frac{d}{ds}\mathcal{A}_{\vec{\omega}_s}^{-1}\right) = \mathcal{A}_{\vec{\omega}_s}\left(\# - \vec{\Omega}_s^{-1}\cdot\vec{\nabla}\#\right)\int_0^1 d\sigma\,\mathcal{A}_{\vec{\omega}_{s,\sigma}}^{-1}\left(\#\vec{\Omega}_{s,\sigma}^{-1}\cdot\vec{\nabla}\#\right)\frac{\partial}{\partial s}\vec{\Omega}_{s,\sigma}^{-1}\cdot\vec{\nabla},$$

$$\Downarrow$$

$$\frac{d}{ds}\mathcal{N}_{\Xi_{0\to s}} = \mathcal{N}_{\Xi_{0\to s}}\left(\left(\mathcal{A}_{\vec{\omega}_s}\vec{F}_s\right)\cdot\vec{\nabla} + \mathcal{A}_{\vec{\omega}_s}\left(-\#\vec{\Omega}_s^{-1}\cdot\vec{\nabla}\#\right)\int_0^1 d\sigma\,\mathcal{A}_{\vec{\omega}_{s,\sigma}}^{-1}\left(\#\vec{\Omega}_{s,\sigma}^{-1}\cdot\vec{\nabla}\#\right)\frac{\partial}{\partial s}\vec{\Omega}_{s,\sigma}^{-1}\cdot\vec{\nabla}\right). \qquad (46)$$

We need to say a few words about the meaning of maps which are "functions" of an adjoint operator such as $\mathcal{A}^{-1}_{\vec{\omega}_{s,\sigma}}\left(\#\vec{\Omega}^{-1}_{s,\sigma}\cdot\vec{\nabla}\#\right)$. They are defined in terms of the formal solution of Eq. (43). Clearly, one can imagine expressing $\mathcal{A}^{-1}_{\vec{\omega}_{s,\sigma}}$ in terms of the functional operator $\vec{\Omega}^{-1}_{s,\sigma}\cdot\vec{\nabla}$. Using Eq. (43), we would get the time-ordered exponential of $\vec{\Omega}^{-1}_{s,\sigma}\cdot\vec{\nabla}$. By the notation $\mathcal{A}^{-1}_{\vec{\omega}_{s,\sigma}}\left(\#\vec{\Omega}^{-1}_{s,\sigma}\cdot\vec{\nabla}\#\right)$, we are referring to the time- ordered exponential of the adjoint operator $\#\vec{\Omega}^{-1}_{s,\sigma}\cdot\vec{\nabla}\#$. We also used a trivial property of the inverse map:

$$\text{if } \frac{d}{d\sigma}\mathcal{A}^{-1}_{\vec{\omega}_{s,\sigma}} = \mathcal{A}^{-1}_{\vec{\omega}_{s,\sigma}}\vec{\Omega}^{-1}_{s,\sigma}\cdot\vec{\nabla} \Rightarrow \frac{d}{d\sigma}\mathcal{A}_{\vec{\omega}_{s,\sigma}} = -\vec{\Omega}^{-1}_{s,\sigma}\cdot\vec{\nabla}\mathcal{A}_{\vec{\omega}_{s,\sigma}}. \qquad (47)$$

In the case of a Hamiltonian system we regain a formula first derived by Dewar [81]:

$$\text{if } \exists\ w_{\vec{\omega}^{-1}_{s,\sigma}} \text{ and } H_s \text{ such that } \begin{cases} \vec{\Omega}^{-1}_{s,\sigma}\cdot\vec{\nabla} =: -w_{\vec{\omega}^{-1}_{s,\sigma}}: \\ \vec{F}_s\cdot\vec{\nabla} =: -H_s: \end{cases}$$

$$\Downarrow$$

$$\mathcal{A}_{\vec{\omega}_s}\left(\frac{d}{ds}\mathcal{A}^{-1}_{\vec{\omega}_s}\right) = $$
$$\mathcal{A}_{\vec{\omega}_s}\left(\#-:-w_{\vec{\omega}^{-1}_s}:\#\right)\int_0^1 d\sigma\,\mathcal{A}^{-1}_{\vec{\omega}_{s,\sigma}}\left(\#:-w_{\vec{\omega}^{-1}_{s,\sigma}}:\#\right)\frac{\partial}{\partial s}:-w_{\vec{\omega}^{-1}_{s,\sigma}}: \qquad (48)$$
$$\text{using the homomorphisms of Eq. (35)}$$
$$\Downarrow$$

$$K_s = \underbrace{\mathcal{A}_{\vec{\omega}_s}H_s}_{\text{Transformed Hamiltonian}} + \underbrace{\mathcal{A}_{\vec{\omega}_s}\int_0^1 d\sigma\,\mathcal{A}^{-1}_{\vec{\omega}_{s,\sigma}}\frac{\partial}{\partial s}w_{\vec{\omega}^{-1}_{s,\sigma}}}_{\text{Time dependent term}}$$

$$\frac{d}{ds}\mathcal{N}_{\vec{\Xi}_{0\to s}} = \mathcal{N}_{\vec{\Xi}_{0\to s}}:-K_s:$$

As one can see from Eq. (48), the expression for K_s appears pretty horrible.* We recognize the two usual terms of perturbation theory. The new Hamiltonian is the old one transformed by the canonical transformation $\vec{\omega}_s$ plus an additional term which depends on the s-dependence of the Lie operator generating the map $\mathcal{A}^{-1}_{\vec{\omega}_s}$. Equation (48) is the starting point of any Deprit-type algorithm. The most inconvenient aspect of Eq. (48) is the necessity for an s derivative in the computation of K_s. This is the main reason for using the map approach.

---

* If one is really serious about normalizing an s-dependent Hamiltonian, then the order by order formulae generated by Eq. (48) are actually simpler than those obtained from a generating function approach (i.e. the standard Hamilton-Jacobi equation). This is why the Lie formalism, first proposed by Hori [22], has supplanted the generating function techniques in astronomical circles.

III.2 Lie Algebras and Charts
III.2.1 Why Are We Stuck with the Lie Operators?

We have seen that the map $\mathcal{M}_{\vec{\xi}_{0 \to s}}$, as well as the normalized map $\mathcal{N}_{\vec{\Xi}_{0 \to s}}$, is determined by the Lie algebra generated by the vector field $\vec{F}_s \cdot \vec{\nabla}$. The real question is: does the Lie group made of the Lie operators $\vec{F}_s \cdot \vec{\nabla}$ have any special properties at all which we want to preserve? The answer is yes. If we study a hadron machine then the underlying motion is Hamiltonian (i.e. $\exists H_s$ such that $\vec{F}_s \cdot \vec{\nabla} =: -H_s :$ and therefore the flow generated by $\vec{\xi}_{0 \to s}$ is symplectic. Hence we would like to preserve this property as we transform the flow by $\vec{\omega}_s$ as explained in Figures 1 and 2. In fact, we would like all our algorithms to be symplectic including the integrations through the patches. This will be insured if we use Lie methods to represent the various maps in the integration process as well as during the computation of $\vec{\omega}_s$ and its associated functional map $\mathcal{A}_{\vec{\omega}_s}$.

But what about electron machines, isn't it true that radiation pulls us away from the symplectic group? The answer is yes but by a **microscopic amount**. This small difference is qualitatively important. However the difference is so small that it is customary to try to study the symplectic system for less than a damping time. But in the end, we must include radiation in the treatment. Classical radiation can be included in the single-particle dynamics. In any mathematically elegant and powerful treatment, we must take advantage of the smallness of this radiative effect. In others words, how do we insure that the operators and the transformations we introduce, such as $\mathcal{A}_{\vec{\omega}_s}$, are as close as possible to the original non-radiative Hamiltonian treatment? The answer is unique: we use Lie operators. This is easy to understand using the formal expressions we derived in Section III.1.3. For an electron machine, the equation for the map, i.e. Eq. (29), has the following form:

$$\vec{F}_s \cdot \vec{\nabla} =: H_s : + \varepsilon \vec{R}_s \cdot \vec{\nabla}. \qquad (49)$$

If we assume, quite correctly, that the term in ε is very small, how can we insure that all the maps introduced in this section (the maps of Figure 2) are Hamiltonian except for terms of order ε? As we said, a Lie operator treatment will automatically insure this property. To see this we simply take the commutator of two vector fields $\vec{F}_s \cdot \vec{\nabla}$ and $\vec{F}_{s'} \cdot \vec{\nabla}$, and we make use of the homomorphisms of Eq. (13) and (35):

$$\left\{ \vec{F}_s \cdot \vec{\nabla}, \vec{F}_{s'} \cdot \vec{\nabla} \right\} = \left\{ : H_s : + \varepsilon \vec{R}_s \cdot \vec{\nabla}, : H_{s'} : + \varepsilon \vec{R}_{s'} \cdot \vec{\nabla} \right\}$$
$$= \left\{ : H_s :, : H_{s'} : \right\} + O(\varepsilon) \ldots$$
$$= \underbrace{: [H_s, H_{s'}] :}_{\text{Hamiltonian vector field}} + O(\varepsilon) \ldots \qquad (50)$$

Since all the operations of the last sections involved commutators only, the result of Eq. (50) will apply to them. Hence, the use of maps derivable from vector fields in our canonical transformations guaranties that, if we are near the symplectic group, we will remain near the symplectic group.

There are occasions when one may want to investigate what happen if non-Hamiltonian transformations are made on the Hamiltonian. For example, in Refs. [82-84], minimal normal forms are discussed. While it is not clear what the virtues of minimal normal forms are in a circular ring, should they be found, they might provide a reason to complicate the theory by destroying its Hamiltonian character.

Finally, in some cases, the algebra of Poisson brackets induces a bracket on a different space. The new bracket, which is called a Lie-Poisson bracket, may have properties different from the Poisson bracket. Unlike the homomorphism between commutators and Poisson brackets, which is actually trivial, in a true Lie-Poisson case, the homomorphism is not trivial. This happens whenever the dynamics is reduced by symmetries or when one studies the dynamics of functionals such as moments. One finds that there are certain elements of the algebra, called Casimirs, which commute with all other elements of the algebra. In the case of symmetries, the Casimirs are associated with the invariants of the original Hamiltonian problem. In the case of moments [87], one can show that there exist an infinite collection of Casimirs if the underlying motion is linear. For example, in one degree of freedom, the so-called emmittance, $\langle x_1^2 \rangle \langle x_2^2 \rangle - \langle x_1 x_2 \rangle^2$, is a Casimir. Needless to say, the preservation of the Poisson bracket automatically imposes the preservation of the Lie-Poisson bracket. In reduced systems, some work has been done on the theory of Lie-Poisson integrators [88].

III.2.2 Charts: the Traveler Analogy

We would like to conclude this section by taking a final look at the mathematical objects of this section from a geometrical angle. It is best to give an analogy. Suppose one is traveling on the earth's surface by car using a road map* (i.e. a chart); the charts that display the road from location A to point B can be very different. They can have different scales, different projections, etc. But, if they are topologically correct, they should permit a traveler to go from A to B correctly. The transport of an observable (i.e. the traveler) from A to B must be chart- independent.

Normally, when we travel by car, we change charts once in a while to ease our journey, for example near state or international borders. It would be very awkward to use the same charts for the whole of Mexico, the United States, and Canada for all sorts of reasons: language, road signs, density of roads, etc. However, changing charts is time-independent (we are not continuously changing road maps as we drive along!). In this case, the equation for the new map has

---

* We remind the reader of the earlier discussion about language. In the English language, the word "map" also refers to a cartographic chart, which is highly unfortunate for our analogy.

a simpler form then that of Eq. (46):

$$\frac{d}{ds}\mathcal{N}_{\vec{\Xi}_{0\to s}} = \mathcal{N}_{\vec{\Xi}_{0\to s}}\left(\mathcal{A}_{\vec{\omega}_s}\overrightarrow{\vec{F}_s}\right)\cdot\vec{\nabla} = \mathcal{N}_{\vec{\Xi}_{0\to s}}\left\{\left(\vec{F}\cdot\vec{\nabla}\xi_k^{-1}\right)\circ\vec{\xi}\right\}\partial_k. \qquad (51)$$

The expression $\left\{\left(\vec{F}\cdot\vec{\nabla}\xi_k^{-1}\right)\circ\vec{\xi}\right\}\partial_k$ for the vector field is a completely geometric object and expresses the chart-independence of the vector field, which can be viewed as a directional derivative in the direction of propagation.

Analogous to the geographical states are the beam line elements. We would like to change charts only at a finite number of boundaries. In our new approach, we should not be seeing the bizarre second term of Eq. (46). This term would represent the invidual who keeps changing charts infinitely often as he travels from A to B.

If we consider a one-turn map, i.e. a round trip from A back to A, where the travel outside A is a black box (taking the train, for example), then the only chart necessary is for A itself. We may want a better city map describing A. Again, the theory is time-independent because it involves a change of chart only at point A.

The local process of integration of Eqs. (17) or (28) as given by the local patches in Eq. (20) is similar to the use of local charts as we move from state to state along the highway. The normalization procedure in the hybrid approach is one where we emphazise the end points of the calculation, which, in a ring, are return trips. Only the chart of the end point is changed by canonical transformation.

III.3 Conclusion

We have described the common trunk between the hybrid and the traditional Hamiltonian approach. It is clear that ultimately the motion in a ring is continuous, and therefore described by the infinitesimal Hamiltonian operator.

Because of the large number of elements in a ring, the parametric dependence of H is extremely complicated. Therefore, the hybrid approach is based on a dual unified strategy:
1. The simulations are done with a local Hamiltonian using only local concepts as explained in Refs. [53,47]. This can be done because tracking is a local numerical operation. As in the real ring, the electron does not need knowledge of global quantities to find its way through a beam line (local interaction).
2. The analytical and conceptual description of a ring requires the introduction of global quantities (tunes, beta functions, island widths, etc.). These concepts are extracted from the one-turn map and understood in a Hamiltonian-free theory.

For a highly complex Hamiltonian, this seems to be the only reasonable strategy. The second part of this strategy is explained in the next section.

IV The Hybrid Approach: Analytic Calculations

In this section, we will assume that we have the ability to compute an approximate map $\vec{\zeta}_s$ as a power series in the state variables $(\vec{x}, \vec{\Delta})$ to some degree k_0 in the Taylor series. Furthermore, it is assumed that we can compute a canonical transformation $\vec{\omega}_s(\vec{x}, \vec{\Delta})$ which totally normalizes the map $\vec{\zeta}_s$ into a map $\vec{\underline{Z}}_s$ to degree k_0. As indicated previously, the tools of automatic differentiation allow us to compute this transformation at any location s around the ring. In this paper, we will not discuss this topic because it is well documented elsewhere [44, 45]. Instead, our purpose is to use this information to perform analytical perturbation theory on the ring using the map $\vec{\underline{Z}}_s$ as our zeroth-order approximation. This is traditionally called "analytical calculations."

At this point we should emphasize the difference between "power series calculations" and "purely analytical calculations." In Eq. (25), the transfer map $\vec{\zeta}_s$ is a Taylor series in the state variables $(\vec{x}, \vec{\Delta})$. As we said, it is possible to write a computer algorithm that will normalize this map, giving us the maps $\vec{\underline{Z}}_s$ and $\vec{\omega}_s$. Using these two maps, we can compute what we call a global quantity in Ref. [53], for example the beta function. Of course, this quantity will come out from the algorithm as a Taylor series in the parameters

$$\beta_x\left(\vec{\Delta}\right) = \beta_x^0 + \beta_x^1\left(\vec{\Delta}\right) + \beta_x^2\left(\vec{\Delta}\right) + \ldots \quad . \tag{52}$$

The functions β_x^k are kth-degree homogeneous polynomial in the variables $\vec{\Delta}$. First, it should be noted that the coefficients of this Taylor series are a pile of numbers. We do not have an analytical formula (let alone understanding) of the dependence of the beta function as a function of the unperturbed global concepts (phase advance between perturbations, beta functions at the perturbation, etc.). Nevertheless, it is extremely important for the reader to realize that, if one can extract a transfer map as a power series in $(\vec{x}, \vec{\Delta})$, it is possible to write a normalization algorithm [45] that will always provide formulae of the type of Eq. (52). This is a great advance due to automatic differentiation [44, 45]. In all fairness, it should be noted that the first power series algorithms in the field of accelerators were written by Dragt, Forest and Neri [68, 69]. They realized that, given a map, one should easily process the global quantities out of it. They used a map where the Lie polynomials where represented by a power series. It is totally equivalent to a power series on the transfer map of one degree less than the degree of the Lie polynomials.

So, what is the difference between a power series calculation and an analytic calculation? In general, one can view any global quantity as a functional of the map. The function inside the functional is just the complex and patchy Hamiltonian of Eq. (17) or its non-symplectic equivalent in Eq. (28). In a power series calculation, we actually evaluate this functional for a given function,* namely

---

* Sorry about confusing functions and functional forms, but it is convenient at this point.

$H_s(\vec{x}, \vec{\Delta})$. In an analytical calculation, we evaluate the global functions for an unperturbed Hamiltonian $H_s(\vec{x})$ and then we compute a functional derivative of the global quantity under a special class of functions (multipoles, beam-beam, etc.). What emerges is an analytic result that depends on the unperturbed global quantities, which in turn are computed by power series normalization. For example, the famous formula for the tune shift of a quadrupole,

$$\delta\nu_i = \frac{(-1)^{i+1}}{4\pi} \oint_0^L k(s)\beta_i ds \quad , \tag{53}$$

is the result of a functional derivative. The inputs to the formula, the betas at the quadrupole positions, are obtained from a power series normalization, i.e. by evaluating the functional for beta on the unperturbed Hamiltonian H_s.

So, in general, a power series normalization is just a matter of computer power. An analytical normalization is a far more complex and ambitious enterprise. But, again, it suffices to say that there is nothing in the traditional approach which is not do-able in the hybrid approach—more easily do-able for complex accelerator systems.

In this section, we will describe this process of functional derivative. In practice, it is done in 2 major steps and several intermediate steps:
1. The power series normalization
(a) The unperturbed power series map is computed using the tracking algorithm.
(b) The unperturbed power series map is normalized.
(c) The appropriate unperturbed global quantites are computed.
2. The analytic normalization
(a) One must first compute the functional derivative of the map, i.e. the perturbed map.
(b) Then one must compute the canonical transformation needed to normalize the perturbed map. In other words, it is the (vector-)functions of the Lie operators associated with these maps which are the functionals under consideration.
(c) Finally, the functional derivative of the global quantity of interest is computed from the functional derivative of the canonical transformation or/and the normalized map. The numerical input to the formulae comes from the power series normalization.

It should be noted that for a kth-order calculation one must evaluate a kth-order functional derivative.

This concept of functional derivatives helps us to see why there is such a strong analogy with quantum mechanics. In quantum mechanics, we have a unitary transformation $\mathcal{U}$ which lives in some abstract Hilbert space. Here, we have the map $\mathcal{M}$ which is also an abstract concept. They both obey a differential equation involving a Hamiltonian operator. More importantly, the observables are obtained from the map through some process involving the map and a type of functional. In quantum mechanics we are interested in probabilities $|\langle f|\mathcal{U}|\psi\rangle|^2$. To compute them we must have a recipe to expand the abstract vector $|\psi\rangle$ in an appropriate basis:

$$|\psi\rangle = \sum_n \langle E_n|\psi\rangle \; |E_n\rangle \; . \tag{54}$$

The quantities $\langle E_n|\psi\rangle$ are the energy wave functions of ordinary quantum mechanics. In classical mechanics, we must extract functions from the abstract map: namely the (vector-)functions of the Lie operators of a certain diagonalized (i.e. normalized) version of the map or/and the action of these operators on specified functions.

The analogy goes all the way down to the form of the normalized map. The appropriate basis to compute an observable may involve a normalization procedure very similar to that of quantum mechanics. In fact, if the reader looks at Section IV.2.3 on the eigenbasis, the analogy with the raising and lowering operator of quantum mechanics is obvious. Classical perturbation theory starts from the point of view that the ring is nearly a harmonic oscillator. In the language of Section IV.2.3, we say that the Hamiltonian is a function of h_+h_- only. The reader will notice that the functions h_+ and h_- have a dependence on the coordinate and momentum identical to that of the quantum harmonic oscillator lowering and raising operators. Of course the functional operator $:h_+h_-:$ plays a role similar to the number operator $a^\dagger a$. In a complete normalization, the new Hamiltonian can depend only on powers of h_+h_-—in a complete diagonalization, the quantum mechanical Hamiltonian is a function of $a^\dagger a$ only.

Not surprisingly, the reader should now expect the coming treatment to look like a quantum mechanical treatment with special emphasis on a periodic unitary transformation. Aside from our conceptualization of the problem in terms of finite time maps, these concepts of functional derivatives can be introduced while working directly on the Hamiltonian. For example, in Ref. [24], Eq. (56), Michelotti introduces a Green's function for the computation of the canonical transformation which normalizes the Hamiltonian. It turns out that Michelotti's Green's functions represent precisely the response of a term in the Lie exponent of the canonical transformation due to the insertion of a localised perturbation somewhere in the ring. This is exactly what we derive from a map point of view in this section. His treatment brings the traditional approach very close to the hybrid approach without actually crossing over. The computation of the Green's function is found in the Appendix. The analogy with quantum mechanics is again remarkable.

The map procedure reproduces Michelotti's results as well as the results of completely intuitive* formalisms such as the "Collins distortion functions"[29]. Forest has called this entire procedure the Hamiltonian-free perturbation theory [55]. It should be said that the word Hamiltonian-free refers to the conceptualization of the ring in the hybrid approach. Unfortunately, some people have been left with the impression that the equations of motion have vanished. As we indicated previously, not only have they not left us, but in the hybrid approach they exist in their original complex and patchy form! This is why we chose the word "hybrid."

Let us start with the computation of the perturbed map, which is done in a true "map" philosophy following a method initially used by Forest but recently expanded and popularized by Irwin in single-pass calculations (final focusing systems of linear accelerators) [59].

IV.1 Computation of the Perturbed Map

IV.1.1 A Localized Perturbation: Effect on the One-Turn Map

Imagine a vector field perturbation $\vec{V}_s \cdot \vec{\nabla}$ which is localized at a position s. The map passing through this perturbation is given by Eq. (16):

$$\mathcal{P}_s = \mathcal{I} + ds\vec{V}_s \cdot \vec{\nabla} + ... = \exp\left(ds\vec{V}_s \cdot \vec{\nabla}\right) = \exp\left(d\hat{V}_s\right).$$

$$\underbrace{\vec{\pi}_s = \mathcal{P}_s \vec{I}}_{\text{Transfer map of the perturbation}} \qquad (55)$$

In Eq. (55) we expressed the perturbation as an exponential. This does no harm since to order ds all the expressions for the map in (55) are equivalent. In fact, the exponential will insure that we stay in or close to the group of interest, i.e. the symplectic group.

The total map from position 0 back to 0 is obtained by propagating from 0 to s, then by applying $\mathcal{P}_s$ and finally by propagating from s to L (of course $L = 0$ modulo L).

$$\mathcal{M}^{perturbed}_{\vec{\zeta}_0} = \mathcal{M}_{\vec{\xi}_{s \to L}} \circ \vec{\pi}_s \circ \vec{\xi}_{0 \to s}, \qquad \mathcal{M}^{perturbed}_{\vec{\zeta}_0} = \mathcal{M}_{\vec{\xi}_{0 \to s}} \mathcal{P}_s \mathcal{M}_{\vec{\xi}_{s \to L}}. \qquad (56)$$

---

* While it is true that some of the results produced by Collins and others are correct to leading order in some simple cases, the highly intuitive formalism based on a largely unjustified generalization of the linear problem ultimately leads to a dead end. Unfortunately, many papers attempt to extend such ideas beyond first order in perturbation.

IV.1.2 A Localized Perturbation: Effect on the One-Turn Floquet Map

By assumption, we stated that it is possible to normalize $\vec{\zeta}_s$ into $\vec{Z}_s$ using the canonical transformation $\vec{\omega}_s$. We now use the functional map $\mathcal{A}_{\vec{\omega}_s}$ associated with $\vec{\omega}_s$ to simplify the perturbed $\mathcal{M}_{\vec{\zeta}_0}^{perturbed}$:

$$\mathcal{M}_{\vec{Z}_0}^{perturbed} = \mathcal{A}_{\vec{\omega}_0} \mathcal{M}_{\vec{\zeta}_0}^{perturbed} \mathcal{A}_{\vec{\omega}_0}^{-1}$$

$$= \mathcal{A}_{\vec{\omega}_0} \mathcal{M}_{\vec{\xi}_{0\to s}} \underbrace{\mathcal{A}_{\vec{\omega}_s}^{-1} \mathcal{A}_{\vec{\omega}_s}}_{\text{Introducing an identity}} \mathcal{P}_s \underbrace{\mathcal{A}_{\vec{\omega}_s}^{-1} \mathcal{A}_{\vec{\omega}_s}}_{\text{Introducing an identity}} \mathcal{M}_{\vec{\xi}_{s\to L}} \mathcal{A}_{\vec{\omega}_0}^{-1}$$

$$= \mathcal{M}_{\vec{\omega}_s^{-1}\circ\vec{\xi}_{0\to s}\circ\vec{\omega}_0} \underbrace{\mathcal{A}_{\vec{\omega}_s} \mathcal{P}_s \mathcal{A}_{\vec{\omega}_s}^{-1}}_{\text{Transformed perturbation}} \mathcal{M}_{\vec{\omega}_0^{-1}\circ\vec{\xi}_{s\to L}\circ\vec{\omega}_s} \quad (57)$$

$$= \mathcal{M}_{\vec{\Xi}_{0\to s}} \underbrace{\exp\left(\mathcal{A}_{\vec{\omega}_s} d\widehat{V}_s \mathcal{A}_{\vec{\omega}_s}^{-1}\right)}_{\text{Transformed perturbation}} \mathcal{M}_{\vec{\Xi}_{s\to L}}$$

$$= \mathcal{M}_{\vec{\Xi}_{0\to s}} \underbrace{\exp\left(d\widehat{V}_s^\dagger\right)}_{\text{Transformed perturbation}} \mathcal{M}_{\vec{\Xi}_{s\to L}}$$

The important result of Eq. (57) is the form of the perturbed operator:

$$d\widehat{V}_s^\dagger = \mathcal{A}_{\vec{\omega}_s} d\widehat{V}_s \mathcal{A}_{\vec{\omega}_s}^{-1} = \left\{\left(ds\vec{V}_s \cdot \vec{\nabla}\omega_{s;k}^{-1}\right) \circ \vec{\omega}_s\right\} \partial_k. \quad (58)$$

It is worth emphasizing that this result is exact for a localized perturbation at position s. In the case of a Hamiltonian perturbation, we simply compose the Lie polynomial of the perturbation by $\vec{\omega}_s$, i.e. "plug in the new variables":

$$d\widehat{V}_s^\dagger = \mathcal{A}_{\vec{\omega}_s} d\widehat{V}_s \mathcal{A}_{\vec{\omega}_s}^{-1} = \underbrace{\mathcal{A}_{\vec{\omega}_s} : -ds\, v_s : \mathcal{A}_{\vec{\omega}_s}^{-1} =: -ds\, v_s \circ \vec{\omega}_s :}_{\text{If it is Hamiltonian } \exists v_s \text{ such that } d\widehat{V}_s =: -ds\, v_s :}. \quad (59)$$

For example, consider a sextupole perturbation:

$$v_s(\vec{x}) = k_s \frac{x_1^3 - 3x_1 x_3^2}{3} \Rightarrow v_s \circ \vec{\omega}_s(\vec{x}) = v_s(\vec{\omega}_s(\vec{x}))$$

a possible choice in the case of midplane symmetry $\left.\begin{array}{c}\\\\\end{array}\right\}$ $\vec{\omega}_s(\vec{x}) = \begin{pmatrix} \sqrt{\beta_{1;s}} x_1 + \eta_{1;s}\delta \\ \frac{-\alpha_{1;s}}{\sqrt{\beta_{1;s}}} x_1 + \frac{1}{\sqrt{\beta_{1;s}}} x_2 + \eta'_{1;s}\delta \\ \sqrt{\beta_{2;s}} x_3 \\ \frac{-\alpha_{2;s}}{\sqrt{\beta_{2;s}}} x_3 + \frac{1}{\sqrt{\beta_{2;s}}} x_4 \end{pmatrix}$,

$$v_s(\vec{\omega}_s(\vec{x})) = k_s \frac{\left(\sqrt{\beta_{1;s}} x_1 + \eta_{1;s}\delta\right)^3 - 3\left(\sqrt{\beta_{1;s}} x_1 + \eta_{1;s}\delta\right)\beta_{2;s} x_3^2}{3}.$$
(60)

The reader will notice that we considered only the 4-dimensional phase space. The time of flight must also be changed. But because we will modify time-independent potentials, this is not important in this discussion. [If δ ($= x_5$) is the momentum deviation then x_6 is the path length. In an electron ring, path length and time of flight are nearly equivalent owing to the very high value of the relativistic γ.]

IV.1.3 Rules for Computation of the Total Map

It is worth noting that the formulae we will now derive can be extracted from Eq. (29) by going into an iteraction picture (see for example Ref. [85]). Instead, in this section, we use a method pioneered by Forest and Irwin for computations in circular rings. Recently, Irwin has used it extensively in final focusing computations where it is useful to introduce transformations into Floquet variables on account of the quasi-periodicity of the system. It is also notable that even Dragt and Forest, when they used the continuous results from Ref. [85] to study electron microscopes [86], applied the results only to individual magnets: to the fringe fields of long solenoid lenses. In the end, they had to use the calculus of discrete maps to analyse the full microscope. Therefore one may as well start with the discrete algebra and skip the continuous discussion altogether in perfect harmony with the philosophy of this paper.

We now display the rules of kth-order perturbation theory for a perturbation due to a single potential $d\widehat{V}_s^\dagger$.

Rules to Compute a Perturbed Map

1. Put k insertions in the ring, which are labelled $d\widehat{V}_{s_1}^\dagger, ..., d\widehat{V}_{s_k}^\dagger$.
2. Compute the one-turn map for those k insertions assuming an arbitrary unperturbed Floquet map $\mathcal{M}_{\underline{\underline{s}}_i \to s_{i+1}}$ between the perturbations $d\widehat{V}_{s_i}^\dagger$ and $d\widehat{V}_{s_{i+1}}^\dagger$.
3. Bring all the perturbations to the left-hand side, which is achieved by "phase advancing" [or parallel transporting] $d\widehat{V}_{s_i}^\dagger$ by $\mathcal{M}_{\underline{\underline{0}} \to s_i}$ into the new operator

$$d\widehat{V}^{\dagger\dagger}_{s_i} = \mathcal{M}_{\vec{\Xi}_{0\to s_i}} d\widehat{V}^{\dagger}_{s_i} \mathcal{M}^{-1}_{\vec{\Xi}_{0\to s_i}}.$$

4. Using the BCH theorem for two or more operators, lump the perturbations into one exponent up to $(k-1)$-fold commutators.

5. Finally, sum/integrate over the actual distribution of the perturbation. The integer label i ranging from $i=1$ to $i=k$ is interpreted as a time-ordering label.

End of Rules

Let us apply this to a second-order calculation $k=2$. Following rules 1, 2 and 3, we can write immediately the "one-turn map" prior to the summation:

$$\mathcal{M}^{perturbed}_{\vec{Z}_0} = \mathcal{M}_{\vec{\Xi}_{0\to s_1}} \exp\left(d\widehat{V}^{\dagger}_{s_1}\right) \mathcal{M}_{\vec{\Xi}_{s_1\to s_2}} \exp\left(d\widehat{V}^{\dagger}_{s_2}\right) \mathcal{M}_{\vec{\Xi}_{s_2\to L}},$$

$$\mathcal{M}^{perturbed}_{\vec{Z}_0} = \exp\left(\mathcal{M}_{\vec{\Xi}_{0\to s_1}} d\widehat{V}^{\dagger}_{s_1} \mathcal{M}^{-1}_{\vec{\Xi}_{0\to s_1}}\right) \exp\left(\mathcal{M}_{\vec{\Xi}_{0\to s_2}} d\widehat{V}^{\dagger}_{s_2} \mathcal{M}^{-1}_{\vec{\Xi}_{0\to s_2}}\right) \mathcal{M}_{\vec{Z}_0},$$

$$\mathcal{M}^{perturbed}_{\vec{Z}_0} = \exp\left(d\widehat{V}^{\dagger\dagger}_{s_1}\right) \exp\left(d\widehat{V}^{\dagger\dagger}_{s_2}\right) \mathcal{M}_{\vec{Z}_0}.$$
(61)

Now following rule 4, we apply the BCH formula as given by Eq. (38); we get

$$\mathcal{M}^{perturbed}_{\vec{Z}_0} = \exp\left(d\widehat{V}^{\dagger\dagger}_{s_1} + d\widehat{V}^{\dagger\dagger}_{s_2} + \frac{1}{2}\left\{d\widehat{V}^{\dagger\dagger}_{s_1}, d\widehat{V}^{\dagger\dagger}_{s_2}\right\} + O\left(|dV|^3\right)\right) \mathcal{M}_{\vec{Z}_0}.$$
(62)

Finally, we apply rule 5:

$$\mathcal{M}^{perturbed}_{\vec{Z}_0} = \exp\left(\sum_{n=1}^{N_{total}} d\widehat{V}^{\dagger\dagger}_{s_n} + \frac{1}{2}\sum_{n=1}^{N_{total}} \sum_{m>n}^{N_{total}} \left\{d\widehat{V}^{\dagger\dagger}_{s_n}, d\widehat{V}^{\dagger\dagger}_{s_m}\right\} + O\left(|dV|^3\right)\right).$$
(63)

Clearly, in the case of a distributed perturbation, we regain the continuous result of Ref. [85]:

$$\mathcal{M}^{perturbed}_{\vec{Z}_0} = \exp\left(\int_0^L d\widehat{V}^{\dagger\dagger}_s + \frac{1}{2}\int_0^L \int_s^L \left\{d\widehat{V}^{\dagger\dagger}_s, d\widehat{V}^{\dagger\dagger}_{s'}\right\} + O\left(|dV|^3\right)\right) \mathcal{M}_{\vec{Z}_0}.$$
(64)

In Eqs. (63) and (64), the commutators will become Poisson brackets in the Hamiltonian case due to the homomorphism of Eq. (13). For example, in a midplane symmetric case where we perturbed a linear problem, the map $\vec{\Xi}_{0\to s}$ is a rotation (i.e. a phase advance):

$$\vec{\Xi}_{0\to s}(\vec{x}) = \begin{pmatrix} R_1 & 0 \\ 0 & R_2 \end{pmatrix} \vec{x},$$
(65)

$$\text{where } R_i = \begin{pmatrix} \cos(\mu_{i;0\to s}) & \sin(\mu_{i;0\to s}) \\ -\sin(\mu_{i;0\to s}) & \cos(\mu_{i;0\to s}) \end{pmatrix}.$$

If we apply this method to the sextupole example, the operator $d\widehat{V}_s^{\dagger\dagger}$ has the following form:

$$d\widehat{V}_s^{\dagger} =: -ds \; v_s \circ \vec{\omega}_s :$$
$$\Downarrow$$
$$d\widehat{V}_s^{\dagger\dagger} =: -ds \; v_s \circ \vec{\omega}_s \circ \vec{\Xi}_{0 \to s} :$$
$$\Downarrow$$

$$v_s\left(\vec{\omega}_s\left(\vec{\Xi}_{0\to s}(\vec{x})\right)\right) = k_s \frac{\left(\sqrt{\beta_{1;s}} x_1^\mu + \eta_{1;s}\delta\right)^3 - 3\left(\sqrt{\beta_{1;s}} x_1^\mu + \eta_{1;s}\delta\right)\beta_{2;s}(x_3^\mu)^2}{3}$$

$$\text{where } \vec{x}^\mu = \begin{pmatrix} R_1 & 0 \\ 0 & R_2 \end{pmatrix} \vec{x} \quad \text{as given by Eq. (65)} \tag{66}$$

We would like to conclude this section with a few comments that enhance the pictorial and diagrammatic aspect of the rules. First, the reader will see from the sextupole example that the perturbation $d\widehat{V}_s$ is of the form $ds \; k(s) \; \widehat{W}_s^{\dagger}$ where $k(s)$ is the strength distribution of the perturbation around the ring. To give a propagator flavor to the discussion, one can imagine that $ds \; k(s)$ represent a relative probability of seeing the effect of the perturbation. The s dependence of $\widehat{W}$ comes from the change of charts at each location induced by $\vec{\omega}_s$.

The dots on Figure 3 refer to successive points in time (i.e. s). The straight lines represent the unperturbed map $\vec{\Xi}_{s_i \to s_j}$ which, as pointed out in the sextupole example, is usually a rotation. The wiggles represent the perturbations. Following the third rule, the wiggles are phase advanced to the beginning of time (i.e. $s=0$). From a geometric point of view, we say that they undergo parallel transport under the unperturbed vector field of $\vec{\Xi}_{0 \to s_j}$. Still we note from our dots that the perturbations act one after the other. [This is why the time ordering label is so important.] The next step is to use the BCH formula of Eq. (38) to lump the two vector fields into one vector field acting at a unique "time." This vector field is not just the sum because the two perturbations do not commute. Finally, under rule 5, we perform a summation over all possible locations in time (i.e. s) of the perturbation. This is given by the distribution $k(s)$.

IV.2 Normalization of the Perturbed Map

The normalization of the perturbed map is an attempt to give the new map, as given by Eq. (63), a form similar to the unperturbed map $\vec{Z}_0$. The word "similar" is defined in terms of the vector fields (i.e. Lie operators) that are needed to generate the functional map $\mathcal{N}_{\vec{Z}_0}$. For example, in the symplectic

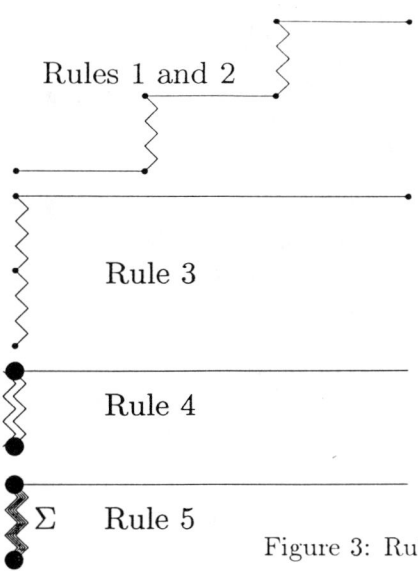

Figure 3: Rules for a 2nd-order computation.

case with no cavity present, it is often useful to *totally* normalize the map into an amplitude-dependent rotation in the transverse plane and into a drift-like motion in the longitudinal plane. In such a case, the approximate map $\mathcal{N}_{\underline{\vec{Z}}_0}$ has the following form:

$$\mathcal{N}_{\underline{\vec{Z}}_0} = \exp\left(:\Gamma\left(\vec{J}\right):\right),$$
$$J_1(\vec{x}) = \frac{x_1^2 + x_2^2}{2} \text{ and } J_2(\vec{x}) = \frac{x_3^2 + x_4^2}{2} \quad (67)$$
$$J_3(\vec{x}) = x_5 = \delta \; .$$

There is also another common type of normalization called "the single-resonance normalization." In this case, the original map $\underline{\vec{Z}}_0$ has Lie exponents that form a closed set under commutation. Of course, the set of all Hamiltonian vector fields is closed by definition. In the resonant case, the set has some very special properties. In particular, each element is invariant under a set of discrete rotations which form a group. For example, in the $3\nu_x$ single-resonance normalization, the Lie polynomials of $\mathcal{N}_{\underline{\vec{Z}}_0}$ are all invariant under a $120°$ rotation in the $x_1 - x_2$ plane.

In general, normalization is a process where one simplifies the map (by changing charts) into a map describable by a subalgebra of the full Lie algebra. For a total normalization, we end up with an Abelian subalgebra. Everything happily commutes, and therefore from Liouville's theorem we automatically have invariant tori in the symplectic case. In a single-resonance normalization, the charts have simple symmetries like snow flakes (3-fold, 4-fold, etc.) and again one can find a set of invariants. However, they are topologically more complex than tori due to the presence of separatrices.

The purpose of this section is to illustrate pictorially the general mechanical rules used in the process of normalization.

IV.2.1 The Canonical Transformation $\mathcal{B}$

We start with the perturbed map to kth order as given by the rules of Section IV.1.3. For example, in a second-order calculation, the map is given by Eq. (63). As implied by Eq. (22), a change of chart is given by a simple similarity transformation on the map:

$$\mathcal{N}^{perturbed} = \mathcal{B} \mathcal{M}_{\underline{Z}_0}^{perturbed} \mathcal{B}^{-1}. \qquad (68)$$

In Eq. (68), the map $\mathcal{B}$ can have various representations. We list a few:

$$\begin{aligned} &\underbrace{\mathcal{B} = \mathcal{B}_k \cdots \mathcal{B}_1 = \exp\left(\vec{B}_k \cdot \vec{\nabla}\right) \cdots \exp\left(\vec{B}_1 \cdot \vec{\nabla}\right)}_{\text{Dragt-Finn, see Eq.(42)}} \\ &\underbrace{\mathcal{B} = \mathcal{B}_{2^n} \cdots \mathcal{B}_4 \mathcal{B}_2 \mathcal{B}_1}_{\text{Superconvergent Dragt-Finn}} \\ &\underbrace{\mathcal{B} = \exp\left(\sum_{n=1}^{k} \vec{B}_n \cdot \vec{\nabla}\right)}_{\text{Single exponent}} \\ &\underbrace{\mathcal{B} = \exp\left(\sum_{n=1}^{\log_2(k)} \vec{B}_{2^n} \cdot \vec{\nabla}\right)}_{\text{Superconvergent single exponent}} \end{aligned} \qquad (69)$$

In numerical calculations using automatic differentiation, the superconvergent methods are easily implemented. In analytic calculations, it is often more convenient to employ methods with normal convergence. In any case, the canonical transformation $\mathcal{B}$ is of two possible families: the factored type (Dragt-Finn) and the single-exponent type. The rules that we will describe apply directly on the single-exponent type and recursively on the factored type. (It is also possible to work directly with a time-ordered exponential which has been expanded out. This is convenient with the Hamiltonian, but not really with the map. Historically, this was the representation used by Deprit. See Refs. [80, 77] for example.) Now we are set for the computation of $\mathcal{N}^{perturbed}$.

IV.2.2 The Manipulation Rules of Normalization

As in the previous section, let us assume that the map $\mathcal{B}$ is in a single-exponent representation $\mathcal{B} = \exp\left(\vec{B} \cdot \vec{\nabla}\right)$. The new map is given by the expression

$$\mathcal{N}^{perturbed} = \exp\left(\vec{B} \cdot \vec{\nabla}\right) \mathcal{M}_{\underline{Z}_0}^{perturbed} \exp\left(-\vec{B} \cdot \vec{\nabla}\right)$$

$$\mathcal{N}^{perturbed} = \exp\left(\vec{B} \cdot \vec{\nabla}\right) \exp\left(\widehat{V}_{total}^{\dagger\dagger}\right) \mathcal{N}_{\underline{Z}_0} \exp\left(-\vec{B} \cdot \vec{\nabla}\right)$$

where $\widehat{V}_{total}^{\dagger\dagger} = \sum_{n=1}^{k} \widehat{V}_n^{\dagger\dagger}$ (70)

$$= \underbrace{\sum_{n=1}^{N_{total}} d\widehat{V}_{s_n}^{\dagger\dagger}}_{n=1} + \underbrace{\frac{1}{2} \sum_{n=1}^{N_{total}} \sum_{m>n}^{N_{total}} \left\{d\widehat{V}_{s_n}^{\dagger\dagger}, d\widehat{V}_{s_m}^{\dagger\dagger}\right\}}_{n=2} + \cdots \quad \text{all the way to order } k.$$

Note that for all practical purposes, this looks like three perturbing potentials. Hence we can use the calculus of Section IV.1.3.

<u>Rules for Normalizing a Perturbed Map</u>

1. Phase advance the map $\exp\left(-\vec{B} \cdot \vec{\nabla}\right)$ by $\mathcal{N}_{\underline{Z}_0}$. The result is as usual $\exp\left(-\mathcal{N}_{\underline{Z}_0} \vec{B} \cdot \vec{\nabla} \mathcal{N}_{\underline{Z}_0}^{-1}\right)$.

2. Using the usual BCH theorem as in rule 4 of Section IV.1.3, lump the perturbations into one exponent.

3. Solve for the functional form of the Lie operator $\vec{B} \cdot \vec{\nabla}$ so that the Lie exponent of the perturbation belongs to the subalgebra which defines the normalization process. In a recursive procedure, this will involve solving the following equation k times, or $\log_2(k)$ times for a superconvergent method:

$$\vec{B} \cdot \vec{\nabla} - \mathcal{N}_{\underline{Z}_0} \vec{B} \cdot \vec{\nabla} \mathcal{N}_{\underline{Z}_0}^{-1} + \widehat{V}_{total}^{\dagger\dagger} \in \text{Subalgebra.} \quad (71)$$

3'. For a recursive process, loop over rules 1, 2, 3, and 4 until reaching order k.

4. Now that the perturbation belongs to the subalgebra of the normalization, redefine the Lie exponent of the unperturbed propagator to include the perturbation. In a complete normalization, this is trivial since all the final Lie exponents commute. In a single-resonance normalization, the process is tricky and will not be discussed here. The reader can consult Ref. [63].

Again a diagram can be used to picture the rules as shown in Fig. 4. In the normalization process, the appearance of the well-known small denominators is a consequence of rule 3. To see this the reader is invited to read Section IV.2.3 on the resonance eigenbasis and try on his own to see how it occurs as a result of Eq. (71). If that fails, the reader can consult Refs. [80, 55, 63]. Connection with the Hamiltonian method can be found in the Appendix.

460 Single-Particle Dynamics for Circular Rings

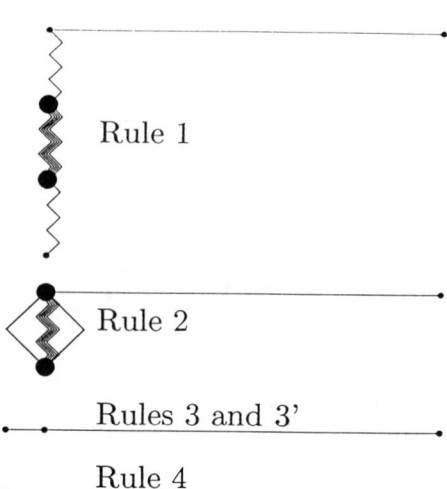

Figure 4: Rules for normalization.

IV.2.3 The Eigenbasis of the Linear Motion
The Symplectic Case

For completeness, we would like to introduce here the so-called "resonance basis." As we said, the normalization is defined in terms of a subalgebra. This subalgebra is in turn defined in terms of properties under rotation in the $x_{2i-1} - x_{2i}$ planes. Given that the operator $: J_i :$ is the generator of rotations in the $x_{2i-1} - x_{2i}$ plane, it would be nice to classify all monomials in terms of their properties under the action of $: J_i :$.

First we define the monomials it the obvious way:

$$R^{2n} \to R^{2n}, \qquad \vec{x} \mapsto P_{\vec{m}}(\vec{x}) = x_1^{m_1} \cdots x_{2n}^{m_{2n}}, \qquad (72)$$

Consider the vector space of linear monomials (i.e. $|\vec{m}| = 1$) where the norm is defined as $|\vec{m}| = \sum_{j=1}^{2n} |m_j|$; of course this vector space has dimension $2n$. It is interesting to find out what are the eigenfunctions of an arbitrary linear combination of the $: J_i :$'s. Here is the answer:

$$: \vec{\mu} \cdot \vec{J} : h_k^{\pm} = i\mu_k h_k^{\pm} \quad \text{where} \quad h_k^{\pm}(\vec{x}) = x_{2k-1} \pm i x_{2k} \equiv q_k \pm i p_k. \qquad (73)$$

Using these eigenfunctions, one builds an eigenbasis for the space of polynomials of any degree:

$$|\vec{m}\rangle = P_{\vec{m}} \circ \vec{h} \Rightarrow P_{\vec{m}} \circ \vec{h} = \left(h_1^+\right)^{m_1} \left(h_1^-\right)^{m_2} \ldots \left(h_n^-\right)^{m_{2n}},$$
$$\vec{h} = \left(h_1^+, h_1^-, \ldots, h_n^-\right). \qquad (74)$$

However, it is more convenient for analysis purposes to separate the vector of integers $\vec{m}$ into two vectors $\vec{n}^+$ and $\vec{n}^-$ which are the powers of $\vec{h}^+$ and $\vec{h}^-$ respectively [more correctly, $\vec{n}^+$ contains the powers of the $(2i-1)$th components while $\vec{n}^-$ contains the powers of the $2i$th components]:

$$|\vec{m}>= |\vec{n}^+, \vec{n}^- >= P_{\vec{m}} \circ \vec{h} \Rightarrow \left(P_{\vec{m}} \circ \vec{h}\right)(\vec{x}) = \left(h_1^+\right)^{n_1^+} \left(h_1^-\right)^{n_1^-} ... \left(h_n^-\right)^{n_n^-}, \quad (75)$$
$$\vec{h} = \left(h_1^+, h_1^-, ..., h_n^-\right).$$

Let us apply our operator $:\vec{\mu} \cdot \vec{J}:$ to the function $|\vec{n}^+, \vec{n}^- >$:

$$:\vec{\mu} \cdot \vec{J}: |\vec{n}^+, \vec{n}^- >= i\vec{\mu} \cdot \left(\vec{n}^+ - \vec{n}^-\right) |\vec{n}^+, \vec{n}^- >. \quad (76)$$

As advertised, the function $|\vec{n}^+, \vec{n}^- >$ is an eigenfunction of the arbitrary rotation generator $:\vec{\mu} \cdot \vec{J}:$.

As an example, consider a rotation of angle $2\pi/3$ in the $x_1 - x_2$ plane. Its functional map is just $\exp(:-(2\pi/3)J_1:)$. Now, let us apply this map to an arbitrary function $|\vec{n}^+, \vec{n}^- >$:

$$\exp\left(:-\frac{2\pi}{3} J_1 :\right) |\vec{n}^+, \vec{n}^- >= \exp\left(\frac{i2\pi \left(n_1^- - n_1^+\right)}{3}\right) |\vec{n}^+, \vec{n}^- >. \quad (77)$$

Clearly, if $n_1^- - n_1^+$ is a multiple of 3, the function is invariant under rotation. Furthermore, the set of functions obeying this relation is invariant under the Poisson bracket. It forms a nontrivial subalgebra of the total algebra.

<u>The General Case: Eigen-Operators</u>

Here we will make use of the tools introduced in the previous section. In a radiation situation, the linear map can be brought into a damped rotation by a real similarity transformation. The approximate normalized transfer map looks as follows:

$$\vec{\underline{Z}}_s\left(\vec{x}, \vec{\Delta}\right) = \vec{\underline{Z}}_s^{[1]} + \vec{\underline{Z}}_s^{[2]} + \cdots + \vec{\underline{Z}}_s^{[No]} + \cdots O\left(\left|\left(\vec{x}, \vec{\Delta}\right)\right|^{No+1}\right). \quad (78)$$

The linear part of this map is

$$\underline{Z}_{2i-1}^{[1]}(\vec{x}) = \exp(-\alpha_i) \{\cos(\mu_i) x_{2i-1} + \sin(\mu_i) x_{2i}\},$$
$$\underline{Z}_{2i}^{[1]}(\vec{x}) = \exp(-\alpha_i) \{\cos(\mu_i) x_{2i} - \sin(\mu_i) x_{2i-1}\}. \quad (79)$$

The constants $\vec{\alpha} = (\alpha_1, ..., \alpha_n)$ are the usual damping rates and of course $\vec{\mu} = (\mu_1, ..., \mu_n)$ are the tunes. The map associated with $\vec{\underline{Z}}^1$, which we denote by $\mathcal{R}$, is just

$$\mathcal{R} = \exp\left(:-\vec{\mu} \cdot \vec{J}: -\vec{\alpha} \cdot \vec{\Delta}^+\right),$$
$$\Delta_i^\pm = I_{2i-1}\partial_{2i-1} \pm I_{2i}\partial_{2i}, \quad (80)$$

where $\vec{I}$ is the identity transfer map $I_k(\vec{x}) = x_k$.

Remarkably, the operators $: J_k :$ and Δ_m commute for all k and m. Now we change charts. In fact we go into the complex basis defined by Eq. (73). Our goal is to express the Lie exponent of $\mathcal{R}$ in this basis. This is done by using the transformational rules of Eq. (40) for the general case. The answer is just

$$\mathcal{R} = \exp\left(-i\vec{\mu}\cdot\vec{\Delta}^- - \vec{\alpha}\cdot\vec{\Delta}^+\right). \tag{81}$$

As before, the eigenbasis is defined in terms of the linear transfer map $\mathcal{R}$. Our goal is to find eigen-operators of the form $\vec{F}\cdot\vec{\nabla}$ which change like scalars under a change of chart induced by $\mathcal{R}$. Following our discussions on homomorphisms, it follows that it suffices that $\vec{F}$ be an eigenvector-function under the application of the vector bracket with $\vec{R}$, where the vector $\vec{R}$ is defined by the relation $\vec{R}\cdot\vec{\nabla} = -i\vec{\mu}\cdot\vec{\Delta}^- - \vec{\alpha}\cdot\vec{\Delta}^+$. Using the definition of the vector bracket found in Eq. (33), the kth component of the bracket is

$$\left\langle \vec{R},\vec{F} \right\rangle_k = R_a\partial_a F_k - F_a\partial_a R_k = \vec{R}\cdot\vec{\nabla}F_k + F_k\left((-1)^{2k}i\mu_{[k]} + \alpha_{[k]}\right) \\ = \left\{ :-\vec{\mu}\cdot\vec{J}: + \left((-1)^{2k}i\mu_{[k]} + \alpha_{[k]}\right)\right\}F_k. \tag{82}$$

The notation $\mu_{[k]}$ points to the tune of the kth plane. The beauty of Eq. (82) resides in the uncoupling of planes. Hence, the eigenvectors are the same as before in this basis:

$$\left\{:-\vec{\mu}\cdot\vec{J}: + \left((-1)^{2k}i\mu_{[k]} + \alpha_{[k]}\right)\right\}|\vec{n}^+,\vec{n}^->$$
$$= \left\{\underbrace{-i\vec{\mu}\cdot(\vec{n}^+ - \vec{n}^-)}_{\text{Change of the function}} + \underbrace{\left((-1)^{2k}i\mu_{[k]} + \alpha_{[k]}\right)}_{\text{Change of the vector}}\right\}|\vec{n}^+,\vec{n}^-> \tag{83}$$

As indicated, there are two components to a chart transformation. One is the change of the functional form of $\vec{F}$ and the other is the change of orientation of the vector. In our basis, these two are neatly decoupled. In the Hamiltonian case, the homomorphism between the vector bracket and the Poisson bracket allows us to use a scalar theory which is much simpler.

Finally, to insure that the original vector field is real, the components of the vector in the eigenbasis must obey the rules:

$$F_{2k} = F^*_{2k-1} \quad \text{for } k = 1,n. \tag{84}$$

From the definition of the eigenmonomial in Eq. (75), the complex conjugate of $|\vec{n}^+,\vec{n}^->$ is just $|\vec{n}^-,\vec{n}^+>$. As an example, let us work out the most general second-degree vector field operator in one degree of freedom:

$$F_1 = \underbrace{A|2,0>}_{3\nu_x} + \underbrace{B|1,1>}_{\nu_x} + \underbrace{C|0,2>}_{}, \quad F_2 = \underbrace{A^*|0,2>}_{3\nu_x} + \underbrace{B^*|1,1>}_{\nu_x} + \underbrace{C^*|2,0>}_{}.$$
$$\tag{85}$$

One notices that in general, the quadratic vector field has two driving terms for the $3\nu_x$ resonance just as it does in the Hamiltonian case. However the ν_x resonance has four driving terms instead of only two. This is why it is important to use Lie operators whenever a map is near the group of symplectic maps; otherwise a flurry of large and unexpected terms will appear to complicate the description. Incidently, it is easy to check that the sympletic condition on the ν_x resonance amounts to $C = 2B^*$.

IV.2.4 The Sextupole Example Revisited

Recently, as a result of the study of quasi-isochronous electron rings, the computation of the second-order "momentum compaction" as a function of sextupole strengths was needed. In the standard way of looking at a circular ring, it is easy to set up pages of equations and get essentially nowhere. This was illustrated during the Brookhaven workshop. Instead, let us use Eq. (67) as our goal, Eq. (60) for the sextupole in Floquet variables, and the rules of Sections IV.1.3 and IV.2.2. Then the map after normalization to leading order in sextupole strengths k_s is just

$$\mathcal{N}^{perturbed} = \mathcal{N}^{perturbed}_{\underline{Z}_0} \exp\left(: K :\right)$$
$$\text{where } K = \int_0^L k_s \left\{ (\beta_{1;s}\eta_{1;s}J_1 - \beta_{2;s}\eta_{1;s}J_2) x_5 + \frac{\eta_{1;s}^3}{3} x_5^3 \right\} ds. \tag{86}$$

Immediately, from Eq. (86), we get the chromaticities,

$$\frac{\partial \nu_n}{\partial \delta} = \frac{(-1)^n}{2\pi} \int_0^L k_s \beta_{n;s} \eta_{1;s}\, ds, \tag{87}$$

as well as the change in the time of flight x_6 due to the energy deviation x_5 (or δ),

$$\frac{1}{L}\Delta x_6 = \frac{1}{L}\frac{\partial K}{\partial x_5} = \underbrace{\int_0^L \frac{k_s}{L}\eta_{1;s}^3 x_5^2 ds}_{\text{Second order momentum compaction}} + \underbrace{\int_0^L \frac{k_s}{L}(\beta_{1;s}\eta_{1;s}J_1 - \beta_{2;s}\eta_{2;s}J_2)\, ds}_{\text{Amplitude dependent momentum compaction}}. \tag{88}$$

IV.2.5 Phase Advance, Matching and Hamiltonian-free Theory

At this time, we will pause to examine some pedagogical consequences of the Hamiltonian- free theory: in particular, the concept of phase advance, matched insertion, and matching.

In the hybrid approach one does not try to normalize the problem by a Courant-Snyder ansatz of the pseudo-harmonic oscillator type [see Bengtsson's Foreword, Eq. (4)]. It is clear that the hybrid approach does not require a pseudo-harmonic representation of the s-dependent motion and therefore one is tempted to assume that the concept of a phase advance is not accessible to the Hamiltonian-free theory. Nothing can be further from the truth. In fact, the Hamiltonian-free approach provides a clear framework for understanding the concept of phase advance and other related concepts such as matched insertions and "matching."

Matched Insertion

We start with the concept of a matched insertion because it involves only a single circular ring. In a circular ring, we say that an insertion from s_1 to s_2 is a matched insertion if and only if the one-turn maps $\mathcal{M}_{s_1}$ and $\mathcal{M}_{s_2}$ are identical. This is an extremely compact and powerful statement. It follows from this definition that the partial maps $\mathcal{M}_{s_1 \to s_2}$, $\mathcal{M}_{s_2 \to s_1}$, if considered as one-turn maps, can be normalized by the same canonical transformation. Furthermore, the total normalized map of the ring, $\mathcal{R}$, is the product of $\mathcal{R}_{s_1 \to s_2}$ and $\mathcal{R}_{s_2 \to s_1}$.

As a trivial example, take a ring containing an arc of identical cells. If the cells are linearly matched into the ring, it follows from the above definition that the total tune (modulo integers) of the ring is the sum of the cell tune and the tune of the rest of the machine — computed by considering these two beam lines as independent rings. This is a powerful design guide and is behind the various "modules" used by lattice designers. There is nothing in this which requires the Courant-Snyder parameterization of the pseudo-harmonic motion.

Matching Two Rings

Another self-contained concept is the matching of two rings. Let us denote the maps of the first ring by $\mathcal{M}$ and those of the second ring by $\mathcal{N}$. Let us assume that particles from ring 1 are allowed to escape into ring 2 at a given time. If we call the one-turn map at the point of escape $\mathcal{M}_1$ and the one-turn map at the point of arrival $\mathcal{N}_2$, then the following equations are true:

$$\mathcal{M}_1 = \mathcal{A}_1^{-1} \mathcal{R}_\mathcal{M} \mathcal{A}_1 \quad \text{and} \quad \mathcal{N}_2 = \mathcal{B}_2^{-1} \mathcal{R}_\mathcal{N} \mathcal{B}_2 . \tag{89}$$

In Eq. (89), we assume that we can totally normalize the maps $\mathcal{M}_1$ and $\mathcal{N}_2$. In addition, we can find immediately expressions for their invariants at these positions:

$$I_i^2 = \mathcal{A}_1^{-1} J_i \quad \text{and} \quad I_i^2 = \mathcal{B}_2^{-1} J_i \quad \text{where} \quad J_i(\vec{x}) = \left(x_{2i-1}^2 + x_{2i}^2\right)/2. \tag{90}$$

In their respective rings, stationary distributions must be unaffected by the one-turn map, hence they must be functions of invariants only. Let us denote by $\mathcal{T}_{12}$ the map linking the rings (i.e. the transfer line) and by ψ_1 or ψ_2 the distributions [again, for clarity we confuse functions and functional form: $\psi_1(\vec{I})$ should be $\psi_1 \circ \vec{\omega}_1^{-1}$ where $\vec{\omega}_1^{-1} = \mathcal{A}_1^{-1} \vec{I}$ ($\vec{I}$ = identity). Also note that distributions are transformed by the inverse of a symplectic map.]:

$$\psi_1\left(\vec{I}^1\right) \underset{T_{12}}{\rightarrow} \psi_2\left(\vec{I}^2\right)$$
$$\Downarrow$$
$$\forall \psi_1 \quad \psi_2\left(\vec{I}^2\right) = T_{12}^{-1}\psi_1\left(\vec{I}^1\right)$$
$$\Downarrow$$
$$\forall \psi_1 \quad \psi_2\left(\mathcal{B}_2^{-1}\vec{J}\right) = T_{12}^{-1}\psi_1\left(\mathcal{A}_1^{-1}\vec{J}\right) \tag{91}$$
$$\Downarrow$$
$$\forall \psi_1 \quad \psi_2\left(\vec{J}\right) = \mathcal{B}_2 T_{12}^{-1}\mathcal{A}_1^{-1}\psi_1\left(\vec{J}\right)$$
$$\Downarrow \text{ (if maps are symplectic)}$$
$$\exists F_{12} \quad \mathcal{A}_1 T_{12}\mathcal{B}_2^{-1} = \exp\left(:F_{12}\left(\vec{J}\right):\right) = \mathcal{F}_{12} \quad \text{and} \quad \psi_1\left(\vec{J}\right) = \psi_2\left(\vec{J}\right).$$

The map $\mathcal{F}_{12}$ is a rotation. It is the "phase advance" of the transfer line. A beam line between two rings is matched if and only if $\mathcal{A}_1 T_{12}$ normalizes the map $\mathcal{N}_2$. All the above statements are true with/without coupling and with/without nonlinearities: clearly the s-dependent pseudo-harmonic parameterization does not enter. In the hybrid approach, the only thing entering is the bare minimum: the point of exit, the point of entry, and the map between them.

Now, since we have introduced the rotation $\mathcal{F}_{12}$ and called it the phase advance, let us conclude with a description of this concept within a single circular ring.

Phase Advance

Let us assume that for every location in the ring we have a recipe to compute the canonical transformation $\mathcal{A}$. Now, we consider two locations 1 and 2. We can do manipulations very similar to those of Eq. (91):

$$\mathcal{M}_1 = \mathcal{M}_{1\to 2}\mathcal{M}_2\mathcal{M}_{1\to 2}^{-1}$$
$$\Downarrow$$
$$\mathcal{A}_1^{-1}\mathcal{R}\,\mathcal{A}_1 = \mathcal{M}_{1\to 2}\mathcal{A}_2^{-1}\mathcal{R}\mathcal{A}_2\mathcal{M}_{1\to 2}^{-1}$$
$$\Downarrow \tag{92}$$
$$\mathcal{R} = \mathcal{A}_1\mathcal{M}_{1\to 2}\mathcal{A}_2^{-1}\mathcal{R}\mathcal{A}_2\mathcal{M}_{1\to 2}^{-1}\mathcal{A}_1^{-1}$$
$$\Downarrow$$
$$\exists F_{12} \quad \mathcal{A}_1\mathcal{M}_{1\to 2}\mathcal{A}_2^{-1} = \exp\left(:F_{12}\left(\vec{J}\right):\right) = \mathcal{F}_{12}.$$

If all the maps are analytic around the origin, symplectic, and nonresonant (the linear tunes are prime amongst each other) then the map $\mathcal{R}$ is unique and $\mathcal{F}_{12}$ is a nonlinear rotation. It is "the" advance. In the paper on the Hamiltonian-free approach (Ref. [55]), a great deal of time is spent on the concept of phase advance. A complete understanding of the freedom one has in selecting the "phase advance" puts in proper perspective the special choice of Courant and Snyder. Also, it is important to realize that there is nothing special about the Courant-Snyder choice. Therefore, let us review briefly the kind of freedom allowed.

Following the language of this paper (Section IV) and of Ref. [55], we can view the canonical transformation $\mathcal{A}$ as a functional of the one-turn map $\mathcal{M}_s$. So, in general, we can write

$$\mathcal{A} = \rho(\mathcal{M}_{s_1}; s) \ . \tag{93}$$

Of course, we must have periodicity in s. If the functional ρ depends explicitly on s, then almost nothing can be said about the resulting phase advance $\mathcal{F}_{s_1 \to s_2}$, except that the one-turn phase advance $\mathcal{F}_{s_1 \to s_1 + L}$ is independent of s_1. In fact, we denoted it by $\mathcal{R}$.

The map $\mathcal{R}$ is the only "physical" phase advance since the angle of rotation of $\mathcal{R}$ corresponds to the locations of the peaks on a Fourier transform of the motion. The phase advance between two points can be made into anything by the explicit s dependence and therefore is not "physical." This disease was present in an early Hamiltonian-free treatment of Dragt [57] which preceded the works of Forest and of Bazzanni et al. [55, 56]. In Ref. [55], Forest studied a more interesting case: what happens if we restrict ourselves to cases where the functional $\mathcal{A} = \rho(\mathcal{M}_{s_1})$ depends on the one-turn map only? In other words, if ρ and ρ' are two algorithms which depend on the one-turn map only, what is the relationship between the two phase advances? The answer is contained in the following theorem:

• Between matched points in the rings, the phase advance is independent of the algorithm used to get the canonical transformation.

So, in general, the phase advance between two arbitrary points is a mathematical construction that is arbitrary. However, between matched points, the phase advance is correlated to a shift observable in a Fourier transform of the motion. To see this, we consider a function f_1 which is observed at position 1 after n turns:

$$f_1^{[n]} = \mathcal{A}_1^{-1} \mathcal{R}^n \mathcal{A}_1 f_1 \ . \tag{94}$$

Next, we compute this function at position 2:

$$f_2^{[n]} = \mathcal{M}_{1 \to 2} f_1^{[n]} = \mathcal{M}_{1 \to 2} \mathcal{A}_1^{-1} \mathcal{R}^n \mathcal{A}_1 f_1 = \mathcal{A}_1^{-1} \underbrace{\mathcal{A}_1 \mathcal{M}_{1 \to 2} \mathcal{A}_2^{-1}}_{\mathcal{F}_{12}} \mathcal{A}_2 \mathcal{A}_1^{-1} \mathcal{R}^n \mathcal{A}_1 f_1$$

$$= \mathcal{A}_1^{-1} \mathcal{F}_{12} \underbrace{\mathcal{A}_2 \mathcal{A}_1^{-1}}_{=\mathcal{I} \text{ if matched}} \mathcal{R}^n \mathcal{A}_1 f_1$$

$$= \mathcal{A}_1^{-1} \exp\left(: F_{12}\left(\vec{J}\right) :\right) \exp\left(: n\Gamma\left(\vec{J}\right) :\right) \mathcal{A}_1 f_1$$

$$= \mathcal{A}_1^{-1} \exp\left(: F_{12}\left(\vec{J}\right) + n\Gamma\left(\vec{J}\right) :\right) \mathcal{A}_1 f_1 \ . \tag{95}$$

The actual phase that enters the Fourier transform is given by

$$\vec{\phi}^{[n]} = \vec{\phi} + \partial_{\vec{J}} F_{12}\left(\vec{J}\right) + n\partial_{\vec{J}} \Gamma\left(\vec{J}\right) = \vec{\phi} + \vec{\Delta}\left(\vec{J}\right) + n\vec{\mu}\left(\vec{J}\right) . \qquad (96)$$

The additional term $\partial_{\vec{J}} F_{12}(\vec{J})$ depends only on the one-turn map and the map $\mathcal{M}_{1\to 2}$ if the insertion is matched. Therefore, it is "physical" and directly measurable from the two Fourier transforms. To see this, we first express Eq. (95) in terms of the action-angles:

$$\mathcal{A}_1 f_1 = \sum_{\vec{m}} A_{\vec{m}}\left(\vec{J}\right) e^{i\vec{m}\cdot\vec{\phi}} . \qquad (97)$$

Then, we substitute this in Eq. (95):

$$f_2^{[n]} = \mathcal{A}_1^{-1} \exp\left(: F_{12}\left(\vec{J}\right) + n\Gamma\left(\vec{J}\right) :\right) \sum_{\vec{m}} A_{\vec{m}}\left(\vec{J}\right) e^{i\vec{m}\cdot\vec{\phi}}$$

$$= \sum_{\vec{m}} A_{\vec{m}}\left(\vec{I^1}\right) \exp\left(i\vec{m}\cdot\vec{\psi}^1 + in\vec{m}\cdot\vec{\mu}\left(\vec{I^1}\right) + i\vec{m}\cdot\vec{\Delta}\left(\vec{I^1}\right)\right) \quad (98)$$

where $\vec{\psi}^1 = \mathcal{A}_1^{-1}\vec{\phi}$ and $\vec{I^1} = \mathcal{A}_1^{-1}\vec{J}$.

Finally, we perform a discrete Fourier transform over turn number:

$$f_1^{[\omega]} = \frac{1}{N} \sum_{n=1}^{N} e^{-i\omega n} f_2^{[n]} = A_{\vec{m}}\left(\vec{I^1}\right) \exp\left(i\vec{m}\cdot\vec{\psi}^1 + i\vec{m}\cdot\vec{\Delta}\left(\vec{I^1}\right)\right) \delta_{\omega,\vec{m}\cdot\vec{\mu}}$$
$$= \exp\left(i\vec{m}\cdot\vec{\Delta}\left(\vec{I^1}\right)\right) f_1^{[\omega]} . \qquad (99)$$

In Eq. (99), the relation between the Fourier peaks at positions 1 and 2 is a function of the (non)linear phase advance only. This is a nice result which follows directly from a canonical transformation that depends on the one-turn map only. It should be added that we do not get this result if the canonical transformation is an arbitrary functional of the Hamiltonian operator. In fact, as shown in the Appendix, the one-turn map is the only object that enters naturally in the computation of $\mathcal{A}$ even in a Hamiltonian setting.

To see examples on how all this connects to the Hamiltonian treatment, we refer the reader to Section VI of Ref. [55]. In that paper, Forest applies the Hamiltonian-free method to linear problems in the limit of infinitesimal maps [i.e. the Hamiltonian: see Eq. (16)]. The standard formulas are rederived in extremely simple terms, and, in addition, the concepts of this section are illustrated on the linear problem.

IV.3 Conclusion

In this section, we have emphasized the analytic part of the new approach. In a theory based on the normalization of a one-turn map at some location s_0, the normalization algorithm involves s_0 only as a parameter. This is a great advantage of the map approach over the traditional approach when applied to complex periodic systems. As an exercise, we invite the reader to normalize the following Hamiltonian into an amplitude-dependent rotation:

$$H = H_0 + \varepsilon H_1 + \tfrac{1}{2}\varepsilon^2 H_2$$
$$\text{where } H_0 = \nu J \, ,$$
$$H_1 = A(s) J^{\frac{3}{2}} \left(\tfrac{3}{4} \sin(\varphi) - \tfrac{1}{4} \sin(3\varphi) \right) \, ,$$
$$H_2 = B(s) J^2 \left(\tfrac{3}{8} - \tfrac{1}{2} \cos(2\varphi) + \tfrac{1}{8} \cos(4\varphi) \right) \, .$$
(100)

This calculation will require the computation and analysis of the map to 2nd order in ε. The formula for the tune shift can be found in Ref. [80], where Michelotti studies this problem with the help of Lie methods in the Hamiltonian formalism (i.e. the traditional approach). The reader is strongly encouraged to test his or her understanding with such problems.

We have completed our short description of our new method with emphasis on the purely analytical aspects. We have left out some essential details which make the *hybrid approach* a complete theoretical framework enveloping modeling, tracking, production and analysis of Taylor series maps and beyond. This framework forms an indivisible whole and therefore we remind the reader of the parts not discussed in this paper:

1. Integration through patches, which requires a full understanding of what we call the LEGO approach. The LEGO approach itself requires the mastering of a few techniques:

   (a) Explicit symplectic integration [47, 72, 67].

   (b) Realization of the Euclidian group with the phase-space variables $(x, p_x, y, p_y, \delta, \ell)$. This is needed to shape the dynamics around the symmetry of the various beam-line elements. Therefore it is of paramount importance when dealing with fringe fields and more realistic models for magnets [47, 53].

2. Automatic differentiation through truncated power series algebra: this technique, first introduced by Berz in our field, permits the extraction of approximate maps as Taylor series [40].

3. A software library, based on automatic differentiation, for the manipulation of Taylor series maps. This allows for the normalization of the map [44, 45, 53] prior to any analytic calculation.

4. The full Hamiltonian-free formalism as described in Refs. [55, 56] should be understood. The practitioner of this trade should not rest until he or she is absolutely convinced that there is nothing provided by ordinary means which is outside the scope of our new hybrid approach.

V New Results

In the beginning of this paper, we listed three reasons for rephrasing the traditional theory. The first two reasons were practical. One ought to have the best tools to do a given job. In practice, those of us in a position of guidance must display at least a minimal knowledge of the state of the art, so that students are not led into a dead end. The concepts of this paper do not and should not be expected to appear in the classic references: time has passed, and we have learned other ways of expressing the problems. (After all, how many scientists read the original papers of Maxwell or Einstein?) That there is not a single good texbook which wraps up these concepts in a complete way means that more papers and monographs need to be written.

The third reason was more abstract, less believable unless one can point to convincing historical evidence. In this last chapter, we review cases where fascinating, new and profound results emerged from the fresh look given by the "Hybrid Approach." We start with explicit symplectic integration.

V.1 Symplectic Integration

V.1.1 Ruth's Original Idea

In 1983, Ronald Ruth proposed a scheme for improving the performamce of the so-called kick codes [78]. These codes are based on a simple Hamiltonian of the form

$$H = A(\vec{p}) + V(\vec{q}). \tag{101}$$

It turns out that the simplified Hamiltonian used by accelerator physicists in most simulations has the form of Eq. (101). [We should point out that a correct treatment produces a totally different Hamiltonian which does not have the form of Eq. (101)]. Using the theory of generating functions to approximate the transfer map for a time s, Ruth was able to compute approximations to the real flow. To be more explicit, let us consider the two following maps:

$$\mathcal{M}_1(s) = \exp(s:-A:), \quad \mathcal{M}_2(s) = \exp(s:-V:). \tag{102}$$

This is to be compared with the exact solution for the Hamiltonian of Eq. (101):

$$\mathcal{M} = \exp(s:-H:). \tag{103}$$

The trick is to use the BCH theorem in conjunction with the maps $\mathcal{M}_1$ and $\mathcal{M}_2$ to approximate the exact map $\mathcal{M}$. For example, one can write immediately a first- and second-order integrator:

$$\mathcal{M}(s) = \mathcal{M}_1(s)\mathcal{M}_2(s) + O(s^2),$$
$$\mathcal{M}(s) = \mathcal{M}_1(s/2)\mathcal{M}_2(s)\mathcal{M}_1(s/2) + O(s^3). \tag{104}$$

Ruth was able, with the help of an algebraic manipulator, to compute a fourth-order integrator. His derivation, again based on the special Hamiltonian of Eq.

(101), contained eight equations with hundreds of terms. Ruth needed to search numerically for the solution. The result for this fourth-order integrator is

$$\mathcal{M}(s) = \mathcal{M}_1(s_1)\mathcal{M}_2(s_2)\mathcal{M}_1(s_3)\mathcal{M}_2(s_4)\mathcal{M}_1(s_3)\mathcal{M}_2(s_2)\mathcal{M}_1(s_1) + O(s^5) \quad (105)$$

where $s_1 = \dfrac{s}{2(2-2^{1/3})}$, $s_2 = \dfrac{s}{2-2^{1/3}}$, $s_3 = \dfrac{s(1-2^{1/3})}{2(2-2^{1/3})}$, $s_4 = \dfrac{-2^{1/3}}{2-2^{1/3}}$.

The Lie methods provide a generalization and a simpler proof of Ruth's integrator. This could not have been forseen by Ruth. This is due entirely to a map approach based on Lie methods.

V.1.2 Beyond Ruth

It is remarkable that these results are totally independent of the actual form of the Hamiltonian. It suffices that the Hamiltonian be separable into two functions, and that the map generated by each function can be obtained explicitly:

$$H = \underbrace{H_1}_{\Downarrow} + \underbrace{H_2}_{\Downarrow} \quad (106)$$
$$\mathcal{M}_1(s) \quad \mathcal{M}_2(s)$$

This result holds for any Lie algebra: the Poisson bracket, the Lie-Poisson, the commutator, etc. This was first noted by Neri and Forest.

This is not the end of the story. Yoshida (actually first Suzuki) [72,73] provided an elegant generalization of Ruth's algorithm. They found that Ruth's algorithm is the beginning of an infinite sequence of higher-order integrators. Later Forest et al. [67] found that the Yoshida-Suzuki formalism can be applied if there exists an approximation of the map $\mathcal{T}_{2n}$ with the property

$$\mathcal{T}_{2n}(s)\mathcal{T}_{2n}(-s) = \mathcal{I}, \quad \mathcal{M}(s) = \mathcal{T}_{2n}(s) + O(s^{2n+1}). \quad (107)$$

It is possible to compute a $2n+2$ approximation of the map using $\mathcal{T}_{2n}$:

$$\mathcal{T}_{2n+2}(s) = \mathcal{T}_{2n}(d_1 s)\mathcal{T}_{2n}(d_2 s)\mathcal{T}_{2n}(d_1 s) + O(s^{2n+3}) \quad (108)$$

where $d_1 = -2^{1/(2n+1)}/\left(2-2^{1/(2n+1)}\right)$ and $d_2 = 1/\left(2-2^{1/(2n+1)}\right)$. The proof of Eq. (108) has two lines! It can be found in Yoshida's paper [72] or in Ref. [67]. As we said, it was found by Suzuki in the context of path integral formalism and recently it was rediscoved by Bandrauk and Shen [74] in a paper on the approximate solution of the Schrödinger equation in chemical application. Ruth's original integrator can be found by using Eq. (108) and the following choice for $\mathcal{T}_2$:

$$\mathcal{T}_4(s) = \mathcal{T}_2(d_1 s)\mathcal{T}_2(d_2 s)\mathcal{T}_2(d_1 s) + O(s^5)$$

where $\begin{cases} \mathcal{T}_2(s) = \mathcal{M}_1(s/2)\mathcal{M}_2(s)\mathcal{M}_1(s/2) \\ d_1 = -2^{1/3}/\left(2-2^{1/3}\right) \text{ and } d_2 = 1/\left(2-2^{1/3}\right) \end{cases}. \quad (109)$

This re-interpretation of the Ruth integrator as a Lie group integrator has led to its application to more complex Hamiltonians. In particular, it is applicable to the complex Hamiltonian of small rings [47], to solar problem integrations [75], and to plasma problems [76]. In all these cases the Hamiltonian does not have the original drift-kick form of Ruth. Finally, Suzuki [73] has completely classified all the integrators based on approximate maps obeying Eq. (107).

All this development is based on the recognition that the underlying equation for the map (or inverse map) is of the form

$$\frac{d}{ds}\mathcal{M} = \mathcal{M}\,\widehat{H}. \tag{110}$$

The example of symplectic integration illustrates with absolute clarity the advantages of a new point of view as forced upon us by our new approach.

V.2 Fitted Maps: Computation and Normalization

V.2.1 The Concept of a Fitted Map

Let us assume for a moment that we are interested in a special region of phase space defined in terms of some preconditioned action-angle variables $(\vec{\Phi}, \vec{J})$ [defined through some power series transformation $\vec{\omega}_s$ as shown in Eq. (112)]:

$$U = \left\{ \vec{\Phi}, \vec{J} \mid 0 < J_{a;i} < J_i < J_{b;i}\,,\ \Phi_i \in [0, 2\pi]\,,\ i = 1, ..., N \right\}. \tag{111}$$

The set U is a product of annuli in phase space. Let us assume that we are interested in the motion within the set U and that we will consider a particle to be lost if it leaves the said set. Then our goal is to represent the map within that set as accurately as possible (see [28]). To do this we simply take tracking data normalized by the map $\vec{\omega}_s$ as indicated in Figure 2. It is important to note that in many applications the transformation $\vec{\omega}_s$ need not be accurate. The action-angle variables are defined in a natural way:

$$\omega^{-1}_{s;2i-1}(\vec{x}) = \sqrt{2J_i}\cos(\Phi_i)\,, \qquad \omega^{-1}_{s;2i}(\vec{x}) = -\sqrt{2J_i}\sin(\Phi_i)\,. \tag{112}$$

The map itself can be expressed as follows:

$$\vec{\Phi}^1 = \vec{\Phi} + \vec{\Theta}\left(\vec{\Phi}, \vec{J}\right), \tag{113a}$$

$$\vec{J}^1 = \vec{J} + \vec{R}\left(\vec{\Phi}, \vec{J}\right). \tag{113b}$$

The fitted map is obtained by expressing the functions $\vec{\Theta}$ and $\vec{R}$ as a Fourier series in the angles and using a spline representation for the actions. However, in long-term tracking applications, we are interested in a symplectic map to computer accuracy. In the case of fitted maps, this can be achieved with generating functions. We now describe this process.

V.2.2 Fitting a Generating Function Map

The map is defined to be a transformation from the "old" variables $(\vec{\Phi}, \vec{J})$ to the "new" variables $(\vec{\Phi}^1, \vec{J}^1)$ as shown in Eq. (113). The generating function in this case will be in terms of old action and new angle variables:

$$G\left(\vec{\Phi}^1, \vec{J}\right) = \sum_{\vec{m}} g_{\vec{m}}\left(\vec{J}\right) e^{i\vec{m}\cdot\vec{\Phi}^1}. \tag{114}$$

The resulting transformation is then just

$$\left(\vec{\Phi}, \vec{J}\right) \mapsto \left(\vec{\Phi}^1, \vec{J}^1\right),$$

$$\vec{\Phi} = \vec{\Phi}^1 + \partial_{\vec{J}} G\left(\vec{\Phi}^1, \vec{J}\right), \tag{115a}$$

$$\vec{J}^1 = \vec{J} + \partial_{\vec{\Phi}^1} G\left(\vec{\Phi}^1, \vec{J}\right). \tag{115b}$$

We start with a "source map," which gives the final variables as an explicit function of the initial variables as given symbolically by Eq. (113). This map will usually be defined as the result of tracking over one turn and preconditioning the data as explained in Eq. (112). The Fourier coefficients are obtained from Eqs. (115b) and (113b) as

$$g_{\vec{m}}\left(\vec{J}\right) = \left(1/(2\pi)^d / im_k\right) \int_0^{2\pi} d\vec{\Phi}^1 \, \partial_{\Phi_k^1} G^1\left(\vec{\Phi}^1, \vec{J}\right) e^{-i\vec{m}\cdot\vec{\Phi}^1},$$

$$= \left(1/(2\pi)^d / im_k\right) \int_0^{2\pi} d\vec{\Phi}^1 R_k\left(\vec{\Phi}\left(\vec{\Phi}^1, \vec{J}\right), \vec{J}\right) e^{-i\vec{m}\cdot\vec{\Phi}^1}. \tag{116}$$

Since one does not know $\vec{R}$ as a function of $\vec{\Phi}^1$, one performs a change of variables in the integral to get an integral over $\vec{\Phi}$:

$$g_{\vec{m}}\left(\vec{J}\right) = \int_0^{2\pi} \frac{d\vec{\Phi}}{(2\pi)^d im_k} R_k\left(\vec{\Phi}, \vec{J}\right) e^{-i\vec{m}\cdot\vec{\Phi}^1} e^{-i\vec{m}\cdot\vec{\Theta}(\vec{\Phi},\vec{J})} \det\left(1 + \partial_{\vec{\Phi}} \vec{\Theta}\left(\vec{\Phi}, \vec{J}\right)\right). \tag{117}$$

The integral is then discretized to obtain

$$g_{\vec{m}}\left(\vec{J}\right) = \frac{1}{im_k \prod_n N_n} \sum_{\vec{j}} R_k\left(\vec{\Phi}_{\vec{j}}, \vec{J}\right) e^{-i\vec{m}\cdot\vec{\Phi}_{\vec{j}}} e^{-i\vec{m}\cdot\vec{\Theta}(\vec{\Phi}_{\vec{j}}, \vec{J})} \det\left(1 + \partial_{\vec{\Phi}} \vec{\Theta}\left(\vec{\Phi}_{\vec{j}}, \vec{J}\right)\right) \tag{118}$$

where N_n is the number of Φ_n mesh points in the n dimension, and the summation is over the integer vectors $\vec{j}$ such that $j_n \in \{0, ..., N_n - 1\}$. The $\vec{m} = \mathbf{0}$ mode must be handled differently. We instead must use $\vec{\Theta}$ values. The resulting summation is

$$g_0\left(\vec{J}\right) = -\frac{1}{\prod_n N_n} \sum_{\vec{j}} \vec{\Theta}\left(\vec{\Phi}_{\vec{j}}, \vec{J}\right) \det\left(1 + \partial_{\vec{\Phi}} \vec{\Theta}\left(\vec{\Phi}_{\vec{j}}, \vec{J}\right)\right). \tag{119}$$

To increase the speed of evaluation of the map, Fourier modes that are smaller than the expected or desired accuracy of the map can be removed from the generating function.

We obtain values of $g_{\vec{m}}(\vec{J})$ for values on a mesh in $\vec{J}$. We then choose a set of basis functions $B^{(k)}_{j_k}(J^k)$ to use in interpolating the coefficients such that

$$g_{\vec{m}}\left(\vec{J}\right) = \sum_{\vec{j}} g_{\vec{m},\vec{j}} \prod_k B^{(k)}_{j_k}(J_k). \qquad (120)$$

The index k labels the different degrees of freedom. For the $\vec{m} \neq \mathbf{0}$ modes, the interpolation is straightforward. For the $\vec{m} = \mathbf{0}$ mode, one must be careful to consider that the derivatives of the basis functions are linearly dependent. Details of this can be found in [28]. It is advantageous to choose B-splines for the basis functions. Because they have a small region where they are nonzero, their use greatly increases the speed of evaluation of the map.

The map is evaluated by performing a Newton iteration to obtain $\vec{\Phi}^1$ and then substituting into Eq. (115b) to get $\vec{J}^1$. An initial guess for the Newton iteration is provided by an explicit map with a small number of modes retained.

The method can be used in any number of dimensions. In a 3-dimensional accelerator problem, however, it is not advantageous to do the third dimension in action-angle variables. Instead, note that most of an accelerator ring is time-independent. One can construct a map for the time-independent part that has the energy deviation as an additional parameter, which is treated on equal footing with the actions. The time-dependent parts (usually cavities) can then be treated separately as the user chooses. Because time-of-flight is canonically conjugate with energy, it is obtained by taking a derivative of the generating function with respect to energy deviation.

Finally, note that since one wants to perform the action interpolation over a finite domain that does not include the origin in each phase-space plane, the plain source map is sometimes not well suited for being directly used by this method. This can be overcome by performing a preliminary canonical transformation on the source map so as to have the new source map take an annulus of initial conditions into a similar (larger) annulus. This can be easily achieved by a linear transformation or a low-order Taylor series mixed-variable generating function, i.e. Eq. (112).

Results of applying this method to a Taylor series map that is intended to approximate the SSC are shown in Figure 5. The "mode cutoff" is a measure of the maximum size of the Fourier modes that are being removed from the generating function. The number of actions indicates the number of mesh points in each dimension of action interpolation. The order refers to the order of B-splines used in action interpolation. The curves have approximately slope 1 when the error is dominated by the number of Fourier modes being thrown away. They begin to level off when the error is dominated by the action interpolation (low actions) or failure of symplecticity of the source map (high actions : dashed lines on Figure 5).

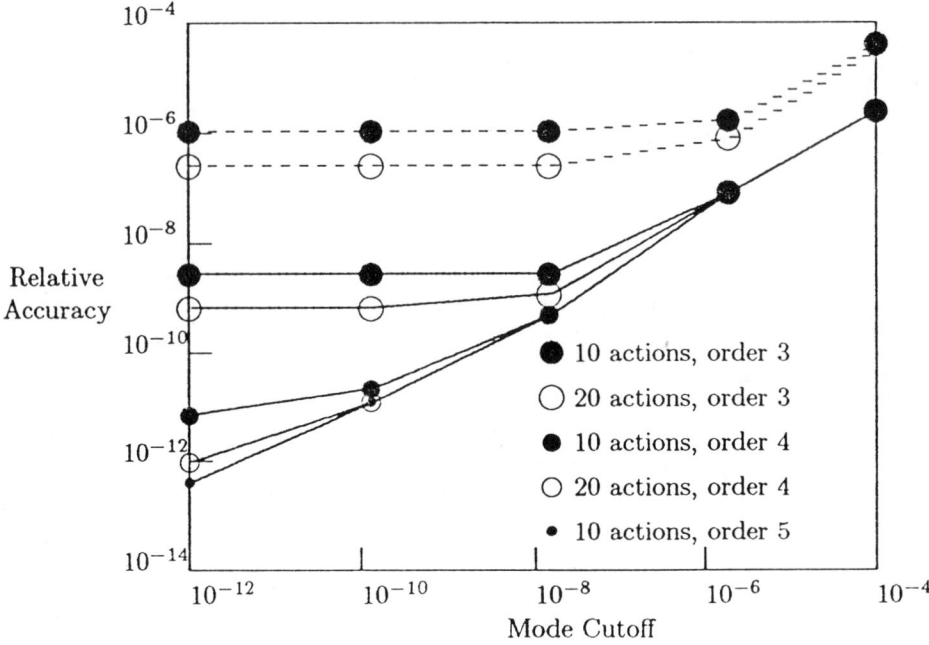

Figure 5: Relative accuracy of a 2-d map.

V.2.3 Conclusion on Fitted Maps

Fitted maps allow us to represent the map accurately in regions far from the origin. With the help of a generating function, they can be used for symplectic tracking and experience has shown they are fast in some regions where the underlying tracking code or an equally good Taylor series is much slower. It should be said that the technique requires only tracking data that have been preconditioned by a power series transformation. Within the hybrid approach, the technique can be used on arbitrarily complex systems. Finally, the map can be put on an integer grid for exact symplectic tracking without truncation error, although there are still some convergence problems associated with the integer grid which have not yet been investigated [90, 89].

V.2.4 Normalizing by Fitting Techniques

Using preconditioned variables as in Eq. (112), one can ask whether or not the motion in the set U is well approximated by the motion on an N-torus or by a single-resonance map [pendulum $\times$ $(N-1)$-torus]. If it is, one can use the approximate invariants to give almost rigorous bounds on the motion.* The technique uses a generating function G of the form

$$\left(\vec{\Phi}, \vec{J}\right) \mapsto \left(\vec{\Psi}, \vec{I}\right),$$

$$\vec{\Psi} = \vec{\Phi} + \partial_{\vec{I}} G\left(\vec{\Phi}, \vec{I}\right), \tag{121a}$$

$$\vec{J} = \vec{I} + \partial_{\vec{\Phi}} G\left(\vec{\Phi}, \vec{I}\right). \tag{121b}$$

An orbit lying on an invariant surface has a constant value of $\vec{I}$, and for such an orbit Eq. (121a) gives $\vec{J}$ as a function of $\vec{\Phi}$. We can compute values of this function by tracking, obtaining $(\vec{\Phi}_i, \vec{J}_i)$, $i = 1, \ldots, n$, by n successive applications of one-turn tracking. The problem is to obtain the Fourier coefficients of G from this information and Eq. (121b). It is not possible to get the coefficients from a simple discrete Fourier transform, since the $\vec{\Phi}_i$ do not lie on a regular mesh; they fall where they may as a result of the dynamics. One could try to solve a system of linear equations for the coefficients, but the system would be ill-conditioned and typically very large.

The way out of this difficulty [96] is to choose a different set of unknowns in place of the Fourier coefficients, namely, the values $\vec{J}(\vec{\Phi})$ on a regular mesh. The matrix relating the tracking values $\vec{J}_i(\vec{\Phi}_i)$ to the values on the mesh is close to the unit matrix, and the system is therefore amenable to iterative solution, provided that we choose the tracking values appropriately. The appropriate data are such that there is one point per cell of the mesh. The orbit is followed long enough to acquire at least one point per cell, and for cells getting more than one, the point closest to a mesh point is chosen. It may happen that one point per cell cannot be achieved in a reasonable number of turns, and that is a convenient signal of a resonant orbit; points lie along "lines" in $\vec{\Phi}$ space. When a resonant orbit is found, the initial condition is automatically shifted, in search of a non-resonant orbit lying on an approximately invariant torus.

After the values of $\vec{J}$ on the mesh are found, an FFT yields the Fourier coefficients. According to Eq. (121b), the invariant action $\vec{I}$ is the constant term of the Fourier series. Fitting orbits for the various initial conditions, and computing the corresponding values of $\vec{I}$, one gets coefficients that can be interpolated in $\vec{I}$, as in the case of maps. Thus a canonical transformation is defined in an open domain of phase space. The interpolation bridges resonances: many narrow, high-order resonances, but occasionally a wide and

---

* Contrary to popular belief, the bounds are actually more rigorous than actual tracking since the word "almost" refers to the impossibility of scanning the entire set U. However, when studying the fluctuation of an invariant, we can scan a greater portion of U than avaible by brute force tracking methods. This is true because to predict k-turn stability, we need to track only a fraction of k. This fraction depends on how well the normalized system approximates the real motion.

significant one. Remarkably, the variables $(\vec{\Psi}, \vec{I})$, obtained by bridging a wide resonance in a somewhat arbitrary way, have been found to follow an isolated resonance model with the mode numbers of that resonance. This is gratifying and useful for studying stability of the motion, but not completely understood; further investigation is in order. [The resonance is wide compared to the distortions of the invariants but still very small. Hence, they are probably well described by the approximate one-resonance model mentioned in Bengtsson's foreword. In other words, if we remove "low"-order distorsions, what is left in the neighborhood of these resonances is a quasi-pendulum. This is not the case if large low-order resonances dominate. They must be treated by the exact single-resonance theory which is not available with fitted maps.]

The canonical transformation that is obtained gives very good results. Often it is easy to obtain a new variable $\vec{I}$ that is constant to 10^{-6} relative error throughout an interpolated region of substantial nonlinearity.

Finally, one can obtain values for the tune as a function of amplitude by this method, using Eq. (121a). This can be useful in locating low- and high-order resonances, even at high amplitudes.

A significant application of this method is to determine long-term bounds for the motion [26]. If one finds that in some region, for some number of turns n, the change in the new variable $\vec{I}$ is less than some value δI, then crossing a region of size ΔI will take at least $N = n\Delta I/\delta I$ turns. While δI cannot be determined rigorously over a region, it can be determined at many points in a region and at expected "bad" points (such as near resonance separatrices), to attempt to obtain some kind of "likely" bound.

Finally, it is also interesting to point out that the use of a Von Zeipel transformation permits the two-step fitting process (i.e. first fitting the Fourier modes on a single trajectory and then interpolating them in the actions). It is one case where the Lie methods, with their explicit transformations, are at a disadvantage.

V.2.5 Conclusion on Fitting Techniques

These techniques can now be used in conjunction with a symplectic integrator and power series normalization (for preconditioning). Again it is interesting to note how one is naturally led to a map-based theory for practical complex calculations. The original work of Ruth, Warnock and Gabella [27] was done using the Hamilton-Jacobi equation. They solved it with a shooting method; that would be like using a differential equation for the lattice functions and getting them through an iteration process. It requires the use of some Floquet variables throughout the ring. This was clearly an overkill!

V.3 Symplectified Taylor Series Maps for Tracking

V.3.1 Generating Functions

Contrary to popular belief, the first kind of symplectic tracking introduced by Dragt and collaborators had <u>absolutely</u> nothing to do with Lie methods. It was based on a generating function representation of the map. Given a power series representation of the map as in Eq. (25), it is possible to find a generating function that will approximate the given Taylor series. In fact, up to the order of truncation, the new map will agree perfectly if the original map is symplectic. Using the tools of automatic differentiation [45, 41], it easy to compute the correct generating function and to track symplectically with it.

Unfortunately, maps produced by generating functions are not necessarily free of poles or branch points even if the original map is entirely free of them. Sometimes, it is convenient to factor the linear part of the map and to represent only its nonlinear part as a generating function as one does in a Dragt-Finn factorization [see Eq. (42)].

It is possible that the resulting nonlinear map may still be too "big" to be used through the region of interest. If that is the case, Lie methods can come to the rescue. Using again automatic differentiation [45], one can compute the modified Dragt-Finn factorization and compute the nth root of the nonlinear part of the map:

$$\mathcal{M}_{\vec{\zeta}_s} = \exp(:h_2:)\exp(:h_{k\geq 3}:)$$
$$\downarrow$$
$$\mathcal{M}_{\vec{\zeta}_s} = \exp(:h_2:)\left\{\exp\left(:\frac{h_{k\geq 3}}{n}:\right)\right\}^n .$$
(122)

The total map is made of a linear map and n identical nonlinear maps:

$$\vec{\zeta}_s = \underbrace{\vec{\zeta}_s^{1/n} \circ \cdots \circ \vec{\zeta}_s^{1/n}}_{n \text{ times}} \circ \vec{\zeta}_s^{1}$$
(123)

where $\vec{\zeta}_s^{1} = \exp(:h_2:)\vec{I}$ and $\vec{\zeta}_s^{1/n} = \exp\left(:\frac{h_{k\geq 3}}{n}:\right)\vec{I}$.

Needless to say this process can be slow. It is not recommended for long-term tracking. Even though Lie methods are used in this type of tracking, it remains tracking with a generating function.

V.3.2 Jolt Tracking

The first type of "Lie algebraic" tracking ever implemented is due originally to Irwin [35, 45]. In its original form it was formulated in terms of "kicks" and rotations. Let us concentrate on the nonlinear part $\mathcal{N}$ of the map since it is the only part that poses some problems. According to Irwin, it can be rewritten as follows:

$$\mathcal{N}_{\vec{\zeta}_s} = \exp\left(: \mathcal{R}_1 \chi^1 :\right) \exp\left(: \mathcal{R}_2 \chi^2 :\right) \quad \exp\left(: \mathcal{R}_{\widehat{N}} \chi^{\widehat{N}} :\right)$$

$$\text{where } \mathcal{R}_i = \exp\left(: -\vec{\alpha}^i \cdot \vec{J} :\right),$$

$$\chi^i = \sum_{\vec{m}} \beta^i_{\vec{m}} \underbrace{q_1^{m_1} \cdots q_N^{m_N}}_{\substack{\text{Position only} \\ \vec{q} = (x_1, x_3, ..)}}.$$
(124)

The minimal value of the integer $\widehat{N}$ is a function of the degree of the power series and of the phase-space dimension. It is also a function of the choice of linear maps $\mathcal{R}_i$, which are rotations in the original treatment of Irwin.

For each individual functional map of Eq. (124), one can find the image of the identity function:

$$\vec{\Xi}^i = \exp\left(: \mathcal{R}_i \chi^i :\right) \vec{I} = \vec{I} + \left[\mathcal{R}_i \chi^i, \vec{I}\right].$$
(125)

The series terminates exacly. Therefore, the one-turn map $\vec{\zeta}_s$ is approximated symplectically as

$$\vec{\zeta}_s \approx \underbrace{\vec{\Xi}^{\widehat{N}} \circ \cdots \circ \vec{\Xi}^1}_{\text{Nonlinear part}} \circ \underbrace{\vec{\zeta}_s^{-1}}_{\text{Linear part}}.$$
(126)

The good points of the Irwin factorization are:
- Exactly symplectic and defined all over phase space.
- Obvious exact inverse.
- Can be put on an integer grid without ambiguity.
- Can be computed by automatic differentiation [45].

Unfortunately many unanswered questions remain with this representation:
- What should be the group chosen for the maps $\mathcal{R}_i$? Irwin picked the group of linear phase advances. One can show that the drifts would suffice.
- Second, given a group, how to choose the elements for a given number $\widehat{N}$. How do they foliate the polynomials into different equivalence classes?
- What is the normal form in each equivalence class leaf?
- Do all the above while minizing the difference between the maps; i.e. $\left\|\vec{\zeta}_s(\vec{x}) - \left(\vec{\Xi}^{\widehat{N}} \circ \ldots \circ \vec{\Xi}^1 \circ \vec{\zeta}_s^{-1}\right)(\vec{x})\right\|$ must remain small within the aperture.

Strangely and unexpectedly, these questions are mathematically very hard to answer. Dragt and Abell have looked into these questions [36] and have found that the goodness of the approximation is very sensitive to the set chosen. In other words, the set of angles $\{\vec{\alpha}^i | \ i = 1, \widehat{N}\}$ is critical in determining the goodness of the approximation.

In conclusion, the topic of factorization in terms of jolts or any other exactly solvable functions is certainly wide open. But, as in symplectic integration, it is bound to a view of the ring based on maps and on the use of Lie methods. However, unlike symplectic integration, it remains to be seen how useful it is.

V.3.3 Monomial Tracking

Besides the jolt factorization, there is another way to do symplectic "Lie" tracking. It is based on the realization that monomial maps are exactly solvable [37]. Consider the map

$$\vec{\kappa}_{\vec{m}} = \exp\left(: P_{\vec{m}} :\right) \vec{I} \tag{127}$$

where $P_{\vec{m}}$ is defined by Eq. (72). It can be shown that the monomial map $\vec{\kappa}_{\vec{m}}$ involves finite Taylor series map (kick), exponential functions as well as roots and ratios of polynomials. So, the map $\mathcal{M}_{\vec{\zeta}_s}$ can be expressed as follows:

$$\mathcal{M}_{\vec{\zeta}_s} = \prod_{n=1}^{\widehat{N}} \exp\left(: P_{\vec{m}_n} :\right) \text{ or equivalently } \vec{\zeta}_s = \vec{\kappa}_{\vec{m}_{\widehat{N}}} \circ \cdots \circ \vec{\kappa}_{\vec{m}_1}. \tag{128}$$

It is easy to check that the various monomial maps have poles, and therefore this technique probably suffers from the combined diseases of the generating functions and the jolts. Nevertheless, it can be very useful for short-term tracking and to represent complex elements symplectically. Again, with the help of automatic differentiation, it has been implemented [45].

V.3.4 Conclusion on Taylor Series Symplectification

We are forced to deal with three methods (generating functions, jolts and monomials) which cannot be absolutely accurate. For example, it is clear that a jolt map cannot have any poles whereas the real map has poles. On the other hand, generating functions and monomials have poles and branch cuts that are spurious. To the extent that these maps are attempting to represent the entire "useful" phase space, one must exercise the same caution as one used in selecting the time step of an integrator. For this purpose, the nth-root trick used in Section V.3.1 can be used to improve the accuracy of any symplectification scheme and to push spurious poles outside the aperture. [While in press, we learned that Yan et al. have combined jolts, monomials and the tools of symplectic integration in an attempt to integrate the Dragt-Finn factorization.]

V.4 Connecting Ray and Wave Optics

It is well known that geometrical optics is governed by a Hamiltonian

$$H = -\sqrt{n\left(\vec{q}; z\right)^2 - p^2} \tag{129}$$

where $n\left(\vec{q}; z\right)$ is the index of refraction and the coordinate z plays the role of time. Normally, it is customary to expand the equation of motion around some ray trajectory as we do in accelerators. Then the functional map of the system can be written as a product of maps following the Dragt-Finn factorization:

$$\mathcal{M} = \underbrace{\mathcal{M}_2}_{\substack{\text{Paraxial} \\ \text{Gaussian} \\ \text{optics}}} \underbrace{\exp\left(: f_3 :\right) \ldots \exp\left(: f_{N_o} :\right)}_{\text{Aberrations}}. \qquad (130)$$

V.4.1 The Equation of Motion for Wave Optics

Consider the problem of trying to deduce wave optics behavior from our knowledge of the symplectic map for ray optics. The work that follows, due to Dragt [91], makes full use of the functional operator. One starts with the scalar wave equation

$$\nabla^2 \phi - \frac{n^2}{c^2} \frac{\partial^2 \phi}{\partial t^2} = 0. \qquad (131)$$

The next step is to assume a monochromatic solution $\phi(\vec{q}, t) = \psi(\vec{q}) e^{-i\omega t}$. Then ψ obeys the "generalized Helmholtz" equation

$$\nabla^2 \psi + \frac{\omega^2 n^2}{c^2} \psi = 0. \qquad (132)$$

Then we convert this equation into a first-order differential equation that neglects backward propagation in the z direction by taking a square root (and hand waving!):

$$\frac{\partial \psi}{\partial z} = i \left\{ \frac{\omega^2 n^2}{c^2} + \left(\frac{\partial}{\partial x}\right)^2 + \left(\frac{\partial}{\partial y}\right)^2 \right\}^{1/2} \psi. \qquad (133)$$

It is absolutely trivial to find an equation for the transfer operator connecting an initial wave front ψ^i to a final wave front ψ^f. Calling this operator $\mathcal{U}$, it obeys the equation:

$$\frac{d\mathcal{U}}{dz} = \frac{i}{\lambda} \left\{ n^2 - P_x^2 - P_y^2 \right\}^{1/2} \mathcal{U},$$

$$P_x = \frac{c}{i\omega} \frac{\partial}{\partial x} = \frac{\lambda}{i} \frac{\partial}{\partial x}, \quad P_y = \frac{\lambda}{i} \frac{\partial}{\partial y} \Rightarrow \{X, P_x\} = \{Y, P_y\} = i\lambda. \qquad (134)$$

By defining the Hamiltonian operator as in the ray optics case, we get

$$\frac{d\mathcal{U}}{dz} = \frac{i}{\lambda} \widehat{H} \mathcal{U}. \qquad (135)$$

The equation for the symplectic map looks very similar to Eq. (135) except for the reverse ordering of the operators. However, instead we can look at the differential equation for the inverse of the symplectic map:

$$\frac{d}{dz} \mathcal{M}^{-1} = : H : \mathcal{M}^{-1}. \qquad (136)$$

In both cases, the final maps $\mathcal{U}$ and $\mathcal{M}^{-1}$ depend on the Lie algebras generated by $\widehat{H}$ and $: H :$ respectively.

V.4.2 Relation between Hamilton and Schrödinger

Rather then looking at the Dragt-Finn factorization, let us look first at the single exponent representation (ignoring matters of convergence). Then, it is possible to rewrite formally the maps $\mathcal{U}$ and $\mathcal{M}^{-1}$ with a single Lie operator using a sligthly modified version of our generalized BCH formula of Eq. (37):

$$\mathcal{U} = \exp\left(\widehat{H}_{Total}\right)$$
$$\text{where } \widehat{H}_{Total} = \frac{1}{i\lambda} \int_0^z \text{inviexp}\left(\#\frac{1}{i\lambda}\widehat{H}_{Total}\#\right) \widehat{H},$$
$$\mathcal{M}^{-1} = \exp\left(: H_{Total} :\right) \qquad (137)$$
$$\text{where } H_{Total} = \int_0^z dz' \, \text{inviexp}\left(: H_{Total} :\right) H .$$

This last equation and the commutation rules of the wave position vector $\vec{Q} = (X, Y)$ with the momentum vector $\vec{P} = (P_x, P_y)$ suggest a redefinition of the commutator of two wave operators:

$$\left[\left[\widehat{f},\widehat{g}\right]\right] = \frac{1}{i\lambda}\left\{\widehat{f},\widehat{g}\right\} \Rightarrow [Q_i, P_j] = \delta_{ij}$$
$$\Downarrow \qquad (138)$$
$$\frac{1}{i\lambda} \# f \# = \left[\left[\widehat{f},\bullet\right]\right] .$$

Using this last equation, the Lie exponent of the wave map $\mathcal{U}$ obeys the equation

$$\widehat{H}_{Total} = \frac{1}{i\lambda} \int_0^z dz' \, \text{inviexp}\left(\left[\left[\widehat{H}_{Total}, \bullet\right]\right]\right) \widehat{H} . \qquad (139)$$

Let us assume for one instant that the Lie algebra of the two brackets ($[\![\bullet , \bullet]\!]$ and $[\bullet , \bullet]$) are exactly the same. Then one can immediately write the solution for the map $\mathcal{U}$ in terms of $\mathcal{M}^{-1}$. The answer is

$$\widehat{H}_{Total} = \frac{1}{i\lambda} H_{Total}. \qquad (140)$$

Of course, the old symplectic H_{Total} is a function of the wave equivalents of $\vec{q}$ and $\vec{p}$ (i.e. $\vec{Q}$ and $\vec{P}$) as defined by Eq. (134). Unfortunately, the situation is not quite this simple. The unitarity of the operators of wave optics requires anti-Hermitian Lie exponents. This implies that polynomials in $\vec{Q}$ and $\vec{P}$ must be suitably ordered. From this, one can show that the two Lie algebras differ by λ^2. Therefore, one must properly order the Lie polynomials of ray optics and then replace them by their wave equivalent.

Finally, we must point out that all these results are also true in the (generalized) Dragt-Finn representation. After all, the single-exponent representation and the Dragt-Finn representation differ only by elements of the Lie algebra. More rigorously, one can use the interaction representation or the rules of map computation of Section IV.1.3 to isolate the nonlinear part of the map whose single Lie exponent will then follow the rules of this section. Finally, one notices that the Lie algebras of quadratic operators, wave and ray, are isomorphic. This is true simply because there is a unique ordering procedure. This is well-known in Fourier optics, the wave analog of paraxial ray optics. The results of this section give a coherent and simple procedure for attempting to compute corrections to the familiar wave optics: it involves some messy integrals, which are essentially Feynman path integrals, and can be approximated by aberration catastrophe integrals [97,98]. Recently, Dragt has shown that for standard aberrations it reproduces the kind of interference patterns one would expect [91].

V.5 Conclusion

One could be bold and imagine using the Irwin factorization discussed in Section V.3.2 to relate the interference patterns of various maps, as prescribed by Section V.4. This may sound funny, but this is precisely what was done independently by Bandrauk and Shen [74], on the Schrödinger equation for molecular optics. Indeed, they derived an exactly unitary propagator, and in the process got the Yoshida-Ruth integrator. As we said, in some cases, integration can be viewed as a jolt factorization on the local map.

We cannot predict the future, we do not have a crystal ball. But, the results of this section are sufficiently impressive in their interdisciplinary breadth to vindicate our claims. There are other areas of physics which overlap with our field: for example, normal forms are used in circuit theory [92] and in dynamical systems connected with the Navier-Stokes equations [93], to name just two. There are no reasons whatsoever to restrict our horizons, especially considering the many aspects of physics that enter into the designs of beam transport systems. Many readers may think that the use of Lie methods for wave optics is an exotic topic, however there are problems in free-electron lasers where one looks beyond ray optics [94, 95].

We cannot afford to ignore advances in the fields of mathematical physics and modern engineering. This paper is to enlighten the reader about the power and direct applicability of these modern methods with the hope of training state-of-the-art practitioners in the field.

Give heed! for noble things I speak;
honesty opens my lips. ...
Sincere are all the words of my mouth,
no one of them is wily or crooked;
all of them are plain to the man of intelligence,
and right to those who attain knowledge.
  Proverbs 8:6, 8-9.

Appendix The Green's Function

We start with Eq. (48)

if we choose $\frac{\partial}{\partial \sigma} w = 0$ in Eq. (48)

$\Downarrow$

$$\begin{cases} H = H_0 + \delta_p (s - s_0) V \\ K_s = \exp(:w:) H + i\exp(:w:) \frac{\partial}{\partial s} w \end{cases} \quad (A.1)$$

$\Downarrow$

$$K_s = H_0 + :-H_0: w + \frac{\partial}{\partial s} w + \delta_p (s - s_0) V + O(w^2).$$

From Eq. (A.1) emerges the term linear in w and V:

$$\left(:-H_0: + \frac{\partial}{\partial s}\right) w + \delta_p (s - s_0) V . \quad (A.2)$$

Outside the delta function, we set the first-order terms to zero:

$$\left(:-H_0: + \frac{\partial}{\partial s}\right) w = 0$$

$\Updownarrow$

$$\frac{\partial}{\partial s} w = :H_0: w . \quad (A.3)$$

We recognize that this equation is for the inverse zeroth-order map (denoted by $\mathcal{M}^{-1}_{\Xi'_{s'\to s}}$ in Section IV.1.3) $\mathcal{M}^{-1}_{s'\to s}$. Therefore we can immediately write a solution for Eq. (A.3):

$$w(s) = \mathcal{M}^{-1}_{s'\to s} w(s') . \quad (A.4)$$

To proceed further we integrate across the delta function from s_0^- to s_0^+:

$$w(s_0^+) - w(s_0^-) = -V . \quad (A.5)$$

Combining Eq. (A.5) and (A.4), we can solve for $w(s_0^+)$:

$$w(s_0^+) - \mathcal{M}^{-1}_{s_0} w(s_0^+) = -V ,$$

$\Downarrow$

$$w(s_0^+) = -\left(\mathcal{I} - \mathcal{M}^{-1}_{s_0}\right)^{-1} R \quad (A.6)$$

where $V = R + T \begin{cases} R \in \text{Range}\left(\mathcal{I} - \mathcal{M}^{-1}_{s_0}\right) \\ T \in \text{Ker}\left(\mathcal{I} - \mathcal{M}^{-1}_{s_0}\right). \end{cases}$

The readers with some knowledge of map normalization will recognize in Eq. (A.6) the small denominator of a map-based theory which involves the one-turn map. The end result for the Green's function w is

$$w(s) = \mathcal{M}^{-1}_{s_0 \to s} \left(\mathcal{I} - \mathcal{M}^{-1}_{s_0}\right)^{-1} R . \quad (A.7)$$

In Eq. (A.7), the variable s is a modulo L number which automaticaly insures the periodicity of the Green's function. Hence, the final form of w for an arbitrary perturbation $V(\vec{x};s)$ is

$$w(s) = -\int_0^L \mathcal{M}_{s_0 \to s}^{-1} \left(\mathcal{I} - \mathcal{M}_{s_0}^{-1} \right)^{-1} R(\vec{x};s_0) \, ds_0. \qquad (A.8)$$

It turns out that $w(s)$ can be changed by an element of $\mathrm{Ker}\left(\mathcal{I}-\mathcal{M}_s^{-1}\right)$. This freedom is intimately connected to the concept of phase advance [55] and free terms [82-84] in normal forms. In fact, the final value of $w(s)$ for different algorithms will differ by elements within this kernel. If H_0 depends only on the actions, then, under nonresonant conditions, the free term of w is also a function only of the actions. Of course, the map $\mathcal{B}$ of Section IV.2.2 is just $e^{:w:}$ to leading order.

Hamiltonian-Free

We write the perturbed one-turn map immediately after the perturbation:

$$\mathcal{M}_{+\to+}^{perturbed} = \mathcal{M}_{s_0} \exp\left(:-V:\right). \qquad (A.9)$$

We then normalize it following the rules of Section IV.2.2:

$$\begin{aligned}
\mathcal{N}_{+\to+}^{perturbed} &= \exp\left(:w_+:\right) \mathcal{M}_{s_0} \exp\left(:-V:\right) \exp\left(:-w_+:\right) \\
&= \mathcal{M}_{s_0} \underbrace{\mathcal{M}_{s_0} e^{:w_+:} \mathcal{M}_{s_0}}_{\exp(:\mathcal{M}_{s_0}^{-1} w_+:)} e^{:-V:} e^{:-w_+:} \\
&= \mathcal{M}_{s_0} \exp\left(:\{\mathcal{M}_{s_0}^{-1} - \mathcal{I}\} w_+ - V:\right) + O\left(w^2\right).
\end{aligned} \qquad (A.10)$$

The Green's function follows from Eq. (A.6). The total w of Eq. (A.7) follows from the application of the summation rules of Section IV.2.2.

Conclusion

In Ref. [24], Michelotti derived this Green's function of Eq. (A.7). He made an effort to connect it to the existing normal-form literature within accelerator theory. [To our knowledge, all our attempts to explain the hybrid approach in terms of the older framework have failed. One does not learn to cross oceans by following the shores.] His treatment, if formulated within a map theoretical framework, conceptually bridges the gap between the traditional and the hybrid approach.

References

[1] J. Moser, Stable and random motion in dynamical systems, Princeton University Press, Princeton, New Jersey (1973).
[2] A. Schoch, CERN 57-21 (1958).
[3] R. Hagedorn, CERN 57-1 (1957).
[4] F. Willeke, Fermilab, FN-422 (1985).
[5] E.D. Courant and H. S. Snyder, Ann. Phys. **3**, 1 (1958).
[6] J. G. Truxal, Control System Synthesis, McGraw-Hill, New York (1955).
[7] Chi-Tsong Chen, Linear System Theory and Design, Holt, Rinehart and Winston, New York (1984).
[8] Proc. 2nd Advanced ICFA Beam Dynamics Workshop, CERN 88-04, 152 (1988).
[9] L. B. Rall, Automatic Differentiation: Techniques and Applications, in Lecture Notes in Computer Science No. 120, Springer-Verlag (1981).
[10] Automatic Differentiation of Algorithms: Theory, Implementation, and Application, SIAM, Philadelphia (1991).
[11] Z. Parsa, S. Tepikian and E. D. Courant, Part. Accel. **22**, 205 (1987).
[12] N. Merminga, Thesis, U. of Michigan (1989).
[13] N. Merminga and K.-Y. Ng, SSC-N-594 (1989).
[14] G. Guignard, CERN 78-11 (1978).
[15] E. D. Courant, R.D. Ruth and W. T. Weng, AIP Conf. Proc. **127**, 294 (1985).
[16] R. D. Ruth, AIP Conf. Proc. **153**, 152 (1987).
[17] H. Von Zeipel, Arkiv Mat. Astron. Fys. **31**, 7 (1916).
[18] A. Ando, Part. Accel. **15**, 177 (1984).
[19] J. Bengtsson, CERN 88-05 (1988).
[20] F. Schmidt and F. Willeke, CERN SPS/88-30 (1988).
[21] R. Nagaoka, K. Yoshida and M. Hara, Part. Accel. **33**, 57 (1990).
[22] G. Hori, Publ. Astr. Soc. Japan **18**, 287 (1966).
[23] A. Deprit, Cel. Mech. **1**, 12 (1969).
[24] L. Michelotti, AIP Conf. Proc. **153**, 236 (1987).
[25] L. Michelotti, Part. Accel. **19**, 202 (1986).
[26] R. L. Warnock and R. D. Ruth, Physica D **56**, 188 (1992).
[27] W. E. Gabella, R. D. Ruth and R. L. Warnock, Phys. Rev. A **46**, 3493 (1992).
[28] J. S. Berg, R. L. Warnock, R. D. Ruth and E. Forest, SLAC-PUB-6037 (1993).
[29] T. L. Collins, Distortion functions, Fermilab-84/114, (Oct. 1984).
[30] B. Autin, AIP Conf. Proc. **153**, 288 (1987).
[31] A. Antillòn, Part. Accel. **23**, 187 (1988).
[32] K. Brown and R. Servranckx, AIP Conf. Proc. **127**, 62 (1985).
[33] M. Berz and H. Wollnik, Nucl. Instr. Meth. **A258**, 364 (1987).
[34] K. Brown and R. Servranckx, AIP Conf. Proc. **153**, 123 (1987).
[35] J. Irwin, SSC-228 (1989).

[36] A. J. Dragt and D. Abell, Jolt Factorization of Symplectic Maps, in Proc. 15th Int. Conf. on High Energy Accelerators, to appear in 1993.
[37] I. Gjaja, Monomial Factorization of Symplectic Maps, Physics Dept., U. of Maryland (1992).
[38] D. R. Douglas, AIP Conf. Proc. **153**, 390 (1987).
[39] M. Berz and H. Wollnik, Nucl. Instr. Meth. **A258**, 364 (1987).
[40] M. Berz, Part. Accel. **24**, 109 (1989).
[41] M. Berz, AIP Conf. Proc. **184**, 961 (1989).
[42] V. Garczynski, BNL AD/AP-47 (1992).
[43] G. Corliss and L. B. Rall, Trans. First Army Conf. on Applied Mathematics and Computing, ARO Report 84-1 (1984).
[44] E. Forest, M. Berz and J. Irwin, Part. Accel. **24**, 91 (1989).
[45] E. Forest and J. Bengtsson have worked out a library based on automatic differentiation which allows power series normalization, various factorizations (Dragt-Finn, Irwin) and the computation of generating functions. It also normalizes maps in the nonsymplectic case, which is useful in electron machines.
[46] L.M. Healy, Thesis, U. of Maryland (1986).
[47] E. Forest, A paper on small machine to appear in Part. Accel.
[48] K. Oide, mostly unpublished material concerning the SAD tracking code at KEK.
[49] P. Krejcik, Proc. IEEE PAC Conf., 1278 (1987).
[50] K. Halbach, Nucl. Instr. Meth. **A187**, 109 (1981).
[51] J. Jäger and D. Möhl, CERN, PS/DL/LEAR/Note 81-7 (1981).
[52] A. J. Dragt, Part. Accel. **12**, 205 (1982).
[53] E. Forest and K. Hirata, KEK Report 92-12 (Aug. 1992).
[54] L. Schackinger and R. Talman, Part. Accel. **22**, 35 (1987).
[55] E. Forest, J. Math. Phys. **31**, 1133 (1990); originally SSC-138 (1987).
[56] A. Bazzanni, P. Mazzanti, G. Servizi and G. Turchetti, Nuovo Cimento **B 102**, 51 (1988).
[57] A. J. Dragt, in Proc. 1984 Study on Design and Utilization of the Superconducting Super Collider, Snowmass, R. Donaldson and J. Morfin, Eds., Am. Phys. Soc. (1985).
[58] S. G. Peggs, SSC-17 (1988); presented at 14th Int. Conf. on High Energy Accelerators, Tsukuba, Japan (Aug. 1989).
[59] J. Irwin, Nucl. Instr. Meth. **A298**, 460 (1990).
[60] A. Jackson, SSC-107 (1987).
[61] E. Forest, SSC-95 (1986).
[62] J. Bengtsson and J. Irwin, SSC-232 (1990).
[63] E. Forest and J. Irwin, in Proc. Workshop, Capri 1990, p. 46, W. Scandale and G. Turchetti, Eds., World Scientific (1990).
[64] G. E. Lee-Whiting, Nucl. Instr. Meth. **83**, 232 (1970).
[65] E. Forest and J. Milutinovic, Nucl. Instr. Meth. **A269**, 474 (1988).
[66] E. Forest and K. Ohmi, KEK Report 92-14 (Sept. 1992).
[67] E. Forest, J. Bengtsson and M. Reusch, Phys. Lett. A **158**, 99 (1991).

[68] E. Forest, Part. Accel. **22**, 15 (1987).
[69] A. J. Dragt et al., Ann. Rev. Nucl. Sci. **38**, 455 (1988).
[70] E. Forest and R.D. Ruth, Physica D **43**, 105 (1990).
[71] E. Forest, J. Comp. Phys. **99** (2), 209 (1992).
[72] H. Yoshida, Phys. Lett. A **150**, 190 (1990).
[73] M. Suzuki, Phys. Lett. A **165**, 387 (1992).
[74] A. Bandrauk and H. Shen, Chem. Phys. Lett. **176**, 428 (1991).
[75] J. Wisdom and M. Holman, Astron. J. **102**, 1528 (1991).
[76] J. R. Cary and I. Doxis, to appear in J. Comp. Phys.
[77] J. R. Cary, Lie Transform Perturbation Theory for Hamiltonian Systems, Phys. Rep. **79**(2), 129 (1981).
[78] R.D. Ruth, IEEE Trans. Nucl. Sci. NS-30, 2669 (1983).
[79] L. Smith, ESG Tech. Note 24, Lawrence Berkeley Lab (1986).
[80] L. Michelotti, Part. Accel. **16**, 233 (1985).
[81] R. L. Dewar, J. Phys. A: Math. Gen. **9**(12), 2043 (1986).
[82] P. B. Kahn and Y. Zarmi, Physica D **54**, 65 (1991).
[83] S.R. Mane and W. T. Weng, Phys. Rev. E. **48**(1), **532** (1993).
[84] E. Forest and D. Murray, submitted to Physica D (1993).
[85] A.J. Dragt and E. Forest, Advances in Electronics and Electron Physics, **24**, 2734 (1983).
[86] A.J. Dragt and E. Forest, Advances in Electronics and Electron Physics, **67**, 65 (1986).
[87] A.J. Dragt, F. Neri and G. Rangarajan, Phys. Rev. A **45**, 2572 (1992).
[88] P. J. Channell and J.C. Scovel, Physica D **50**, 80 (1991).
[89] D.J.D. Earn and S. Tremaine, Physica D **56**, 1 (1992).
[90] J. C. Scovel, Phys. Lett. A **159**, 396 (1992).
[91] A. J. Dragt, Unpublished work and talk given in fall 1992 at Stony Brook; manuscript in preparation.
[92] M. Ashkenazi and S.-H. Chow, IEEE Trans. Circuits and Systems **35**, 850 (1988).
[93] P. Manneville, Dissipative Structures and Weak Turbulence, Academic Press, Boston (1990).
[94] K. J. Kim, Nucl. Instr. Meth. **A246**, 71 (1986).
[95] Y. H. Chin, K. J. Kim and M. Xie, Phys. Rev. A **46**, 6662 (1992).
[96] R. L. Warnock, Phys. Rev. Lett. **66**(14), 1803 (1991).
[97] R. Gilmore, Catastrophe Theory for Scientists and Engineers, Wiley, New York (1981).
[98] M.V. Berry and C. Upstill, Progress in Optics **18**, (1980).

Towards C++ Object Libraries for Accelerator Physics

Leo Michelotti
Fermi National Accelerator Laboratory

This will be telegraphic. I have spoken and written before about creating libraries of reusable objects in the language C++ for doing accelerator design and analysis [1]. To avoid repeating the same material too many times, I shall here merely state a few issues without discussion.

We must write robust, flexible, portable software that is easier to understand, maintain, modify, reuse, and extend. These attributes are more than mere buzzwords. They are important goals that computer scientists strive to achieve by refining their programming models and devising languages to support them. An important breakthrough was achieved by the introduction of object-oriented programming (OOP) as the computing model behind such languages as Smalltalk, ADA, Eiffel, Objective-C, and C++ . OOP is *not* a "fad": it is arguably the most significant development in programming since the invention of FORTRAN and the way that the best software will be written well into the next century. An "object" comprises structures of data, the functions that manipulate them, and rules for bringing them into and out of scope. OOP is a methodology for realizing and fully utilizing this abstract concept, an extension to programming of the basic technique that has advanced mathematics for centuries.

Each of the OOP languages can undoubtedly boast advantages compared to the others, but C++ possesses features which lend themselves to writing portable, scientific software. The two libraries of C++ classes (objects) which have been under development are: (a) MXYZPTLK, which implements automatic differentiation, and (b) BEAMLINE, which provides objects for modelling beamline and accelerator components. The principal classes in the MXYZPTLK library are:

- class DA : public dlist   ...is the fundamental object which carries data about functions and their derivatives and propagates them through arithmetic operations. The notation "public dlist" means that DA is derived from class dlist,[1] which is a doubly linked list. Each link of the list contains the index and value of a specific derivative.

- class coord : public DA   ...is the DA object which serves as the starting point for calculations; it is the DA implementation of a function which projects onto a coordinate. An important feature distinguishing it from a normal DA variable is that it cannot be changed by placing it to the left of an equal sign.

---

[1] More accurately, it "inherits from" dlist.

`class LieOperator` ...implements vector field operators of the form $\underline{v}(\underline{z}) \cdot \partial/\partial\underline{z}$. Among other things, it is (to be) used in creating exponential maps.

while those in the BEAMLINE library include:

`class bmlnElmnt` ...is the abstract base class containing data applicable to all possible beamline elements, including the geometric information determining its position and orientation in space. No `bmlnElmnt` declaration should ever appear in an application; rather, the beamline elements which do appear are all derived from this class.

`class drift, quadrupole rbend, sector, ...:public bmlnElmnt` ...are the primitive beamline elements which can be instanced in application programs. Not all objects appearing in MAD are presently available; for now, objects are created as needed by the author or upon request from others.

`class beamline : public bmlnElmnt, public dlist` ...is the composite object comprising a sequential, geometric arrangement of beamline elements. It inherits from `bmlnElmnt` as well as `dlist` to make it trivial to insert `beamlines` into `beamlines`, allowing for their hierarchical construction. This feature also facilitates certain recursive procedures in the class.

`class proton, DAproton, bunchOfProtons, ...` ...are the objects that can be propagated through beamline elements. Using a `proton` object performs tracking; using a `DAproton` object creates polynomial Poincaré maps, as its state is modelled with the DA variables from MXYZPTLK.

`class beamlineImage, circuit, beamlineOverseer, ...` ...are utility classes which, it is hoped, will make writing interactive programs easier. For example, `circuit` objects enable attributes of beamline elements to be changed in a correlated manner. The `beamlineOverseer` (not yet written) will perform communication between beamline elements, keep track of particles within the elements, and so forth. It will be useful for modelling such things as slow extraction of a bunch using septa or for connecting monitors to kickers in a stochastic cooling system.

Information describing a beamline element's location and orientation in space is contained in a "geometry" struct within the base class, `bmlnElmnt`. This struct comprises two frames, each consisting of a `threeVector` point of origin (yet another object) and three `threeVectors` which are unit vectors for a local right-handed coordinate system. The plane spanned by the first two unit vectors, and passing through the frame's point of origin, is a face of the beamline element; the two frames thus model the element's in-face and the out-face. Every beamline element which inherits from `bmlnElmnt` possesses a member function called `.propagate` whose argument is a particle type (e.g, proton) and whose primary action is to change the state of the particle so as to reflect its passage from the in-face to the out-face. In this way, the beamline element classes implement the "lego" concept discussed by Forest [2].

The methods employed by the .**propagate** function are contained in a collection of small files of C++ source code, the "physics files," one for each element type and particle class of interest. (The default provided by the base class, bmlnElmnt, is drift physics.) In fact, there will be a number of such collections reflecting varying levels of sophistication. Simple applications – such as calculating lattice functions for an uncoupled, design lattice – could use the most basic physics files, while more complicated applications – such as simulating stochastic cooling in a ring with alignment errors and using symplectic numerical integration through nonlinear focussing elements – would use the more elaborate physics files. (Of course, the latter could be used in the simpler calculations as well, but there is some value in doing simple calculations in a simple manner.) A collection of different physics files amount to what Iselin descriptively calls "pluggable modules" [3]. Because the physics has been isolated within these files it will be easy to change from one to another, or even to modify them in order to try different ideas.

It was mentioned that the .**propagate** functions alter the state of the particle in passing from the in-face to the out-face of a beamline element. They can also change the element itself: for example, it is possible to write the functions so that a bunch of protons sets up a mode in an RF cavity which then modifies the cavity's action on subsequent protons. Using OOP, no further complexity need be added to the programming paradigm in order to achieve these things.

MXYZPTLK exists in a reasonably usable state; BEAMLINE is still in its infancy. The most ambitious goal is to put together a library of objects for use (and reuse) in programs at *all* levels of accelerator design and analysis, from the zeroth-order task of placing dipoles so as to establish the correct design orbit to the problem of calculating normal forms of 137th degree polynomial maps. This goal is not something that can be accomplished by one person (at least, not *this* person) or in a short amount of time; although one can learn enough C++ to *use* objects fairly quickly, it takes years to become proficient at *creating* classes well. However, it is achievable provided that we start from the correct programming paradigm so that the library can evolve in a coherent, controllable manner ("organic programming") from its earliest stage all the way through, if desired, the development of complete, graphics-intensive accelerator CAD programs. I am proceeding under the beliefs that (a) object-oriented programming is that paradigm and (b) C++ is the correct language in which to implement it. Those who wish to play around with the MXYZPTLK and BEAMLINE libraries in whatever their current state of existence may obtain them via ftp from calvin.fnal.gov, in the subdirectories pub/outgoing/michelotti/beamline and pub/outgoing/michelotti/mxyzptlk. Regrettably, the only documentation included at this point is a three-year-old MXYZPTLK User's Guide, but a few simple demo programs are provided in the package.

References

[1] L. Michelotti. Exploratory orbit analysis. In F. Bennett & J. Kopta, Eds., *Proc. 1989 IEEE Part. Accel. Conf.*, March 1989. IEEE Cat. No. 89CH2669-0.
 — A C++ hacker's implementation of automatic differentiation. In *Automatic Differentiation of Algorithms: Theory, Implementation, and Application*. SIAM, Philadelphia, PA, 1991.
 — MXYZPTLK: A practical, user-friendly C++ implementation of differential algebra: User's guide. Fermi Note FN-535, Jan. 1990.
 — MXYZPTLK and BEAMLINE: C++ objects for beam physics. In *Advanced Beam Dynamics Workshop on Effects of Errors in Accelerators, Their Diagnosis and Correction. (Corpus Christi, Oct. 1991)*. AIP Conf. Proc. 255, 1992.
 — Accelerator physics analysis with an integrated toolkit. Technical Report Fermilab-Conf-92/219, July 1992.
 — A note on the automated differentiation of implicit functions. TM-1742, Fermilab, June, 1991.

[2] E. Forest and K. Hirata, KEK Report 92-12, August 1992.

[3] F. Christoph Iselin. This conference.

© 1994 American Institute of Physics

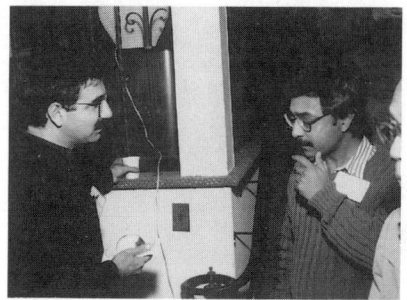

Photographs 495

List of Participants

Abell, Dan T.
Physics Dept., U. of Maryland
College Park, MD 20742

Amiry, Ali
UCLA
Marina Del Rey, CA 90292

Bengtsson, Johan
Lawrence Berkeley Laboratory
Berkeley, CA 94720

Berg, J. Scott
SLAC
Stanford, CA 94309

Bhatia, T.S.
U.S. Dept. of Energy, DHEP
Washington, DC 20585

Blind, Barbara
Los Alamos National Laboratory
Los Alamos, NM 87545

Bourianoff, George I.
SSC Laboratory
Dallas, TX 75237

Bruening, Oliver
DESY
2 Hamburg 52, Germany

Cai, Yunhai
SSC Laboratory
Dallas, TX 75237

Cappi, R.
CERN PS Division
CH 1211 Geneva 23, Switzerland

Caussyn, David
Indiana U. Cyclotron
Bloomington, IN 47405

Channell, Paul
Los Alamos National Laboratory
Los Alamos, NM 87545

Chao, Alex
SSC Laboratory
Dallas, TX 75237

Cole, Ben
SSC Laboratory
Dallas, TX 75237

Courant, Ernest D.
Brookhaven National Laboratory
Upton, NY 11973

Dell, G. Fritz
Brookhaven National Laboratory
Upton, NY 11973

Dome, Georges
CERN SL Division
CH-1211 Geneva 23, Switzerland

Dragt, Alex J.
Physics Dept., U. of Maryland
College Park, MD 20742

Dutt, Samir K.
SSC Laboratory
Dallas, TX 75237

Ellison, Michael J.
Indiana U. Cyclotron
Bloomington, IN 47405

Ellison, James
SSC Laboratory
Dallas, TX 75237

Forest, Etienne
Lawrence Berkeley Laboratory
Berkeley, CA 94720

Garczynski, Vlodek
Brookhaven National Laboratory
Upton, NY 11973

Gerasimov, Andrei
Fermilab
Batavia, IL 60510

Giovannozzi, Massimo
CERN and U. of Bologna
CH-1211 Geneva 23, Switzerland

Gjaja, Ivan
Physics Dept., U. of Maryland
College Park, MD 20742

List of Participants

Goderre, Glenn
Fermilab
Batavia, IL 60510

Hahn, Harald
Brookhaven National Laboratory
Upton, NY 11973

Harrison, Michael A.
Brookhaven National Laboratory
Upton, NY 11973

Iselin, Christoph
CERN SL Division
CH-1211 Geneva 23, Switzerland

Koga, James K.
U. of Texas, Inst. for Fusion
Austin, TX 78712

Koul, Rabinder K.
Argonne National Laboratory
Argonne, IL 60439

Krinsky, Samuel
Brookhaven National Laboratory
Upton, NY 11973

Lee, S.Y.
Dept. of Physics, Indiana U.
Bloomington, IN 47405

Lessner, Eliane S.
Argonne National Laboratory
Argonne, IL 60439

Li, Derun
Indiana U. IUCF
Bloomington, IN 47405

Litvinenko, Vladimeir
Duke U. FEL Lab
Durham, NC 27708

Luccio, Alfredo
Brookhaven National Laboratory
Upton, NY 11973

Maletic, Dusan
Brookhaven National Laboratory
Upton, NY 11973

Mane, Sateesh
Brookhaven National Laboratory
Upton, NY 11973

Meddahi, Malika
Lawrence Berkeley Laboratory
Berkeley, CA 94720

Mehta, Naresh C.
SSC Laboratory
Dallas, TX 75237

Michelotti, Leo
Fermilab
Batavia, IL 60510

Milutinovic, Janko
Brookhaven National Laboratory
Upton, NY 11973

Mishra, C. Shekhar
Fermilab
Batavia, IL 60510

Month, Melvin
Brookhaven National Laboratory
Upton, NY 11973

Murphy, James
Brookhaven National Laboratory
Upton, NY 11973

Neri, Filippo
Los Alamos National Laboratory
Los Alamos, NM 87544

Neuffer, David
CEBAF
Newport News, VA 23606

Parzen, George
Brookhaven National Laboratory
Upton, NY 11973

Peggs, Stephen G.
Brookhaven National Laboratory
Upton, NY 11973

Pilat, Fulvia
SSC Laboratory
Dallas, TX 75237

Pusterla, Modesto
Physics Dept., U. of Padova
35131 Padova, Italy

Rhoades-Brown, Mark
Brookhaven National Laboratory
Upton, NY 11973

Roser, Thomas
Brookhaven National Laboratory
Upton, NY 11973

Ruggiero, A.G.
Brookhaven National Laboratory
Upton, NY 11973

Satogata, Todd
Fermilab/Northwestern U.
Batavia, IL 60510

Scandale, Walter
CERN
1211 Geneva 23, Switzerland

Schmidt, Frank
CERN
CH-1211 Geneva 23, Switzerland

Sen, Tanaji
SSC Laboratory
Dallas, TX 75237

Servranckx, Roger
TRIUMF
B.C. VOR 1X0, Canada

Sessler, Andrew
Lawrence Berkeley Laboratory
Berkeley, CA 94720

Shi, Jack
Physics Dept., U. of Houston
Houston, TX 77204

Shih, Hsiuan-Jeng
SSC Laboratory
Dallas, TX 75237

Stupakov, Gennady
SSC Laboratory
Dallas, TX 75237

Tajima, T.
U. of Texas Inst. for Fusion
Austin, TX 78712

Talman, Richard
SSC Laboratory
Dallas, TX 75237

Teng, Lee
Argonne National Laboratory
Argonne, IL 60439

Tepikian, Steven
Brookhaven National Laboratory
Upton, NY 11973

Trbojevic, Dejan
Brookhaven National Laboratory
Upton, NY 11973

Tzenov, Stephan
CERN SL/AP
CH-1211 Geneva 23, Switzerland

Wei, Jie
Brookhaven National Laboratory
Upton, NY 11973

Weng, Wu-Tsung
Brookhaven National Laboratory
Upton, NY 11973

Willeke, Ferdinand
DESY
D-2000 Hamburg 52, Germany

Wu, Ying
Duke U. FEL Lab
Durham, NC 27708

Xu, Jianming
Brookhaven National Laboratory
Upton, NY 11973

Yan, Yiton
SSC Laboratory
Dallas, TX 75237

Zimmerman, Frank
DESY
2000 Hamburg 52, Germany

AUTHOR INDEX

A

Abell, D. T., 230

B

Ball, M., 163, 170
Bazzani, A., 337
Bengtsson, J., 417
Berg, J. S., 417
Bourianoff, G., 13, 102
Brabson, B., 163, 170
Brüning, O., 289
Budnick, J., 163, 170

C

Caussyn, D. D., 163, 170
Chao, A. W., 163, 170
Collins, J., 163, 170
Curtis, S., 170

D

Dell, G. F., 397
Derenchuk, V., 163, 170
Dragt, A. J., 230, 417
Dutt, S., 163, 170

E

East, G., 163, 170
Ellison, M., 163, 170
Ellison, T., 163, 170

F

Forest, É., 58, 417
Friesel, D., 163, 170

G

Gabella, W., 163, 170
Garczynski, V., 93
Gerasimov, A., 375
Giovannozzi, M., 385
Goderre, G. P., 36

H

Hamilton, B., 163, 170
Harfoush, F., 208
Holt, J., 36
Huang, H., 163, 170
Hunt, S., 102

I

Iselin, C., 73

J

Jones, W. P., 163, 170

K

Koul, R. K., 217

L

Lamble, W., 170
Lee, S. Y., 163, 170, 260
Li, D., 163, 170

M

Maletic, D., 345
Mane, S. R., 307, 310
Mathieson, D., 102

Michelotti, L., 417, 488
Milutinovic, J., 182
Minty, M. G., 163, 170
Mishra, C. S., 208
Morpurgo, G., 102

N

Nagaitsev, S., 163, 170
Ng, K. Y., 163, 170, 260
Nosochkov, Y., 135

O

Ohnuma, S., 325

P

Parzen, G., 48, 409
Peggs, S., 85, 361
Pei, X., 163, 170
Pilat, F., 102, 135
Pusterla, M., 337

R

Rondeau, G., 163, 170
Ruggiero, A. G., 90, 182, 345

S

Satogata, T., 361

Scandale, W., 3
Sen, T., 135
Shi, J., 325
Sloan, T., 163, 170
Stupakov, G. V., 267
Syphers, M., 163, 170

T

Talman, R., 102
Tepikian, S., 163, 170
Trbojevic, D., 260

V

Venturini, M., 337

W

Wang, Y., 163
Weng, W. T., 310
Willeke, F., 33

Y

Yan, Y. T., 163, 170, 177

Z

Zhang, P. L., 163
Zimmerman, F., 273